AI AND BIG DATA SCIENCES FOR BIOINFORMATICS AND SYSTEMS MEDICINE

AI AND BIG DATA SCIENCES FOR BIOINFORMATICS AND SYSTEMS MEDICINE

Editors

Jiajia Liu

Kexin Huang

Xiaobo Zhou

The University of Texas Health Science Center at Houston, USA

World Scientific

NEW JERSEY • LONDON • SINGAPORE • BEIJING • SHANGHAI • HONG KONG • TAIPEI • CHENNAI • TOKYO

Published by

World Scientific Publishing Co. Pte. Ltd.
5 Toh Tuck Link, Singapore 596224
USA office: 27 Warren Street, Suite 401-402, Hackensack, NJ 07601
UK office: 57 Shelton Street, Covent Garden, London WC2H 9HE

British Library Cataloguing-in-Publication Data
A catalogue record for this book is available from the British Library.

AI AND BIG DATA SCIENCES FOR BIOINFORMATICS AND SYSTEMS MEDICINE

ISBN 978-981-98-1428-2 (hardcover)
ISBN 978-981-98-1429-9 (ebook for institutions)
ISBN 978-981-98-1430-5 (ebook for individuals)

For any available supplementary material, please visit
https://www.worldscientific.com/worldscibooks/10.1142/14342#t=suppl

Desk Editor: Joy Quek

Contents

Preface

The convergence of artificial intelligence (AI) and big data sciences has revolutionized bioinformatics and systems medicine, paving the way for new paradigms in precision medicine, drug discovery, and disease modeling. The exponential growth of biomedical data, from electronic medical records (EMRs) and genomic sequencing to high-resolution medical imaging and single-cell analyses, has necessitated the development of computational tools that can extract meaningful insights from complex datasets. This book, *AI and Big Data Sciences for Bioinformatics and Systems Medicine*, serves as a comprehensive resource for researchers, clinicians, and computational biologists seeking to harness the power of AI-driven analytics in biomedical sciences.

With the rapid advancements in machine learning, deep learning, and network-based modeling, bioinformatics has transitioned from traditional statistical approaches to AI-powered methodologies. These approaches have enabled breakthroughs in understanding genomic variation, predicting protein structures, and modeling disease progression. The book is structured to provide a logical progression from fundamental AI models to their applications in biomedical data analysis and systems biology.

The first few chapters introduce core machine learning and AI models, setting the foundation for their applications in biomedical big data analysis. Fundamental models such as statistical modeling approaches, machine learning, and deep learning are discussed, along with their implementations in healthcare and biological data interpretation. Subsequent chapters delve into specific data modalities, including genomics, transcriptomics, proteomics, and imaging data, outlining computational frameworks and best practices for their analysis. A strong emphasis is placed on functional annotation and knowledgebase integration, which are essential for translating multi-omics data into biological insights. Functional annotation databases play a crucial role in linking genomic variations to disease phenotypes, facilitating biomarker discovery, and improving personalized treatment strategies.

As precision medicine gains traction, integrating multimodal data has become a critical challenge. This book addresses this need by presenting strategies for single-cell sequencing, spatial transcriptomics, and proteomics, along with multimodal integration techniques. These methodologies enable the reconstruction of complex cellular processes, shedding light on cancer evolution, drug resistance mechanisms, and immune system interactions. Understanding single-cell and spatial omics data is particularly important in unraveling tumor heterogeneity, cell fate decisions, and tissue microenvironments, providing new avenues for targeted therapies and regenerative medicine.

Furthermore, the book explores systems biology approaches to model disease progression, from precancerous states to metastasis, offering computational frameworks for studying drug resistance and immunotherapy responses. By incorporating mathematical modeling techniques such as ordinary differential equations (ODEs) and partial differential equations (PDEs), researchers can simulate disease dynamics and predict therapeutic outcomes. Systems biology approaches are instrumental in deciphering the interactions between genetic, epigenetic, and environmental factors, ultimately leading to better disease classification and prognosis.

In addition to disease modeling, the book covers emerging topics such as synthetic biology, mRNA optimization, and small-molecule generation. Advances in AI-driven drug discovery, including sequence optimization and virtual screening, highlight the transformative potential of computational methodologies in therapeutic development. The integration of AI with synthetic biology accelerates the design of novel therapeutics, including mRNA-based vaccines and gene-editing strategies, which have shown great promise in combating infectious diseases and genetic disorders.

This book aims to bridge the gap between computational and biomedical sciences, providing a unified perspective on how AI and big data can be leveraged to tackle pressing challenges in healthcare. We hope that researchers, bioinformaticians, and medical professionals alike will find this resource valuable in their pursuit of innovation and discovery in bioinformatics and systems medicine.

We extend our gratitude to the contributors who have shared their expertise and to the broader scientific community whose ongoing research continues to push the boundaries of AI-driven biomedical discovery. Their dedication to advancing the field has been instrumental in shaping this book, and we hope it will serve as a foundation for future breakthroughs in biomedical data science and precision medicine.

Contributors

Jiajia Liu • Center for Computational Systems Medicine, McWilliam School of Biomedical Informatics, The University of Texas Health Science Center at Houston, Houston, TX, USA

Kexin Huang • Center for Computational Systems Medicine, McWilliam School of Biomedical Informatics, The University of Texas Health Science Center at Houston, Houston, TX, USA

Xiaobo Zhou • Center for Computational Systems Medicine, McWilliam School of Biomedical Informatics, McGovern Medical School, School of Dentistry, The University of Texas Health Science Center at Houston, Houston, TX, USA

Ruihan Luo • Center for Computational Systems Medicine, McWilliam School of Biomedical Informatics, The University of Texas Health Science Center at Houston, Houston, TX, USA

Jianguo Wen • Center for Computational Systems Medicine, McWilliam School of Biomedical Informatics, The University of Texas Health Science Center at Houston, Houston, TX, USA

Tiangang Wang • Center for Computational Systems Medicine, McWilliam School of Biomedical Informatics, The University of Texas Health Science Center at Houston, Houston, TX, USA

Zhiwei Fan • West China Biomedical Big Data Center, West China Hospital, Sichuan University, Chengdu, Sichuan, China

Xiaona Liu • Center for Computational Systems Medicine, McWilliam School of Biomedical Informatics, The University of Texas Health Science Center at Houston, Houston, TX, USA

Sijia Wu • School of Life Science and Technology, Xidian University, Xi'an, Shaanxi, China

Yankai Yu • School of Computing and Artificial Intelligence, Southwest Jiaotong University, Chengdu, Sichuan, China

Yanfei Wang • Department of Health Outcomes and Biomedical Informatics, College of Medicine, University of Florida, Gainesville, FL, USA

Wentao Li • McWilliam School of Biomedical Informatics, The University of Texas Health Science Center at Houston, Houston, TX, USA

Mengyuan Yang • Department of Cell Biology and Genetics, School of Basic Medical Sciences, Xi'an Jiaotong University Health Science Center, Xi'an, China

Yuzhou Feng • Medical College of Shihezi University, Shihezi University, Shihezi, Xinjiang, China

CHAPTER 1

Basic Statistical Modeling and Machine Learning Approaches

In this chapter, we will introduce the basic statistical modeling and machine learning concepts, which will be applied to the later chapters. We do not try to cover all models, as there are other relevant books readers can read.

1.1 Basic Statistical Modeling Approaches and Information Theory

In this section, we will introduce probability, probability density function (PDF), cumulative distribution function, Bayesian theories, entropy, divergence, mutual information, linear regression model, maximum likelihood estimation, Bayesian approach, and Gaussian mixture model (GMM).

1.1.1 Probability, Probability Density Function, and Cumulative Distribution Function

A random variable is a quantity that takes on various values depending on chance. In mathematical terms, there are two types of random variables: discrete random variables and continuous random variables. (1) Discrete random variables form a countable set of outcomes; we will take binomial random variables as an example of a type of discrete random variable. (2) Continuous random variables form a continuum of possible outcomes; we will take normal or Gaussian random variables as a way to familiarize ourselves with continuous random variables.

There are many types of discrete random variables. Here, we introduce the binomial family. Binomial random variables are discrete random variables of "counts" that describe the number of "successes" (X) in n independent Bernoulli trials, where each Bernoulli trial has a probability of success designated as p. In other words, binomial random variables have two parameters: n and p. n is the number of independent "Bernoulli" trials. p is the probability of success for each trial (which does not change from trial to trial). Let b represent "binomial distribution" and "~" represent "distributed as." Then, $X \sim b(n, p)$ is read as "random variable X is distributed as a binomial random variable with parameters n and p."

Let $Pr(X = x)$ represent "the probability that random variable X takes on a value of x." Let $Pr(X \leq x)$ represent "the probability random variable X takes on a value less than or equal to x." This is the cumulative probability of the event. The probability mass function (PMF) assigns probabilities for all possible outcomes of a discrete random variable.

Continuous random variables form a continuum of possible outcomes. There are many different types of continuous random variables. These random variable types occur in families (e.g., uniform random variables, normal random variables, and chi-squared random variables).

PDFs assign probabilities for all possible outcomes for continuous random variables. PDFs cannot be shown in tabular form. They can be represented with integral functions. The normal distribution represents a family of continuous random variables. Each family member is characterized by two parameters, μ and σ, where μ is defined as the PDF's mean or expected value and σ is defined as the PDF's standard deviation. Let $X \sim N(\mu, \sigma)$ represent a normal random variable with mean μ and standard deviation σ.

A cumulative distribution function (CDF) is a function that provides the probability that a random variable X takes a value less than or equal to a given x. The CDF is represented by a step function, where each step corresponds to a possible value of the random variable. The height of each step represents the cumulative probability. For a random variable X, the CDF is denoted as $F(X)$, and its formula is given by $F(X) = Pr(X \leq x)$.

1.1.2 Bayes' Theorem

The conditional probability of an event is a probability obtained with the additional information that some other event has already occurred. We used $P(B|A)$ to denote the conditional probability of event B occurring, given that event A has already occurred. The following formula was provided for finding $P(B|A)$: $P(B|A) = P(A \text{ and } B)/P(A)$.

A prior probability is an initial probability value originally obtained before any additional information is obtained. A posterior probability is a probability value that has been revised by using additional information that is later obtained. Now, we can give the Bayes' theorem. The probability of event A, given that event A has subsequently occurred, is given by

$$P(A|B) = \frac{P(A) \cdot P(B|A)}{P(A) \cdot P(B|A) + P(\bar{A}) \cdot P(B|\bar{A})}, \tag{1.1}$$

where $\bar{A}$ refers to the event A does not happen.

Bayes' theorem for probability densities

There is also a version of Bayes' theorem for continuous distributions. It is somewhat harder to derive since probability densities, strictly speaking, are not probabilities, so Bayes' theorem has to be established by a limit process. Bayes' theorem for probability densities is formally similar to the theorem for probabilities. Suppose we have data x which we model using the PDF $f(x|\theta)$, which depends on parameter θ. Once we have observed the data, $f(x|\theta)$ is the likelihood function for θ and is a function of θ (for fixed x) rather than of x (for fixed θ).

Furthermore, suppose we have prior probability about likely values of θ expressed by a PDF $\pi(\theta)$. The resulting distribution for θ is called the posterior distribution for θ as it expresses our beliefs about θ given the data. It summarizes all the knowledge about the parameter θ. The posterior PDF for θ is

$$\pi(\theta|x) = \frac{\pi(\theta) \cdot f(x|\theta)}{f(x)}. \tag{1.2}$$

Bayes' theorem can be rewritten as $\pi(\theta|x) \propto \pi(\theta) \cdot f(x|\theta)$. That is, posterior $\propto$ prior $\times$ likelihood. Therefore, to obtain the posterior distribution, we need (i) data, from which we can form the likelihood $f(x|\theta)$, and (ii) a suitable distribution, $\pi(\theta)$, that represents our prior beliefs about θ.

1.1.3. Information Theorem

We use a probability space (Ω, F, P) to describe randomness, where Ω is the sample space (the set of all possible outcomes), F is a σ-algebra over Ω (the collection of events for which probabilities are defined), and P is the probability measure that assigns a probability to each event in F. A random variable is a measurable function from (Ω, F) to $(R, B(R))$, where R is the set of real numbers and $B(R)$ is the Borel σ-algebra on R. We will often omit these notations and only talk about random variables. Furthermore, we will focus on discrete random variables that take values in a discrete subset $\aleph \subset R$, whose distributions can be described by their PMFs, for example, for a random variable X, its PMF is $p(x) = P(X = x)$ for $x \in \aleph$.

Entropy

The entropy $H_b(X)$ in base b of a discrete random variable X is defined as:

$$H_b(X) = -\sum_{x \in \aleph} P(X = x) \log_b P(X = x). \tag{1.3}$$

For $b = 2$, we usually write $H(X)$ instead of $H_2(X)$ and write $\log(q)$ instead of $\log_2(q)$. Thus, $H(X) = -E[\log p(X)]$. $H(X)$ serves as a measure of the average uncertainty we have about the value of X: the less certain we are, the bigger $H(X)$.

Divergence

Let p and q be PMFs on X. The divergence is defined as:

$$D(p\|q) = \sum_{x \in \aleph} p(x) \log \frac{p(x)}{q(x)} = E\left[\log \frac{1}{q(X)}\right] - H(X). \tag{1.4}$$

Divergence is also known as information divergence, Kullback–Leibler (KL) divergence, and relative entropy. We can think of divergence as the cost we incur if we use the distribution q to encode a random variable X with distribution p. Divergence is always nonnegative.

Mutual information

Let X and Y be discrete random variables taking values in X and Y, respectively. The mutual information $I(X; Y)$ between X and Y is defined as

$$I(X;Y) = \sum_{x \in \aleph} \sum_{y \in \aleph} P(X = x; Y = y) \log \frac{P(X = x; Y = y)}{P(X = x)P(Y = y)}, \tag{1.5}$$

notice that $I(X; Y) = D(p_{X;Y} \| P_x P_Y)$.

Hence, we can regard mutual information as a measure of how much dependence there is between two random variables.

1.1.4. Basic Statistical Modeling Approaches

Here, we will briefly introduce the linear regression model, maximum likelihood estimation, Bayesian approach, and GMM.

Linear regression model

The simplest deterministic mathematical relationship between two variables, x and y, is a linear relationship:

$$y = \beta_0 + \beta_1 x. \tag{1.6}$$

We can also call it a linear probabilistic model. If the two (random) variables are probabilistically related, then for a fixed value of x, there is uncertainty in the value of the second variable. Hence, we assume

$$Y = \beta_0 + \beta_1 x + \varepsilon, \tag{1.7}$$

where ε is a random variable, following an independent and identically distributed distribution with the mean being 0 and standard deviation being σ^2. The values of β_0, β_1, and σ^2 will almost never be known to an investigator. Instead, sample data consists of n observed pairs (x_1, y_1), …, (x_n, y_n), from which the model parameters and the true regression line itself can be estimated. The data (pairs) are assumed to have been obtained independently of one another. The "best fit" line is motivated by the principle of the least squares or can be optimized by the maximum likelihood estimation, which is given by

$$\underset{\beta 0,\, \beta 1}{\operatorname{argmin}} \sum_{i=1}^{n} [y_{\mathrm{i}} - x_{\mathrm{i}}]^2 . \tag{1.8}$$

Maximum likelihood estimation

Whenever we have a distribution $p(X;\theta)$ parametrized by some parameter θ. We may use data to find a value of θ that maximizes the likelihood that the data would have been generated by a distribution with this choice of parameter. For instance, assume that we observe a set of temperature measurements $X = \{x_{1,\ \ldots,}\ x_m\}$. In this case, we could try finding a normal distribution such that the likelihood $p(X;\theta)$ of the data under the assumption of a normal distribution is maximized. Note that this does not imply in any way that the temperature measurements are actually drawn from a normal distribution. Instead, it means that we are attempting to find the Gaussian that fits the data in the best fashion. While this distinction may appear subtle, it is critical: we do not assume that our model accurately reflects reality. Instead, we simply try doing the best possible job at modeling the data given a specified model class. Later, we will encounter alternative approaches at estimation, namely Bayesian methods, which make the assumption that our model ought to be able to describe the data accurately. For a given model or PDF $p(.;\theta)$ parametrized by θ and observations X, the maximum likelihood estimator (MLE) is given by

$$\hat{\theta}_{ML} = \arg\max_{\theta} p(X;\theta) . \tag{1.9}$$

Given different kinds of PDF of $p(.;\theta)$, then we can optimize this model.

Bayesian approach: Maximum a posteriori estimator

The analysis of the maximum likelihood method might suggest that inference is a solved problem. After all, in the limit, MLE is unbiased and exhibits as small variance as possible. Empirical results use a finite amount of data. While not making any assumptions can lead to interesting and general theorems, it ignores the fact that, in practice, we almost always have some idea about what to expect of our solution. It would be foolish to ignore such additional information. Consequently, such prior knowledge should be incorporated into the estimation process. In fact, the use of side information to guide estimation turns out to be a tool for building estimators that work well in high dimensions.

Recall Bayes' rule which states that $p(\theta|x) \propto p(\theta) \cdot p(x|\theta)$. In our context, this means that if we are interested in the posterior probability of θ assuming a particular value, we may obtain this using the likelihood (often referred to as evidence) of x having been generated by θ via $p(x|\theta)$ and our prior belief $p(\theta)$ that θ might be chosen in the distribution generating x. Observe the subtle but important difference to MLE: instead of treating θ as a parameter of a density model, we treat θ as an unobserved random variable that we may attempt to infer given the observations X. This can be done for a number of different purposes: we might want to infer the most likely value of the parameter given the posterior distribution $p(\theta|X)$. This is achieved by

$$\hat{\theta}_{MAP}(X) = \arg\max_{\theta} p(\theta|X) = \arg\max_{\theta} p(X|\theta) \cdot p(\theta). \tag{1.10}$$

The second equality follows since $p(X)$ does not depend on θ. This estimator is also referred to as the maximum a posteriori (MAP) estimator. It differs from the MLE by adding the negative log, before the optimization problem. For this reason, it is sometimes also referred to as penalized MLE. Effectively, we are penalizing unlikely choices θ via $p(\theta)$.

Note that using $\hat{\theta}_{MAP}(X)$ as the parameter of choice is not quite accurate. After all, we can only infer a distribution over θ, and in general, there is no guarantee that the posterior is indeed concentrated around its mode. A more accurate treatment is to use the distribution $p(\theta|X)$ directly via

$$p(x|X) = \int p(x|\theta) \cdot p(\theta|X)d\theta. \tag{1.11}$$

In other words, we integrate the unknown parameter θ and obtain the density estimate directly. As we will see, it is generally impossible to solve this equation exactly, an important exception being conjugate priors. In other cases, one may resort to sampling from the posterior distribution to approximate the integral. That is why we often see the Markov chain and Monte Carlo (MCMC) approach and the variational Bayesian inference approach being employed to approximate the MAP solution.

1.2 Basic and Popular Machine Learning Approaches

Here, we will introduce the basic and popular machine learning approaches to be used in this book.

1.2.1 Classification and Clustering Methods

k-Nearest neighbor

The nearest neighbor (NN) method is a fundamental approach in machine learning and pattern recognition. It is primarily used for classification tasks. The key idea is to predict the label or value of a new data point based on the labels or values of the closest data points in the training set. For classification, the k-nearest neighbor (k-NN) algorithm is often used, where the label of a query point is determined by the majority vote among its k closest neighbors. The distance metric used, such as Euclidean or Manhattan distance, plays a critical role in determining "closeness."

K-Means

The K-means algorithm is an unsupervised machine learning method used for clustering, where the goal is to group data into k clusters based on similarity. Unlike the NN method, which requires labeled data for classification or regression, K-means identifies natural groupings in unlabeled datasets.

Key steps in K-means:

1. Initialization: Randomly select k initial centroids.
2. Cluster assignment: Assign each data point to the nearest centroid using a distance metric, usually the Euclidean distance. This assignment is expressed as:

$$c_i = \arg \min_{j \in \{1,2,\ldots,k\}} \|x_i - \mu_j\|^2, \tag{1.12}$$

   where c_i is the cluster assigned to data point x_i, and μ_j is the centroid of cluster j.
3. Centroid update: Recompute the centroid of each cluster as the mean of all points assigned to that cluster:

$$\mu_j = \frac{1}{|C_j|} \sum_{x_i \epsilon C_j} x_i. \tag{1.13}$$

4. Iteration: Repeat the assignment and update steps until the centroids stabilize or a convergence criterion is met.

Support vector machines

Support vector machine (SVM) is a powerful and versatile supervised learning algorithm used for classification, regression, and outlier detection tasks.

SVM is particularly well-suited for high-dimensional data and problems where the decision boundary between classes is not linear. At its core, SVM seeks to find the optimal hyperplane that separates data points of different classes with the maximum margin. The margin is the distance between the hyperplane and the nearest data points of each class, known as support vectors. The goal of SVM is to maximize this margin, ensuring robust classification.

In the case of linearly separable data, SVM tries to find a hyperplane $w^T x + b = 0$ such that:

$$w^T x_i + b \geq 1 \text{ for } y_i = +1, \tag{1.14}$$

$$w^T x_i + b \leq -1 \text{ for } y_i = -1, \tag{1.15}$$

where w is the normal vector to the hyperplane, b is the bias term, y_i is the class label (+1 or −1), and x_i is a data point. The margin is $\frac{2}{\|w\|}$, and the optimization problem becomes:

$$\min_{w,b} \frac{1}{2}\|w\|^2, \tag{1.16}$$

subject to the constraints $y_i(w^T x + b) \geq 1 \forall i$.

Random Forests

Random forest (RF) builds a collection of decision trees during training, where each tree is grown on a randomly selected subset of the training data and features. The algorithm operates in the following steps:

1. Bootstrap sampling (bagging): A random subset of the training data is selected with replacement to train each tree. This process ensures that each tree sees slightly different data, introducing diversity. This technique is known as bagging (bootstrap aggregating).
2. Feature subset selection: At each split in the tree, a random subset of features is considered instead of evaluating all available features. This further introduces randomness and reduces correlation between trees.
3. Tree construction: Each decision tree is grown to its maximum depth or until a stopping criterion (e.g., minimum number of samples per leaf node) is met. Unlike traditional decision trees, RF does not prune trees, allowing them to capture intricate patterns.

4. Aggregation: For classification, the output is determined by majority voting among all the trees. For regression, the output is the average of the predictions from all the trees.

RF is a powerful and versatile supervised learning algorithm used for both classification and regression tasks. It can reduce overfitting, increase generalization, and improve prediction accuracy compared to individual decision trees.

Gaussian mixture model and expectation-maximization algorithm

A GMM is a probabilistic model used to represent a dataset as a mixture of multiple Gaussian distributions. Each component in the mixture is a multivariate Gaussian distribution characterized by its mean, covariance, and a mixing coefficient that reflects the proportion of data points belonging to that component (Fig. 1.1). GMMs are widely used in clustering, density estimation, and pattern recognition due to their ability to model data with complex, multimodal distributions.

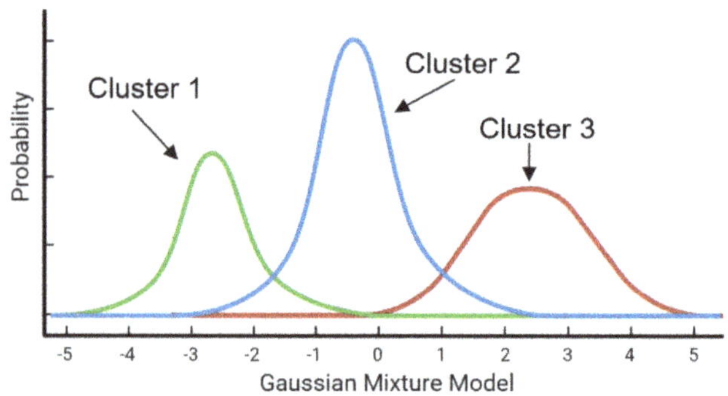

Fig. 1.1 Gaussian mixture model with three components.

Mathematically, the PDF of a GMM is given as:

$$p(x) = \sum_{k=1}^{K} \pi_k \mathcal{N}\left(x|\mu_k, \sum_k\right), \tag{1.17}$$

where K is the number of Gaussian components, π_k is the mixing coefficient for the k-th component ($\sum_{k=1}^{K} \pi_k = 1$ and $0 \leq \pi_k \leq 1$), and $\mathcal{N}\left(x|\mu_x, \sum_k\right)$ is the Gaussian density function with mean μ_x and covariance matrix $\sum_k$.

Unlike algorithms such as K-means, which assign each data point to a single cluster, GMMs provide a soft clustering approach. Each data point has a probabilistic association with each cluster, represented by the posterior probability:

$$\gamma_{ik} = \frac{\pi_k \mathcal{N}\left(x_i | \mu_k, \sum_k\right)}{\sum_{j=1}^{K} \pi_j \mathcal{N}\left(x_i | \mu_j, \sum_j\right)}, \tag{1.18}$$

where γ_{ik} represents the probability that data point belongs to cluster k.

The expectation-maximization (EM) algorithm is a powerful optimization technique used to estimate the parameters of a GMM in the presence of incomplete or hidden data (e.g., cluster assignments). It iteratively improves the model parameters by alternating between two steps:

1. Expectation (E-step): Compute the posterior probabilities γ_{ik} of the latent variables (cluster assignments) based on the current parameters (π_k, μ_k, $\sum_k$).
2. Maximization (M-step): Update the parameters to maximize the likelihood of the data given the posterior probabilities:

   Update the mixing coefficients: $\pi_k = \frac{N_k}{N}$, where $N_k = \sum_{i=1}^{N} \gamma_{ik}$ is the effective number of points assigned to cluster k and N is the total number of data points.

   Update the means: $\mu_k = \frac{\sum_{i=1}^{N} \gamma_{ik} x_i}{N_k}$.

   Update the covariance matrices: $\sum_k = \frac{\sum_{i=1}^{N} \gamma_{ik} (x_i - \mu_k)(x_i - \mu_k)^T}{N_k}$.

The algorithm alternates between the E-step and M-step until convergence.

1.2.2 Dimension Reduction Methods

Principal component analysis

Principal component analysis (PCA) is a widely used dimensionality reduction technique that transforms a high-dimensional dataset into a lower-dimensional one while retaining as much of the original variance as possible. PCA achieves this by identifying new orthogonal axes, called principal components, along which the variance of the data is maximized. These components capture the most significant patterns or directions in the data, enabling efficient data compression and noise reduction.

The PCA algorithm involves several key mathematical steps:

1. Centering the data: To ensure the data is centered around the origin, the mean of each feature is subtracted from the corresponding data points.
2. Computing the covariance matrix: The covariance matrix of the centered data captures the relationships between features.
3. Eigenvalue decomposition: The covariance matrix is decomposed into eigenvalues and eigenvectors. The eigenvalues represent the amount of variance explained by each eigenvector (or principal component).
4. Selecting principal components: Eigenvectors are ranked based on their eigenvalues, with the largest eigenvalues corresponding to the principal components that explain the most variance. If d components are desired, the top d eigenvectors are chosen.
5. Projecting data: The original data is projected onto the selected principal components to obtain the lower-dimensional representation.

t-Distributed stochastic neighbor embedding

t-Distributed stochastic neighbor embedding (t-SNE) is a popular machine learning algorithm for visualizing high-dimensional data in lower-dimensional spaces, typically 2D or 3D. Developed by Laurens van der Maaten and Geoffrey Hinton [1], t-SNE excels at preserving local structures in the data, making it particularly well-suited for visualizing clusters and patterns in complex datasets. Unlike linear dimensionality reduction methods such as PCA, t-SNE is nonlinear, focusing on capturing relationships between similar data points while separating dissimilar points.

For each data point i, t-SNE calculates the conditional probability $p_{j|i}$ that another point j is a neighbor of i in the high-dimensional space:

$$p_{j|i} = \frac{\exp(-\|x_i - \mathrm{x}_j\|^2/2\sigma_i^2)}{\sum_{k \neq i} \exp(-\|x_i - x_k\|^2/2\sigma_i^2)}, \tag{1.19}$$

where σ_i is the bandwidth of a Gaussian distribution centered on x_i, adjusted to ensure that the perplexity of the distribution matches a user-defined value.

The joint probability p_{ij} is symmetrized:

$$p_{ij} = \frac{p_{j|i} + p_{i|j}}{2n}, \tag{1.20}$$

where n is the number of data points.

In the lower-dimensional space, t-SNE models the similarity between points i and j using a Student's t-distribution:

$$q_{ij} = \frac{(1 + \|y_i - y_j\|^2)^{-1}}{\sum_{k \neq l}(1 + \|y_k - y_l\|^2)^{-1}}, \tag{1.21}$$

where y_i and y_j represent the low-dimensional coordinates of points i and j.

The algorithm minimizes the KL divergence between the probability distributions $P = \{p_{ij}\}$ and $Q = \{q_{ij}\}$:

$$KL(P\|Q| = \sum_{i \neq j} p_{ij} \log(\frac{p_{ij}}{q_{ij}}). \tag{1.22}$$

Gradient descent is used to optimize the positions of points in the low-dimensional space.

Uniform manifold approximation and projection

Uniform manifold approximation and projection (UMAP) is a state-of-the-art dimensionality reduction algorithm that excels in preserving both the local and the global structures of high-dimensional data when mapping it to a lower-dimensional space. Developed by McInnes, Healy, and Melville [2], UMAP is grounded in manifold learning and fuzzy topological principles, making it highly effective for visualization, clustering, and data compression. Compared to similar techniques such as t-SNE, UMAP is computationally faster and better at capturing global structures.

UMAP works by modeling the high-dimensional data as a weighted graph and optimizing a lower-dimensional representation to preserve the relationships in the original data.

1. Construction of a high-dimensional graph
   Manifold assumption: UMAP assumes that the data lies on a Riemannian manifold embedded in a high-dimensional space. It aims to approximate the manifold's structure in the lower-dimensional space.

NN: For each data point, UMAP identifies its k-NN using a distance metric (e.g., Euclidean distance) to define local connectivity.
Fuzzy simplicial complex: The relationships between data points are represented as probabilities, forming a weighted graph where the edge weights reflect the probability of one point being connected to another in the manifold.

2. Construction of a low-dimensional graph
UMAP initializes a low-dimensional representation of the data and constructs a graph in this space. The weights of edges in this graph represent the relationships between points in the lower-dimensional space.
3. Optimization via cross-entropy
The two graphs (high-dimensional and low-dimensional) are aligned by minimizing a cross-entropy loss function, ensuring that the local and global structures are preserved:

$$L = \sum_{(i,j)} w_{i,j} \log\left(\frac{p_{ij}}{q_{ij}}\right) + (1 - w_{i,j}) \log(\frac{(1 - p_{ij})}{(1 - q_{ij})}), \tag{1.23}$$

where p_{ij} and q_{ij} are the edge weights in the high-dimensional and low-dimensional graphs, respectively, and $w_{i,j}$ is the importance of the edge.

1.2.3 Deep Learning Methods

Neural networks and backpropagation

Neural networks are a class of machine learning models inspired by the structure and function of biological neural systems. They consist of layers of interconnected nodes, called neurons, organized into an input layer, one or more hidden layers, and an output layer. Each neuron processes information by applying a weighted sum of its inputs, adding a bias term, and passing the result through an activation function, which introduces nonlinearity. Neural networks are highly versatile and have been successfully applied to tasks such as image recognition, natural language processing (NLP), and time-series prediction.

At the core of neural network training is the concept of forward propagation, where input data is passed through the network layer by layer. Each layer computes an output based on its weights and biases, which are adjusted during the training process. The final output is compared to the true labels using a loss function, which quantifies the error between the predicted and actual outputs.

Common loss functions include mean squared error for regression tasks and cross-entropy loss for classification problems.

The backpropagation algorithm is a cornerstone of neural network training, enabling the network to learn by iteratively updating its weights and biases. Backpropagation works by computing the gradient of the loss function with respect to each weight and bias in the network using the chain rule of calculus. This process involves two main steps: a backward pass and an optimization step. In the backward pass, the algorithm propagates the error signal from the output layer back through the network, calculating how much each weight contributed to the error. The gradients obtained are then used in the optimization step, where weights are updated to minimize the loss. This is typically achieved using an optimization algorithm such as stochastic gradient descent (SGD) or its variants such as Adam.

One of the challenges in backpropagation is the vanishing gradient problem, where gradients diminish as they are propagated backward through deep networks, making it difficult for earlier layers to learn. This issue has been mitigated through advancements such as activation functions such as rectified linear unit (ReLU) and architectural innovations such as residual networks (ResNets).

Autoencoders

Autoencoders are a type of artificial neural network used primarily for unsupervised learning, particularly in tasks related to dimensionality reduction, feature extraction, and data denoising. The architecture of an autoencoder consists of two main parts: an encoder and a decoder (Fig. 1.2). The encoder maps the input data into a lower-dimensional representation, often referred to as the latent space or bottleneck layer. The decoder then attempts to reconstruct the original input data from this compact representation. The goal of an autoencoder is to learn an efficient representation of the input by minimizing the reconstruction error, typically measured using a loss function such as mean squared error.

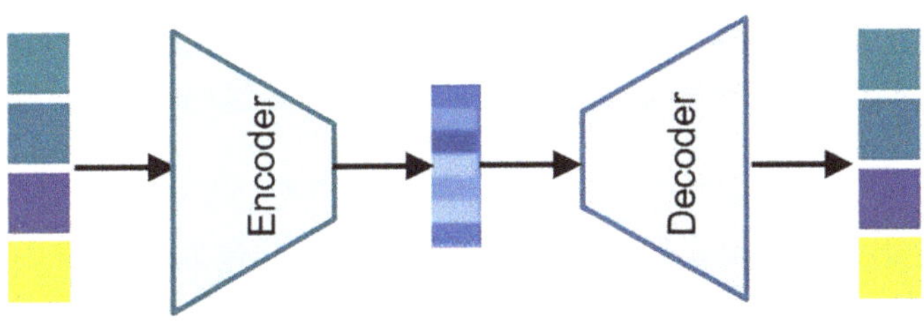

Fig. 1.2 Structure of autoencoder.

The encoder part of an autoencoder compresses the input data into a latent space representation. This is done by passing the input through one or more layers of neurons, with each layer applying weights, biases, and activation functions. The number of neurons in the latent space is typically smaller than in the input layer, which forces the network to learn a compressed, meaningful representation of the data. This reduced-dimensional representation often captures the most salient features or patterns in the input data while discarding noise or irrelevant information.

The decoder part of an autoencoder is responsible for mapping the latent space representation back to the original input space. It consists of layers that gradually increase the dimensionality of the data, using the learned features to reconstruct the input as accurately as possible. The final output of the decoder should ideally be a close approximation of the original input, although in practice, perfect reconstruction is not always achieved, especially in the cases of lossy compression or noisy data.

Autoencoders are trained using a supervised learning approach, with the input data serving as both the input and the target. The network is trained to minimize the reconstruction error, often using gradient descent and backpropagation. Over time, the autoencoder learns to encode and decode the data in a way that captures the essential structures and patterns, which can then be useful for various downstream tasks such as anomaly detection, clustering, or generating new data samples.

One of the main advantages of autoencoders is their ability to learn efficient data representations without needing labeled data. This makes them particularly useful in unsupervised learning scenarios. In addition, autoencoders have been widely applied for tasks such as denoising, where the network is trained to reconstruct clean data from noisy inputs, and image compression, where the encoder learns to represent high-dimensional images in a compact form.

Variational autoencoders

Variational autoencoders (VAEs) are a class of generative models that extend the traditional autoencoder architecture by introducing probabilistic elements, enabling them to generate new data samples that resemble the input data distribution. Unlike regular autoencoders, which directly learn a deterministic mapping from inputs to a lower-dimensional latent space, VAEs model the latent space probabilistically, allowing for more flexible and powerful generative capabilities (Fig. 1.3). The core idea of a VAE is to learn a distribution over the latent variables, which can then be sampled to generate new data points.

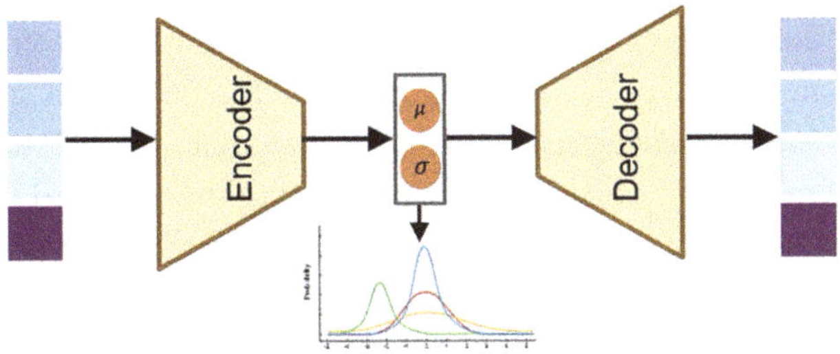

Fig. 1.3 Structure of variational autoencoder.

The architecture of a VAE consists of two main parts: an encoder and a decoder, similar to standard autoencoders. However, instead of outputting a single deterministic encoding, the encoder in a VAE outputs the parameters of a probability distribution, typically a Gaussian distribution, for each latent variable. These parameters are the mean and variance (or log-variance) of the distribution. By sampling from this distribution, the VAE introduces stochasticity into the encoding process, allowing for a more flexible and continuous latent space.

The decoder part of the VAE reconstructs the input data from the sampled latent variables, just like in a standard autoencoder. The goal of training a VAE is to maximize the likelihood of the data under the model, which involves two key components: the reconstruction loss and the KL divergence. The reconstruction loss measures how well the decoder can reconstruct the input data from the latent space representation, usually using a loss function such as mean squared error or binary cross-entropy. The KL divergence, on the other hand, regularizes the latent space by encouraging the learned latent distribution to be close to a prior distribution, typically a standard Gaussian (with zero mean and unit variance). The KL divergence term ensures that the latent space does not become overfitted to the training data, helping to maintain a smooth and well-structured latent space.

Mathematically, the training objective of a VAE is to maximize the variational lower bound (also known as the evidence lower bound [ELBO]), which is a combination of the reconstruction loss and the KL divergence. The ELBO is given by:

$$\mathbb{E}_{q(z|x)}[\log p(x|z)] - D_{KL}[q(z|x) \| p(z)], \tag{1.24}$$

where $q(z|x)$ is the approximate posterior distribution (the encoder's output), $p(x|z)$ is the likelihood function (the decoder), and $p(z)$ is the prior distribution over the latent variables (usually a standard Gaussian).

One of the key advantages of VAEs over traditional autoencoders is their ability to generate new data by sampling from the learned latent space. Since the latent variables are modeled probabilistically, new data points can be generated by sampling from the prior distribution and passing the sampled latent variables through the decoder. This makes VAEs useful in a variety of generative tasks, such as generating images, text, or other complex data structures. In addition, because VAEs enforce a well-structured latent space, they allow for smooth interpolation between data points in the latent space, enabling meaningful variations in the generated samples.

Long short-term memory

Long short-term memory (LSTM) is a type of recurrent neural network (RNN) architecture specifically designed to address the limitations of traditional RNNs when modeling sequential data. RNNs are well-suited for tasks such as time-series forecasting, NLP, and speech recognition because they have the ability to maintain information from previous time steps. However, traditional RNNs suffer from the problem of vanishing gradients, where gradients become very small as they are backpropagated through time, making it difficult for the network to learn long-range dependencies. LSTM addresses this problem by introducing a more sophisticated architecture that can capture long-term dependencies more effectively.

The key innovation in LSTM networks is the introduction of memory cells and gates. Each LSTM unit consists of a memory cell, an input gate, a forget gate, and an output gate. These gates control the flow of information into, out of, and within the memory cell. The memory cell serves as a storage unit for information that can be carried across multiple time steps, allowing the network to remember information for longer durations than a traditional RNN. The gates are responsible for regulating the updating, forgetting, and outputting of information within the cell, making it possible for the network to selectively retain or forget information over time.

The input gate controls how much of the incoming information (from the current input and the previous hidden state) should be stored in the memory cell. The forget gate determines how much of the previous memory should be forgotten, effectively allowing the model to discard irrelevant information. The output gate controls how much of the information in the memory cell should be passed to the next time step as the output of the current LSTM unit. These gates are all governed by learned weight parameters, and the decisions made by the gates are based on the current input and the previous hidden state, which makes the LSTM network capable of dynamically adjusting its memory over time.

The training of an LSTM network involves a process known as backpropagation through time (BPTT), which is an extension of the standard backpropagation algorithm applied to sequential data. During training, the model learns to adjust the weights of the gates to minimize the loss function, typically using a variant of gradient descent such as SGD or Adam. This allows the LSTM to learn complex temporal dependencies by adjusting how information is passed between time steps and selectively updating its memory.

LSTMs have been shown to be particularly effective in tasks involving long-range dependencies, where traditional RNNs might fail. They are widely used in fields such as speech recognition, machine translation, text generation, and time-series forecasting.

Adversarial learning

Adversarial learning is a machine learning paradigm that involves training models through the interaction of two competing networks, typically referred to as the generator and the discriminator. The core idea behind adversarial learning is to create models that can produce high-quality outputs by learning from their interactions with a discriminator model that evaluates the authenticity of these outputs (Fig. 1.4). This type of learning has gained significant attention due to its success in generative models, particularly in the case of generative adversarial networks (GANs), which were introduced by Ian Goodfellow in 2014 [3].

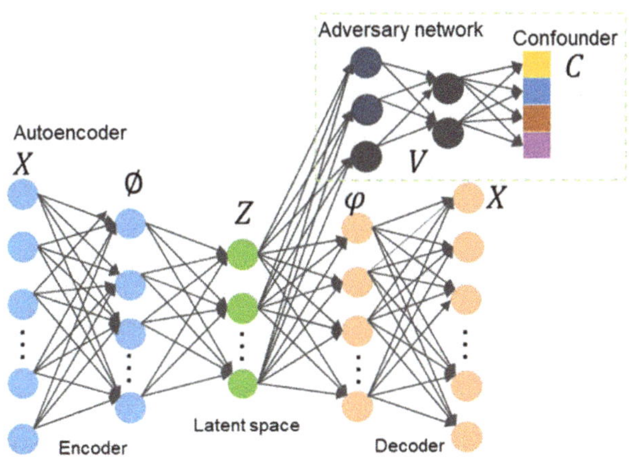

Fig. 1.4 Structure of adversary network.

In an adversarial learning setup, the generator network attempts to create data samples that resemble real data, while the discriminator network's task is to distinguish between real and fake data samples. The generator and discriminator are trained simultaneously but with opposing objectives. The generator learns to generate increasingly realistic data, while the discriminator improves its ability to correctly classify data as real or fake. The training process is akin to a game, where the generator aims to "fool" the discriminator, and the discriminator aims to detect whether the data is real or generated. This setup is often described as a zero-sum game, where the success of one network comes at the expense of the other.

The loss function in adversarial learning reflects the interaction between the generator and the discriminator. In a typical GAN setup, the discriminator's goal is to maximize its ability to correctly classify real and fake data, which is typically achieved through a binary cross-entropy loss. The generator, on the other hand, seeks to minimize the probability that the discriminator can identify its generated data as fake. This is done by minimizing a loss that penalizes the generator when the discriminator successfully distinguishes its output from real data. The overall objective is for the generator to produce data that is indistinguishable from real data and for the discriminator to become more adept at detecting subtle differences.

One of the key challenges in adversarial learning is maintaining the balance between the generator and the discriminator. If the discriminator becomes too powerful, the generator may fail to learn how to produce realistic data, as it will always be outclassed by the discriminator's ability to distinguish fake data. Conversely, if the generator becomes too powerful, it may begin producing data that the discriminator cannot distinguish from real data, but this could result in poor-quality or unrealistic outputs. Techniques such as minibatch discrimination, feature matching, and Wasserstein loss have been proposed to address these challenges and stabilize the training process. Adversarial learning has found widespread applications in generative models, particularly in the creation of GANs.

1.2.4 Large Language Models

The emergence of large language models (LLMs) has brought milestone progress to NLP. These LLMs, particularly exemplified by bidirectional encoder representations from transformers (BERT) [4] and generative pretrained transformer (GPT) [5], usually use the "pretraining + fine-tuning" approach (Fig. 1.5). Pretraining is mainly to train a general/foundation model with strong generalization ability on a large-scale text corpus. Fine-tuning relies on the pretrained model, undergoing additional training for a specific task. This process allows the model to adjust to the unique data and demands of the task at hand. The goal of fine-tuning is to improve the performance of the model on specific tasks, such as sentiment analysis, question answering systems, abstract text summarization, machine translation, and creative text generation in NLP.

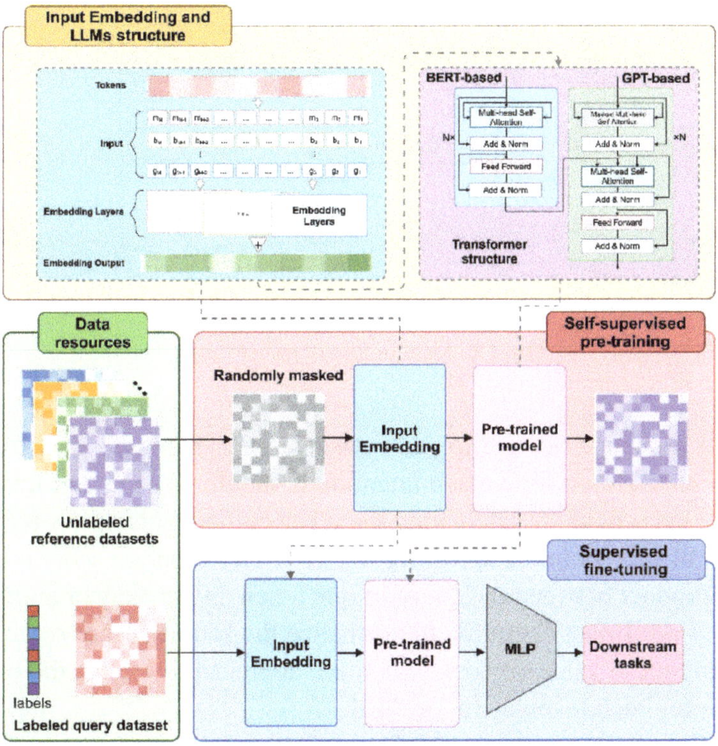

Fig. 1.5 Schematic diagram of the large language model pretraining and fine-tuning process.

Transformers are the foundational architecture in LLMs and consist of two main components: the encoder and the decoder (Fig. 1.6). The encoder takes the input data and processes it in parallel across multiple layers to capture relationships within the sequence. The decoder, on the other hand, generates output sequences based on the encoder's processed information, typically used in tasks such as translation or text generation. Each component is built on layers of multi-head attention, add and norm layer, and feed-forward layer.

Attention mechanism

A key innovation of the transformer is the attention mechanism, particularly self-attention [6], which allows the model to weigh the importance of different

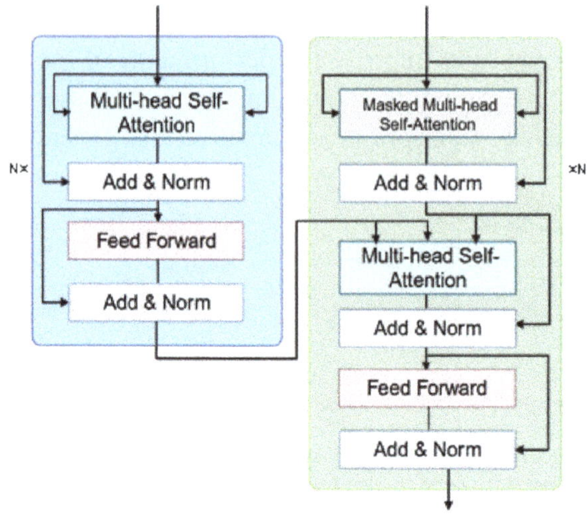

Fig. 1.6 Architecture of transformer.

tokens in a sequence relative to each other. In self-attention, each token computes a score based on how much attention it should pay to other tokens in the sequence. This is done by calculating three key components: query (Q), key (K), and value (V) vectors for each token (Fig. 1.7). The attention score is computed as the dot product between the query of one token and the key of another token, followed by a softmax operation to normalize the scores. These scores are then used to weight the value vectors, which are aggregated to form the output representation for each token as follows [6]:

$$Attention(Q, K, V) = softmax(\frac{QK^T}{\sqrt{d_k}})V. \tag{1.25}$$

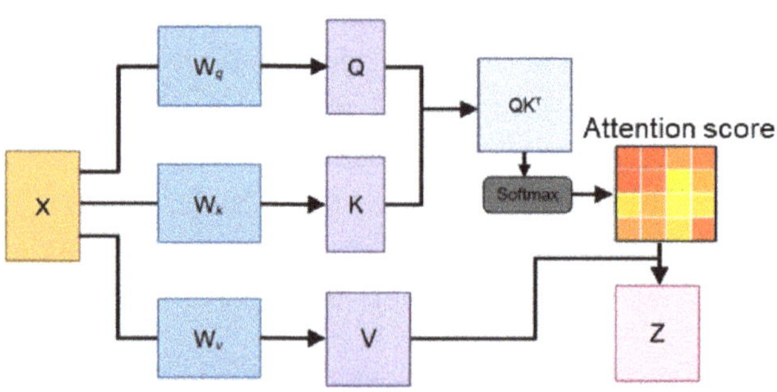

Fig. 1.7 Attention mechanism.

Multi-head attention

Multi-head attention extends this idea by running multiple attention mechanisms (or "heads") in parallel. Each attention head processes the input tokens in a slightly different way by using different sets of learned weights for the Q, K, and V vectors. The results of all heads are concatenated and linearly transformed, allowing the model to capture different aspects of relationships between tokens simultaneously. This mechanism enables the model to focus on various parts of the input sequence at once, learning different types of interactions between tokens. For example, in single-cell foundation models, self-attention can help identify important gene interactions by determining which genes (tokens) should focus on each other during processing. In this way, multi-head attention allows the model to capture complex relationships between genes in single-cell RNA-seq data, where multiple aspects of gene expression (such as co-expression patterns or functional relationships) need to be captured simultaneously.

Add and norm layer

The add and norm layer performs layer normalization and residual connections, which help stabilize training by ensuring that the output from each layer is added to the input before being normalized. This allows for smoother gradient flow and avoids the vanishing gradient problem.

Feed-forward layer

After the attention mechanism, the feed-forward network is a fully connected neural network (Fig. 1.8), helping the model learn complex mappings and capture more abstract representations of the input data.

BERT and GPT stand as two exceptional language models. Both BERT and GPT leverage the transformer architecture, employing attention mechanisms to grasp dependencies within input data.

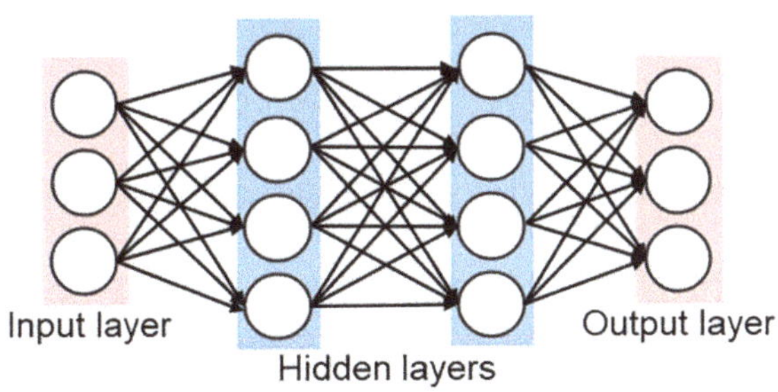

Fig. 1.8 Feed-forward network.

Bidirectional encoder representations from transformers

BERT stands as a revolutionary model in the realm of NLP, reshaping the landscape of deep learning. Introduced by Google in 2018, BERT is engineered to comprehend the context and subtleties of human language in unprecedented ways [4]. At its core, BERT employs a transformer architecture, which enables it to capture the relationships between words in both directions of a sentence, making it "bidirectional." This implies that BERT considers not only the words preceding a given word but also those following it, allowing it to grasp the full meaning of a sentence. One of BERT's most remarkable features is its pretraining process. It learns from an extensive corpus of text from the internet, effectively absorbing vast amounts of knowledge. The pretrained model can subsequently undergo fine-tuning for diverse natural language understanding tasks, including text classification, question answering, and language translation. BERT has consistently demonstrated state-of-the-art performance across a broad spectrum of NLP tasks and benchmarks, thanks to its ability to handle context, polysemy, and long-range dependencies effectively. Its versatility and accuracy have made it a pivotal tool in various applications, from chatbots and virtual assistants to sentiment analysis and content recommendation systems.

BERT, originally developed for NLP, has emerged as a powerful tool in bioinformatics, enabling new ways to analyze and interpret complex biological data. BERT is typically pretrained using a masked language model (MLM) task, where random tokens in a sequence are masked, and the model is tasked with predicting them. At its core, BERT's bidirectional transformer architecture allows it to capture intricate contextual relationships within sequences (Fig. 1.9), making it particularly well-suited for modeling the sequential and contextual nature of biological data such as DNA, RNA, and proteins.

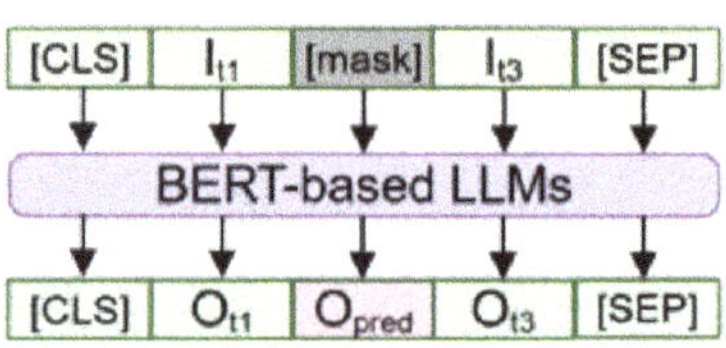

Fig. 1.9 BERT-based bidirectional prediction pretraining process.

In bioinformatics, BERT has been adapted for a variety of tasks, ranging from genomic sequence analysis to protein structure prediction and single-cell transcriptomics. For example, models such as DNABERT [7] and ProteinBERT [8] leverage BERT's architecture to analyze nucleotide sequences and predict protein functions, respectively. By treating genomic sequences or amino acid chains as "sentences" and individual bases or residues as "words," BERT-based models can identify patterns, dependencies, and functional relationships embedded in biological sequences.

In single-cell studies, adaptations such as scBERT [9] have shown great promise. By applying BERT's MLM approach to single-cell RNA-seq data, scBERT can learn the co-expression patterns and complex transcriptional relationships between genes. This enables it to capture the full transcriptional context of individual cells, improving downstream analyses such as cell type classification and trajectory inference. Similarly, transformer-based models inspired by BERT have been applied to tasks such as enhancer-gene prediction, functional annotation of noncoding regions, and drug discovery, showcasing their versatility across diverse areas in bioinformatics.

The use of BERT in bioinformatics represents a paradigm shift, demonstrating how deep learning models originally designed for text data can be effectively adapted to biological challenges. By leveraging BERT's ability to learn from large-scale datasets and capture contextual information, researchers can uncover novel insights into biological systems, paving the way for advancements in precision medicine, systems biology, and functional genomics.

Generative pretrained transformer

GPT stands as a remarkable achievement in the realm of NLP. Introduced by OpenAI, GPT is a sophisticated language model designed to generate human-like text, making it a pivotal component in a wide array of applications [5]. At its core, GPT is built upon the transformer architecture, renowned for its attention mechanisms that allow it to comprehend the nuances of context and language structure. What sets GPT apart is its "pretraining" phase, where it's exposed to an extensive corpus of text from the internet, enabling it to absorb an immense amount of linguistic knowledge. The real brilliance of GPT lies in its capacity to produce text that is both coherent and contextually relevant. Given a prompt, it can produce everything from creative stories and natural-sounding dialog to summaries and translations. This versatility has found applications in chatbots, content generation, language translation, and even code generation. GPT's knack for understanding context, handling ambiguity, and generating human-like text has earned it a place of honor in various domains, from creative writing and customer support to academic research and data analysis. Its adaptability, powered by fine-tuning, makes it a potent tool for diverse language-related tasks.

Similar to BERT, GPT has been increasingly adapted for bioinformatics, leveraging its ability to generate, understand, and analyze sequential data across various innovative applications. Unlike BERT, GPT employs a unidirectional training approach, processing input sequences from left to right (Fig. 1.10). It is trained using an autoregressive learning method, where each token is predicted based on its preceding tokens, making it particularly effective for generative tasks. A standout feature of GPT is its proficiency in zero-shot learning, enabling it to perform tasks without requiring task-specific training data. With its transformer-based architecture, GPT excels in modeling long-range dependencies and capturing contextual relationships within sequences, making it highly suited for analyzing complex biological datasets, such as genomic sequences, protein structures, and cellular pathways.

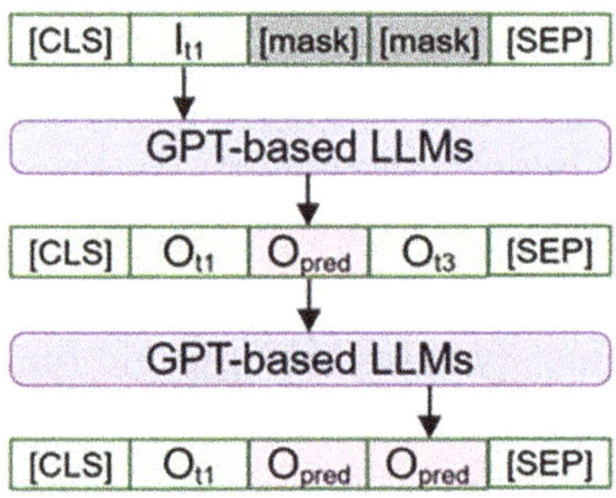

Fig. 1.10 GPT-based bidirectional prediction pretraining process

In bioinformatics, GPT has been leveraged to address challenges in areas such as genomic sequence generation, protein design, and single-cell transcriptomics. For instance, models inspired by GPT, such as DNAGPT [10], use its autoregressive capabilities to predict the next nucleotide or amino acid in a sequence, enabling tasks such as de novo DNA or protein sequence generation. This is particularly valuable in synthetic biology and drug discovery, where designing novel biological molecules is a critical goal.

In single-cell biology, scGPT [11] has been utilized to model gene expression profiles and predict cellular states by learning from large-scale single-cell RNA-seq datasets. scGPT can generate realistic transcriptional profiles and predict cell-type-specific gene expression patterns, providing valuable insights into cellular heterogeneity and differentiation trajectories. These capabilities significantly enhance the study of complex cellular processes and improve our understanding of single-cell dynamics.

The generative capabilities of GPT also extend to biomedical text generation and knowledge extraction. Models fine-tuned on biomedical literature, such as PubMed articles, are employed for tasks such as summarizing research findings, generating hypotheses, or even creating training datasets for other bioinformatic tasks [12–18].

GPT's application in bioinformatics exemplifies the transformative potential of generative artificial intelligence (AI) in biological research. By harnessing GPT's ability to model and generate biological data, researchers can accelerate discovery processes, uncover hidden biological patterns, and push the boundaries of what is achievable in computational biology and precision medicine.

Comparison between BERT and GPT

BERT and GPT stand as two exceptional language models that have transformed the landscape of NLP. While they share some common traits, they also exhibit significant differences in their design and applications. Both BERT and GPT leverage the transformer architecture, employing attention mechanisms to grasp contextual information and dependencies within text data. This architecture has demonstrated significant effectiveness in comprehending and generating text that resembles human language. Furthermore, both models go through a pretraining phase, during which they are exposed to extensive amounts of text data sourced from the internet. This unsupervised learning process helps them learn language structure, grammar, and a broad range of linguistic patterns. Furthermore, both BERT and GPT excel at transfer learning. These models undergo pretraining on an extensive text corpus and can be fine-tuned for particular downstream tasks, including text classification, sentiment analysis, or language generation. This makes them versatile and applicable across various NLP tasks.

However, there are notable differences between these models. BERT employs a bidirectional approach, considering both left and right contexts when learning language representations. It aims to create deep contextual embeddings for words. In contrast, GPT uses a left-to-right approach, generating text autoregressively by predicting the next word in a sequence based on the preceding words. This design makes GPT more suitable for generation tasks. Moreover, their architectures differ, with BERT using a bidirectional transformer encoder and an MLM objective during pretraining, while GPT uses a unidirectional transformer decoder and a causal language model objective. BERT undergoes training through an MLM task and next sentence prediction, whereas GPT

(GPT-1) is trained to predict the succeeding word in a sequence. In practice, BERT is often preferred for tasks requiring deep contextual understanding, such as question answering, sentiment analysis, and text classification. It is widely used for tasks where understanding the context of words is crucial. On the other hand, GPT shines in text generation tasks, including creative writing, language translation, and dialog generation. It is an excellent choice for tasks that prioritize generating human-like text.

1.3 Conclusion

The convergence of AI and big data sciences with bioinformatics and systems medicine has ushered in a new era of precision and scale in biomedical research. This chapter has emphasized the foundational statistical modeling approaches, including probability theory, Bayes' theorem, and information theory, as critical tools for processing and interpreting the vast amounts of data generated in life sciences. These methods enable researchers to extract meaningful insights from complex biological systems and create predictive models tailored to systems medicine applications.

The incorporation of machine learning approaches, particularly in the context of bioinformatics, highlights their transformative impact on big data analytics. Classification and clustering methods, such as SVMs and k-means, are pivotal for identifying biological patterns, defining molecular subtypes, and organizing multidimensional datasets. Dimensionality reduction techniques, such as PCA and t-SNE, simplify high-dimensional biological datasets, such as single-cell multi-omic or proteomic data, enabling intuitive visualization and discovery of novel biological phenomena.

Deep learning and its application in bioinformatics and systems medicine have emerged as game-changers. Neural networks, including autoencoders and LSTM, have redefined tasks such as protein structure prediction, drug discovery, and medical imaging. Furthermore, the integration of LLMs, such as GPT and BERT, has extended these capabilities to biomedical text mining, enabling automated literature reviews, hypothesis generation, and knowledge graph construction. These advancements underline the potential of AI to bridge the gap between data generation and actionable insights in medicine.

In conclusion, the synergy between AI, big data sciences, and bioinformatics drives innovation in systems medicine, creating unprecedented opportunities to decipher complex biological systems. By leveraging advanced computational approaches alongside statistical foundations, researchers are better equipped to address challenges in personalized medicine, drug discovery, and disease modeling. As these technologies continue to evolve, their potential to accelerate breakthroughs in systems medicine and improve patient outcomes remains unparalleled.

References

1. Van der Maaten, L. and G. Hinton. Visualizing data using t-SNE. *J Mach Learn Res*. 2008; **9**(11): p. 2579–2605.
2. McInnes, L., J. Healy, and J. Melville. UMAP: uniform manifold approximation and projection for dimension reduction. *arXiv preprint arXiv:1802.03426*, 2018.
3. Goodfellow, I., et al. Generative adversarial nets. *Adv Neural Inf Process Syst*. 2014; **27**.
4. Devlin, J., et al. BERT: pre-training of deep bidirectional transformers for language understanding. *arXiv preprint arXiv:1810.04805*, 2018.
5. Radford, A., et al. Improving language understanding by generative pre-training. Preprint 2018.
6. Vaswani, A., et al. Attention is all you need. *Adv Neural Inf Process Syst*. 2017; **30**.
7. Ji, Y., et al. DNABERT: pre-trained bidirectional encoder representations from transformers model for DNA-language in genome. *Bioinformatics*. 2021; **37**(15): p. 2112–2120.
8. Brandes, N., et al. ProteinBERT: a universal deep-learning model of protein sequence and function. *Bioinformatics*. 2022; **38**(8): p. 2102–2110.
9. Yang, F., et al. scBERT as a large-scale pretrained deep language model for cell type annotation of single-cell RNA-seq data. *Nat Mach Intell*. 2022; **4**(10): p. 852–866.
10. Zhang, D., et al. DNAGPT: a generalized pretrained tool for multiple DNA sequence analysis tasks. *bioRxiv*. 2023: p. 2023.07. 11.548628.
11. Cui, H., et al. scGPT: toward building a foundation model for single-cell multi-omics using generative AI. *Nat Methods*. 2024; **21**(8): p. 1470–1480.
12. Hou, W. and Z. Ji. Assessing GPT-4 for cell type annotation in single-cell RNA-seq analysis. *Nat Methods*. 2024; **21**(8): p. 1462–1465.
13. Chen, Y. and J. Zou. GenePT: a simple but effective foundation model for genes and cells built from ChatGPT. *bioRxiv*. 2024: p. 2023.10. 16.562533.
14. Zhao, S., et al. LangCell: language-cell pre-training for cell identity understanding. *arXiv preprint arXiv:2405.06708*, 2024.
15. Lu, Y.-C., et al. scChat: a large language model-powered co-pilot for contextualized single-cell RNA sequencing analysis. *bioRxiv*. 2024: p. 2024.10. 01.616063.
16. Liu, T., et al. scELMO: embeddings from language models are good learners for single-cell data analysis. *bioRxiv*. 2023: p. 2023.12. 07.569910.
17. Heimberg, G., et al. Scalable querying of human cell atlases via a foundational model reveals commonalities across fibrosis-associated macrophages. *bioRxiv*. 2023: p. 2023.07. 18.549537.
18. Levine, D., et al. Cell2Sentence: teaching large language models the language of biology. *bioRxiv*. 2023: p. 2023.09. 11.557287.

CHAPTER 2

Biomedical Big Data in Electronic Medical Records (EMR), Hematoxylin and Eosin (H&E), and Medical Imaging

The explosive expansion of biomedical big data provides enormous opportunities and challenges for the knowledge discovery of biomedical big data and translational applications in personalized medicine. Efficient management, analysis, and interpretation of big data can transform big data sciences by opening new avenues for personalized medicine. However, unprecedented advances in the automated collection of large-scale molecular and clinical data have presented significant challenges for data analysis and interpretation, requiring the development of new computational approaches. Some potential challenges include the curse of dimensionality, data heterogeneity, missing data, class imbalance, and scalability issues. In this chapter, we review the recent progress and breakthroughs in the application of big data and summarize the content, sources, technologies, tools, challenges, gaps, and opportunities of big data research to improve and advance big data applications in healthcare. We also discuss strategies for overcoming the challenges.

2.1 Introduction

The explosive expansion of biomedical big data [1–3] provides enormous opportunities and challenges for knowledge discovery on biomedical big data and translational applications in personalized medicine. Big data processing in healthcare refers to generating, collecting, analyzing, and holding clinical data that is too vast or complex to be inferred by classical data processing methods. Healthcare big data sources include electronic medical records (EMRs), electronic health records (EHRs), and imaging. These big data contain patients' medical histories, diagnoses, medications, treatment plans, allergies, laboratory test results, genomic sequencing, medical imaging, insurance providers, and other clinical data. In this chapter, we discuss different machine learning algorithms applied to various types of healthcare data, as well as the challenges faced in processing and handling big data and their applications.

Figure 2.1 depicts a general flowchart of big data science for personalized and precision medicine (PPM). It consists of two components: a cloud-based learning module and cloud-based artificial intelligence (AI)-aided diagnosis and treatment. In the training module, patient data are collected and deposited into a data warehouse when they come to the hospital, including EMR data, EHR data, claims data, biomedical imaging data, and omics data generated from tissue/blood species. These types of datasets will require quality control, data imputation, preprocessing, normalization, and batch effects removal through their domain knowledge and data mining approaches, respectively. For example, the new generation of sequencing technologies enables the processing of billions of DNA sequence data per day, and the application of EHRs is documenting large amounts of patient data. Imaging informatics is now more rapidly integrated with cloud platforms to share medical image data and workflows. Public health informatics leverages big data techniques for predicting and monitoring infectious disease outbreaks, such as Ebola. Specifically, in bioinformatics, high-throughput experiments facilitate the research of new genome-wide association studies of diseases. With clinical informatics, the clinical field benefits from the vast amount of collected patient data for making intelligent decisions. In Fig. 2.1, once we process each type of data, we can then project them into a signature space where we could potentially integrate the heterogenous data and stratify patients into different subgroups for personalized medicine to predict potential optimal drugs, drug combinations, and other treatment strategies. In the AI-aided diagnosis and treatment stage, clinicians can incorporate various data of a new patient into the trained PPM model. The model can recommend an optimal treatment strategy for this patient. In this chapter, we mainly discuss the following topics related to big data science for PPM.

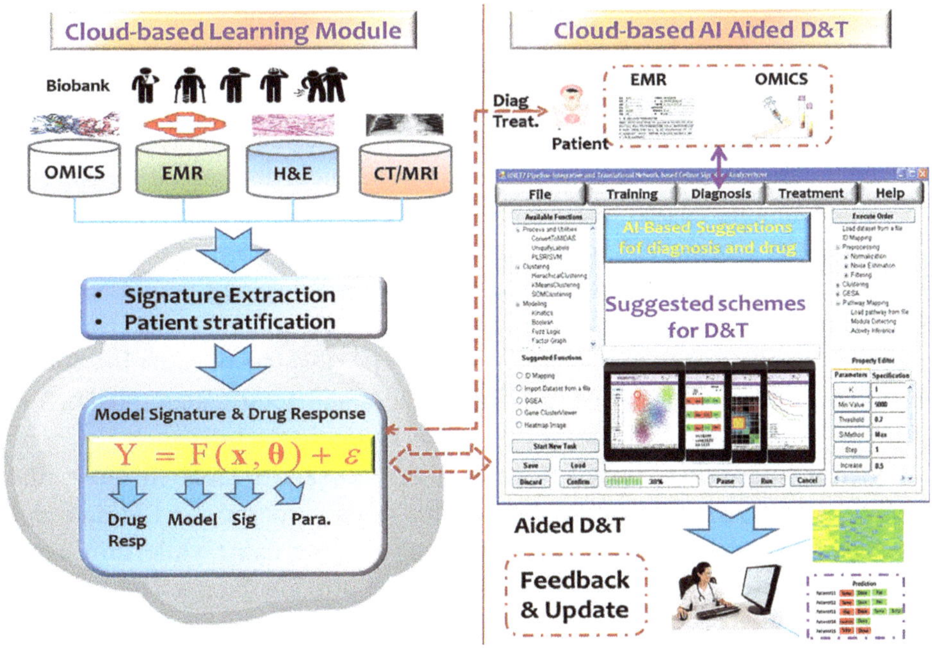

Fig. 2.1 Big data science flowchart for personalized and precision medicine (PPM). Left panel: Training process. Right panel: Prediction of PPM.

2.2 Data Warehouse and Data Management

Clinicians and data scientists often ask where and how to extract the data from a hospital's database when deciding to study a specific disease. To answer this question, we need to understand structured data in different tables and unstructured data such as clinical notes in text format. In the healthcare industry, the sources of big medical data include hospital and pharmaceutical enterprise records, medical records of patients, results of medical examinations, medical costs, and social security records. Figure 2.2 illustrates the overall architecture and data flow in the data management system. The core of the system is the translational data warehouse (TDW) [4] based on i2b2 (Informatics for Integrating Biology & the Bedside) core technology [5]. TDW represents a centralized data repository that aggregates data from nonclinical and clinical data sources. Nonclinical data resources include preclinical centers, nonclinical imaging, and shared resource cores. Clinical and research data sources include clinical EMRs such as EPIC and Public Health Sciences' research databases, clinical research imaging, clinical registries, and data from various clinical centers [4]. The i2b2 framework [5] and open-source toolsets can be used to structure these disparate data sources systematically and integrate them into one central data warehouse, TDW. Hospital databases such as EPIC system have a certain search function, but this function is designed for general searches. Hence, it may not be enough for complex criteria. Therefore, we often need to design an algorithm to meet the needs of clinicians. This can be done in the clinical database. The i2b2 team developed another search and visualization tool called "shrine" on top of the i2b2 core. Once using this tool to find the appropriate patients for certain criteria, doctors can further validate those patients by checking their records and then determine whether they are good candidates for the next step. It is worth pointing out that other tools such as REDCap, Hospital Information System (HIS), and Observational Health Data Sciences and Informatics (OHDSI) are also very good.

Investigator Portal provides researchers access to tools to query data and request services, including datasets for research purposes. Big data mining provides researchers with tools to analyze their data, while informatics services can use developed tools to create and provide datasets for researchers to use. To ensure good data governance, "Data Control Agents (DCAs)" areused to facilitate a centralized, secure, streamlined process for managing clinical research data requests through Institutional Review Board (IRB) and Health Insurance Portability and Accountability Act (HIPAA) processes. The Clinical Research Management System, "Click Commerce," is also accessible via the Investigator Portal. It helps researchers manage their protocols, patient scheduling, study-specific data collection, and analysis for Food and Drug Administration (FDA)-mandated data submission and reporting.

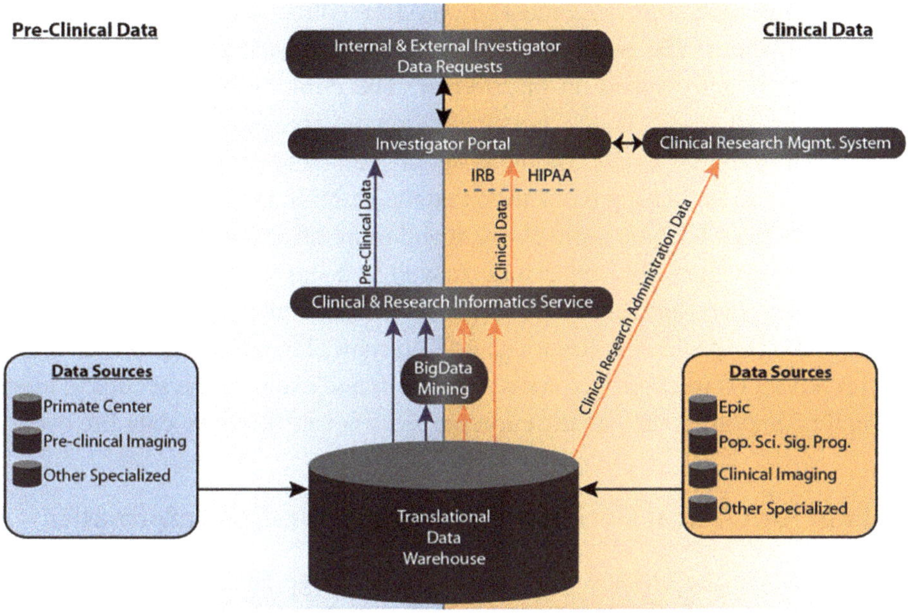

Fig. 2.2 The framework of data warehouse.

Here, we just present an example of TDW with an i2b2 core. Overall, there are over thousands of patients' clinical notes from 2001 to 2020 for a middle-level hospital. These records are usually stored in five tables, including "Patient_dimension," "Visit_dimension," "Observation_fact," "Concept_dimension," and "Provider_dimension". "Patient_dimension" provides the demographic and medical enrollment information of each patient. "Visit_dimension" stores physicians' notes related to each clinical encounter (visit). "Observation_fact" gives detailed records related to all diagnoses, medical procedures, lab tests, and drug information. Every row in the "Observation_fact" table shows a specific clinical operation description. "Concept_dimension" explains various medical items, considered the clinical dictionary for our EMR indexes. "Provider_dimension" stores a mapping table between each provider ID and the related department annotation. International Classification of Diseases, Tenth Revision (ICD-10) is adopted as a diagnosis code. Medical procedure notes are expressed using Healthcare Common Procedure Coding System (HCPCS) [6]. The depiction of drugs and lab tests is given by Current Procedural Terminology (CPT) [7]. All EMR datasets are organized by PostgreSQL database schemas [8].

The biomedical research community generates large volumes of digital data in different formats using different modalities. These data should be findable, accessible, interoperable, and reusable (FAIR). Thus, scientists can take full advantage of them. The ability to reuse, integrate, or add to the existing datasets can significantly speed up discoveries. The curation of digital data assets is critical for all researchers. Digital curation involves annotating, managing, and preserving digital assets. As the costs of managing digital data continue to climb, there is an increasing realization that the curation processes for biomedical research data must become more efficient and more effective. The generation and curation of biomedical datasets can be viewed as a single process involving people at different steps and levels, including scientists at all career levels, technical support staff, IT specialists, curators, and librarians. Therefore, there is an urgent need for advanced methods and tools to monitor and evaluate the accuracy, integrity, quality, and efficiency of the curation process for the entire data life cycle.

2.3 EMR for Clinical Decision-Making and Medical Informatics

Rapid growth in the clinical implementation of large EHRs, fueled by recent federal provisions in the HITECH Act [9, 10], has led to an unprecedented expansion in the availability of large-scale, dense, and longitudinal clinical datasets [9, 11]. Huge efforts are already underway to link EHRs across institutions to facilitate clinical and translational research, for example, to study disease onset and treatment outcomes [12–14]. Numerous studies have reported EMR-based findings [15–17]. EHR plays a crucial role in the newly developed concept of Learning Healthcare System (LHS) [18]. LHS aims to efficiently convert data about care and operations into knowledge, which is further translated into evidence-based clinical practice and health system change. However, challenges exist when utilizing EHR data for research and operation. The heterogeneity and complex nature of EHR data make it difficult to be directly used for analysis. First, approximately 80% of EHR data (and big data in general) consists of unstructured data [19]. The same clinical information can be expressed in different formats with inherent ambiguity. Second, EHR systems are, by nature, highly diverse [20]. Relevant data could be scattered among different documents, databases, and registries in different formats. These data need to be integrated to ensure a profound analysis. Before EHR data can be used for direct analysis, intensive manual data extraction and normalization processes are needed, which becomes a bottleneck for the secondary use of EHR data. Fortunately, recent advances in informatics research have provided alternative solutions to automate data extraction and normalization. Natural language processing technologies for extracting structured clinical data from the unstructured narrative text have shown promising results [21–23]. In addition, ontology-based approaches have also been widely applied to EHR data. For example, the Strategic Health IT Advanced Research Project for Normalization [24], funded by the National Coordinator for Health Information Technology, aims to develop a framework of open-source services that can be dynamically configured to transform EHR data into standards-conforming and comparable information suitable for large-scale analysis.

Despite the success of biomedical informatics research, current informatics tools still have very limited capacity for handling big data. For example, MetaMap [25], a widely used biomedical natural language processing system developed at the National Library of Medicine, takes about 15 days to process 20 million biomedical abstracts, even when using a powerful server enabling parallel processing. EHR systems in large academic centers have accumulated decades of data, often from several million patients, each with potentially thousands of clinical records. To achieve the goal of the LHS, such large-scale EHR data must be manipulated in real time. However, current biomedical informatics software face challenges in storing, querying, and analyzing such large-scale EHR data. Therefore, there is an urgent need to introduce parallel computing frameworks for big data, such as Hadoop, into the biomedical domain, thus enabling existing informatics tools to handle EHR data. The latest research focuses on bridging the gap between raw EHR data and the needs of ontological concept–based big data integration and harmonization by developing new methods and software platforms that allow parallel computing of clinical data in EHR; specifically, topics include (i) developing novel natural language processing- and ontology-based approaches to extract and normalize clinical data in EHR and deliver them in standard ontological formats and (ii) developing an open-source distributed parallel framework to speed clinical data processing and retrieval. As a platform with strong advantages in unstructured data processing, Hadoop can play a key role in more medical big data application scenarios in the future. For example, Hadoop can help doctors to formulate personalized treatment plans for patients, assist in diagnosis, monitor patients' vital characteristics, and so on.

2.3.1 EMR and Clinical Decision

Clinical decision support is the core value of EMR data. Kidd et al. [26] studied 229 kinds of EMR clinical decision support tools (CDST) and assessed 3,786 obstetric inpatients with Obstetric venous thromboembolism evaluation. They found that using CDST can significantly improve the compliance of hospitalized obstetric patients to complete Obstetric venous thromboembolism risk assessments.

A classic application of EMR data is for better healthcare management of patients. By designing an appropriate AI diagnostic model, it is possible to obtain diagnostic results that are practically accurate or even better than those of clinicians. Although clinical diagnosis by AI cannot guarantee completely correct results, nor can it completely replace clinicians, for a considerable number of diseases, such as pneumonia, meningitis, sinusitis, and chickenpox, its accuracy rate can reach more than 90% and even exceed the detection level of doctors. For example, Brown et al. [27] proposed an algorithm for the automatic diagnosis of retinopathy based on deep learning. They used a dataset with 5,511 retinal photos to train a deep convolutional neural network and achieved diagnostic accuracy equivalent to or even better than human experts. Adem et al. [28] designed a SoftMax classification model with a stacked autoencoder method for the automatic diagnosis of cervical cancer. The model was trained and tested on a dataset containing 668 samples and 30 attributes in the UCI database and achieved a classification accuracy of 97.8%.

EMR can also promote and improve compliance with clinical practice guidelines for patient management. Clinical guidelines are a medical standardization standard based on "evidence-based medicine", which is used to help doctors provide appropriate medical and healthcare services under specific circumstances. The clinical practice based on EMR has a good role in promoting traditional clinical guidelines. By analyzing patient data from different sources, a knowledge base is formed to provide a new basis for the improvement of clinical guidelines or clinical pathways. Knowledge sharing through diagnosis and treatment behavior and EMR systems is conducive to achieving safer and more efficient health services. Terasaki et al. [29] have discussed the role of the structured approach of EMR in the outpatient management of chronic obstructive pulmonary disease (COPD). Through experimental comparative analysis, they found that after the implementation of EMR intervention, the diagnostic screening indicators of some COPD patients were significantly improved. Jessica et al. [30] also studied a clinical pathway of obesity hypoventilation syndrome (OHS) through EMR intervention. By comparing and analyzing the success rate of OHS patients treated with bilevel positive airway pressure (BiPAP) before and after EMR intervention, it was confirmed that EMR is possible to help promote and improve BiPAP treatment.

Another classic application of EMR is risk analysis. A risk assessment is the process of identifying potential hazards and analyzing what could happen if a hazard occurs. A risk factor is a variable associated with an increased risk of disease or infection. Linear regression [31], logistic regression [32], and Cox regression [33] are often used to analyze risk and to evaluate the importance of risk factors. For example, based on EMR data, many scholars established different risk prediction models to predict the readmission risk of patients [34, 35]. Mirzaei [36] compared and evaluated the agreement between cardiovascular disease (CVD) prediction of the Iran Package of Essential Non-communicable Disease (IraPEN) and the Framingham risk score (FRS) to estimate the risk of CVD. EMR-based risk analysis overcomes the limitations of traditional medical statistical methods, provides a new data source selection for disease risk assessment, and makes up for the shortcomings of traditional statistical methods in model construction.

2.3.2 EMR and Medical Quality Assessment

Medical quality is a standard to measure the professional level of medical workers, as well as the comprehensive strength of medical institutions and the treatment effect of specific conditions. Traditional medical quality evaluation methods are usually based on expert judgment or peer review, which are strongly subjective and cannot objectively reflect the real level of medical quality. The modern clinical medical evaluation system is based on EMR. With the wide application of EMR, the diagnosis and treatment processes of various diseases have been completely recorded and stored, providing a good basis for medical quality evaluation. The EMR system integrates patient information from various medical systems, allowing data analysts to quickly access and analyze it for improved medical quality evaluation. For example, Arnold et al. used EMRs and diabetes registry data to assess the relationship between timely treatment intensification and hemoglobin A1C levels, controlling quality-of-care performance measures among patients with uncontrolled type 2 diabetes [37]. To evaluate the use quality of antibiotics, Broek [38] used EMR data from three hospitals to assess the adaptability of antibiotic prescriptions, which indirectly illustrates the key role of EMR in various quality evaluations within and between hospitals. Besides, to further verify the importance of EMR to medical quality, Lin [39] assessed the relationship between the degree of EMR adoption and patient outcomes. They divided the levels of EMR utilization into no EMRs, partial EMRs, and full EMRs and designed three primary healthcare quality indicators, including inpatient mortality, readmission within 14 days, and 48-hour postoperative mortality. Finally, it was proved that full EMR adoption could improve healthcare quality in medical institutions treating severely ill patients.

2.3.3 EMR and Health System Performance

EMR plays a key role in improving the performance of the medical and health system [40, 41] because EMR-based health information-sharing platforms can provide all kinds of medical records in time and conduct good management and sharing. Recent studies have also confirmed that EHRs can leverage health information to improve the performance of the entire medical and health service system.

Moreover, in order to strengthen the informatization level of medical institutions and build smart hospitals, many countries have taken the application level of the EMR system as an important indicator of hospital performance appraisal. In China, the application level of the EMR system has been graded, evaluated, and incorporated into the informatization requirements for performance evaluation of level III public hospitals and level II public hospitals. This also shows that EMRs have been in the core position in the construction of hospital informatization.

2.4 Security and Quality of Biomedical Big Data

The Health Information Trust Alliance (HITRUST) is the organization responsible for creating the Common Security Framework (CSF). CSF incorporates and cross-references security requirements from the HIPAA [42], Payment Card Industry (PCI) [43], National Institute of Standards Technology (NIST) [44], Centers for Medicare and Medicaid Services (CMS) [45] Information Security Acceptable Risk Safeguards, and CMS Minimum Security Requirements, so we can easily evaluate our compliance with each of these regulations or standards. Using CSF, we have a flexible and scalable "defense-in-depth" strategy, which allows us to remain current with cybersecurity technology and standards and address emerging threats. Features of this defense-in-depth strategy include (1) security and privacy policy and procedures; (2) new employee and annual training on security and privacy; (3) architecture and infrastructure which implements security at multiple layers from the network perimeter to the desktop; (4) industry-leading products such as TippingPoint Intrusion Prevention Systems, Palo Alto Next-Generation Firewalls and Checkpoint Full Disk Encryption; (5) monitoring of activity and possible incidents using RSA enVision security events and incident monitoring (SEIM) and FairWarning event monitor for HIPAA compliance; (6) risk assessment and vulnerability analysis using the Nessus scanner and HP WebInspect; and (7) commitment to security and privacy demonstrated by hiring a nationally known Chief Information Security Officer (CISO) [46] and a formal Privacy and Security Committee with members from Legal, IRB, Compliance, Privacy, IT Security, and HR.

DCAs [47] can facilitate a centralized, secure, streamlined process for the management of research data requests (and subsequent datasets) made by researchers across individual hospitals. The primary functions of the data control agents are to verify that clinical data requested by researchers from the Investigator Portal is IRB approved and to verify that the datasets generated for the researchers contain only IRB-approved data elements before they are released. Clinical research datasets generated during this process are stored on secure servers at the medical center. Access is restricted to research staff, Informatics Service Core programmers, and DCAs. Informatics Service Core programmers have access to 'working' folders that are not accessible by researchers. Once research datasets are created, they are verified by the DCAs and copied to a secure folder accessible by the researcher.

This process includes cohort identification and data requests for IRB-approved research protocols. Before IRB submission, researchers can use i2b2 software to query de-identified data in the TDW to see if there are adequate numbers of patients for a research proposal. At this time, this tool does not provide a set of identified data, only total number of patients, and a de-identified list of patients meeting the criteria chosen. The Informatics Service Core can use this de-identified list and the search criteria to query the identified data for the full set of data elements needed. Training is required to access the i2b2 platform to ensure users understand data privacy and know how to navigate the system properly.

Researchers with an identified cohort can request specific data elements. Each request must be associated with an approved IRB protocol or qualify for expedited or exempted IRB status and provide information about the security of the data, privacy, Protected Health Information (PHI) and HIPAA, and annual compliance and HIPAA training of medical center employees. These requests are submitted using the online Service Request form (included in the Investigator Portal). When completed, the DCAs receive an automated email and can access the request form via a service request tracking system.

2.5 Biomedical Imaging Informatics for Translational Research and Clinical Diagnosis and Treatment

2.5.1 Radiomics and Radiogenomics

In contrast to the traditional practice of using medical images solely for visual interpretation, AI has demonstrated remarkable progress in image-recognition tasks, especially at recognizing complex patterns in imaging data and providing quantitative assessment in an automated fashion. The rapid development of radiogenomics shows great promise for many different malignancies and represents the "evolution of radiology-pathology correlation from the anatomical-histological level to the genetic level" [48]. In order to improve clinical decision-making, radiomics focuses on the extraction and analysis of quantitative imaging features of intratumoral and intertumoral heterogeneity by providing diagnostic, prognostic, and predictive data in a noninvasive and repeatable way. To obtain the radiomic signatures as shown in Fig. 2.3, image processing, tracking, registration, segmentation, quantification, and classification are crucial for a series of downstream analyses such as diagnosis, disease staging, and treatment. The emergence of AI has led to a seamless imaging workflow by increasing efficiency, reducing errors, and achieving objectives with minimal manual input. Because of the exponential growth in computing power, AI is used to construct prediction models that can learn from existing datasets and analyze and perform predictions on related but new datasets.

2.5.2 Medical Imaging Analysis

Medical imaging includes cellular imaging, hematoxylin and eosin stain (H&E), and other pathological imaging and x-ray/CT (Computed Tomography), MRI (Magnetic Resonance Imaging), PET (Positron Emission Tomography), SPECT (Single-Photon Emission Computed Tomography)/ultrasound imaging. These technologies are beyond the scope of this review. Here, we focus on the following various aspects. For cellular imaging analysis, we refer to review articles [49–51]; these technologies can be used to study drug signatures by looking into the cell phenotype and cell phase changes before and after treatment [52]. For H&E and other pathological image analyses, we refer to previous studies [50, 53], where cell population signatures can be explored in response to drug treatment. There are numerous approaches developed for x-ray/CT/MRI/PET/SPECT/ultrasound imaging; here, we refer to the study by Parmar et al. [51] where imaging features can be investigated to define signatures for drug treatment.

Traditional image analysis includes basic steps such as preprocessing, registration, alignment, segmentation, quantification, tracking, and classification. The latest progress in deep learning bypasses some of these steps and directly inputs the raw imaging data to the different networks and output classification labels. In recent years, deep learning algorithms, including convolutional networks, transformer-based networks, and graph-based networks, have rapidly become a methodology of choice for analyzing medical images, and deep learning algorithms have achieved state-of-the-art performance while avoiding manual feature-extracting algorithms [49–52]. From the technical aspect, there are various deep learning approaches in segmentation, classification, active learning, and tracking. From the application aspect, many promising applications have been reported in neuro, retinal, pulmonary, digital pathology, breast, cardiac, abdominal, and musculoskeletal fields. However, data label quality and data diversity and volume are major factors affecting performance. Collecting such large amounts of data remains challenging. Deep learning-based algorithms are still challenged by some critical issues, such as data imbalance, model robustness, interpretability, sparse and noisy labels, federated learning, and uncertainty quantification. Some promising techniques have been proposed to address these challenging issues, such as shape value techniques to make deep learning models explainable [58], domain adaptation techniques [59] to enhance model robustness, and self-supervised deep learning techniques [60] to ease the requirement of labor-intensive data labeling work. If these challenges are addressed, the next era of AI medical analysis will arrive and bring dramatic changes to clinical applications.

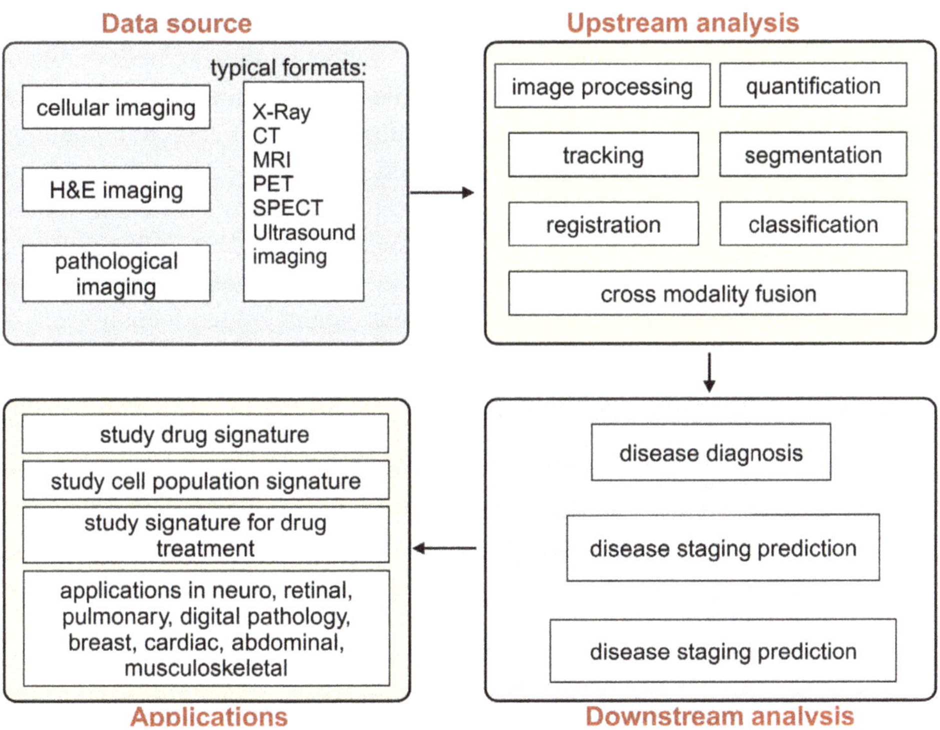

Fig. 2.3 Medical image analysis routines.

A single image source has its own advantages and disadvantages. Although most studies are based on single-modality data, cross-modality fusion techniques are able to further improve the performance of downstream tasks in medical image analysis if multiple modalities of image resources are properly fused [61, 62].

2.6 Big Data to Knowledge (BD2K), Knowledge Graph, and Functional Annotations

Here, we mainly introduce the medical knowledge annotation [63] and knowledge graph [64]. Medical knowledge graph [65] is a particular knowledgebase that organizes itself around the relationship between the nodes or entities that make up the graph. Articulating this data as a graph allows for graph-based approaches to querying and working with the data. For example, three types of nodes in the lexigraph are drugs, anatomy parts, and diseases. Disease nodes are also connected to each other; for example, diabetes type I (node 1) and diabetes type II (node 2) are both types of diabetes (node 3).

Most of the existing medicine recommendation systems are based mainly on EMRs [66, 67], which significantly assist doctors in making better clinical decisions, benefiting both patients and caregivers. Even though the growth of EMRs is at a lightning-fast speed in the era of big data, the content limitations in EMRs restrain the existing recommendation systems for reflecting relevant medical facts, such as drug-drug interactions (DDIs). Many medical knowledge graphs containing drug-related information, such as DrugBank [68], may give hope to the recommendation systems. However, the direct use of these knowledge graphs in systems suffers from robustness caused by the incompleteness of the graphs.

To address these challenges, we stand on recent advances in graph embedding learning techniques [69] and propose a novel framework called Safe Medicine Recommendation (SMR) [70] in this chapter. Specifically, SMR first constructs a high-quality heterogeneous graph by bridging EMRs (MIMIC-III) and medical knowledge graphs (ICD-9 ontology and DrugBank). Then, SMR jointly embeds diseases, medicines, patients, and corresponding relations into a shared lower-dimensional space. Finally, SMR uses the embeddings to decompose the medicine recommendation into a link prediction process while considering the patient's diagnoses and adverse drug reactions. Extensive experiments on real datasets are conducted to evaluate the effectiveness of the proposed framework.

With the increasing emergence of knowledge graphs [71], many world-leading researchers have successfully extracted information from huge volumes of medical databases to build up giant heterogeneous graphs that reflect medical facts about medicines and diseases. For instance, DrugBank [68] is a rich source of medical information. It contains a wide range of entities (drugs, drug targets, chemistry, etc.) and relationships (enzymatic pathways, DDIs, etc.). The ICD-9 ontology [14] represents a knowledge base of human diseases and can be used to classify patient diagnoses. Leveraging a well-built medical knowledge graph in the EMRs-based medicine recommendation system [72] may enable a reinvented system to provide appropriate prescriptions for particular patients and alerts to possible side effects and serious DDIs.

As shown in Fig. 2.4, linking EMRs and medical knowledge graphs to generate a large and high-quality heterogeneous graph is a promising avenue for medicine recommendations on a wider scope, but it is no easy task. Specifically, the newly designed system confronts the following challenges: (1) Computational efficiency. Querying specialized medical entities and relationships based on conventional graph-based algorithms has limitations in portability and scalability. Computational complexity becomes unfeasible when the heterogeneous graph reaches a considerable scale. (2) Data incompleteness. The medical knowledge graphs also follow the long-tail distribution like other types of large-scale knowledge bases. Data incompleteness is another serious problem among entities and relationships in such a distribution. For example, since DDIs are often not identified at the clinical trial phase, the lack of significant DDIs in DrugBank does not support a comprehensive precaution to the medication. Last but not least, medicine recommendation suffers from Cold Start. As conventional systems normally recommend medicines based on historical records, the pace of recommendation changes cannot keep up with the frequent updates of new therapies and treatments in medical practices. Little information on adverse reactions to the newly updated medicines in historical EMRs or even in well-built knowledge graphs makes the evidence-based recommendation model hardly support the new medicines as updated recommendations. Taking all the challenges above into account, we can have a novel medicine recommendation framework based on graph embedding techniques inspired by the idea of link prediction. This framework is named SMR [70]. The recommendation process mainly includes the following: (i) A large heterogeneous graph is constructed from EMRs and medical knowledge graphs. The nodes are entities (medicines, diseases, and patients), and the edges (or links) represent various relationships between entities, such as DDIs. (ii) Different parts of the generated heterogeneous graph (patient-medicine bipartite graph, patient-disease bipartite graph, medicine knowledge graph, and disease knowledge graph) are embedded into a shared low-dimension space based on graph-based embedding models. Afterward, a joint learning algorithm is proposed to optimize the integrated graph simultaneously. (iii) Based on the learned embeddings, a new patient, represented by the vectors of his/her diagnoses, is modeled as an entity in the disease–patient graph. Recommending medicines for the patient is translated to predicting links from the patient to medicines.

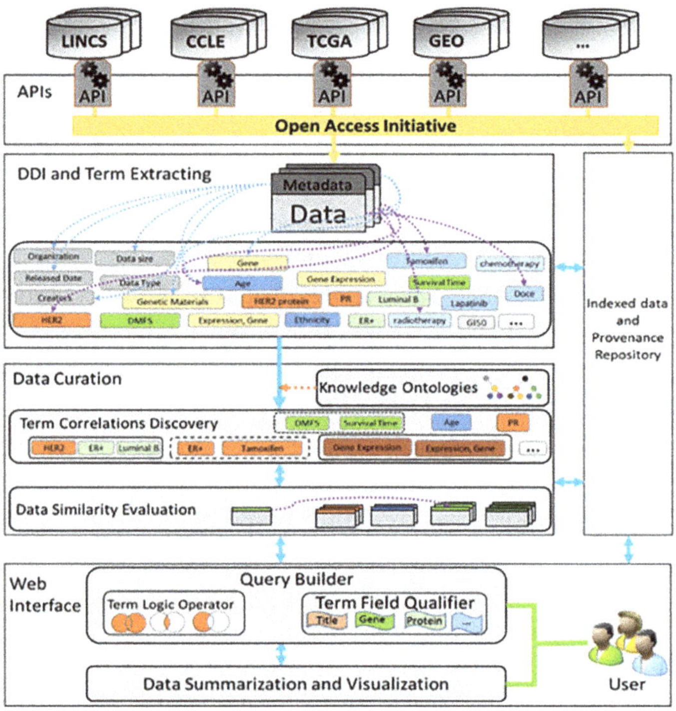

Fig. 2.4 Flowchart of development of knowledge environment.

As described, the knowledgebase contains signatures of various types and at different conceptual levels, from the low-level molecular and cellular levels to higher levels, such as individual patients or diseases. For a given signature type, each extracted signature is essentially a set of features; a given signature can be viewed as a point in a multidimensional feature space. Different methods have been proposed for indexing in multidimensional feature spaces [73–75], and most of these methods work well for specific features, distance measures, or query types.

2.7 Conclusion

With the continuous development and application of big data-related technologies in the medical field, many medical fields, such as medical and health undertakings, pharmacy, and clinical medicine, have benefited from it. However, in the face of increasingly complex large-scale data, how to conduct more effective management, analysis, and interpretation requires further developing big data-related analysis tools and technologies to achieve multidisciplinary integration and progress so that biomedicine can better serve humanity and promote social development.

References

1. Whetzel, P. L., et al. BioPortal: enhanced functionality via new web services from the National Center for Biomedical Ontology to access and use ontologies in software applications. *Nucleic Acids Res.* 2011; **39**(suppl 2): p. W541–W545.
2. Schadt, E. E., et al. Computational solutions to large-scale data management and analysis. *Nat Rev Genet.* 2010; **11**(9): p. 647–657.
3. Rosenthal, A., et al. Cloud computing: a new business paradigm for biomedical information sharing. *J Biomed Inform.* 2010; **43**(2): p. 342–353.
4. Johns, M., et al. *A Comprehensive Portal for Clinical and Translational Data Warehouses.* Public Health and Informatics; 2021.
5. Ganslandt, T., et al. Unlocking data for clinical research—the German i2b2 experience. *Appl Clin Inform.* 2011; **02**(01): p. 116–117.
6. *HCPCS - General Information.* https://www.cms.gov/medicare/coding/medhcpcsgeninfo.
7. Association, A. M. *Current Procedural Terminology: CPT.* Current Procedural Terminology: CPT; 2011.
8. Juba, S., A. Vannahme, and A. Volkov. *Learning PostgreSQL.* Packt Publishing Ltd; 2015.
9. Shea, S., and G. Hripcsak. Accelerating the use of electronic health records in physician practices. *N Engl J Med.* 2010; **362**(3): p. 192–195.
10. Price, L. H., S. Hendricks, and C. Cook. Incorporating computer-aided language sample analysis into clinical practice. *Lang Speech Hear Serv Sch.* 2010; **41**(2): p. 206–222.
11. Jha, A. K., et al. Use of electronic health records in U.S. hospitals. *N Engl J Med.* 2009; **360**(16): p. 1628–1638.
12. McCarty, C. A., and R. A. Wilke. Biobanking and pharmacogenomics. *Pharmacogenomics.* 2010; **11**(5): p. 637–641.
13. Ritchie, M. D., et al. Robust replication of genotype-phenotype associations across multiple diseases in an electronic medical record. *Am J Hum Genet.* 2010; **86**(4): p. 560–572.
14. Pace, W. D., et al. An electronic practice-based network for observational comparative effectiveness research. *Ann Int Med.* 2009; **151**(5): p. 338.
15. Schildcrout, J. S., et al. An analytical approach to characterize morbidity profile dissimilarity between distinct cohorts using electronic medical records. *J Biomed Inform.* 2010; **43**(6): p. 914–923.

16. McCarty, C. A., et al. The eMERGE Network: a consortium of biorepositories linked to electronic medical records data for conducting genomic studies. *BMC Med Genom.* 2011; **4**: p. 13.
17. Kho, A. N., et al. Electronic medical records for genetic research: results of the eMERGE consortium. *Sci Transl Med.* 2011; **3**(79): p. 79re1.
18. Institute of Medicine (US) Roundtable on Evidence-Based Medicine. *The Learning Healthcare System: Workshop Summary*. In: Olsen L, Aisner D, McGinnis JM, editors. Washington (DC): National Academies Press (US); 2007. PMID: 21452449.
19. Petro, J. *Natural Language Processing in Electronic Health Records*. KevinMD.com; 2011.
20. Chute, C. G., et al. The Enterprise Data Trust at Mayo Clinic: a semantically integrated warehouse of biomedical data. *J Am Med Inform Assoc.* 2010;**17**(2): p. 131–135.
21. Friedman, C. A broad-coverage natural language processing system. *Proc AMIA Symp.* 2000: p. 270–274.
22. Aronson, A. Effective mapping of biomedical text to the UMLS Metathesaurus: the MetaMap program. *Proc AMIA Symp.* 2001: p. 17–21.
23. Chapman, W., et al. A simple algorithm for identifying negated findings and diseases in discharge summaries. *J Biomed Inform.* 2001; **34**(5): p. 301–310.
24. Office_of_the_National_Coordinator. *Strategic Health IT Advanced Research Projects (SHARP) Program*. 2011 [13 March 2011]. http://healthit.hhs.gov/portal/server.pt/community/healthit_hhs_gov__sharp_program/1806.
25. Aronson, A. R., and F. M. Lang. An overview of MetaMap: historical perspective and recent advances. *J Am Med Inform Assoc.* 2010; **17**(3): p. 229–236.
26. Kidd, J., et al. 229 EMR clinical decision support tools improve compliance with venous thromboembolism risk assessment in obstetrical patients. *Am J Obstet Gynecol.* 2021; **224**(2): p. S151–S152.
27. Brown, J. M., et al. Automated diagnosis of plus disease in retinopathy of prematurity using deep convolutional neural networks. *JAMA Ophthalmol.* 2018; **136**(7): p. 803–810.
28. Adem, K., S. Kiliçarslan, and O. Cömert. Classification and diagnosis of cervical cancer with softmax classification with stacked autoencoder. *Expert Systems with Applications.* 2019; **115**.
29. Terasaki, J., et al. Using EMR to improve compliance with clinical practice guidelines for management of stable COPD. *Respir Med.* 2015; **109**(11): p. 1423–1429.
30. Jessica, C., and G. Katherine. 0777 The OHS pathway: an EMR-integrated clinical pathway to facilitate BiPAP on discharge for hospitalized patients with obesity hypoventilation syndrome. *Sleep.* 2022; **45**(Supplement_1): p. A338–A338.
31. Kemp, F. Applied multiple regression/correlation analysis for the behavioral sciences. *J R Stat Soc Series D: The Statistician.* December 2003; **52**(4): p. 691. https://doi.org/10.1046/j.1467-9884.2003.t01-2-00383_4.x
32. Menard, S. Logistic regression. *Am Stat.* 2004; **58**(4): p. 364.
33. Cox, D. R. Regression models and life-tables. *J R Stat Soc: Series B (Methodological).* 1972; **34**(2): p. 187–202.
34. Ansari, M. S., et al. Predictive model based on health data analysis for risk of readmission in disease-specific cohorts. *Perspect Health Inf Manage.* 2021; **18**(Spring): p. 1j.

35. Mahmoudi, E., et al. Use of electronic medical records in development and validation of risk prediction models of hospital readmission: systematic review. *BMJ (online)*. 2020; **369**: p. m958.
36. Mirzaei, M., and M. Mirzaei. Agreement between Framingham, IraPEN and non-laboratory WHO-EMR risk score calculators for cardiovascular risk prediction in a large Iranian population. *J Cardiovasc Thorac Res*. 2020; **12**(1): p. 20–26.
37. Arnold, R. J. G., et al. Assessment of the relationship between diabetes treatment intensification and quality measure performance using electronic medical records. *PLoS One*. 2018; **13**(6): p. e0199011.
38. van den Broek, A. K., et al. A mandatory indication-registration tool in hospital electronic medical records enabling systematic evaluation and benchmarking of the quality of antimicrobial use: a feasibility study. *Antimicrob Resist Infect Control*. 2021; **10**(1): p. 103.
39. Lin, H. L., et al. Association between electronic medical records and healthcare quality. *Medicine (Baltimore)*. 2020; **99**(31): p. e21182.
40. Dagroso, D., et al. Implementation of an obstetrics EMR module: overcoming user dissatisfaction. *J Healthc Inf Manag*. 2007; **21**(1): p. 87–94.
41. Amalia, N., et al. The implementation of electronic medical record (EMR) in the development health care system in Indonesia. *Int J Adv Life Sci Res*. 2021; **4**(3): p. 8–12.
42. Sollins, H. Health Insurance Portability and Accountability Act (HIPAA): must there be a trade-off between privacy and quality of health care, or can we advance both? *Circulation*. 2003; **108**(8): p. 915–918.
43. Payment, C. I. *Payment Card Industry (PCI) Data Security Standard*. Data Security; 2004.
44. Averill, J. D., P. Reneke, and R. D. Peacock. *NPA Plan, NIST Public Access*. National Institute of Standards and Technology (NIST); 2004.
45. Iglehart, J. K. The centers for medicare and medicaid services. *N Engl J Med*. 2001; **345**(26): p. 1920–1924.
46. Armstrong, I. Is the CISO gaining influence? (Chief Information Security Officer) (Editorial). *Sc Magazine*. 2009; **20**(6): p. 4–4.
47. *OPS Data Collection Agent (DCA)*. https://www.konicaminolta.com.au/KonicaMinolta/media/KonicaMinolta/Support/Documents/Guide-DCA-System-Requirements.pdf
48. Segal, E., et al. Decoding global gene expression programs in liver cancer by noninvasive imaging. *Nat Biotechnol*. 2007; **25**(6): p. 675–680.
49. Xiong, G., X. Zhou, and L. Ji. Automated segmentation of Drosophila RNAi fluorescence cellular images using deformable models. *IEEE Trans Circuits Syst I: Regul Pap*. 2006; **53**(11): p. 2415–2424.
50. Gurcan, M. N., et al. Histopathological image analysis: a review. *IEEE Rev Biomed Eng*. 2009; **2**: p. 147–171.
51. Parmar, C., et al. Data analysis strategies in medical imaging. *Clin Cancer Res*. 2018; **24**(15): p. 3492–3499.
52. Shao, H., et al. Systematically studying kinase inhibitor induced signaling network signatures by integrating both therapeutic and side effects. *PLoS One*. 2013; **8**(12): p. e80832.
53. Barisoni, L., et al. Digital pathology and computational image analysis in nephropathology. *Nat Rev Nephrol*. 2020; **16**(11): p. 669–685.

54. Zhou, S. K., et al. A review of deep learning in medical imaging: imaging traits, technology trends, case studies with progress highlights, and future promises. *Proc IEEE*. 2021; **109**(5): p. 820–838.
55. Salvi, M., et al. The impact of pre- and post-image processing techniques on deep learning frameworks: a comprehensive review for digital pathology image analysis. *Comput Biol Med*. 2021; **128**: p. 104129.
56. Deng, S., et al. Deep learning in digital pathology image analysis: a survey. *Front Med*. 2020; **14**(4): p. 470–487.
57. Litjens, G., et al. A survey on deep learning in medical image analysis. *Med Image Anal*. 2017; **42**: p. 60–88.
58. Molnar, C. *Interpretable Machine Learning: A Guide for Making Black Box Models Explainable*. Lulu.com; 2022.
59. Wang, R., P. Chaudhari, and C. Davatzikos. Embracing the disharmony in medical imaging: a simple and effective framework for domain adaptation. *Med Image Anal*. 2022; **76**: p. 102309.
60. Xu, J. A review of self-supervised learning methods in the field of medical image analysis. *Int J Image Graph Signal Process*. 2021; **10**(4): p. 33.
61. Huang, B., et al. A review of multimodal medical image fusion techniques. *Comput Math Methods Med*. 2020; **2020**(1): p. 8279342.
62. Rajalingam, B., and R. Priya. Multimodal medical image fusion based on deep learning neural network for clinical treatment analysis. *Int J ChemTech Res*. 2018; **11**(06): p. 160–176.
63. Dingli, A. *Knowledge Annotation: Making Implicit Knowledge Explicit*. Heidelberg: Springer; 2011.
64. Zhang, L. Knowledge graph theory and structural parsing. [PhD Thesis - Research UT, graduation UT, University of Twente]. Twente University Press (TUP); 2002.
65. Desarkar, et al. Med-Tree: a user knowledge graph framework for medical applications. *Bioinform Bioeng*. 2013; 1–4.
66. Wang, W., X. Zhang, and D. Q. Dai. springD2a: capturing uncertainty in disease–drug association prediction with model integration. *Bioinformatics*. 2021; **38**: p. 1353–1360.
67. Chen, R., et al. Machine learning for drug-target interaction prediction. *Molecules*. 2018; **23**(9): p. 2208.
68. Wishart, D. S., et al. DrugBank: a knowledgebase for drugs, drug actions and drug targets. *Nucleic Acids Res*. 2008; **36**(suppl_1): p. D901–D906.
69. Zhang, X., et al. Graph-embedding-based learning for robust object tracking. *IEEE Trans Ind Electron*. 2013; **61**(2): p. 1072–1084.
70. Wang, M., et al. Safe medicine recommendation via medical knowledge graph embedding. *arXiv preprint arXiv:1710.05980*, 2017.
71. Kejriwal, M. *Domain-Specific Knowledge Graph Construction*. Cham: Springer International Publishing; 2019.
72. Gurawa, P. and M. Nickles. Drug similarity and link prediction using graph embeddings on medical knowledge graphs. *arXiv preprint arXiv:2110.13047*, 2021.

73. Bohm, C., S. Berchtold, and D. Keim. Searching in high-dimensional spaces—index structures for improving the performance of multimedia Databases. *ACM Comput Surv.* 2001; **33**(3): p. 322–373.
74. Grebhahn, A., et al. Challenges in finding an appropriate multi-dimensional index structure with respect to specific use cases. In *24th G1-Workshop on Foundations of Databases (Grundlagen von Datenbanken)*, 2012: p. 77–82.
75. Samet, H. Foundations of multidimensional and metric data structures. In: The Morgan Kaufmann series in computer graphics and geometric modeling. Amsterdam; Boston: Elsevier/Morgan Kaufmann; 2006, p. xxvii, 993.

CHAPTER 3

Functional Annotation and Knowledgebase

3.1 Introduction

The rapid advancement of sequencing technologies and large-scale biological experiments has produced an overwhelming amount of genomic and proteomic data. However, the true value of this data lies in its interpretation and transformation into meaningful biological knowledge. Functional annotation, as the process of assigning biological significance to genes, proteins, and regulatory elements, forms the foundation for understanding complex biological systems [1]. By predicting gene functions, identifying molecular pathways, and integrating experimental evidence, functional annotation helps bridge the gap between raw data and actionable insights in biology.

In parallel, biological knowledgebases serve as repositories of curated information, providing researchers with access to validated data, including gene sequences, protein structures, molecular interactions, and pathway relationships [2]. These knowledgebases play an indispensable role in bioinformatics by offering standardized frameworks for data interpretation and integration. Computational tools for functional annotation rely heavily on these knowledgebases using algorithms and machine learning approaches to infer gene and protein functions, identify conserved domains, and explore biological networks.

The integration of computational tools with knowledgebases has transformed bioinformatics research, enabling the systematic study of biological systems at unprecedented scales [3, 4]. Together, they allow researchers to uncover functional relationships, investigate disease mechanisms, and develop innovative solutions in fields such as precision medicine, synthetic biology, and agriculture. This chapter will delve into the methodologies, tools, and resources used in functional annotation and explore the critical role of knowledgebases in advancing biological research. By examining their applications and challenges, we aim to provide a comprehensive understanding of how these tools empower scientists to interpret biological data and drive discoveries.

3.1.1 Definition of Functional Annotation in Multiomics Data

Functional annotation is a critical process in genome research and extends beyond to encompass the transcriptome, proteome, microbiome, metabolome, and even integrated multiomics datasets [5]. It involves identifying, categorizing, and characterizing the biological roles of genes, transcripts, proteins, and other molecular components to transform raw high-throughput sequencing data into biologically meaningful insights. Functional annotation serves as the backbone for understanding the molecular mechanisms driving cellular functions, biological processes, and phenotypic outcomes across diverse biological systems.

1. **Functional Annotation in the Genome:** In genomic research, functional annotation focuses on identifying and characterizing genes and other genomic elements within an organism's DNA. It begins with the identification of coding sequences (CDS), noncoding RNAs (ncRNAs), and regulatory regions, such as promoters, enhancers, and insulators [6–8]. Gene prediction tools, such as Augustus [9], Glimmer [10], or MAKER [11], are used to identify open reading frames (ORFs) and noncoding elements within the genome. Once these features are mapped, functional annotations are inferred through a combination of computational methods and experimental validations.

   Sequence similarity searches, such as Basic Local Alignment Search Tool (BLAST) [12] and HMMER [13], allow researchers to assign potential functions to genes based on homology to known sequences in curated databases such as NCBI GenBank [14] and UniProt [15]. Beyond sequence alignments, annotation often incorporates structured vocabularies such as gene ontology (GO) [16], which categorizes genes into molecular functions, biological processes, and cellular components. Additionally, pathway databases such as KEGG [17] and Reactome [18] enable mapping of genes to metabolic, signaling, or disease pathways.

   Noncoding regions are equally significant in genomic annotation. Regulatory elements, including enhancers, silencers, and ncRNAs such as microRNAs (miRNAs) and long noncoding RNAs (lncRNAs), are annotated using tools like FANTOM [19] and Rfam [20]. These elements play key roles in gene regulation, epigenetic modifications, and transcript stability, providing a more complete understanding of genome functionality. Genome annotation thus underpins downstream analyses such as comparative genomics, population genetics, and synthetic biology applications.
2. **Functional Annotation in the Transcriptome:** Transcriptome annotation focuses on the characterization of all RNA molecules, including coding and noncoding transcripts, expressed in a specific tissue, condition, or developmental stage. High-throughput RNA sequencing (RNA-seq) is central to transcriptome studies, enabling researchers to measure gene expression levels, identify splice variants, and discover novel transcripts. Functional annotation of the transcriptome begins with the identification

and quantification of expressed genes using tools such as STAR [21] and HISAT2 [22], followed by assembly with pipelines like StringTie [23] or Cufflinks [24].

Once transcripts are assembled, they are functionally annotated by integrating gene expression data with existing databases. Tools such as Trinotate [25] combine sequence homology searches, domain prediction, and GO annotations to infer the biological roles of transcripts. Transcript-level annotations also reveal functional isoforms, posttranscriptional regulation, and alternative splicing events that add complexity to gene function. ncRNAs, including lncRNAs, miRNAs, and circular RNAs (circRNAs), are annotated using tools like miRbase [26] or LncBook [27], enabling insights into their regulatory roles.

Transcriptome annotation is particularly powerful for identifying condition-specific expression patterns. By analyzing differentially expressed genes (DEGs) and integrating GO enrichment and pathway analyzes, researchers can link transcripts to specific biological processes, disease mechanisms, or stress responses. This approach is essential for functional genomics, biomarker discovery, and therapeutic target identification.

3. **Functional Annotation in the Proteome**: Proteome annotation focuses on identifying and characterizing the set of proteins expressed by a genome or transcriptome. It provides insights into protein functions, structures, post-translational modifications (PTMs), and interactions that govern cellular processes. Proteomics studies typically involve mass spectrometry (MS)-based technologies, which allow for the large-scale identification of proteins and their modifications. Tools such as MaxQuant [28], Mascot [29], and Proteome Discoverer [30] analyze MS data to identify protein sequences and quantify their abundance.

   Functional annotation of proteins relies heavily on domain and motif analyzes to detect conserved functional units. Tools like InterProScan [31] and MEME [32] enable the identification of protein families, domains, and active sites. Annotation also includes mapping proteins to biochemical pathways using databases like KEGG and BioCyc, revealing their roles in metabolic networks and signaling cascades. Structural prediction tools, such as AlphaFold [33] and I-TASSER [34], further enhance functional annotation by linking protein sequences to 3D structures, providing insights into enzymatic activity, ligand binding, and interactions.

Proteome annotation extends beyond individual proteins to include functional protein-protein interactions (PPIs), which can be identified using databases such as STRING [35] and BioGRID [36]. By integrating PPIs with pathway and GO enrichment analyses, researchers can reconstruct protein interaction networks that reveal functional modules and key regulatory hubs within the cell.

4. **Functional Annotation in Single-Cell Sequencing:** Single-cell sequencing represents a revolutionary advancement in functional annotation, enabling the study of gene expression and genome function at the resolution of individual cells. Unlike bulk sequencing, which averages signals across heterogeneous populations, single-cell RNA sequencing (scRNA-seq) reveals transcriptional heterogeneity, rare cell types, and dynamic changes in gene expression across tissues or conditions. Functional annotation at the single-cell level involves clustering cells based on gene expression profiles, assigning cell types, and identifying marker genes associated with specific biological functions. Tools like Seurat [37], Scanpy [38], and CellRanger [39] help annotate single-cell data and uncover cell-type-specific pathways and regulatory mechanisms. Single-cell multiomics technologies, such as CITE-seq (RNA and surface proteins) and scATAC-seq (chromatin accessibility), provide further functional insights by linking gene expression to epigenomic and proteomic changes at single-cell resolution.
5. **Functional Annotation in Spatial Sequencing:** Spatial sequencing combines transcriptomic and genomic data with spatial context, enabling researchers to study gene expression patterns within intact tissues. Technologies like 10x Genomics Visium, Slide-seq, and MERFISH preserve spatial information while measuring transcript abundance, allowing functional annotation of genes within their native tissue architecture. Spatial sequencing reveals how cells interact within their microenvironments, identifies tissue-specific functional zones, and maps signaling gradients that drive biological processes.

By integrating spatial and single-cell sequencing data, functional annotation gains unprecedented depth, uncovering the interplay between cellular functions, spatial organization, and tissue-level phenotypes. These technologies are particularly transformative in fields like developmental biology, neuroscience, and cancer research, where understanding the spatial relationships between genes and cell types is critical.

6. **Functional Annotation by Using Multiomics Data:** With advancements in high-throughput technologies, functional annotation has expanded to include integrated multiomics datasets, combining genomic, transcriptomic, proteomic, metabolomic, and epigenomic information. Multiomics annotation provides a comprehensive view of molecular processes by linking changes in gene expression, protein function, and metabolite levels to biological outcomes. For example, integrating genomics and transcriptomics allows for identifying functional variants that influence gene expression, such as single-nucleotide polymorphisms (SNPs) in regulatory regions. Tools like eQTL analysis and databases like GTEx facilitate the linking of genetic variants to expression changes across tissues.

   When combined with proteomics, transcriptomic data can be used to resolve discrepancies between mRNA levels and protein abundance, revealing posttranscriptional and translational regulation. Further integration with metabolomics enables researchers to annotate metabolic pathways, linking gene functions and protein activities to changes in cellular metabolites. Epigenomic data, such as histone modifications and DNA methylation patterns, add another layer of functional annotation by highlighting regulatory mechanisms controlling gene expression. Tools like ChIP-seq and ATAC-seq are used to annotate regulatory elements that influence genome and transcriptome functions. To enable integrated multiomics annotation, tools like OmicsNet [40], Pathway Tools [41], and Galaxy [42] offer computational platforms that combine various omics layers to derive functional insights. These approaches allow researchers to build systems biology models, identify biomarkers, and explore disease mechanisms in a holistic manner.

3.1.2 Overview of Knowledgebases in Biological Research

Biological knowledgebases are structured repositories that play a pivotal role in bioinformatics research by storing, organizing, and disseminating large volumes of curated biological data. These resources serve as the backbone for functional annotation, providing researchers with reliable, accessible, and interoperable datasets that can be queried and analyzed to generate new biological insights. Knowledgebases act as essential tools for transforming raw experimental and computational data into meaningful biological knowledge by organizing diverse data types such as genomic sequences, protein structures, metabolic pathways, and disease-related mutations.

One of the distinguishing features of knowledgebases is their ability to aggregate heterogeneous biological data from multiple sources and harmonize it into standardized formats. By applying controlled vocabularies and ontologies, such as GO [43] for gene function descriptions or the enzyme commission (EC) system [44] for enzyme classification, knowledgebases ensure consistency across studies and enable cross-species and cross-platform comparisons. This standardization is crucial for facilitating reproducible research and integrating data from high-throughput experimental technologies.

Biological knowledgebases can be broadly categorized based on their scope and focus. General repositories, such as NCBI resources [45–47], and UniProt [15], provide foundational data on gene and protein sequences, annotations, and evolutionary relationships. These databases serve as comprehensive resources for researchers across various fields of biology, offering access to extensive datasets for comparative genomics, functional annotation, and evolutionary studies. For instance, UniProt is widely recognized as the golden standard and reference for protein sequence and functional information, linking proteins to structural features, functions, and disease relevance.

In addition to general knowledgebases, pathway, and interaction-focused repositories such as KEGG [17], Reactome [18], and STRING [35] provide invaluable insights into the biological networks and molecular interactions that govern cellular processes. KEGG maps genes and proteins to pathways involved in metabolism, signal transduction, and diseases, enabling researchers to contextualize gene functions within broader biological systems. Reactome, on the other hand, focuses on manually curated pathways, integrating experimental evidence to build high-quality models of biochemical reactions. Similarly, STRING aggregates known and predicted PPIs, supporting researchers in exploring functional relationships among biomolecules.

Knowledgebases dedicated to disease and clinical research, such as OMIM [48], COSMIC [49], and ClinVar [50], are indispensable for understanding the genetic basis of diseases and identifying potential therapeutic targets. OMIM provides a comprehensive catalog of human genes and genetic phenotypes, serving as a critical resource for linking genetic variations to clinical outcomes. COSMIC focuses specifically on somatic mutations in cancer, offering curated data on mutation frequencies, gene alterations, and their impact on cancer progression. ClinVar complements these resources by curating clinically significant variants, enabling clinicians and researchers to interpret the role of genetic mutations in human health and disease.

Building a biological knowledgebase is a complex, multistep process that requires careful planning, robust computational infrastructure, and ongoing curation. The first step involves defining the scope and objectives of the knowledgebase, such as the types of data it will include and the biological questions it aims to address. Data collection follows, where information is gathered from experimental results, published literature, existing databases, and computational predictions. Once collected, the data undergoes rigorous curation to ensure accuracy, consistency, and completeness. This process often combines automated pipelines for large-scale data processing with manual curation by experts to validate and refine annotations.

Data integration is a critical aspect of knowledgebase construction, where heterogeneous datasets are harmonized into a unified framework using standardized ontologies and formats. The database infrastructure is then designed to support efficient data storage, retrieval, and querying. User interfaces and APIs are developed to provide researchers with easy access to the data, while visualization tools are incorporated to enable the exploration of pathways, networks, and annotations in an intuitive manner. Maintenance and regular updates ensure that the knowledgebase remains current, incorporating new discoveries and addressing errors as needed.

The importance of biological knowledgebases cannot be overstated. They enable researchers to analyze complex biological systems, uncover functional relationships, and generate testable hypotheses that drive scientific discovery. By providing access to curated and validated data, knowledgebases accelerate progress in fields ranging from fundamental biology to translational medicine. As biological data continues to grow in volume and complexity, the future of knowledgebases will be shaped by advances in artificial intelligence, machine learning, and cloud computing. These technologies will enhance data integration, curation, and analysis, ensuring that knowledgebases remain at the forefront of modern bioinformatics research.

In summary, biological knowledgebases are indispensable tools that support the organization, interpretation, and dissemination of biological data. They provide the foundation for functional annotation, comparative analysis, and systems biology research, enabling scientists to unlock the full potential of large-scale genomic and proteomic datasets.

3.2 Computational Approaches for Functional Annotation

Functional annotation is driven by a diverse array of computational tools that help assign biological significance to genes, proteins, and other genomic elements. These tools employ sequence analysis, structural predictions, and integrative multidimensional approaches to explore functional relationships, identify pathways, and uncover interactions within biological systems. Each computational approach plays a unique role in functional annotation, and their effective use relies on integrating biological data with advanced algorithms.

3.2.1 Sequence-Based Methods

Sequence-based methods are among the most widely used approaches for functional annotation, as they leverage the principle that sequence similarity often reflects functional conservation. These methods operate on the premise that proteins or genes with similar sequences are likely to share similar functions, structures, and evolutionary origins. Over time, the accumulation of large, annotated biological databases has made this principle even more powerful, as researchers can query unknown sequences and draw inferences based on well-annotated homologs. These methods are essential for identifying gene functions, predicting evolutionary relationships, and revealing the molecular mechanisms of biological systems.

At the core of sequence-based annotation lies BLAST [12], a foundational and widely used program for pairwise sequence comparisons. BLAST allows researchers to compare nucleotide or protein sequences against massive databases such as NCBI GenBank or UniProt [15]. By calculating sequence alignments and similarity scores, BLAST identifies homologous sequences that share significant matches, enabling functional predictions based on shared annotations. For example, a newly discovered gene can be queried against these databases to find functionally characterized genes with high similarity, providing strong evidence of its likely role. Different flavors of BLAST cater to specific research needs: BLASTn for nucleotide sequence searches, BLASTp for protein sequence comparisons, and BLASTx for translating nucleotide sequences and searching against protein databases, which is particularly helpful when analyzing genomic sequences of unknown coding regions [51].

While traditional BLAST is highly effective, its sensitivity can diminish when searching for homologs that are distantly related. To address this limitation, tools like Position-Specific Iterated BLAST (PSI-BLAST) have been developed [52]. PSI-BLAST enhances sequence sensitivity by building position-specific scoring matrices (PSSMs) after the initial search. These PSSMs capture conserved regions of the sequence and iteratively refine the search, making it possible to detect distant homologs that share functional or structural conservation but may exhibit weak sequence similarity overall. This iterative approach significantly expands the search scope and is particularly useful in identifying remote evolutionary relationships or conserved functional motifs.

Beyond BLAST, other tools like DIAMOND [53] and MMseqs2 [54] have emerged as faster and more scalable solutions for large-scale sequence searches, particularly in metagenomic and high-throughput sequencing applications. DIAMOND, for example, is a high-speed alignment tool optimized for massive datasets, providing rapid sequence comparisons without compromising accuracy [53]. Similarly, MMseqs2 allows clustering and searching of vast sequence datasets efficiently, making it ideal for annotating the growing number of sequences generated by next-generation sequencing technologies [54].

Sequence-based methods also benefit from the integration of specialized databases that extend functional insights. For example, the KEGG database links sequences to metabolic and signaling pathways, providing functional context for annotated genes. Similarly, databases like InterPro consolidate information on conserved protein domains from multiple sources to offer a comprehensive annotation of functional motifs and structures [31]. These domain-based approaches complement traditional sequence alignments, especially when analyzing proteins with modular architectures.

In recent years, advancements in machine learning have further expanded the capabilities of sequence-based functional annotation. Deep learning-based tools, such as DeepGO [55] and DeepFam [56], analyze complex sequence features to predict gene or protein functions based on models trained with large annotated datasets. These methods allow researchers to capture subtle sequence patterns that may elude traditional alignment-based tools, enhancing the prediction of functional roles in sequences with minimal similarity to known genes.

Sequence-based methods are increasingly applied in metagenomic and environmental sequencing studies, where vast datasets are generated from mixed microbial communities. Tools like MEGAN [57] and Kaiju [58] allow researchers to annotate metagenomic reads by comparing them against reference databases, thereby identifying taxonomic and functional information in microbial ecosystems. Additionally, combining sequence-based annotation with structural prediction tools such as AlphaFold [33] or I-TASSER [34] can provide functional insights by linking sequence data to predicted 3D protein structures, especially in cases where homology searches yield limited information. Overall, sequence-based functional annotation has evolved into a versatile and powerful toolkit, combining traditional alignment approaches, profile-based methods, machine learning models, and structural predictions. These tools, supported by ever-growing databases, continue to play a pivotal role in genomics, proteomics, and metagenomics, enabling the discovery and functional characterization of genes and proteins across diverse biological systems.

Motif-based methods complement traditional sequence alignment by focusing on the identification of conserved sequence motifs or domains that are often associated with specific biological functions. This type of method is particularly valuable for detecting functional signatures within sequences, even when overall sequence similarity is low. Motifs represent short, conserved regions of amino acid or nucleotide sequences that are often critical for biological activities such as enzymatic catalysis, PPIs, or DNA/RNA binding. By identifying these conserved elements, motif-based methods enable researchers to infer protein functions, classify sequences into functional families, and reveal structural and biochemical properties.

InterProScan [31], as we mentioned before, integrates multiple domain and motif databases to provide comprehensive functional insights. Multiple EM for Motif Elicitation (MEME) [32] is another popular tool for motif discovery and analysis. MEME identifies statistically significant, ungapped motifs within a set of input sequences, making it ideal for discovering novel functional motifs or binding sites shared among related sequences. Once motifs are identified, they can be compared against databases like TomTom [59] or JASPAR [60], which house collections of known motifs, to predict their functions. This process is particularly powerful for understanding transcription factor (TF) binding sites, regulatory elements, or conserved functional regions in DNA and protein sequences.

PROSITE [61], a key database for motif-based analysis, focuses on manually curated functional motifs and patterns that are often associated with specific biological functions. It includes regular expressions and profiles that describe these conserved regions, enabling precise identification of enzymatic active sites, binding motifs, and PTM sites. For example, PROSITE motifs can help identify serine/threonine phosphorylation sites, glycosylation patterns, and ATP-binding motifs, which are critical for understanding protein regulation and activity.

Motif-based approaches are also employed in tools like MotifScan [62], which focuses on scanning sequences for known functional motifs derived from curated motif libraries. Similarly, tools like Discriminative Regular Expression Motif Elicitation (DREME) [63] and Find Individual Motif Occurrences (FIMO) [64] are part of the MEME Suite and are used to detect short, significant sequence patterns that can reveal binding sites and functional signatures. FIMO, in particular, searches sequences for specific motifs using probabilistic scoring, allowing researchers to identify motif occurrences across entire genomes or transcriptomes with statistical confidence.

In the context of structural and functional annotation, motif-based methods can provide insights into enzymatic activities, metal ion or ligand binding sites, and PPI regions. For example, motifs like the Walker A and B motifs are associated with nucleotide-binding sites in ATPases, while the EF-hand motif is indicative of calcium-binding domains. Similarly, zinc finger motifs, identified through conserved cysteine and histidine residues, are characteristic of DNA-binding proteins that regulate gene expression. By identifying such motifs, researchers can make functional predictions for uncharacterized proteins with high confidence.

Motif-based methods are also critical for studying PTMs. Tools like ScanProsite [65] and PhosphoSitePlus [66] identify conserved motifs that serve as sites for phosphorylation, glycosylation, acetylation, and other regulatory modifications. These functional annotations are essential for understanding protein regulation, signaling pathways, and disease mechanisms, as PTMs often play key roles in modulating protein activity and interactions.

For large-scale analyses, motif-based methods can be integrated with machine learning models to identify novel motifs and functional domains. Tools like DeepBind [67] and BindSpace [68] use deep learning techniques to predict protein-DNA and protein-RNA interactions by learning motif features from large datasets. These approaches enable the discovery of motifs associated with TFs, RNA-binding proteins, and regulatory elements in a high-throughput manner.

3.2.2 Structure-Based Methods

In cases where sequence similarity alone cannot resolve functional roles, structure-based methods offer a powerful alternative by focusing on the three-dimensional (3D) organization of proteins. While sequence-based methods rely on homology at the amino acid level, protein structures are often more conserved across evolutionary time because functional properties are inherently tied to a protein's shape and biochemical interactions. Proteins with highly similar structures can perform similar biological functions even when their sequences have diverged significantly. This makes structure-based annotation particularly valuable for characterizing novel, poorly understood, or divergent proteins where sequence comparisons yield little to no useful results.

The recent development of advanced protein structure prediction tools has revolutionized this field. One of the most prominent tools is AlphaFold [33], developed by DeepMind, which uses deep learning algorithms trained on large databases like the Protein Data Bank (PDB) [69] to predict protein structures with atomic-level accuracy. AlphaFold's success lies in its ability to model protein backbones, side-chain orientations, and flexible regions, providing highly reliable structural information that can be used for downstream functional annotation. Other tools like MODELLER [70] and Rosetta [71] further contribute to this process by offering structure refinement capabilities and modeling protein complexes.

Once a protein structure is predicted, it can be compared to known structures in public repositories such as the PDB [69]. Structural alignment tools like DALI [72], TM-align [73], and VAST [74] are used to identify proteins with similar folds, domains, or active site geometries. For instance, DALI [72] compares 3D structures by calculating similarities in spatial distances between amino acids, enabling researchers to detect conserved structural features even when sequence homology is minimal. TM-align [73] provides quantitative metrics to assess the degree of structural similarity, while tools like structural classification of proteins (SCOP) [75] and class, architecture, topology, and homology (CATH) [76] classify proteins into hierarchical families based on their fold architectures. These alignments often reveal functional relationships that would remain hidden using sequence-based approaches alone, particularly for proteins with divergent evolutionary histories.

Structural annotations offer profound insights into key functional features that govern protein activity. For instance, catalytic sites, where enzymatic reactions occur, are often defined by highly conserved 3D arrangements of amino acid residues. Tools like CASTp [77] and ProBis [78] analyze protein structures to identify functional pockets, active sites, and potential ligand-binding regions, which are crucial for understanding enzymatic mechanisms and substrate specificity. Binding sites for small molecules, cofactors, or metal ions can also be predicted using tools such as COACH-D [79] and SiteMap [80], which map surface cavities and assess their likelihood of interacting with specific ligands. These insights are invaluable for annotating proteins with potential biochemical roles, particularly in the context of enzyme function and receptor-ligand interactions.

Furthermore, structure-based methods play a significant role in identifying PPIs, which are fundamental to cellular processes such as signaling, regulation, and complex assembly. Docking tools like HADDOCK [81], ClusPro [82], and

ZDOCK [83] predict how proteins interact with each other, modeling interaction surfaces and identifying residues critical for binding. By mapping these interaction interfaces, researchers can annotate proteins as components of larger functional complexes or pathways. Structural features such as conserved folds, repeating motifs, and binding domains, like the Rossmann fold for nucleotide binding or zinc finger motifs for DNA interactions, also provide clues about the protein's biological role.

Incorporating experimental validation is a key step in structure-based annotation, ensuring the reliability of computational predictions. Experimental techniques like X-ray crystallography and cryo-electron microscopy (Cryo-EM) provide high-resolution structural data that validate computational models and capture protein conformations under physiological conditions [84]. For smaller proteins, nuclear magnetic resonance (NMR) spectroscopy is used to obtain structural information in solution, offering additional insights into protein dynamics and flexibility [85]. Combined with functional assays, such as site-directed mutagenesis or enzymatic activity tests, these experimental approaches confirm predicted catalytic residues, binding interfaces, and structural domains.

Structure-based annotation has expanded further through integration with multiomics datasets, bridging the gap between genomics, transcriptomics, and proteomics. For example, genomic data identifying novel CDS can be annotated using predicted structures to infer protein functions. Similarly, transcriptomic data from RNA-seq experiments can be linked to specific protein isoforms, whose structures reveal functional differences arising from alternative splicing. In proteomics, MS data mapping PTMs, such as phosphorylation or acetylation, can be superimposed onto protein structures to assess their functional impacts on protein stability, interactions, and enzymatic activity. Structural data also complement metabolomics studies, where enzymes involved in metabolic pathways are annotated based on substrate-binding sites and catalytic mechanisms.

Structure-based methods have found widespread applications in biotechnology, evolutionary biology, and drug discovery. In drug design, for instance, structural models allow researchers to identify active sites and develop small-molecule inhibitors that bind specifically to target proteins. In evolutionary studies, structural comparisons can reveal conserved folds or motifs shared across distant species, shedding light on the evolution of protein functions. These methods are particularly transformative for understanding disease-related proteins, as structural anomalies caused by mutations are often linked to pathologies such as cancer, neurodegenerative disorders, and genetic diseases.

Despite these advances, challenges remain in structure-based annotation. While computational tools have made significant strides, predicting the structures of protein complexes, intrinsically disordered regions, or membrane proteins remain difficult due to their dynamic and flexible nature. Additionally, functional annotation based purely on structural similarity can sometimes be ambiguous, as proteins with similar folds do not always share identical functions. Integrating structure-based predictions with sequence data, experimental validation, and multiomics analyses helps address these limitations, creating a more holistic annotation framework.

In summary, structure-based methods provide a robust approach for functional annotation when sequence similarity alone is insufficient. By predicting and analyzing protein 3D structures, these methods uncover functional clues such as catalytic sites, binding regions, and interaction surfaces that drive biological processes. Combining computational tools with experimental techniques and multiomics integration enhances annotation accuracy, offering a powerful platform to advance our understanding of proteins and their roles across cellular systems, tissues, and organisms.

3.3 Common Knowledgebases and Resources

Biological research is driven by a vast array of knowledgebases and resources that serve as repositories for curated, reliable, and comprehensive biological data. These resources enable researchers to annotate genes and proteins, understand biological pathways, study diseases, and develop therapeutics. Knowledgebases are typically divided into comprehensive databases and specialized repositories, each playing a specific role in biological discovery.

3.3.1 Comprehensive Databases

Comprehensive databases form the cornerstone of biological research by offering integrative and expansive datasets that encompass a wide range of biological entities, including genes, proteins, pathways, and diseases. These knowledgebases are meticulously curated, providing researchers with reliable, up-to-date, and interconnected biological information that drives discoveries in genomics, proteomics, systems biology, and drug development. Comprehensive databases often serve as starting points for large-scale analyses, offering tools for data retrieval, annotation, and integration to help researchers contextualize biological functions and molecular mechanisms.

A prime example is UniProt [15], a leading protein-centric knowledgebase that aggregates extensive information on protein sequences, structures, functions, and PTMs. UniProt serves as a comprehensive platform for protein annotation, linking sequence data to functional insights such as enzymatic activities, subcellular localization, and PPIs. It is divided into two major sections: UniProtKB/Swiss-Prot, a manually curated database that ensures high-quality functional annotations, and UniProtKB/TrEMBL, which contains automatically annotated entries [86]. Researchers use UniProt to identify conserved protein domains, predict functional sites, and study evolutionary relationships, making it invaluable for functional genomics and proteomics research.

Similarly, NCBI Reference Sequences (RefSeq) [87] and Ensembl [88] are critical for genomic and transcriptomic data integration. RefSeq, maintained by the National Center for Biotechnology Information, provides a centralized hub for gene-specific information, including gene nomenclature, genomic sequences, mRNA and protein products, functional descriptions, and related references. It connects to other NCBI tools such as BLAST, GenBank, and PubMed, offering a robust ecosystem for genome annotation and exploration. Ensembl, on the other hand, specializes in genome annotation and sequence analysis across multiple species, enabling comparative genomic studies and population-level analyses. Ensembl integrates data on gene models, sequence variations, and regulatory elements, providing insights into gene evolution, structural variations, and genetic diversity.

Beyond sequence data, the GO [16] knowledgebase plays a pivotal role in standardizing the description of gene products and their functions. GO organizes gene functions into three interconnected ontologies: molecular function, which describes the biochemical activities of gene products; biological process, which defines the pathways and processes genes contribute to; and cellular component, which identifies the subcellular localization of gene products. By providing a consistent vocabulary for gene function, GO enables large-scale analyses such as gene enrichment studies, functional clustering, and network reconstruction. Tools like GOrilla [89], goseq [90], and PANTHER [91] utilize GO annotations to perform pathway enrichment and analyze gene lists for overrepresented biological processes, helping researchers interpret experimental data from RNA-seq, proteomics, and other omics studies.

In summary, comprehensive biological databases provide the foundational resources necessary for functional annotation, systems-level analyses, and biological discovery. From sequence and pathway mapping to multiomics integration, databases play central roles in advancing our understanding of genes, proteins, and pathways. The integration of manual curation, automated predictions, and AI-driven tools ensures these knowledgebases remain essential for genomics, proteomics, and biomedical research, driving progress across diverse areas of biology, from fundamental research to clinical and agricultural applications.

3.3.2 Specialized Knowledgebases

While comprehensive databases provide foundational, broad-scale data for biological research, specialized knowledgebases focus on specific biological domains, offering in-depth, high-quality insights into areas such as diseases, molecular interactions, pathways, regulatory mechanisms, pharmacology, and organism-specific biology. These targeted repositories serve as essential tools for addressing particular research questions, facilitating the precise annotation of genes, proteins, and pathways, and enabling discoveries in clinical biology, systems biology, and translational medicine.

One of the most critical areas covered by specialized knowledgebases is disease biology, where repositories link genetic variations, molecular features, and clinical outcomes to uncover the basis of diseases. Databases like ClinVar [50] curate clinically significant genetic variants, assigning pathogenicity scores and linking mutations to disease phenotypes. This data is invaluable for both clinical diagnostics and research into inherited genetic disorders. Similarly, OMIM [48] focuses on the genetic and phenotypic details of Mendelian diseases, providing curated descriptions of genes and their associated phenotypes. OMIM remains a cornerstone for understanding monogenic disorders and serves as a vital resource for clinicians and researchers working on rare diseases. Another key resource, DisGeNET [92], integrates gene-disease association data from multiple sources, combining experimental evidence, literature mining, and curated annotations to map genes to complex diseases. These databases, alongside cancer-focused platforms such as COSMIC [49], support studies on disease mechanisms, biomarker discovery, and precision medicine by connecting molecular alterations to clinical manifestations.

In the domain of drug discovery and pharmacology, specialized knowledgebases provide essential information on drug mechanisms, therapeutic targets, and pharmacogenomics. DrugBank [93] is a widely used repository that integrates chemical, genomic, and clinical data on drugs, offering detailed annotations on drug targets, mechanisms of action, and therapeutic applications. Researchers rely on DrugBank to identify candidate molecules, predict drug interactions, and explore drug repurposing opportunities. Complementary databases like

ChEMBL [94] and PubChem [95] focus on bioactive small molecules, cataloging chemical structures, screening data, and target interactions. PharmGKB [96] adds a pharmacogenomic dimension by linking genetic variations to drug responses, enabling studies into personalized medicine where treatments are tailored to an individual's genetic profile. In oncology, platforms like Genomics of Drug Sensitivity in Cancer (GDSC) [97] link genomic features to drug sensitivity profiles in cancer cell lines, helping researchers identify biomarkers that predict therapeutic responses and resistance mechanisms. Together, these knowledgebases drive innovation in drug development, precision medicine, and pharmacology.

Specialized knowledgebases also extend to organism-specific biology, offering curated genomic, transcriptomic, and functional data for specific model organisms, agricultural species, and microbial communities. Resources like FlyBase [98] for *Drosophila melanogaster* and Zebrafish Information Network (ZFIN) [99] for Danio rerio provide comprehensive datasets that are critical for studying developmental biology, genetics, and disease modeling in these widely used research organisms. Similarly, WormBase [100] focuses on *Caenorhabditis elegans*, a model organism for studying cellular processes and aging. Microbial-focused resources such as MGnify [101] and Human Microbiome Project (HMP) [102] provide functional annotations for microbial genomes and metagenomes, offering insights into host-microbe interactions, environmental adaptations, and microbiome functions in human health and disease. Organism-specific databases not only support basic research but also facilitate translational studies in agriculture, medicine, and environmental biology.

In conclusion, specialized knowledgebases provide detailed, domain-specific insights into diseases, pathways, molecular interactions, gene regulation, pharmacology, and organism-specific biology. They complement comprehensive databases by enabling targeted functional annotation, offering researchers the tools to explore specific aspects of biology with depth and precision. As these resources continue to evolve and integrate with computational tools, machine learning, and multiomics approaches, they will play an increasingly central role in unraveling the complexities of biological systems, supporting discoveries across fundamental biology, translational medicine, and biotechnology.

3.4 Guidance for Knowledgebase Construction

Constructing a biological knowledgebase is a multifaceted process that involves careful planning, strategic data management, and the integration of advanced computational tools to ensure accessibility, reliability, and utility. A well-designed knowledgebase serves as a central resource for researchers, clinicians, and educators, enabling them to access curated, comprehensive, and high-quality biological data. The process unfolds in a series of well-defined steps, beginning with clear objectives, systematic data collection, careful tool selection, and the creation of user-friendly platforms that allow intuitive data visualization and retrieval.

The first and most critical step in constructing a biological knowledgebase is defining its objectives and scope. This involves identifying the research goals, the target audience, and the specific area of focus. A knowledgebase must be tailored to address well-defined research needs and knowledge gaps in biology or medicine. For example, a knowledgebase might focus on a particular disease domain, such as cancer mutations or rare genetic disorders, to help clinicians and researchers understand the molecular basis of disease progression and identify therapeutic targets. Alternatively, the scope could involve mapping PPI networks, regulatory mechanisms such as TF binding sites, or pathways associated with specific cellular functions. Another approach may center on an organism-specific resource, such as Drosophila melanogaster or Arabidopsis thaliana, where the aim is to consolidate genomic, proteomic, and phenotypic data for fundamental and translational research.

Choosing the topic requires a careful evaluation of the scientific relevance and data availability. Researchers must determine whether the knowledgebase will fill a critical void in the current understanding of the chosen domain. For example, a knowledgebase focusing on microbial genomes for host-microbiome studies can address growing interest in microbiome contributions to human health and disease. Similarly, creating a resource for drug-target interactions could provide essential data for therapeutic development and drug repurposing studies. Assessing data availability is equally vital; this includes considering existing datasets from public resources like NCBI and emerging high-throughput technologies. Data gaps must be identified, and strategies for filling those gaps—such as incorporating experimental data, literature curation, or computational predictions—must be outlined early in the planning phase.

Once the focus is defined, the next step is systematic data collection and integration, which forms the backbone of the knowledgebase. Data sources can include experimental results, high-throughput sequencing data, literature mining, and existing public repositories. For instance, genomic and transcriptomic data can be gathered from platforms such as GEO [103], EMBL [104], TCGA (https://www.cancer.gov/tcga), or GTEx [105], while protein data may come from ProteomicsDB [106] and Proteomics Identification Database (PRIDE) [107]. Specialized tools like text-mining algorithms and natural language processing (NLP) approaches are often employed to extract biological information from research articles and clinical reports. Curating this information requires a combination of manual curation by domain experts, who ensure data accuracy, and automated pipelines that streamline large-scale data extraction and standardization. Standardization is essential for data integration, and researchers must use controlled vocabularies such as GO terms and other ontologies to harmonize annotations across datasets.

Following data collection, tool and platform selection is crucial for enabling efficient analysis, management, and user accessibility. This step often involves identifying tools for data processing, annotation, and visualization. Advanced computational pipelines must be established to clean, format, and integrate heterogeneous datasets. Databases must be designed to allow fast query execution, ensuring that researchers can retrieve information with ease, even when working with large-scale datasets. Technologies like structured query language (SQL) databases, Neo4j for graph-based data (https://neo4j.com/), or NoSQL solutions (e.g., MongoDB [https://www.mongodb.com/]) are commonly used to organize and store biological data efficiently.

Equally important is data visualization, which transforms raw data into interpretable insights through intuitive visual outputs. Effective visualization tools allow users to explore complex relationships within biological systems, such as gene interactions, regulatory pathways, or disease associations. Modern visualization tools like Shiny (https://shiny.posit.co/) or Plotly (https://plotly.com/) support interactive dashboards that allow users to filter and analyze data based on custom parameters. Integration of web-based platforms using frameworks like Django (https://www.djangoproject.com/), Flask (https://flask.palletsprojects.com/), or Node.js (https://nodejs.org/) ensures that the knowledgebase is accessible to a global audience via a user-friendly interface.

Ensuring data quality and reliability is another critical consideration in knowledgebase construction. Researchers must establish clear guidelines for data inclusion, emphasizing curated and validated datasets to maintain confidence in the resource. Annotation confidence levels should be provided, especially for computationally predicted data, to help users distinguish between experimental evidence and theoretical inferences. Continuous updates and versioning are also important to accommodate new findings and technological advancements. Implementing automated workflows for data updates and integrating application programming interfaces (APIs) with external databases enables seamless synchronization of the latest information.

Finally, maintaining a successful biological knowledgebase involves fostering an active user community and promoting collaboration. Knowledgebases must be designed to support both novice and expert users, offering easy navigation, robust search capabilities, and clear documentation. Providing user-friendly tools for data submission and feedback encourages community-driven contributions, improving data coverage and quality over time. Community platforms, such as forums, workshops, and webinars, help engage users, ensuring that the knowledgebase remains dynamic, widely adopted, and aligned with evolving research needs.

In summary, constructing a biological knowledgebase is a complex yet rewarding process that requires defining clear research objectives, curating diverse and reliable datasets, selecting appropriate computational tools, and prioritizing user accessibility through effective visualization. Careful planning and rigorous quality control are essential to ensure that the knowledgebase fulfills its role as a valuable resource for biological discovery. As technologies evolve, integrating machine learning, automated curation, and interactive platforms will further enhance the utility, scalability, and impact of biological knowledgebases in addressing critical research questions across life sciences and medicine.

3.5 Case Studies

3.5.1 Functional Annotation Knowledgebases for Cancer Research

Cancer knowledgebases are critical tools in the field of oncology, offering a systematic way to compile and interpret the vast array of data generated by cancer research. These resources focus on cataloging genomic, transcriptomic, proteomic, and clinical data related to various cancer types, enabling researchers to uncover the molecular mechanisms underlying cancer initiation, progression, and treatment resistance. Our team has constructed various cancer knowledgebases, including the functional annotations of lncRNA [108], alternative splicing [109], alternative polyadenylation (APA) [110] in cancer, and functional annotations for genes related to cancer metastasis [111] and drug resistance [112].

lncRNAfunc [108] is a knowledgebase for decoding lncRNAs in cancer. It annotated over 20,000 cancer-related lncRNAs collected from multiple public resources, including TCGA and ENCODE [113], with a focus on their roles in tumorigenesis, metastasis, and therapy resistance. These annotations were based on integrated data from gene expression profiles, chromatin states, TF binding sites, and lncRNA-microRNA interactions. For transcriptomic data, lncRNAfunc contains differential expression analysis to identify lncRNAs with tumor-specific expression patterns across multiple cancer types and stages. This revealed 1,800 differentially expressed lncRNAs (DElncRNAs), with specific subsets showing strong tissue-specific expression and functional enrichment in biological processes such as epithelial-mesenchymal transition (EMT), immune regulation, and DNA damage repair.

To elucidate the potential mechanisms of lncRNAs, lncRNAfunc used functional enrichment analysis and mapped lncRNAs to cancer-related pathways such as p53 signaling, Wnt signaling, and immune checkpoint regulation. Additionally, lncRNAfunc constructed lncRNA-TF-mRNA regulatory networks to uncover how lncRNAs mediate transcriptional regulation in tumor cells. Co-expression network analysis revealed key hub lncRNAs like MALAT1 and HOTAIR, which play pivotal roles in metastasis and chromatin remodeling. For structural and mechanistic insights, lncRNAfunc incorporated lncRNA-microRNA interaction maps, showing how specific lncRNAs sponge miRNAs to modulate oncogene expression. The resource also includes cancer subtype-specific lncRNA annotations, offering insights into lncRNA diversity across tumors. These annotations provide a valuable foundation for understanding how lncRNAs contribute to cancer biology and for developing potential lncRNA-based biomarkers or therapeutics.

The other knowledgebase, CAFuncAPA [110], annotated 50,000 APA) events from cancer-associated fibroblasts (CAFs) across 15,000 tumor samples collected from TCGA. For transcriptomic analysis, CAFuncAPA identified differential APA events between tumor and normal samples, focusing on APA patterns that influence transcript stability, translation efficiency, and isoform diversity. This analysis revealed 1,200 tumor-specific APA sites, which were functionally enriched in biological processes such as extracellular matrix remodeling, angiogenesis, and immune suppression (Fig. 3.1).

Cancer metastasis is the process by which tumor cells migrate and colonize distant organs from a primary site, and is responsible for the majority of cancer-related deaths. Understanding the cellular and molecular mechanisms underlying this complex process is essential for developing effective metastasis prevention and therapy strategies. To this end, metsDB systematically analyzed 1,786 bulk tissue samples from 13 cancer types, 988,463 single

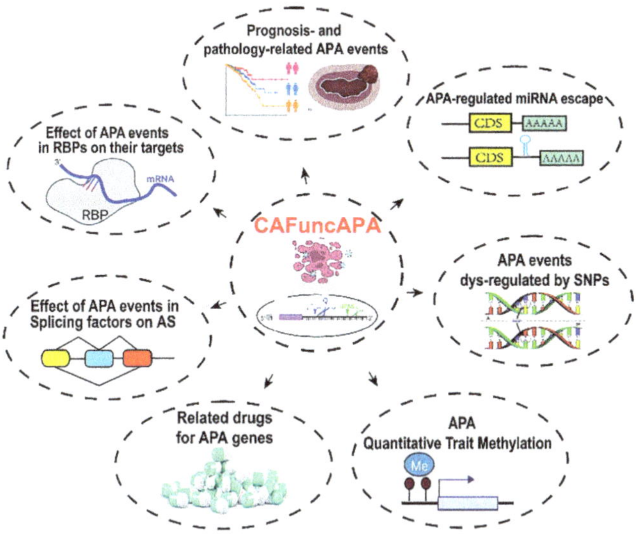

Fig. 3.1 The overview of CAFuncAPA database.

cells from 17 cancer types, and 40,252 spots from 45 spatial slides across 10 cancer types [114]. This database provides insights into alterations in cell constitutions, cell relationships, biological pathways, molecular biomarkers, and drug responses during cancer metastasis at bulk, single-cell, and spatial levels. Users can perform cell or gene searches to obtain multi-view and multi-scale metastasis-related data. This comprehensive resource is invaluable for understanding the metastasis process and for designing molecular therapies.

Drug resistance poses a significant challenge in cancer treatment. Despite the initial effectiveness of therapies such as chemotherapy, targeted therapy, and immunotherapy, many patients eventually develop resistance. Currently, there are several drug-related databases available. For example, the DRESIS [115] database collects validated drug resistance molecules belonging to different drug resistance mechanisms by retrieving published literature and explicitly describes clinically/experimentally verified resistance data for a large number of drugs. The CTR-DB database utilizes pretreatment bulk RNA-seq data to characterize cancer drug response [116]. CeDR Atlas is primarily based on scRNA-seq data to utilize the Connectivity Map resource for drug response prediction [117]. However, there is still a lack of comprehensive characterization of drug resistance mechanisms using single-cell data obtained from drug treatment settings. To this end, our team built the DRMref [112], a reference atlas of drug resistance mechanisms based on a collection of single-cell datasets, to elucidate resistance mechanisms across different cell types. DRMref reveals the intricate landscape of drug resistance, facilitates a deep understanding of resistance complexity at the single-cell level, and aids in the development of improved therapeutic interventions.

DRMref annotated and analyzed 42 single-cell datasets with drug resistance information from 30 studies, 14 of which have both pre and post-treatment samples, encompassing 26 cancer subtypes obtained from GEO database and PubMed queries. These datasets are accompanied by crucial drug response information under three major drug categories: chemotherapy, targeted therapy, and immunotherapy. For those single-cell datasets, DRMref performed rigorous preprocessing steps, including quality control and the elimination of batch effects. Then, DRMref identified resistance-related differentially expressed genes (R-DEGs) by comparing resistant cells with sensitive cells in each of the cell types. To gain insights into the drug resistance mechanisms of these R-DEGs, we conducted functional annotations using enrichment analysis, covering six well-known drug resistance mechanisms: "Unusual Activation of Pro-survival Pathway," "Irregularity in Drug Uptake and Drug Efflux," "Aberration of the Drug's Therapeutic Target," "Epigenetic Alteration of DNA, RNA, or Protein," "Drug Inactivation by Structure Modification," and "Regulation by the Disease Microenvironment" [115]. We also performed enrichment analysis on hallmark [118], KEGG [17], and GO [16] biological process (BP) pathways, as well as TF and miRNA regulatory analysis, for a more in-depth annotation of R-DEGs. Moreover, to provide a deep understanding of the malignant cell activity, we compared intratumoral heterogeneity (ITH) and epithelial–mesenchymal transition (EMT) scores between the resistant and sensitive cells. Given that tumor microenvironment (TME) has been indicated to be associated with drug resistance, we further examined alterations in the TME during drug resistance by comparing the compositions of all cell types. In-depth analysis of cell-cell interactions was performed to investigate the impact of the surrounding microenvironment on cellular responses to drugs. As explained here, DRMref provides a comprehensive knowledge of drug resistance mechanisms.

3.5.2 Functional Annotation Knowledgebases for Aging

Aging is defined as the gradual deterioration of functional characteristics in living organisms [119]. The aging process is also among the most well-known risk factors for most human diseases [120]. Currently, the aging population is growing at an unprecedented pace, and there is an urgent need to initiate detailed studies and develop strategies for healthy aging. Abundant evidence indicates that aging is accompanied by complex molecular and cellular alterations, such as dysregulated intercellular communication, epigenetic and transcriptional alterations, genomic instability, and defects in the telomere maintenance machinery [121]. For example, forkhead box O3 (FOXO3) is recognized as one of the key regulators of aging-related processes. The expression level of FOXO3 decreases with age [122]. The interaction of FOXO3 with NR2E1, SNX3, and CEP5 plays a critical role in oxidative stress response and DNA damage during aging [123]. High expression of FOXO3 regulated by the chromatin accessibility of its enhancer region may have a protective effect on lifespan [124]. Furthermore, FOXO3 is a potential target of resveratrol, which is a candidate agent for the prevention and treatment of aging and aging-related diseases [125]. As shown here, a systematic analysis of individual aging genes and the development of a functional reference annotation will help to gain knowledge of the underlying mechanisms of aging and to identify novel therapeutic targets for anti-aging and aging-related diseases.

There are several available aging databases, such as Human Aging Genomic Resources, Aging Atlas, the Digital Ageing Atlas, AGEMAP, and Open Genes (https://open-genes.com/) [126–128]. These databases mainly focused on providing integrated biological datasets or a simple collection of aging-associated genes. In the era of personalized medicine, with the accumulated genomic data of individuals, an intensive knowledge base annotating the potential mechanisms of individual aging genes is still lacking. Establishing a representative resource on aging will provide a foundation for the individual genomic differences in terms of aging. The high heterogeneity of molecular and cellular changes during aging pushes bulk sequencing techniques to their limits. Thanks to the advancement in single-cell sequencing technologies, single-cell sequencing data can refine our understanding of cellular heterogeneity and molecular alteration at the single-cell level in different tissues during aging. Huang et al. built AgeAnno [129], a knowledgebase of single-cell annotation of human aging, with the goal of providing comprehensive and intensive characterizations for human aging-related genes across diverse tissues and cell types by using single-cell RNA and single-cell ATAC sequencing data (scRNA and scATAC) (Fig. 3.2).

AgeAnno annotated 5,580 aging-related genes collected from six public resources based on 1,678,610 cells from 28 healthy tissue samples with ages ranging from 0 to 110 years. For scRNA data, first, AgeAnno performed differential gene expression analysis for aging-related genes (DEGs) in each cell type and identified tissue- and cell-type-specific DEGs between different age groups. Then, AgeAnno annotated aging-related DEGs into three aging-related biological categories, including immune genes, telomere maintenance genes, and circadian genes. To better understand the potential mechanisms of aging-related genes, AgeAnno performed enrichment analysis and transcriptomic variation analysis to determine aging effect on gene transcriptional noise; cell-cell communication analysis based on the coordinated ligand-receptor (L-R) expression; and transcript factor (TF) regulatory network analysis. To explore the immune microenvironment alteration in the aging context, AgeAnno investigated the immune cell proportions between different age groups. For scATAC data, AgeAnno identified differential accessible regions (DARs) between age groups, and provided detailed genetic annotation information such as enhancers, promoters and introns. Then, AgeAnno performed motif/TF enrichment analysis, footprint analysis, and co-accessibility peak analysis to provide functional annotations for aging-related DARs. AgeAnno also identified gene-drug or gene-chemical interactions and related diseases for aging genes. As described above, AgeAnno provides intensive multiple functional annotations of aging context from single-cell data. We believe that AgeAnno will be a valuable resource that can advance the understanding of the underlying mechanisms of aging and help aging research communities develop anti-aging and aging-related disease treatment strategies.

Moreover, the conservation of well-known aging regulators and their interactions across species highlights the relevance of aging-related research in model organisms to human aging. These crucial evolutionarily conserved aging regulatory factors may serve as potential targets for the development of anti-aging drugs. Since the goal of the animal model research is not solely to comprehend the specific species or the certain genes under investigation, but rather to derive broader insights and draw general conclusions regarding the process of human aging. To this end, Huang et al. have developed AgeAnnoMO [130], a comprehensive knowledgebase encompassing aging-related functional annotations from diverse representative species in aging and longevity research on multiomics level, including genetic and mitochondrial genomes, bulk and single-cell transcriptomes, epigenomes, proteomes, immunomes, microbiomes, and metabolomes. AgeAnnoMO annotated the aging-related functions for 90,972 genes, 16,244 proteins, 841 metabolites, 13 mitochondrial genes, 422 microbiotas,

229,413 age-specific TCR sequences and 1,391 age-specific BCR sequences related to 10 aging hallmarks across multiple species. The current version of AgeAnnoMO houses 136 datasets, including 50 species, 61 tissues, 8,596 samples, and 1,044,141 cells. It identified aging-related mutations, DNA methylations, immune repertoires, and proteostasis associated with primary aging hallmarks. AgeAnnoMO also identified aging-related metabolites and mitochondrial mutations associated with antagonistic aging hallmarks. Moreover, it identified DEGs, intercellular interaction networks, and dysbiosis associated with integrative aging hallmarks. Figure 3.3 shows the construction and the function annotation of AgeAnnoMO. This database provides user-friendly ways to browse and search the results of functional annotations in the database.

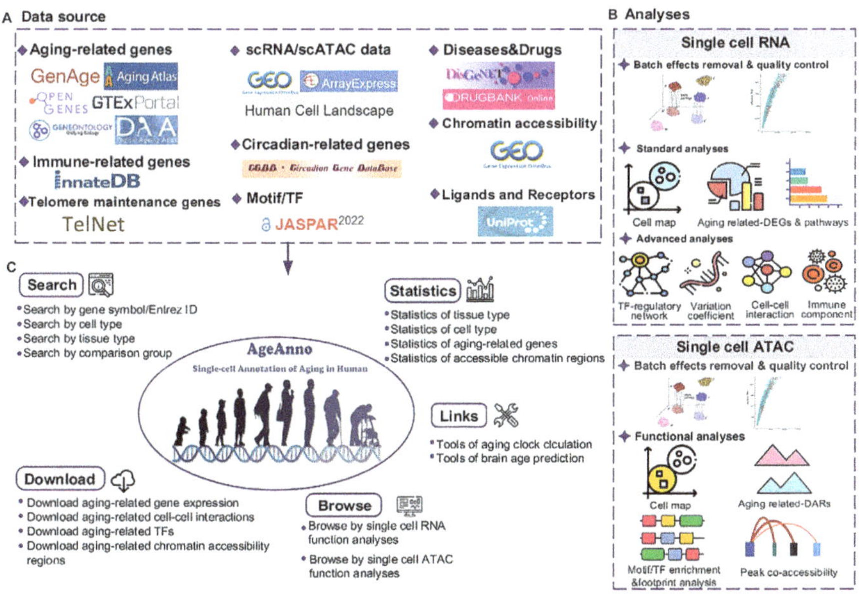

Fig. 3.2 The overview of AgeAnno.

3.5.3 Functional Annotation Knowledgebases for Infectious Diseases

Our team also constructed several databases for infectious diseases, such as SARS-CoV-2. To date, there are multiple resources related to SARS-CoV-2 mutation annotation (i.e., COVID-19 CG [131], CovMT [132], CoV-GLUE [133], Coronapp [134], GESS [135], and Outbreak.info [136]). However, these resources provide a list of mutations detected in SARS-CoV-2 variants without intensive functional annotations and new insights into human health (Table 3.1). Mutations in SARS-CoV-2 genomes led to increased transmissibility, severe disease outcomes, evasion of the immune response, changes in clinical manifestations, and reduced efficacy of vaccines or treatments. Deep understanding the relationship between mutation sites and these factors is crucial for advancing our knowledge of SARS-CoV-2 and for developing effective responses. To address this gap, Feng et al. performed comprehensive bioinformatics analyses for over 1.3 million SARS-CoV-2 genome sequences and associated metadata.

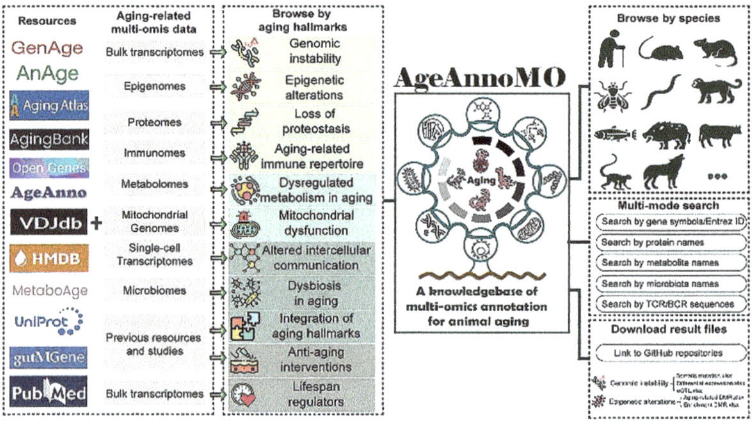

Fig. 3.3 Database content and construction of AgeAnnoMO.

COV2Var [137] is a functional annotation knowledgebase dedicated to SARS-CoV-2 genetic variation. This database plays a critical role in advancing our understanding of the genetic diversity within SARS-CoV-2 and its implications for viral behavior, including infectivity, transmission dynamics, and immune response efficacy. The primary objective of this section is to guide readers through the foundational steps involved in constructing a functional annotation database tailored to SARS-CoV-2 genetic variations.

To construct a functional annotation database for SARS-CoV-2 genetic variation like COV2Var, it is essential to follow a systematic approach that includes several critical steps: first, systematic data collection from publicly available genomic databases to gather high-quality SARS-CoV-2 genome sequences and associated metadata; then, employing robust bioinformatics tools for mutation

Table 3.1 Comparison of information in COV2Var with other mutation-related databases.

Category	COV2Var	Coronapp	GESS	COVID-19 CG	CovMT	CoV-GLUE	Outbreak.info
Mutation type	SNV, Indel	SNV, Indel	SNV	SNV	SNV	SNV, Indel	SNV, Indel
Mutations frequency	✓	✓		✓		✓	
Geographic region distribution	✓	✓	✓	✓	✓	✓	✓
Mutations over time	✓	✓	✓	✓	✓	✓	✓
Co-mutation patterns	✓		✓	✓			
Lineage distribution	✓	✓		✓	✓	✓	✓
Animal hosts	✓						
Patient age	✓						
Patient gender	✓						
Patient status	✓				✓		
Natural selection	✓						
Protein physicochemical properties	✓						
Protein stability and function	✓						
Antigenicity and immunogenicity	✓						
Affinity between RBD and ACE2 receptor	✓						
Affinity between RBD and antibody/serum	✓						
Mutation-related literature	✓						✓
Data source	GISAID	GISAID	GISAID	GISAID	GISAID	GISAID	GISAID

detection, such as variant calling, to identify single nucleotide polymorphisms (SNPs), insertions, and deletions; subsequently, annotating these mutations functionally by integrating sequence data with detailed metadata to assess their impact on viral properties like protein function, stability, and immune escape; designing a database structure using a relational database management system (RDBMS) to store complex relationships between genetic data and epidemiological information; and developing an intuitive user interface with visualization tools like heatmaps and tables to facilitate easy exploration and interpretation of genetic variations. Finally, these tools and the database structure enable applications such as tracking the emergence of new variants, evaluating the impact of mutations on vaccine effectiveness and clinical outcomes, and informing public health interventions. This integrated approach allows researchers to understand the genetic diversity of SARS-CoV-2 and its implications for viral behavior, providing critical insights for vaccine development and public health planning.

The overall schema of COV2Var is represented in Fig. 3.4. As depicted in Fig. 3.4, Panel A, we conducted a comprehensive analysis of over 1.3 million SARS-CoV-2 genome sequences and their associated metadata. Through this extensive analysis, SARS-CoV-2 identified 9,832 common mutations with

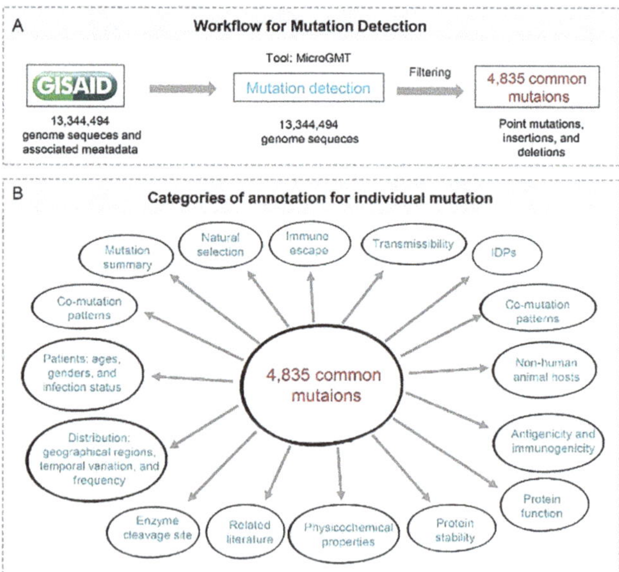

Fig. 3.4 COV2Var Pipeline Overview. (A) Mutation analysis workflow leading to 9832 common mutations. (B) Main categories of common mutations among 9832 Individuals.

competitive advantages—mutations that exhibit greater fitness and spread compared to others. These mutations highlight their evolutionary benefits and biological significance, providing insights into how SARS-CoV-2 adapts and evolves over time. In Fig. 3.4 Panel B, SARS-CoV-2 performed a detailed functional annotation by integrating sequence data and metadata for all 9,832 commonly mutated sites. This process not only identifies the presence of these mutations in different variants of SARS-CoV-2 but also evaluates their functional impact. COV2Var offers a valuable resource for in-depth functional annotations of these mutations across various variants, helping researchers understand their role in viral behavior, such as infectivity, transmissibility, and immune evasion.

3.6 Conclusion

3.6.1 Challenges in Database Integration and Consistency

As the number and diversity of biological knowledgebases continue to grow, integrating data from multiple sources poses significant challenges. One major issue is the lack of standardized formats and terminologies across databases, which complicates data harmonization and interoperability. For instance, variations in how gene identifiers, ontologies, or metadata are represented can result in inconsistencies that hinder accurate comparisons and analyses. To address this, efforts to adopt universal standards such as Findable, Accessible, Interoperable, Reusable (FAIR) principles must be prioritized, ensuring that knowledgebases can effectively share and integrate their datasets.

Another challenge lies in the reliability and completeness of data. Knowledgebases often rely on automated curation processes that, while efficient, may introduce errors or fail to capture the full context of experimental findings. Manual curation by domain experts is essential but resource-intensive, requiring significant time and expertise. Ensuring the quality and consistency of data while maintaining scalability is a critical hurdle for the future of biological knowledgebases. Moreover, the rapid pace of scientific discovery necessitates frequent updates to knowledgebases, which can result in version control issues and discrepancies between datasets. Researchers need tools and workflows that allow seamless tracking of updates and reconciliation of outdated data to maintain the integrity of integrative analyses.

3.6.2 Implications for Future Research

Despite these challenges, biological knowledgebases remain indispensable tools for advancing our understanding of complex biological systems. The development of integrative platforms that combine multiomics data, clinical annotations, and experimental results will open new avenues for research. For example, integrating genomics, transcriptomics, and proteomics data with clinical outcomes can accelerate the discovery of biomarkers and therapeutic targets in precision medicine.

The incorporation of artificial intelligence and machine learning into knowledgebase design and curation processes holds great promise. AI-driven algorithms can enhance data mining, improve error detection, and predict functional relationships, significantly reducing the workload for manual curation. Furthermore, machine learning models can uncover hidden patterns in large-scale datasets, enabling novel insights into biological processes and disease mechanisms.

Future research should also prioritize the development of user-friendly tools for accessing and analyzing knowledgebase data. Interactive visualization platforms, customizable query interfaces, and robust APIs will make knowledgebases more accessible to researchers across disciplines, facilitating interdisciplinary collaborations.

Finally, the expansion of community-driven curation initiatives will ensure that knowledgebases remain relevant and accurate. Engaging the global research community to contribute annotations, validate findings, and share data will foster the continuous growth and refinement of these resources. By addressing these challenges and leveraging emerging technologies, biological knowledgebases will continue to drive innovation in biology, medicine, and biotechnology, paving the way for transformative discoveries in the years to come.

References

1. Stein, L. Genome annotation: from sequence to biology. *Nat Rev Gen*. 2001; **2**(7): p. 493–503.
2. Kumar, A. Introduction to biological databases. *Rec Adv Chem Sci Biotechnol*: p. 25.
3. Krasnogor, N., et al. Web and grid technologies in bioinformatics, computational and systems biology: a review. *Curr Bioinform*. 2008; **3**(1): p. 10–31.
4. Rastogi, S., N. Mendiratta. and P. Rastogi. Bioinformatics: methods and applications:(Genomics, proteomics and drug discovery). 2013: PHI Learning Pvt. Ltd.
5. Misra, B.B., et al. Integrated omics: tools, advances and future approaches. *J Mol Endocrinol*. 2019; **62**(1): p. R21–R45.

6. Srijyothi, L., et al. Roles of non-coding RNAs in transcriptional regulation. *InTech*. 2018. https://doi.org/10.5772/intechopen.76125
7. Itzkovitz, S., E. Hodis, and E. Segal. Overlapping codes within protein-coding sequences. *Genome Res*. 2010; **20**(11): p. 1582–1589.
8. Riethoven, J.-J.M. Regulatory regions in DNA: promoters, enhancers, silencers, and insulators. *Computat Biol Transcript Fact Bind*. 2010: p. 33–42.
9. Stanke, M., et al. AUGUSTUS: Ab initio prediction of alternative transcripts. *Nucleic Acids Res*. 2006; **34**(suppl_2): p. W435–W439.
10. Delcher, A.L., et al. Identifying bacterial genes and endosymbiont DNA with Glimmer. *Bioinform*. 2007; **23**(6): p. 673–679.
11. Cantarel, B.L., et al. MAKER: An easy-to-use annotation pipeline designed for emerging model organism genomes. *Genome Res*. 2008; **18**(1): p. 188–196.
12. Altschul, S.F., et al. Basic local alignment search tool. *J Mol Biol*. 1990; **215**(3): p. 403–410.
13. Finn, R.D., J. Clements, and S.R. Eddy, HMMER web server: interactive sequence similarity searching. *Nucleic Acids Res*. 2011; **39**(suppl_2): p. W29–W37.
14. Benson, D.A., et al. GenBank. *Nucleic Acids Res*. 2009; **37**(suppl_1): p. D26–D31.
15. Consortium, U., UniProt: A worldwide hub of protein knowledge. *Nucleic Acids Res*. 2019; **47**(D1): p. D506–D515.
16. Consortium, G.O., The Gene Ontology (GO) database and informatics resource. *Nucleic Acids Res*. 2004; **32**(suppl_1): p. D258–D261.
17. Kanehisa, M. and S. Goto, KEGG: Kyoto encyclopedia of genes and genomes. *Nucleic Acids Res*. 2000; **28**(1): p. 27–30.
18. Fabregat, A., et al. The reactome pathway knowledgebase. *Nucleic Acids Res*. 2018; **46**(D1): p. D649–D655.
19. Nobusada, T., et al. Update of the FANTOM web resource: enhancement for studying non-coding genomes. *Nucleic Acids Res*. 2024: p. gkae1047.
20. Kalvari, I., et al. Rfam 13.0: shifting to a genome-centric resource for non-coding RNA families. *Nucleic Acids Res*. 2018; **46**(D1): p. D335–D342.
21. Dobin, A., et al. STAR: ultrafast universal RNA-seq aligner. *Bioinform*. 2013; **29**(1): p. 15–21.
22. Kim, D., et al. Graph-based genome alignment and genotyping with HISAT2 and HISAT-genotype. *Nat Biotechnol*. 2019; **37**(8): p. 907–915.
23. Pertea, M., et al. StringTie enables improved reconstruction of a transcriptome from RNA-seq reads. *Nat Biotechnol*. 2015; **33**(3): p. 290–295.
24. Trapnell, C., et al. Differential gene and transcript expression analysis of RNA-seq experiments with TopHat and Cufflinks. *Nat Protocols*. 2012; **7**(3): p. 562–578.
25. Bryant, D.M., et al. *A tissue-mapped axolotl de novo transcriptome enables identification of limb regeneration factors. Cell Rep*. 2017; **18**(3): p. 762–776.
26. Kozomara, A., M. Birgaoanu, and S. Griffiths-Jones, miRBase: from microRNA sequences to function. *Nucleic Acids Res*. 2019; **47**(D1): p. D155–D162.
27. Li, Z., et al. LncBook 2.0: Integrating human long non-coding RNAs with multi-omics annotations. *Nucleic Acids Res*. 2023; **51**(D1): p. D186–D191.

28. Cox, J. and M. Mann, MaxQuant enables high peptide identification rates, individualized ppb-range mass accuracies and proteome-wide protein quantification. *Nat Biotechnol.* 2008; **26**(12): p. 1367–1372.
29. Hirosawa, M., et al. MASCOT: multiple alignment system for protein sequences based on three-way dynamic programming. *Bioinform.* 1993; **9**(2): p. 161–167.
30. Orsburn, B.C., Proteome discoverer—a community enhanced data processing suite for protein informatics. *Proteomes.* 2021; **9**(1): p. 15.
31. Jones, P., et al. InterProScan 5: genome-scale protein function classification. *Bioinform.* 2014; **30**(9): p. 1236–1240.
32. Bailey, T.L., et al. MEME SUITE: tools for motif discovery and searching. *Nucleic Acids Res.* 2009; **37**(suppl_2): p. W202–W208.
33. Jumper, J., et al. Highly accurate protein structure prediction with AlphaFold. *Nature.* 2021; **596**(7873): p. 583–589.
34. Yang, J., et al. The I-TASSER Suite: protein structure and function prediction. *Nat Methods.* 2015; **12**(1): p. 7–8.
35. Mering, C.v., et al. STRING: a database of predicted functional associations between proteins. *Nucleic Acids Res.* 2003; **31**(1): p. 258–261.
36. Oughtred, R., et al. The BioGRID interaction database: 2019 update. *Nucleic Acids Res.* 2019; **47**(D1): p. D529–D541.
37. Butler, A., et al. Integrating single-cell transcriptomic data across different conditions, technologies, and species. *Nat Biotechnol.* 2018; **36**(5): p. 411–420.
38. Wolf, F.A., P. Angerer, and F.J. Theis, SCANPY: large-scale single-cell gene expression data analysis. *Genome Biol.* 2018; **19**: p. 1–5.
39. Zheng, G.X., et al. Massively parallel digital transcriptional profiling of single cells. *Nature* communications. 2017; **8**(1): p. 14049.
40. Zhou, G., et al. OmicsNet 2.0: a web-based platform for multi-omics integration and network visual analytics. *Nucleic Acids Res.* 2022; **50** (w1): p. W527–W533.
41. Karp, P.D., et al. Pathway Tools version 23.0 update: software for pathway/genome informatics and systems biology. *Brief Bioinform.* 2021; **22**(1): p. 109–126.
42. Boekel, J., et al. Multi-omic data analysis using Galaxy. *Nat Biotechnol.* 2015; **33**(2): p. 137–139.
43. Ashburner, M., et al. Gene ontology: tool for the unification of biology. *Nat Genet.* 2000; **25**(1): p. 25–29.
44. Matsuta, Y., M. Ito, and Y. Tohsato, ECOH: an enzyme commission number predictor using mutual information and a support vector machine. *Bioinform.* 2013; **29**(3): p. 365–372.
45. Geer, L.Y., et al. The NCBI biosystems database. *Nucleic Acids Res.* 2010; **38**(suppl_1): p. D492–D496.
46. Federhen, S., The NCBI taxonomy database. *Nucleic Acids Res.* 2012; **40**(D1): p. D136–D143.
47. Schoch, C.L., et al. NCBI Taxonomy: a comprehensive update on curation, resources and tools. *Database.* 2020; **2020**: p. baaa062.
48. Amberger, J.S., et al. OMIM.org: online Mendelian Inheritance in Man (OMIM®), an online catalog of human genes and genetic disorders. *Nucleic Acids Res.* 2015; **43**(D1): p. D789–D798.

49. Bamford, S., et al. The COSMIC (Catalogue of Somatic Mutations in Cancer) database and website. *Br J Cancer*. 2004; **91**(2): p. 355–358.
50. Landrum, M.J., et al. ClinVar: public archive of interpretations of clinically relevant variants. *Nucleic Acids Res*. 2016; **44**(D1): p. D862–D868.
51. Zhang, J. and T.L. Madden, PowerBLAST: a new network BLAST application for interactive or automated sequence analysis and annotation. *Genome Res*. 1997; **7**(6): p. 649–656.
52. Altschul, S.F., et al. Gapped BLAST and PSI-BLAST: A new generation of protein database search programs. *Nucleic Acids Res*. 1997; **25**(17): p. 3389–3402.
53. Buchfink, B., C. Xie, and D.H. Huson, Fast and sensitive protein alignment using DIAMOND. *Nat Methods*. 2015; **12**(1): p. 59–60.
54. Steinegger, M. and J. Söding, MMseqs2 enables sensitive protein sequence searching for the analysis of massive data sets. *Nat Biotechnol*. 2017; **35**(11): p. 1026–1028.
55. Kulmanov, M., M.A. Khan, and R. Hoehndorf, DeepGO: predicting protein functions from sequence and interactions using a deep ontology-aware classifier. *Bioinform*. 2018; **34**(4): p. 660–668.
56. Seo, S., et al. DeepFam: deep learning based alignment-free method for protein family modeling and prediction. *Bioinform*. 2018; **34**(13): p. i254–i262.
57. Huson, D.H., et al. MEGAN analysis of metagenomic data. *Genome Res*. 2007; **17**(3): p. 377–386.
58. Menzel, P., K.L. Ng, and A. Krogh. Fast and sensitive taxonomic classification for metagenomics with Kaiju. *Nat Commun*. 2016;**7**(1): p. 11257.
59. Gupta, S., et al. Quantifying similarity between motifs. *Genome Biol*. 2007; **8**: p. 1–9.
60. Khan, A., et al. JASPAR 2018: update of the open-access database of transcription factor binding profiles and its web framework. N*ucleic Acids Res*. 2018; **46**(D1): p. D260–D266.
61. Sigrist, C.J., et al. PROSITE, a protein domain database for functional characterization and annotation. *Nucleic Acids Res*. 2010; **38**(suppl_1): p. D161–D166.
62. Sun, H., et al. Quantitative integration of epigenomic variation and transcription factor binding using MAmotif toolkit identifies an important role of IRF2 as transcription activator at gene promoters. *Cell Discov*. 2018; **4**(1): p. 38.
63. Bailey, T.L., DREME: motif discovery in transcription factor ChIP-seq data. *Bioinform*. 2011; **27**(12): p. 1653–1659.
64. Grant, C.E., T.L. Bailey, and W.S. Noble, FIMO: scanning for occurrences of a given motif. *Bioinform*. 2011; **27**(7): p. 1017–1018.
65. De Castro, E., et al. ScanProsite: detection of PROSITE signature matches and ProRule-associated functional and structural residues in proteins. *Nucleic Acids Res*. 2006; **34**(suppl_2): p. W362–W365.
66. Hornbeck, P.V., et al. PhosphoSitePlus: a comprehensive resource for investigating the structure and function of experimentally determined post-translational modifications in man and mouse. *Nucleic Acids Res*. 2012; **40**(D1): p. D261–D270.
67. Hassanzadeh, H.R. and M.D. Wang. DeeperBind: enhancing prediction of sequence specificities of DNA binding proteins. in 2016 IEEE International conference on bioinformatics and biomedicine (BIBM). 2016. IEEE.
68. Yuan, H., et al. BindSpace decodes transcription factor binding signals by large-scale sequence embedding. *Nat Methods*. 2019; **16**(9): p. 858–861.

69. Burley, S. K., et al. Protein data bank (PDB): the single global macromolecular structure archive. In: Wlodawer, A., Z. Dauter, M. Jaskolski, editors. *Protein Crystallography. Methods in Molecular Biology*. Vol. 1607. New York, NY: Humana Press; 2017. https://doi.org/10.1007/978-1-4939-7000-1_26
70. Webb, B. and A. Sali. Comparative protein structure modeling using MODELLER. *Curr Prot Bioinform*. 2016; **54**(1): p. 5.6. 1–5.6. 37.
71. Rohl, C. A., et al. Protein structure prediction using Rosetta. *Methods in Enzymology*. Vol. 383. Academic Press; 2004; p. 66–93.
72. Holm, L. Using Dali for protein structure comparison. In: Gáspári, Z., editor. *Structural Bioinformatics. Methods in Molecular Biology*. Vol. 2112. New York, NY: Humana; 2020. https://doi.org/10.1007/978-1-0716-0270-6_3
73. Zhang, Y. and J. Skolnick, TM-align: a protein structure alignment algorithm based on the TM-score. *Nucleic Acids Res*. 2005; **33**(7): p. 2302–2309.
74. Khafif, M., et al. Identification and phylogenetic analyses of VASt, an uncharacterized protein domain associated with lipid-binding domains in Eukaryotes. *BMC Bioinform*. 2014; **15**: p. 1–12.
75. Murzin, A.G., et al. SCOP: a structural classification of proteins database for the investigation of sequences and structures. *J Mol Biol*. 1995; **247**(4): p. 536–540.
76. Orengo, C.A., et al. CATH—a hierarchic classification of protein domain structures. *Structure*. 1997; **5**(8): p. 1093–1109.
77. Tian, W., et al. CASTp 3.0: computed atlas of surface topography of proteins. *Nucleic Acids Res*. 2018; **46**(W1): p. W363–W367.
78. Konc, J. and D. Janežič, ProBiS algorithm for detection of structurally similar protein binding sites by local structural alignment. *Bioinform*. 2010; **26**(9): p. 1160–1168.
79. Wu, Q., et al. COACH-D: improved protein–ligand binding sites prediction with refined ligand-binding poses through molecular docking. *Nucleic Acids Res*. 2018; **46**(W1): p. W438–W442.
80. Halgren, T., New method for fast and accurate binding-site identification and analysis. *Chem Biol Drug Des*. 2007; **69**(2): p. 146–148.
81. Dominguez, C., R. Boelens, and A.M. Bonvin, HADDOCK: a protein–protein docking approach based on biochemical or biophysical information. *J Am Chem Soc*. 2003; **125**(7): p. 1731–1737.
82. Comeau, S.R., et al. ClusPro: a fully automated algorithm for protein–protein docking. *Nucleic Acids Res*. 2004; **32**(suppl_2): p. W96–W99.
83. Chen, R., L. Li, and Z. Weng, ZDOCK: an initial-stage protein-docking algorithm. *Proteins: Struct Funct Bioinform*. 2003; **52**(1): p. 80–87.
84. Srivastava, A., et al. Role of computational methods in going beyond X-ray crystallography to explore protein structure and dynamics. *Int J Mol Sci*. 2018; **19**(11): p. 3401.
85. Kovermann, M., P. Rogne, and M. Wolf-Watz, Protein dynamics and function from solution state NMR spectroscopy. *Q Reviews Biophys*. 2016; **49**: p. e6.
86. Schneider, M., et al. The UniProtKB/Swiss-Prot knowledgebase and its plant proteome annotation program. *J Proteomics*. 2009; **72**(3): p. 567–573.
87. Pruitt, K.D., et al. NCBI Reference Sequences (RefSeq): current status, new features and genome annotation policy. *Nucleic Acids Res*. 2012; **40**(D1): p. D130–D135.

88. Martin, F.J., et al. Ensembl 2023. *Nucleic Acids Res.* 2023; **51**(D1): p. D933–D941.
89. Eden, E., et al. GOrilla: a tool for discovery and visualization of enriched GO terms in ranked gene lists. *BMC Bioinform.* 2009; **10**: p. 1–7.
90. Young, M.D., et al. goseq: Gene Ontology testing for RNA-seq datasets. *R Bioconductor.* 2012; **8**: p. 1–25.
91. Mi, H. and P. Thomas, PANTHER pathway: an ontology-based pathway database coupled with data analysis tools. *Protein Netw Pathway Anal.* 2009: p. 123–140.
92. Piñero, J., et al. The DisGeNET knowledge platform for disease genomics: 2019 update. *Nucleic Acids Res.* 2020; **48**(D1): p. D845–D855.
93. Knox, C., et al. DrugBank 6.0: the DrugBank knowledgebase for 2024. *Nucleic Acids Res.* 2024; **52**(D1): p. D1265–D1275.
94. Gaulton, A., et al. ChEMBL: a large-scale bioactivity database for drug discovery. *Nucleic Acids Res.* 2012; **40**(D1): p. D1100–D1107.
95. Kim, S., et al. PubChem 2019 update: improved access to chemical data. *Nucleic Acids Res.* 2019; **47**(D1): p. D1102–D1109.
96. Barbarino, J.M., et al. PharmGKB: a worldwide resource for pharmacogenomic information. *WIREs Syst Biol Med.* 2018; **10**(4): p. e1417.
97. Yang, W., et al. Genomics of Drug Sensitivity in Cancer (GDSC): a resource for therapeutic biomarker discovery in cancer cells. *Nucleic Acids Res.* 2012; **41**(D1): p. D955–D961.
98. Thurmond, J., et al. FlyBase 2.0: the next generation. *Nucleic Acids Res.* 2019; **47**(D1): p. D759–D765.
99. Sprague, J., et al. The Zebrafish Information Network (ZFIN): the zebrafish model organism database. *Nucleic Acids Res.* 2003; **31**(1): p. 241–243.
100. Harris, T.W., et al. WormBase: a comprehensive resource for nematode research. *Nucleic Acids Res.* 2010; **38**(suppl_1): p. D463–D467.
101. Richardson, L., et al. MGnify: the microbiome sequence data analysis resource in 2023. *Nucleic Acids Res.* 2023; **51**(D1): p. D753–D759.
102. Peterson, J., et al. The NIH human microbiome project. *Genome Res.* 2009; **19**(12): p. 2317–2323.
103. Barrett, T., et al. NCBI GEO: archive for functional genomics data sets—update. *Nucleic Acids Res.* 2012; **41**(D1): p. D991–D995.
104. Kanz, C., et al. The EMBL nucleotide sequence database. *Nucleic Acids Res.* 2005; **33**(suppl_1): p. D29–D33.
105. Lonsdale, J., et al. The genotype-tissue expression (GTEx) project. *Nat Gen.* 2013; **45**(6): p. 580–585.
106. Schmidt, T., et al. ProteomicsDB. *Nucleic Acids Res.* 2018; **46**(D1): p. D1271–D1281.
107. Martens, L., et al. PRIDE: the proteomics identifications database. *Proteomics.* 2005; **5**(13): p. 3537–3545.
108. Yang, M., et al. lncRNAfunc: a knowledgebase of lncRNA function in human cancer. *Nucleic Acids Res.* 2022; **50**(D1): p. D1295–D1306.
109. Kim, P., et al. ExonSkipDB: functional annotation of exon skipping event in human. *Nucleic Acids Res.* 2020; **48**(D1): p. D896–D907.
110. Huang, K., et al. CAFuncAPA: a knowledgebase for systematic functional annotations of APA events in human cancers. *NAR cancer.* 2023; **5**(1): p. zcad004.

111. Wu, S., et al. metsDB: a knowledgebase of cancer metastasis at bulk, single-cell and spatial levels. *Nucleic Acids Res.* 2024: p. gkae916.
112. Liu, X., et al. DRMref: comprehensive reference map of drug resistance mechanisms in human cancer. *Nucleic Acids Res.* 2024; **52**(D1): p. D1253–D1264.
113. Sloan, C.A., et al. ENCODE data at the ENCODE portal. *Nucleic Acids Res.* 2016; **44**(D1): p. D726–D732.
114. Wu, Sijia., et al. metsDB: a knowledgebase of cancer metastasis at bulk, single-cell and spatial levels. *Nucleic Acids Res.* 2025; **53**(D1): p. D1427–D1434.
115. Sun, X., et al. DRESIS: the first comprehensive landscape of drug resistance information. *Nucleic Acids Res.* 2023; **51**(D1): p. D1263–D1275.
116. Liu, Z., et al. CTR-DB, an omnibus for patient-derived gene expression signatures correlated with cancer drug response. *Nucleic Acids Res.* 2022; **50**(D1): p. D1184–D1199.
117. Wang, Y.-Y., et al. CeDR Atlas: a knowledgebase of cellular drug response. *Nucleic Acids Res.* 2022; **50**(D1): p. D1164–D1171.
118. Subramanian, A., et al. Gene set enrichment analysis: a knowledge-based approach for interpreting genome-wide expression profiles. *Proc Natl Acad Sci.* 2005; **102**(43): p. 15545–15550.
119. Strihler, B. Times, cells, and aging. 2012: Elsevier.
120. Niccoli, T. and L. Partridge. Ageing as a risk factor for disease. *Curr Biol.* 2012; **22**(17): p. R741–R752.
121. López-Otín, C., et al. The hallmarks of aging. *Cell.* 2013; **153**(6): p. 1194–1217.
122. Zhao, Y. and Y.-S. Liu, Longevity factor FOXO3: a key regulator in aging-related vascular diseases. *Front Cardiovas Med.* 2021; **8**: p. 778674.
123. Donlon, T.A., et al. FOXO 3 longevity interactome on chromosome 6. *Aging Cell.* 2017; **16**(5): p. 1016–1025.
124. Eijkelenboom, A., et al. FOXO3 selectively amplifies enhancer activity to establish target gene regulation. *Cell Rep.* 2013; **5**(6): p. 1664–1678.
125. Lin, C.-H., et al. Resveratrol enhanced FOXO3 phosphorylation via synergetic activation of SIRT1 and PI3K/Akt signaling to improve the effects of exercise in elderly rat hearts. *Age.* 2014; **36**: p. 1–10.
126. Tacutu, R., et al. Human ageing genomic resources: new and updated databases. *Nucleic Acids Res.* 2018; **46**(D1): p. D1083–D1090.
127. Craig, T., et al. The Digital Ageing Atlas: integrating the diversity of age-related changes into a unified resource. *Nucleic Acids Res.* 2015; **43**(D1): p. D873–D878.
128. Zahn, J.M., et al. AGEMAP: a gene expression database for aging in mice. *PLoS Genet.* 2007; **3**(11): p. e201.
129. Huang, K., et al. AgeAnno: a knowledgebase of single-cell annotation of aging in human. *Nucleic Acids Res.* 2023; **51**(D1): p. D805–D815.
130. Huang, K., et al. AgeAnnoMO: a knowledgebase of multi-omics annotation for animal aging. *Nucleic Acids Res.* 2024; **52**(D1): p. D822–D834.
131. Chen, A.T., et al. COVID-19 CG enables SARS-CoV-2 mutation and lineage tracking by locations and dates of interest. *Elife.* 2021; **10**: p. e63409.
132. Alam, I., et al. CovMT: an interactive SARS-CoV-2 mutation tracker, with a focus on critical variants. *The Lancet Infect Dis.* 2021; **21**(5): p. 602.

133. Singer, J., et al. CoV-GLUE: a web application for tracking SARS-CoV-2 genomic variation. *Preprints*. 2020.2020060225 (2020).
134. Mercatelli, D., et al. Coronapp: a web application to annotate and monitor SARS-CoV-2 mutations. *J Med Virol*. 2021; **93**(5): p. 3238–3245.
135. Fang, S., et al. GESS: a database of global evaluation of SARS-CoV-2/hCoV-19 sequences. *Nucleic Acids Res*. 2021; **49**(D1): p. D706–D714.
136. Gangavarapu, K., et al. Outbreak.info genomic reports: scalable and dynamic surveillance of SARS-CoV-2 variants and mutations. *Nat Methods*. 2023; **20**(4): p. 512–522.
137. Feng, Y., et al. COV2Var, a function annotation database of SARS-CoV-2 genetic variation. *Nucleic Acids Res*. 2024; **52**(D1): p. D701–D713.

CHAPTER 4

General Introduction to Omics Data Analysis for Precision Medicine

4.1 Overview of Omics Technologies: Genomics, Transcriptomics, Proteomics, Metabolomics, and Microbiomics

Omics technologies represent a transformative leap in the field of biological and biomedical research, enabling the large-scale study of molecular components that define the structure, function, and dynamics of biological systems [1]. These technologies encompass diverse disciplines, including genomics, transcriptomics, proteomics, metabolomics, and microbiomics, each contributing unique insights into the intricate workings of cells, tissues, and organisms [2]. Together, they provide a comprehensive framework for understanding the molecular basis of health and disease.

Genomics, the study of an organism's entire DNA sequence, forms the foundation of omics sciences. By analyzing genetic variations, structural modifications, and epigenetic marks, genomics provides insights into heritable traits and predispositions to diseases [3]. Advances in high-throughput sequencing, such as whole-genome and exome sequencing, have accelerated the identification of disease-associated mutations and genetic markers, making genomics indispensable in both research and clinical settings.

Transcriptomics delves deeper into gene expression, focusing on RNA molecules transcribed from DNA. It captures the dynamic nature of gene regulation by analyzing how genes are expressed across different tissues, conditions, or time points [3]. Techniques like RNA sequencing (RNA-seq) reveal transcriptional changes associated with diseases, uncovering key regulatory networks and identifying biomarkers for diagnostic and therapeutic purposes.

Proteomics shifts the focus to proteins, the functional molecules in cells that carry out biological processes. By studying the proteome, researchers can explore protein expression levels, modifications, interactions, and functions. Mass spectrometry (MS)-based proteomics has become a powerful tool for investigating complex biological systems, enabling the discovery of novel protein biomarkers and therapeutic targets [3].

Metabolomics complements these approaches by profiling small molecules and metabolites that reflect cellular biochemical activity. Metabolomics provides a snapshot of the metabolic state of an organism, linking genotype to phenotype through metabolic pathways [4]. This field has been instrumental in identifying disease-specific metabolic signatures and elucidating mechanisms underlying metabolic disorders.

Microbiomics examines the collective genomes and functional contributions of microbial communities residing within the human body. Advances in microbiome research have highlighted its critical role in maintaining health and influencing disease outcomes [5]. Studies of gut microbiota, for example, have uncovered connections to immunity, metabolism, and neurological disorders, making microbiomics an integral part of systems biology.

4.2 Computational Genomics, Genetics, and Epigenetics Analyses

4.2.1 Functional genomics and genetic variants analysis

Advances in next-generation sequencing (NGS) technologies have enabled the precise identification of genetic alterations, including single nucleotide variants (SNVs) and copy number variants (CNVs). These alterations play critical roles in disease susceptibility, progression, and therapy response, making their analysis central to precision medicine. This section outlines key computational tools and techniques for analyzing SNVs, CNVs, and other genetic alterations, emphasizing their applications and limitations in clinical and research contexts.

4.2.1.1 Tools and techniques for analyzing SNVs, CNVs, and other genetic alterations

SNVs represent single-base changes in the genome, often associated with disease phenotypes. Their detection relies on accurate variant calling pipelines, which typically involve preprocessing, alignment to a reference genome, variant calling, and annotation. Genome analysis toolkit (GATK) [6] is a gold standard for SNV detection, offering highly sensitive and specific variant calling workflows. It employs sophisticated statistical methods to identify SNVs and indels (insertions and deletions) from NGS data. SAMtools and BCFtools [7, 8] are lightweight tools for alignment and variant calling, offering fast processing with minimal computational requirements. DeepVariant [9] is an AI-powered tool leveraging deep learning to improve the accuracy of SNV detection. Post-detection, annotation tools like ANNOVAR [10] and SnpEff [11] link SNVs to known databases (e.g., ClinVar [12], dbSNP) to predict their functional impact, aiding in the interpretation of pathogenicity. CNVs involve duplications or deletions of large genomic regions and have profound implications in cancer, developmental disorders, and other diseases. Detecting CNVs requires tools tailored for structural variant discovery. CNVkit [13] identifies CNVs in targeted and whole-exome sequencing (WES) datasets, providing comprehensive visualization of copy number profiles.

Both single-cell RNA sequencing (scRNA-seq) and single-cell DNA sequencing (scDNA-seq) allow for the precise examination of genetic variations at the individual cell level, offering insights into evolutionary dynamics. Nonetheless, challenges like sequencing errors, contamination, and amplification biases in single-cell sequencing lead to potential false positive and negative results. Various tools have been developed to address these issues. Monovar [14] utilizes a probabilistic method for somatic SNV identification, revealing co-occurrence and mutual exclusivity of driver mutations. LiRA [15] employs phased sequence reads with germline mutants as a reference, distinguishing somatic mutants from amplification artifacts. SCcaller [16] employs a likelihood ratio test to differentiate true nonsynonymous single nucleotide variants (nsSNVs) from artifacts, aiding discrimination based on aging and smoking status. Strategies for identifying CNV in single cells include Alleloscope [17] for allele-specific CNVs in scDNA-seq, while InferCNV [18] and CONICS [19] use scRNA-seq data to estimate copy number states, offering insights into intratumoral heterogeneity (ITH) across cancer types. Additionally, Numbat [20] detects allele-specific CNV from spatial transcriptomics and scRNA-seq. Understanding single-cell-based somatic mutations aids clonal lineage reconstruction and inference of the order and timing of genomic events. Advanced methods like B-SCITE [21], integrating WES/whole-genome sequencing (WGS) and scDNA-seq data, and SCARLET [22], leveraging somatic nsSNVs and CNVs from scDNA-seq, facilitate tracking cell population evolution and inferring tumor phylogenies. Beyond SNVs and CNVs, genetic analysis increasingly focuses on structural variants (e.g., inversions, translocations), fusion genes, and epigenomic changes. Manta [23] analyzes structural variants like large deletions, inversions, and translocations with high sensitivity.

4.2.1.2 Cases for identifying disease-associated genetic variants, driver mutations, and genetic biomarkers for precision medicine

Genomic alterations and signatures

Variant calling tools such as GATK and Mutect2 [24] enable the identification of genomic alterations and gene signatures. Notable findings include the impact of microsatellite instability (MSI) on immunotherapy outcomes and the correlation between PD-L1 overexpression and intra-tumoral inflammatory environment [25, 26]. Mutant genes such as TP53 [27], KRAS [27], ARID1A [28], CDK12 [29], PBRM1 [30], and SMARCA4 [31] have been associated with enhanced responses to ICB. Moreover, genomic features associated with exogenous mutagens, such as ultraviolet (UV) light exposure, aflatoxin B1, tobacco smoke, and oncogenic bacteria or viruses, can modulate the effectiveness of immune checkpoint blockade (ICB) therapies [32]. Conversely, alterations in genes such as B2M, JAK, STK11, KEAP1, and EGFR can predict poor outcomes or resistance to immunotherapy [33–36]. Furthermore, amplification of the MDM2 family and EGFR mutations have been correlated with hyper-progression following ICB treatment [37].

Mutational signatures represent patterns of DNA mutations found in the genome, often resulting from specific mutagenic processes such as UV exposure. They can affect specific genomic alterations, tumor mutation burden status, and neo-antigen immunogenicity. For instance, apolipoprotein B mRNA-editing cytidine-deaminase (APOBEC) signatures, characterized by $C > G$ and $C > T$ mutations, result from viral activity and strongly correlate with immune activation [38]. Conversely, the aging process signature shows a nonsignificant correlation with immune activation [38]. APOBEC-related mutagenesis, positioned at the intersection of anti-viral immunity and endogenous nucleic acid editing, also strongly correlates with immune marker expression (e.g., PD-L1 and interferon-gamma in cancer cells), leading to increased neo-peptide hydrophobicity [38]. UV exposure has been associated with specific effects on neo-peptide characteristics, including an increase in hydrophobicity, and this has been linked to improved immunotherapeutic response in certain cancer types [32].

Intratumor heterogeneity

Genomic intratumor heterogeneity (ITH) is a prevalent occurrence in most tumors, encompassing the clonal composition of tumor cells and genomic distinctions among individual subclones [39]. The clinical ramifications of tumor heterogeneity are significant, as tumors characterized by intricate subclonal structures tend to exhibit heightened aggressiveness, making them more prone to the development of treatment resistance, disease recurrence, and immune escape [40]. Bulk sequencing methods like WGS and WES often fail to capture rare subclonal populations, particularly when their distribution is uneven. Emerging techniques, such as circulating tumor DNA (ctDNA) and cell-free DNA (cfDNA) derived from liquid biopsies, enable the detection of minor subclones, including those associated with treatment resistance.

Advanced computational tools address these challenges. For instance, copy number callers like Sclust [41], FACETS [42], Sequenza [43], and TITAN [44] analyze copy number alterations, while algorithms like Battenberg [45], SciClone [46], PyClone [47], and FastClone [48] deconvolve subclonal structures using Bayesian models. Single-cell sequencing methods provide finer resolution, revealing genetic and transcriptomic variations at the cellular level. Benchmark studies [49, 50] suggest that pipelines combining Mutect2, FACETS, and PyClone achieve optimal accuracy, though scalability issues persist for larger datasets [39, 49–51]. Together, these methods advance our understanding of ITH, offering opportunities to refine diagnostics and develop tailored therapeutic strategies.

4.3 Functional Transcriptomics Analyses for Precision Medicine

Functional transcriptomics is a rapidly evolving field that explores the functional roles of RNA molecules within the cell, particularly gene expression regulation and its impact on disease mechanisms. By examining how various forms of RNA, such as messenger RNA (mRNA), noncoding RNA (ncRNA), and small RNA, interact and influence cellular processes, functional transcriptomics provides invaluable insights into the molecular basis of diseases, including cancer, neurodegenerative disorders, and autoimmune diseases [52].

Functional transcriptomics goes beyond traditional transcriptomics, which involves the large-scale study of RNA species expressed in a particular cell or tissue. Functional transcriptomics emphasizes understanding how these RNA molecules contribute to cellular functions and how their expression and regulation can be altered in disease states. It involves both gene expression profiling (such as RNA-seq or microarrays) and the functional characterization of RNA molecules, which include RNA splicing, alternative polyadenylation (APA), RNA editing, and ncRNAs (e.g., miRNAs, lncRNAs). Functional transcriptomics are critical mechanisms of RNA processing that contribute to the complexity of gene expression regulation. These processes can be dysregulated in various diseases, offering novel opportunities for precision medicine—the tailoring of medical treatment to individual genetic and molecular profiles. Below is a detailed exploration of these mechanisms, their roles in disease, and their potential in RNA-based precision medicine [53–57].

4.3.1 RNA splicing

4.3.1.1 RNA splicing in precision medicine

RNA splicing is a process by which noncoding regions (introns) are removed from precursor mRNA (pre-mRNA), and coding regions (exons) are joined together. Alternative splicing (AS) allows the production of multiple mRNA isoforms from a single gene by varying which exons are included or excluded. This increases protein diversity and allows cells to generate different protein isoforms from the same gene. Aberrant AS can produce oncogenic protein isoforms that promote cancer development, progression, and metastasis. For example, mutations or mis-splicing of tumor suppressor genes like p53 can lead to a loss of function or gain of function in oncogenic isoforms. AS also defects are implicated in diseases like spinal muscular atrophy (SMA) [58] and frontotemporal dementia (FTD) [59], where splicing errors produce nonfunctional or toxic proteins. In Duchenne muscular dystrophy (DMD) [60], exon skipping therapies (modulating AS) are used to bypass defective exons and restore functional dystrophin production.

Exon skipping, where specific exons are "skipped" to correct defective splicing, is a well-studied approach in SMA (using antisense oligonucleotides like Spinraza) [61]. Small molecules or gene therapies targeting splicing regulators (like SR proteins or hnRNPs) can restore normal splicing patterns, as in the case of splicing modulators for neurodegenerative diseases [62, 63]. RNA therapeutics can be designed to correct or modulate AS in specific tissues or diseases, providing a promising avenue for precision treatments.

4.3.1.2 Key tools and methodologies in RNA splicing detection

The identification of RNA splicing events, particularly AS, is a critical aspect of transcriptomic analysis. One of the most widely used technologies for detecting splicing events is RNA-seq [64]. This high-throughput technique maps sequencing reads to a reference genome or transcriptome, enabling the detection of exon-exon junctions and the quantification of splicing isoforms. The versatility of RNA-seq allows researchers to examine splicing patterns across different conditions, tissues, and disease states. Single-cell RNA-seq [65] enables the study of gene expression and splicing at the resolution of individual cells. This technique is particularly useful for investigating cellular heterogeneity and understanding how splicing events may vary across different cell types or states. Key features include high-resolution analysis and the ability to explore splicing in specific cell populations.

To focus specifically on splicing events, several RNA-seq protocols are tailored to enhance the detection of spliced transcripts. Exon-exon junction-specific protocols, which focus on capturing reads spanning exon-exon junctions to improve splicing detection. mRNA and poly-A enriched protocols, which enrich for mRNA and polyadenylated transcripts, increase the sensitivity for detecting splicing events. Capture-based techniques, which use probes to target specific regions of interest in the genome, allow for more efficient detection of spliced isoforms.

Long-read sequencing technologies, such as PacBio [66] and Oxford Nanopore [67], offer several advantages over short-read sequencing for studying RNA splicing. These technologies can directly capture full-length transcripts, improving accuracy in isoform identification and enabling the detection of complex AS patterns that may be missed with short reads.

RNA splicing detection has become a key area of focus in transcriptomics due to its importance in gene expression regulation and cellular function. The combination of RNA-seq, single-cell RNA-seq, splicing-specific protocols, and long-read sequencing technologies provides a powerful toolkit for understanding splicing at various levels of complexity. Using tools such as TopHat [68], STAR [69], Cufflinks [70], rMATS [71], LeafCutter [72], SplAdder [73], and MAJIQ [74], researchers can accurately analyze splicing events, identify novel isoforms, and explore how AS contributes to gene regulation and disease processes.

The RNA splicing disease database, in conjunction with specialized resources like ExonSkipDB [75], ExonSkipAD [76], and ASCancer [77], serves as a valuable tool for advancing our understanding of the role of RNA splicing in human diseases. By providing detailed information on splicing mutations and their disease connections, these databases help accelerate research into novel therapeutic strategies and contribute to the development of personalized medicine.

4.3.2 Alternative polyadenylation

4.3.2.1 Alternative polyadenylation in precision medicine

Polyadenylation is the addition of a poly(A) tail to the 3′ end of an mRNA molecule, which affects mRNA stability, translation efficiency, and nuclear export. APA occurs when different polyadenylation sites within a gene's 3′ untranslated region (UTR) are selected, leading to transcripts with varying 3′ UTRs. This process can influence gene expression by affecting the stability and localization of the mRNA [78].

In many cancers, APA patterns are altered, which can lead to changes in the stability or translation of oncogenes and tumor suppressors. For example, truncated forms of mRNA can evade degradation, leading to the expression of proteins that contribute to cancer progression [79]. APA has been shown to regulate critical genes involved in heart function. Dysregulated APA could contribute to diseases like heart failure or cardiac hypertrophy by altering gene expression patterns in the heart. In the brain, APA can affect the expression of genes involved in neuronal differentiation and synaptic plasticity, and APA alterations are linked to diseases like Alzheimer's and autism spectrum disorders (ASD) [80].

Understanding how polyadenylation factors, such as cleavage and polyadenylation specificity factor (CPSF) and cleavage stimulation factor (CstF), regulate APA could lead to therapeutic strategies aimed at correcting APA dysregulation in various diseases [81]. Targeting specific 3′ UTRs may offer a means to restore normal mRNA stability or translation, particularly in conditions like cancer and neurodegenerative diseases. Additionally, RNA-based interventions that modify polyadenylation signals or employ RNA-binding proteins to restore proper APA patterns hold the potential to finely regulate gene expression, providing a promising approach in precision medicine.

4.3.2.2 Key tools and methodologies in alternative polyadenylation detection

APA is a crucial mechanism that regulates gene expression by generating mRNA isoforms with different 3′ UTRs using APA sites. This process can influence mRNA stability, translation efficiency, and localization. The detection of APA events involves a combination of high-throughput sequencing technologies and specialized computational methods that can identify and quantify the use of different polyadenylation sites across genes [82].

RNA-seq [64] remains the primary tool for detecting APA events. By mapping sequencing reads to the genome or transcriptome, RNA-seq can identify the precise locations of polyadenylation sites by detecting reads that extend to the 3′ end of mRNA transcripts. The detection of these polyadenylation sites allows for the identification of APA isoforms, which differ in the length of their 3′ UTRs. RNA-seq data can also be used to quantify the relative use of different polyadenylation sites in each gene, providing insights into how APA contributes to gene regulation in different biological contexts.

PolyA-seq [83] is a specialized variant of RNA-seq specifically designed to focus on the polyadenylation sites. It enriches polyadenylated RNA species, thereby increasing the sensitivity for detecting APA events. PolyA-seq often employs techniques such as oligo(dT) selection to specifically capture mRNA molecules with polyA tails, allowing for more accurate identification of polyadenylation sites across the transcriptome.

3′ end sequencing like EnD-seq and AppEnD are other powerful approaches to detect APA [84]. This method involves sequencing the 3′ ends of mRNA transcripts, enabling the identification of precise polyadenylation sites and the quantification of the different isoforms produced by APA. Targeted 3′ end sequencing (e.g., 3′Tag-seq) and cap analysis of gene expression (CAGE) provide additional resolution for mapping and quantifying APA sites across the genome [85, 86].

Long-read sequencing technologies, such as PacBio and Oxford Nanopore, also play a role in APA detection by capturing full-length transcripts, including the 3′ UTRs. These technologies provide a more comprehensive view of polyadenylation events compared to short-read sequencing. Long reads can help resolve complex polyadenylation patterns that span large regions and detect APA events that are difficult to resolve with short-read sequencing due to their proximity to other polyadenylation sites or repetitive sequences.

In addition to sequencing technologies, various computational tools have been developed to analyze APA events. These include DaPars [87], APAtrap [88], TAPAS [89]. Combining RNA-seq, PolyA-seq, 3′-end sequencing, long-read sequencing, and computational tools offers a powerful toolkit for detecting and analyzing APA events. These methodologies allow researchers to gain insights into the regulatory mechanisms controlling APA, how it impacts gene expression, and its roles in various biological processes and diseases.

Several databases focus on functional aspects of APA events. It aims to understand how APA events contribute to gene expression regulation and their functional roles in various biological processes, including disease mechanisms.

3aTWAS [80], CAFuncAPA [90], and other APA databases are crucial tools for studying the role of APA in diseases. They provide functional insights into how changes in polyadenylation events contribute to gene regulation and disease pathogenesis. Understanding these mechanisms may offer new avenues for diagnosing, treating, or preventing various diseases, including cancers, neurological disorders, and cardiovascular diseases.

4.3.3 RNA editing

4.3.3.1 RNA editing in precision medicine

RNA editing involves the posttranscriptional modification of RNA molecules, which can change the nucleotide sequence of the RNA. The most common type of RNA editing is adenosine-to-inosine (A-to-I) editing, in which adenosine is converted to inosine (which is interpreted as guanosine by the cell). RNA editing can alter protein-coding sequences (CDS), affect RNA splicing, or influence mRNA stability and translation [56].

RNA editing is particularly important in the brain, where it regulates neurotransmitter receptors, ion channels, and other proteins involved in neuronal function [91]. Dysregulation of RNA editing has been linked to diseases like Alzheimer's disease (AD), Parkinson's disease, and schizophrenia. Abnormal RNA editing can lead to changes in protein function, contributing to tumorigenesis and metastasis. For example, altered editing of K-ras can affect its oncogenic potential. In autoimmune diseases, abnormal RNA editing can result in the production of proteins that trigger inappropriate immune responses, as seen in systemic lupus erythematosus and rheumatoid arthritis [92, 93].

The enzymes responsible for RNA editing, such as adenosine deaminases acting on RNA (ADARs), can be targeted to correct or enhance RNA editing at specific loci [94]. This is particularly relevant in treating neurological disorders where RNA editing is critical for normal brain function. Altered RNA editing patterns can serve as diagnostic biomarkers for diseases like cancer or neurodegenerative disorders. This could help with early detection and prognosis monitoring [95]. Exogenous RNA editing or CRISPR-based RNA editing technologies offer the potential to correct pathogenic mutations or generate therapeutic RNA isoforms in diseases [96].

4.3.3.2 Key tools and methodologies in RNA editing detection

RNA editing is a posttranscriptional process that alters RNA sequences, most commonly through the conversion of specific nucleotides, such as A-to-I and cytosine-to-uracil (C-to-U), which can impact protein function, RNA stability, and splicing. Detecting RNA editing events is essential to understanding how RNA modifications contribute to cellular function and disease. Various tools and methodologies have been developed to identify and characterize RNA editing sites, with high-throughput sequencing technologies and computational approaches playing crucial roles.

RNA-seq is the most used technique for detecting RNA editing events, as it enables the comprehensive sequencing of the transcriptome. By mapping RNA-seq reads to a reference genome, researchers can detect differences between the reference DNA sequence and the RNA sequence, which may indicate RNA editing. A-to-I editing is detected by comparing RNA reads to the genome, identifying positions where the RNA sequence differs from the genomic reference (where A is substituted by I in the RNA, which is read as G). Targeted RNA-seq allows for the sequencing of specific regions of interest in the transcriptome, increasing the sensitivity of RNA editing detection. By focusing sequencing efforts on regions known to be prone to editing, such as CDS or 3′ UTRs, researchers can more efficiently identify edited sites. Long-read sequencing technologies, such as PacBio and Oxford Nanopore, allow for the direct sequencing of full-length RNA molecules. These technologies enable more accurate detection of RNA editing because they capture full transcript isoforms, including regions where editing events are prevalent. In particular, long reads are helpful for resolving complex or overlapping editing sites that may be challenging to detect with short-read sequencing. Bisulfite sequencing is widely used for detecting DNA methylation but can also be adapted for RNA editing detection, especially for C-to-U editing. This method works by treating RNA with sodium bisulfite, which deaminates cytosines to uracils, allowing for the identification of editing events in the transcriptome through comparison with untreated RNA samples.

The detection of RNA editing events relies on a combination of advanced sequencing technologies and specialized computational tools. RNA-seq remains the cornerstone technique for identifying RNA editing, but targeted approaches and long-read sequencing offer enhanced sensitivity and resolution. Computational tools such as REDItools [97] facilitate the identification of editing events in large-scale datasets. Collectively, these methodologies provide comprehensive insights into the functional impact of RNA editing in various biological processes and diseases.

The databases ADeditome [98], NeuroEdQTL [99], CAeditome [100], REDIdb [101], and REIA [102] are focused on studying RNA editing events and their implications in various diseases, particularly in the contexts of neurological, cardiovascular, and genetic disorders. These resources catalog RNA editing sites, highlight their functional roles, and explore how alterations in RNA editing contribute to diseases like AD, cancer, and autoimmune conditions. By providing comprehensive data on RNA editing patterns and their genetic associations, these databases help researchers uncover the molecular mechanisms underlying disease pathogenesis and facilitate the development of new therapeutic targets and diagnostic biomarkers.

4.3.4 Noncoding RNAs

4.3.4.1 Noncoding RNAs in precision medicine

ncRNAs are RNA molecules that do not code for proteins but play crucial regulatory roles in gene expression, RNA processing, and cellular function. They are classified into several categories, with the most studied being. MicroRNAs (miRNAs) are small (~22 nucleotides) RNA molecules that regulate gene expression posttranscriptionally by binding to mRNAs and inhibiting their translation or inducing degradation. Long noncoding RNAs (lncRNAs) are RNA molecules longer than 200 nucleotides that can regulate gene expression at the transcriptional, posttranscriptional, or epigenetic level by interacting with chromatin, transcription factors, and other RNA molecules. Circular RNAs (circRNAs) are a type of lncRNA with a circular structure formed by back-splicing of exons, involved in gene regulation, serving as miRNA sponges or interacting with RNA-binding proteins. Small nucleolar RNAs (snoRNAs) are involved in the modification and maturation of other RNAs, particularly ribosomal RNAs (rRNAs) and small nuclear RNAs (snRNAs). Small interfering RNAs (siRNAs) are double-stranded RNAs that mediate RNA interference (RNAi), regulating gene expression by promoting the degradation of target mRNAs [103, 104].

ncRNAs are critical regulators of cellular processes, and their dysregulation is implicated in a wide range of diseases, including cancer, neurological disorders, cardiovascular diseases, and autoimmune conditions. Their roles in disease can be categorized as follows. Many ncRNAs regulate gene expression either directly, by interacting with DNA or mRNA, or indirectly, by modulating chromatin structure or transcription factor activity. When this regulation is disrupted, it can lead to overexpression or underexpression of genes, contributing to disease. For example, in cancer, altered miRNA expression can lead to the silencing of tumor suppressor genes or the activation of oncogenes. For example, the overexpression of miR-21 in tumors contributes to oncogenesis. Certain lncRNAs, such as XIST (involved in X-chromosome inactivation) or HOTAIR (linked to cancer metastasis), can influence cancer progression and metastasis by regulating chromatin remodeling or gene silencing [105, 106].

ncRNAs can interact with chromatin-modifying complexes, leading to changes in DNA methylation, histone modification, and chromatin remodeling. These epigenetic changes are associated with various diseases. lncRNAs and miRNAs have been implicated in conditions like Alzheimer's and dysregulated Parkinson's diseases, where they affect synaptic plasticity, neuronal survival, and amyloid plaque formation [107, 108]. miRNAs are involved in regulating heart development, vascular tone, and the response to injury [109]. Abnormal expression of miRNAs such as miR-1 and miR-133 is associated with heart failure, atherosclerosis, and myocardial infarction [110]. ncRNAs such as miRNAs or siRNAs influence mRNA stability and translation [111]. Dysregulation of these processes can result in abnormal protein synthesis, contributing to diseases such as miRNAs that target mRNAs for degradation can alter tumor suppressor expression, facilitating tumorigenesis. Aberrant regulation of specific lncRNAs involved in muscle cell differentiation or apoptosis can contribute to the progression of muscle diseases [112]. ncRNAs, especially miRNAs and lncRNAs, play significant roles in regulating immune system responses. Dysregulation of these ncRNAs can lead to autoimmune diseases, chronic inflammation, and immune system dysfunction. For example, altered miRNA expression can influence inflammation by regulating immune cells like T-cells and macrophages. Specific lncRNAs are involved in regulating the immune response, and their dysregulation contributes to the development of autoimmune diseases [113].

4.3.4.2 Key tools and methodologies in noncoding RNAs detection

ncRNAs are a diverse group of RNA molecules that do not encode proteins but play crucial roles in cellular regulation, gene expression, and various biological processes. The study of ncRNAs, including miRNAs, lncRNAs, circRNAs, and others, has become increasingly important in understanding gene regulation, disease mechanisms, and cellular function. The detection and analysis of ncRNAs require specialized tools and methodologies due to their complexity, size variability, and the fact that many of them are not well-annotated in existing genome databases.

RNA-seq remains the most widely used and powerful tool for discovering and profiling ncRNAs. It enables high-throughput sequencing of the entire transcriptome, providing a detailed snapshot of both coding and ncRNA species. Whole transcriptome RNA-seq provides an unbiased approach to detect all

RNA species, including both known and novel ncRNAs, by sequencing all RNA molecules in a sample. RNA-seq can also be used to compare the expression levels of ncRNAs across different conditions, providing insights into their role in disease and cellular processes. In addition to detecting ncRNAs, RNA-seq can identify splicing events involving ncRNAs or miRNA precursors. The small RNA-seq [114] protocol is specifically designed to enrich and sequence small RNAs (typically between 18 and 40 nucleotides), including miRNAs, piwi-interacting RNAs (piRNAs), and snoRNAs. Long RNA-seq focuses on sequencing longer RNA molecules like lncRNAs and other long noncoding species (typically longer than 200 nucleotides), offering a detailed view of their structure and function.

Detecting ncRNAs, particularly novel ones, often requires bioinformatics tools that integrate RNA-seq data with genome annotations. These tools help predict and annotate ncRNA genes, identify regulatory elements, and perform functional analysis. miRDeep is the tool for the identification and quantification of miRNAs from RNA-seq data. miRDeep can discover novel miRNAs and predict their target genes [115]. RNAFold/ViennaRNA are for predicting the secondary structure of RNA molecules, which is particularly useful for analyzing ncRNAs like miRNAs and lncRNAs, where structure often determines function [116, 117]. CIRCExplorer is a tool specifically designed to identify and analyze circRNAs from RNA-seq data [118].

Detecting and analyzing ncRNAs requires a combination of high-throughput sequencing technologies, traditional molecular biology techniques, and advanced bioinformatics tools. RNA-seq remains the cornerstone for comprehensive ncRNA profiling, while techniques like qRT-PCR, Northern blotting, and in situ hybridization are critical for validating and visualizing specific ncRNAs. Bioinformatics tools play an essential role in predicting novel ncRNAs, understanding their structures, and analyzing their functional roles in gene regulation and disease processes. By combining these methodologies, researchers can gain a deeper understanding of the functional roles and regulatory mechanisms of ncRNAs in various biological systems.

RNAcentral [119], NONCODE [120], NPInter [121], and lncRNAfunc [122] are key databases that provide valuable resources for studying lncRNAs and their functions. RNAcentral consolidates RNA sequences from various sources, offering a comprehensive resource for RNA sequence information, including lncRNAs. NONCODE focuses specifically on lncRNAs, providing a curated collection of verified and predicted sequences across species, along with annotations related to their function, expression, and disease involvement.

NPInter is dedicated to RNA-protein and RNA-RNA interactions, offering insights into how lncRNAs interact with other molecules to regulate biological processes. lncRNAfunc provides functional annotations of lncRNAs, highlighting their roles in gene regulation, chromatin remodeling, and disease pathways. Together, these databases are essential tools for advancing our understanding of the biology of lncRNAs and their roles in cellular regulation and disease mechanisms.

4.4 Functional Transcriptomics Applications in Precision Medicine

Functional transcriptomics is a key tool in precision medicine, which aims to tailor medical treatment to individual patients based on their unique genetic, molecular, and environmental profiles. By examining the full spectrum of RNA molecules expressed in cells or tissues, whether coding or noncoding, functional transcriptomics provides deep insights into the molecular mechanisms underlying disease, therapeutic responses, and patient-specific characteristics. This allows for the development of more precise diagnostic tools, therapeutic strategies, and drug treatments. The applications of functional transcriptomics in precision medicine are extensive, with significant implications for diagnosis, therapy, and drug development.

4.4.1 Diagnostic biomarkers

4.4.1.1 Identification of disease-specific biomarkers

Functional transcriptomics plays a crucial role in the identification of disease-specific biomarkers by profiling RNA expression patterns associated with various conditions. This approach not only facilitates early diagnosis but also provides deeper insights into disease severity, subtypes, and progression. By analyzing gene expression levels and specific ncRNAs, such as long noncoding RNAs (lncRNAs) and miRNAs, researchers can pinpoint biomarkers that not only signal the presence of disease but also reflect its stage and type.

In cancer, for example, transcriptomic signatures help differentiate between various cancer types and their subtypes, enabling more precise diagnosis and personalized treatment strategies. The expression of certain lncRNAs and miRNAs can be linked to specific cancer stages or even predict patient prognosis [123]. Moreover, functional transcriptomics enables the use of liquid biopsy, an emerging, noninvasive technique for detecting cancer-related biomarkers in bodily fluids, such as blood or urine. This approach allows for continuous monitoring of disease progression, enabling early detection of metastasis or recurrence, without the need for invasive tissue biopsies [124].

Beyond cancer, functional transcriptomics also shows promise in diagnosing neurological diseases like Alzheimer's, Parkinson's, and multiple sclerosis [125]. In these conditions, specific miRNAs and lncRNAs have been identified as potential biomarkers for early diagnosis and ongoing monitoring of disease progression. Furthermore, RNA profiling techniques are being applied to detect viral infections, including COVID-19 [126], HIV [127], and hepatitis [128]. By analyzing the host's RNA expression, these platforms offer a deeper understanding of the immune response to infections, which can inform treatment decisions.

Moreover, transcriptomic biomarkers play a significant role in monitoring therapeutic responses. Changes in the expression of particular genes or ncRNAs can provide valuable information about how well a patient is responding to treatment, allowing for timely adjustments to therapy based on real-time data. This capability can improve treatment outcomes by ensuring that patients receive the most effective interventions tailored to their specific molecular profile.

4.4.1.2 Transcriptomic signatures for early diagnosis

Transcriptomic signatures, which refer to the patterns of gene expression and the presence of specific RNA molecules, play an essential role in the early detection of diseases. By profiling RNA at various stages of disease development, functional transcriptomics allows for the identification of biomarkers that can signal the presence of a disease even before clinical symptoms appear. These signatures provide valuable insights into the underlying molecular mechanisms, helping to detect diseases at a stage when they are more treatable and potentially less advanced.

In cancer, for example, early detection is critical for improving patient outcomes, and transcriptomic profiling has proven to be a powerful tool for this purpose. By analyzing the expression of specific genes, lncRNAs, and miRNAs, researchers can identify unique RNA signatures that are characteristic of early-stage cancers, often before any physical manifestations are visible. This can lead to the identification of cancers at more treatable stages, allowing for less aggressive treatments and better prognosis. Furthermore, the ability to distinguish between different cancer subtypes through transcriptomic analysis supports personalized medicine, where treatments are tailored to the specific genetic and molecular profile of each tumor.

Beyond oncology, transcriptomic signatures are also being utilized for the early diagnosis of a wide range of diseases, including neurological disorders such as AD, Parkinson's disease, and multiple sclerosis. In these cases, specific miRNAs and lncRNAs have been identified as potential biomarkers that can be detected in blood or cerebrospinal fluid before overt symptoms appear. This could pave the way for earlier intervention, potentially slowing or even preventing disease progression in at-risk populations. In infectious diseases, transcriptomic signatures can also play a significant role in early diagnosis. By analyzing host RNA expression in response to infection, researchers can detect the presence of pathogens like viruses (e.g., HIV, COVID-19, and hepatitis) at very early stages. Early viral detection through RNA profiling can provide important information about the host's immune response, enabling more precise and timely treatment strategies, reducing the risk of complications, and improving patient outcomes.

Moreover, transcriptomic biomarkers can be particularly useful in diseases where traditional diagnostic methods are either invasive or insufficiently sensitive. Noninvasive techniques such as liquid biopsy, which involves analyzing RNA from blood samples, have emerged as an exciting application for early diagnosis. These technologies are poised to revolutionize the way we detect and monitor diseases, providing a more comprehensive, cost-effective, and accessible approach to diagnostics.

In summary, transcriptomic signatures offer a powerful, noninvasive means of diagnosing diseases at their earliest stages. By analyzing gene expression and ncRNA patterns, these signatures provide a deeper understanding of disease pathophysiology, facilitate early intervention, and support the development of personalized treatment strategies. The continued advancement of transcriptomic technologies holds great promise for improving diagnostic accuracy and patient outcomes across a wide range of diseases.

4.4.2 Therapeutic targets

4.4.2.1 Discovery of novel therapeutic targets using transcriptomics

RNA splicing, RNA editing, APA, and ncRNAs are increasingly recognized as important regulators of gene expression and play critical roles in disease mechanisms. These processes provide novel therapeutic opportunities, particularly in diseases where gene expression regulation is altered. Altered splicing can produce isoforms that drive tumorigenesis, metastasis, and resistance to apoptosis. For instance, the CD44 gene, which plays a key role in cell-cell interactions and tumor progression, undergoes AS to generate isoforms that facilitate metastasis [129]. Targeting these splicing variants with antisense oligonucleotides or small molecules could specifically inhibit the production of tumor-promoting isoforms. Additionally, diseases like SMA and amyotrophic lateral sclerosis are linked to splicing defects in crucial genes such as SMN1 and TDP-43, respectively [130, 131]. Splicing-modulating therapies, including nusinersen for SMA, aim to correct these defects and restore functional protein production. In cardiac diseases, splicing alterations in genes like TTN, which encodes titin, can lead to conditions such as dilated cardiomyopathy [132]. Correcting these splicing defects using small molecules or gene-editing technologies holds promise for treatment. Antisense oligonucleotides can be designed to bind to pre-mRNA and modulate splicing patterns, while small molecule splicing modulators can interact with splicing regulatory factors to either promote or inhibit specific splicing events, offering potential therapeutic strategies for a range of diseases.

RNA editing plays a crucial role in maintaining normal brain function, particularly in disorders like epilepsy and AD, where alterations in the editing of neurotransmitter receptors, such as glutamate receptors, can disrupt receptor activity and lead to neuronal dysfunction. Therapeutic strategies aimed at restoring normal RNA editing, either by enhancing the activity of ADAR enzymes or using gene-editing tools like CRISPR/Cas9, offer potential treatments for these conditions [133]. In cancer, RNA editing can lead to the production of modified proteins that promote tumor progression, resistance to therapies, and immune evasion. For example, alterations in RNA editing of TGF-β receptors or p53 can influence tumor behavior [134]. Targeting RNA editing enzymes or correcting specific RNA edits may provide novel therapeutic approaches, such as using small molecules to activate ADAR enzymes and restore normal editing patterns or employing CRISPR/Cas9 systems to correct editing defects in vivo, offering new avenues for treatment.

Altered APA patterns in many cancers lead to the generation of shorter 3′ UTRs, which may evade posttranscriptional regulation by miRNAs or RNA-binding proteins, thus promoting oncogene expression. For instance, genes like EGFR undergo APA, enhancing cancer cell proliferation and survival [135]. Therapeutic strategies targeting APA could potentially restore normal regulation of these oncogenes. Similarly, in heart failure, APA in genes associated with cardiac function, such as BNP, may produce unstable mRNA isoforms that contribute to disease progression [136]. Modulating the APA machinery could offer a way to regulate gene expression and slow down disease development. Small molecules or antisense oligonucleotides can influence APA by targeting polyadenylation factors like CPSF, and editing 3′ UTRs to restore proper gene regulation may represent a promising therapeutic approach for both cancer and heart failure [137].

miRNAs are small RNA molecules that regulate gene expression by binding to the 3′ UTRs of target mRNAs, leading to their degradation or repression. In cancer, miR-21 is often upregulated, promoting cell proliferation, migration, and resistance to apoptosis [138]. Restoring the expression of tumor-suppressive miRNAs or inhibiting oncogenic miRNAs offers significant therapeutic potential. Long noncoding RNAs regulate gene expression at both the transcriptional and posttranscriptional levels, and dysregulated lncRNAs, such as HOTAIR, which is involved in chromatin remodeling and cancer metastasis, have been identified as key players in cancer and other diseases [139]. Therapeutic strategies targeting lncRNAs, such as antisense oligonucleotides or small molecules, could help restore normal gene expression. circRNAs are stable, covalently closed RNAs that act as sponges for miRNAs or regulators of transcription. Certain circRNAs, like circRNA-MYLK, are implicated in cancer progression, and modulating their expression could provide new therapeutic avenues [140]. Potential treatments include restoring tumor-suppressive miRNAs or inhibiting oncogenic miRNAs with miRNA mimics or antagomirs, using antisense oligonucleotides to block disease-associated lncRNAs, and developing drugs or gene-editing tools to regulate the expression or function of circRNAs. Together, these strategies represent promising approaches for the treatment of diseases driven by aberrant RNA expression.

Splicing, RNA editing, APA, and ncRNAs represent essential layers of gene regulation that influence a wide range of diseases. Targeting these RNA-based mechanisms opens new avenues for therapeutic intervention, especially in cancers, neurological disorders, and other complex diseases. Advances in RNA-based therapies, such as small molecules, antisense oligonucleotides, and CRISPR-based gene-editing technologies, offer promising strategies to restore normal gene expression and correct molecular defects. The development of such therapies could lead to more personalized, effective treatments with fewer side effects.

4.4.2.2 Targeting RNA modifications for disease treatment

Functional transcriptomics provides valuable insights into RNA molecules and pathways that can be targeted for therapeutic intervention. By identifying dysregulated genes and pathways in diseases, it enables researchers to pinpoint potential therapeutic targets and correct gene expression abnormalities at the transcriptomic level. In cancer, for example, oncogenes and tumor suppressor genes are often regulated by ncRNAs like miRNAs and lncRNAs. Modulating these RNA molecules could inhibit tumor growth or enhance the effectiveness of existing therapies, such as using miRNA mimics or inhibitors to restore the normal expression of tumor-suppressing genes. RNAi therapies, including siRNAs and short hairpin RNAs (shRNAs), offer a strategy to silence specific genes, and functional transcriptomics can help identify aberrantly expressed genes as targets for RNAi treatments. These therapies are being explored for a range of conditions, from viral infections to genetic disorders like Huntington's disease and cystic fibrosis. Additionally, with gene-editing technologies like CRISPR/Cas9, functional transcriptomics is used to map gene expression profiles before and after editing, ensuring more precise targeting for optimal therapeutic outcomes. Transcriptomic data also aids in the design of immunotherapies by identifying immune-related pathways and markers, such as the RNA expression profiles of immune checkpoint molecules like PD-1 and CTLA-4, which can guide the development of more effective immune checkpoint inhibitors for cancer therapy.

4.5 Proteomics and Protein Function Analysis

Proteomics, the large-scale study of proteins and their functions, has emerged as a cornerstone in modern biomedical research. Proteins are the primary executors of biological processes, serving as enzymes, structural components, signaling molecules, and more. Understanding protein dynamics, modifications, and interactions is essential for unraveling disease mechanisms and identifying therapeutic targets. This section explores the tools and techniques central to proteomics, with a focus on their applications in disease biology and drug discovery.

4.5.1 Proteomics techniques

Proteomics encompasses a variety of techniques designed to identify, quantify, and characterize proteins in complex biological samples. These techniques enable researchers to gain insights into protein expression levels, posttranslational modifications (PTMs), and protein-protein interactions, all of which are critical for understanding cellular processes and their dysregulation in disease.

1. **Protein identification and quantification:**
   High-resolution MS is the cornerstone of modern proteomics, enabling the identification and quantification of proteins with remarkable precision.

Proteins are enzymatically digested into peptides, which are separated by liquid chromatography (LC) and analyzed by tandem mass spectrometry (MS/MS). The resulting spectra are matched against protein databases to identify the proteins present in the sample.

2. **Key tools for protein identification include:**
   MaxQuant [141]: A leading software for quantitative proteomics, MaxQuant integrates label-free and isotopic labeling techniques, such as SILAC and TMT, to provide robust protein quantification. It also supports advanced statistical analysis for high-throughput datasets.
   Comet [142] and MSFragger [143]: These open-source tools excel in database searching, rapidly matching MS/MS spectra to peptide sequences. MSFragger, in particular, is optimized for large datasets, providing unmatched speed and sensitivity in protein identification.
   Quantification techniques in proteomics are critical for comparing protein expression levels across conditions. Label-free quantification (LFQ) methods analyze the intensity of peptide signals, while isotope-based approaches provide higher accuracy by incorporating internal standards [144]. These methods are invaluable for identifying differentially expressed proteins in disease states, offering potential biomarkers for diagnosis or targets for therapy.
3. **Posttranslational modifications**
   PTMs, such as phosphorylation, acetylation, glycosylation, and ubiquitination, regulate protein function and localization [145]. Dysregulation of PTMs is often associated with diseases, including cancer, neurodegeneration, and metabolic disorders [145]. Advanced proteomics tools are now capable of identifying and quantifying PTMs with high sensitivity.
   CLUMPS-PTM [146]: This tool focuses on clustering PTM sites to identify patterns and pathways associated with specific cellular functions or diseases. It facilitates the discovery of regulatory networks involving modified proteins.
   DeepKla [147]: A deep learning-based platform, DeepKla predicts lysine acetylation sites from proteomic datasets, offering insights into epigenetic regulation and metabolic control. Its integration with MS data accelerates the discovery of functionally relevant PTMs.
   Comprehensive PTM analysis provides a deeper understanding of how protein modifications influence signaling pathways, enabling researchers to identify potential intervention points for therapeutic development.

4. **Protein-protein interactions and functional analysis**
   Protein-protein interactions (PPIs) are fundamental to nearly all biological processes, from signal transduction to metabolic regulation [148]. Mapping these interactions provides insights into cellular networks and disease mechanisms. Techniques such as co-immunoprecipitation (Co-IP), yeast two-hybrid screens, and affinity purification-mass spectrometry (AP-MS) are widely used for studying PPIs [149]. Proteomics tools also enable large-scale analysis of interaction networks. MSFragger [143] supports cross-linking mass spectrometry (XL-MS) workflows, capturing spatial relationships between interacting proteins. This approach is particularly valuable for understanding protein complexes and their roles in cellular architecture. Bioinformatics platforms like Cytoscape [150] visualize PPI networks, integrating proteomics data with genomic and transcriptomic datasets to provide a holistic view of cellular function. Proteomics plays a pivotal role in understanding disease biology by revealing alterations in protein expression, PTMs, and interaction networks. In cancer research, proteomics has identified dysregulated kinases and signaling pathways, leading to the development of targeted therapies such as tyrosine kinase inhibitors. In neurodegenerative diseases, proteomics has uncovered aberrant protein aggregation and misfolding, providing insights into pathogenesis and potential biomarkers for early diagnosis.
   Therapeutic target identification is another key application of proteomics. By integrating proteomics with systems biology approaches, researchers can pinpoint druggable proteins and pathways. For example, proteomics has identified overexpressed receptors in tumor cells, such as HER2 in breast cancer, which are now targeted by monoclonal antibodies like trastuzumab. Similarly, in infectious diseases, proteomics has revealed host-pathogen interactions, guiding the development of antiviral and antimicrobial agents.

4.5.2 Sequence prediction, optimization, and AI-based drug discovery

Genome sequence optimization forms the foundation of modern drug discovery by revealing genetic variations and molecular targets critical for therapeutic interventions. AI and machine learning algorithms have revolutionized this process, analyzing complex datasets to identify patterns, predict drug-target interactions, and streamline the drug design process. Among the most impactful tools in this domain is AlphaFold2 [151], an AI-driven system that predicts protein structures with unprecedented accuracy.

Proteins are essential to nearly all biological processes, and their three-dimensional structures are key to understanding their function and interaction with potential drugs. AlphaFold2, developed by DeepMind, has addressed one of the most significant challenges in structural biology: predicting a protein's 3D structure from its amino acid sequence. This technology leverages advanced deep learning techniques and extensive training on experimentally determined protein structures to generate highly accurate predictions.

At the heart of AlphaFold2 is the Evoformer module, a transformer-based architecture that processes both sequence data and structural features simultaneously. It utilizes multiple sequence alignments (MSAs) to capture evolutionary information, highlighting conserved residues and structural motifs critical for folding. The Evoformer also generates a pair representation matrix that encodes spatial relationships between residues, a vital component in modeling protein folding dynamics. AlphaFold2's iterative refinement process further distinguishes it from traditional approaches. The system employs gradient-based optimization to minimize energy functions, ensuring that the predicted structures are physically plausible and biologically relevant. Attention mechanisms allow the model to focus on specific regions of the sequence, enhancing its ability to capture long-range interactions necessary for accurate folding predictions. The result is a high-resolution atomic structure with confidence scores assigned to each residue, providing insights into protein stability and function.

The implications of AlphaFold2 in drug discovery are profound. It facilitates structure-based drug design (SBDD) by identifying druggable pockets and guiding the design of molecules that bind with high specificity and affinity. For example, AlphaFold2 has been instrumental in targeting previously undruggable proteins, offering new opportunities for therapeutic development. Additionally, the predicted structures are used in virtual screening platforms, where millions of candidate compounds are computationally evaluated for binding efficacy, significantly accelerating hit identification and lead optimization.

Beyond AlphaFold2, AI and machine learning methods play a broader role in drug design. Generative adversarial networks (GANs) and reinforcement learning algorithms are employed to design novel compounds with desired properties, optimizing pharmacokinetics, toxicity, and efficacy. These approaches reduce the trial-and-error cycles traditionally associated with drug discovery, saving both time and resources.

Despite its groundbreaking capabilities, AlphaFold2 has limitations. The model is less effective in predicting the structures of proteins with high conformational flexibility or those influenced by PTMs. Additionally, its focus on apo structures (proteins without bound ligands) requires supplementary modeling to account for ligand-induced conformational changes. Future developments are likely to integrate molecular dynamics simulations and ligand-protein interaction models, further bridging the gap between computational predictions and experimental validation.

In conclusion, sequence optimization and AI-driven approaches like AlphaFold2 are reshaping the landscape of drug discovery. By providing high-resolution structural insights and streamlining compound design, these technologies have unlocked new possibilities for therapeutic innovation. As tools like AlphaFold2 continue to evolve, their integration with virtual screening platforms and advanced computational models promises to accelerate the development of safe and effective drugs, transforming healthcare on a global scale.

4.6 Microbiome and Metagenomics Data Analysis

4.6.1 Microbial and host sequencing data analysis

The human gastrointestinal tract hosts over 10^{14} microorganisms, impacting the host's health by providing infection protection, immune regulation, and energy harvesting. Changes in intestinal ecology can lead to the expansion of pathobionts, potentially causing infections, senescence, and cancers [152]. Approximately 15%-20% of human cancers can be attributed to oncogenic viruses [153]. Although the role of bacteria in causing cancer is controversial, extensive research demonstrates that certain bacteria contribute to the onset of cancer [154].

However, the experimental approaches limited insights into the function of the large-scale microbiome. To address these limitations, computational models have emerged, enabling sophisticated analyses of complex microbial systems. These models start with preprocessing steps such as filtering, denoising, and normalization of sequencing data to ensure data quality and consistency. Measurement of microbial diversity, such as alpha and beta diversity, is performed to understand the richness and variation in microbial communities. Downstream functional annotation, such as pathway enrichment, helps interpret the microbiome's role and functional capacity. Advanced computational techniques are then employed to uncover the underlying mechanisms of microbiome-host interactions, particularly in immune responses and cancer immunotherapy. This section provides insights into the measurement techniques and computational approaches currently used to study the intricate crosstalk between the microbiome, immune response, and therapeutic interventions for cancer.

4.6.2 Microbiome characterization

4.6.2.1 Tissue-resident microbiome

Tools like PICRUSt and Tax4Fun can be used to predict the functional potential of microbial communities from marker gene sequencing data. PICRUSt relies on reference genomes, while Tax4Fun infers functions based on taxonomic maps [155, 156]. Previous studies compared the stability of several microbial functional annotation tools via multiple aspects, including technical biases, differences in measuring the differential abundance of health-related functional categories, and normalization based on 16S rRNA gene copy number. The results showed a considerable prediction discrepancy between different tools, and PICRUSt2 shows the most robust prediction ability through several datasets. Despite their cost-effectiveness, 16S rRNA data have limitations, including indirect function estimation and lower prediction accuracy. Understanding microbial functions and pathways is essential for grasping the microbiome's impact on host health [157].

4.6.2.2 Host-based microbiome identification

The functions and the interactions of the microbiome in host tissues can be identified using metagenomic sequencing data or host DNA/RNA-seq data. The methods to identify strains using metagenomics data can be grouped into two categories: marker gene-based methods and genome-wide methods. Marker gene-based methods, such as StrainPhlAn [158], can identify the dominant strain of a given species in each sample. Other methods, such as MIDAS and strain, identify microbial species based on genome-wide genotyping rather than focusing on marker genes.

Various methods have been developed to extract intra-tumoral microbiota using bulk RNA-seq data from databases like TCGA. Tools, such as Kraken2 and MetaPhlAn2, are available to aid the identification and analysis of tumor-associated microbiota using bulk RNA-seq data [159–163]. Single-cell analysis of host-microbiome interactions (SAHMI) and CSI-microbes enable the quantification of intra-tumor microbial profiling using scRNA sequencing data [164, 165]. Despite progress, challenges persist in intra-tumor microbiome analysis. Variations in reference genomes and annotation across databases pose obstacles, while the precise localization of microbial genomes from single-cell data remains elusive [166]. Efforts are ongoing to elucidate the relationship between microbial and host genome sequences (Fig. 4.1).

Machine learning offers powerful tools for identifying cancer-associated microbiotas and understanding their contributions after we obtain the microbiome information. Linear regression models are commonly used to identify microbial signatures linked to cancer progression, metastasis, and prognosis [167, 168].

Researchers have also developed multiple resources for the cancer-associated microbiome. ColPortal is a platform that integrates omics data for colorectal cancer, including clinical information, histology slides, molecular features, and microbiome data [169]. There are also several pan-cancer microbiome resources, such as microbioTA [170] and BIC [171]. These databases provide the comparison of microbial profiles of various cancer tissues.

4.6.2.3 Whole-genome shotgun sequencing for microbial community analysis

Whole-genome shotgun (WGS) sequencing offers a high-resolution view of microbial communities by sequencing and reconstructing the complete genomes of organisms present in a sample. Unlike 16S rRNA or ITS sequencing, which targets specific marker genes, WGS captures the total genetic content of a microbial ecosystem, encompassing bacteria, archaea, fungi, viruses, and even host DNA. This comprehensive approach provides insights into both the taxonomic composition and functional potential of the microbiome.

WGS has advanced the study of diseases such as inflammatory bowel disease (IBD), cancer, and obesity by linking microbial genes to metabolic pathways and immune regulation. For instance, genes associated with short-chain fatty acid (SCFA) production, which are linked to anti-inflammatory properties, have been identified in microbiomes of healthy individuals but are diminished in IBD patients. Similarly, genes involved in secondary bile acid metabolism have been implicated in colorectal cancer progression, revealing potential biomarkers and therapeutic targets.

4.6.2.4 Tools for contamination detection and microbiome profiling

Contamination is a major challenge in metagenomic studies, as it can distort microbial profiles and confound downstream analyses. Tools designed for contamination detection have become essential to ensure the reliability of metagenomic datasets:

Decontam: An R-based tool widely used to identify and remove contaminants in sequencing datasets. Decontam applies statistical models based on DNA concentration and prevalence across samples. For example, contaminants typically show a linear relationship with DNA concentration, allowing Decontam to distinguish true microbial sequences from extraneous noise. Its integration with tools like QIIME2 ensures streamlined workflows for contamination-free data analysis.

SourceTracker: This Bayesian tool tracks the sources of microbial contamination by comparing sample profiles to reference databases. It is particularly effective for distinguishing environmental contaminants from true sample-derived sequences. For instance, SourceTracker has been applied in clinical metagenomics to detect reagent-derived contaminants, ensuring accurate profiling of patient microbiomes.

PhyloFlash: Designed to reconstruct near-complete small subunit (SSU) rRNA sequences from metagenomic data, PhyloFlash excels in identifying potential contaminants early in the analysis pipeline. Its ability to visualize taxonomic distributions in real-time provides an additional layer of quality control, making it an invaluable tool for large-scale studies.

MicroDecon: A contamination removal tool that uses a statistical framework to model the distribution of microbial taxa across samples. By preserving the integrity of true microbial signals, MicroDecon is particularly suited for environmental microbiome studies where contamination levels are often high.

For microbiome profiling, tools have evolved to provide robust taxonomic and functional insights into microbial communities: MetaPhlAn2 is a widely used tool that relies on clade-specific marker genes for high-resolution taxonomic profiling. MetaPhlAn2 can accurately profile microbial abundance even in mixed and complex communities, making it a preferred choice for large-scale studies. Its updated version, MetaPhlAn3, integrates additional marker genes to improve accuracy and computational efficiency. HUMAnN3 regards as a complementary to MetaPhlAn2. HUMAnN3 performs functional profiling by mapping microbial genes to metabolic pathways and biological functions. It leverages databases like UniRef90 and KEGG, enabling researchers to link microbial activities to host phenotypes. For example, HUMAnN3 has been instrumental in identifying microbial pathways associated with immune modulation and inflammation in cancer patients. Kraken2 is a k-mer-based taxonomic classifier that assigns sequences to microbial taxa with high sensitivity and speed. Kraken2's database flexibility allows users to customize their analyses, making it an ideal choice for studies involving underrepresented or novel microbial taxa.

QIIME2: While traditionally associated with amplicon-based analyses, QIIME2 has expanded to support metagenomic workflows. Its plugins for taxonomic and functional annotation provide an all-in-one solution for microbiome studies. QIIME2's visualization tools, including interactive plots and diversity metrics, enhance data interpretation and communication.

MEGAN6: A comprehensive tool for visualizing and analyzing metagenomic datasets, MEGAN6 integrates taxonomic, functional, and phylogenetic data in an intuitive interface. It supports comparisons across multiple samples, making it invaluable for longitudinal and multi-cohort studies.

4.6.2.5 Microbial diversity

After obtaining OTUs or microbial abundance, alpha and beta diversity are used to assess microbiome differences between tumor and healthy tissue. Alpha diversity measures within-sample diversity using indices like Chao1, Shannon, and Simpson, while beta diversity evaluates between-sample dissimilarity with metrics like Bray-Curtis and Jaccard [172, 173]. These analyses help explore differences in community composition. Studies have linked microbiota diversity to cancer treatment outcomes, notably showing that reduced diversity from pre-immunotherapy antibiotic use correlates with poorer treatment responses [174–177].

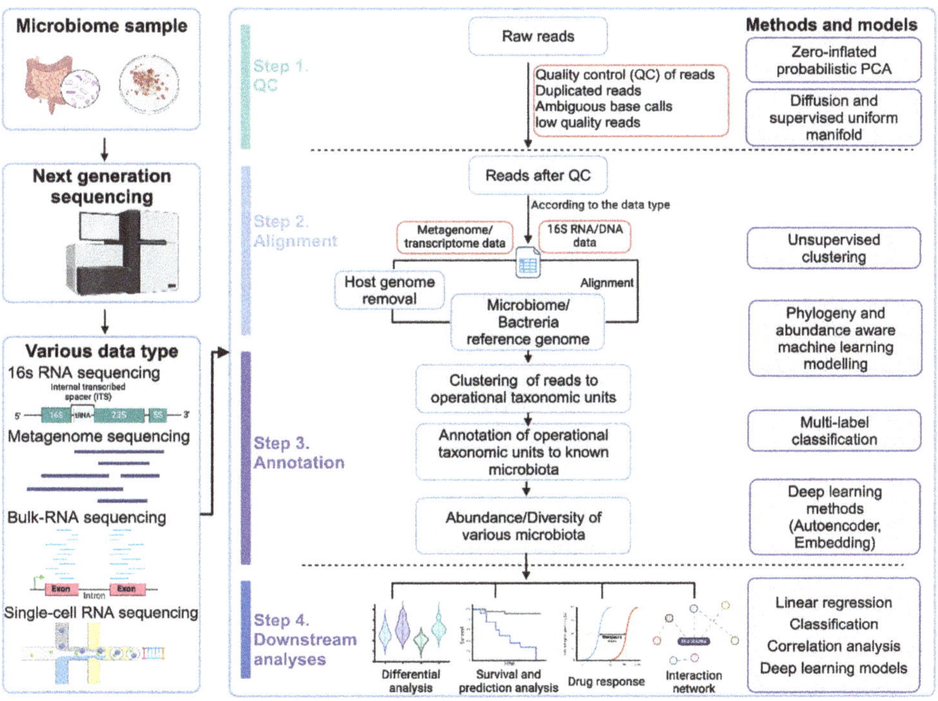

Fig. 4.1 The identification of the microbiome from NGS data.

4.6.2.6 Host–microbial Interactions

Interactions between the microbiome and host play crucial roles in tumorigenesis and immune response and can be categorized into microbiota co-expression networks, microbial metabolite networks, PPIs, and RNA-mediated interactions [178]. Various tools have been developed to infer these interactions. Both SparCC and CCLasso are tools for interfering with microbial co-abundance networks [179, 180]. Assembly of gut organisms through reconstruction and analysis (AGORA) and AGORA2 enable genome-scale metabolic reconstructions of human gut microbes [181]. MDiNE incorporates time-series information from longitudinal microbiome data to infer dynamic networks that capture temporal dependencies and changes in microbial interactions [182]. MiMeNet employs neural network-based approaches to predict relationships between microbial community composition and the metabolome [183]. ViRBase is a comprehensive database that provides experimental virus-host RNA interaction data [184].

Currently, computational studies in microbiome and cancer immunotherapy primarily focus on identifying drug-microbiota interactions and predicting treatment responses based on microbial signatures. In the realm of drug-microbiota interactions, various computational models have emerged. GACNNMDA employs a two-layer graph attention network to predict microbiota-drug interaction pair scores [185]. A data-driven approach that uses machine learning models like the RF, supported vector machine, and logistic regression to forecast drug impacts on the human microbiome [186]. Additionally, multiple computational methods were developed to predict the treatment response of immunotherapy by using microbiome signatures. Coxnet is a regularized Cox regression model for prognostication and drug response prediction [187]. In the realm of drug-microbiota interactions, various computational models have emerged. GACNNMDA employs a two-layer graph attention network to predict microbiota-drug interaction pair scores [185]. A data-driven approach that uses machine learning models like the RF, supported vector machine, and logistic regression to forecast drug impacts on the human microbiome [186]. Additionally, multiple computational methods were developed to predict the treatment response of immunotherapy by using microbiome signatures. Coxnet is a regularized Cox regression model for prognostication and drug response prediction [187].

Computational methods offer several advantages for analyzing interactions between the host and individual microbes or microbial communities. These advantages include (1) enhanced scalability: computational approaches can efficiently process large datasets, enabling inferences to be made across large numbers of microbiotas and samples; (2) improved reproducibility: when complemented by interoperability, automation, robust version control, and comprehensive documentation, computational methods can improve the reproducibility of research; and (3) performance assessment: model performance can be assessed using a variety of metrics. Moreover, computational methods can select and prioritize interactions under different conditions, such as various diseases and treatments.

Numerous tools and databases are available for gut microbiome analysis, reflecting the diverse perspectives from which microbiome research is conducted. With the ongoing development of advanced computational tools and pipeline management platforms, the field of host-microbiome interaction analysis is expected to expand further in the future.

4.7 Microbiome and ICB Adverse Events

In a phase three clinical trial, the incidence of irAEs among lung cancer patients was similar regardless of PD-L1 expression levels (≥1% or <1%), suggesting that molecular markers might not be reliable predictors for irAEs [188]. In contrast, the microbiota seems to play a more substantial role in forecasting irAEs. Previous studies identified the reduction of diversity and relative abundance in the gut microbiome can be used to predict both irAEs and their severity in liver cancer [189]. Linear models can be used to identify microbial markers for irAE and non-irAE patients [190]. Despite these findings, the development of computational tools specifically designed to study the association between the microbiome and irAEs remains limited. Expanding computational methodologies in this area could represent a valuable future direction, enabling more precise prediction models and improving our understanding of the microbiome's role in irAEs.

4.8 Translational Challenges

The microbiome has gained attention in clinical trials and FDA-approved therapies for its ability to modulate treatment responses. Multiple clinical trials are underway to evaluate the role of the microbiome in enhancing immunotherapy. For instance, Vedanta Biosciences' VE800 is a live bacterial product in clinical trials to improve cancer immunotherapy outcomes by activating CD8+ T cells, essential for antitumor immunity. An example of microbial therapy in cancer treatment undergoing FDA consideration is MRx0518 by 4D Pharma [191]. MRx0518, a live biotherapeutic derived from Bacillus bacteria, is in clinical trials to enhance immune responses to ICBs in advanced cancers [192]. The microbiome's role in cancer treatment is expanding, supported by computational models that identify promising drug targets and evaluate drug-target binding affinities. Microbiome profiling is also being explored as a biomarker for disease risk, progression, and therapy response. However, challenges like population variability, causality, and regulatory hurdles persist. Solutions include diverse cohort studies, region-specific databases, multi-omics integration, and standardized protocols. These advancements aim to translate microbiome research into reliable clinical applications.

4.9 The Integration of Multi-Omics Data in Precision Medicine for a Holistic Understanding of Disease

While individual omics technologies offer valuable insights, the integration of multi-omics data has emerged as a game-changer in precision medicine. By combining genomic, transcriptomic, proteomic, metabolomic, and microbiomic data, researchers can gain a holistic understanding of disease mechanisms and uncover complex interactions between molecular pathways.

The integration of multi-omics data enables the construction of detailed molecular networks that reveal the hierarchical relationships between genes, proteins, metabolites, and microbiota. For example, integrating transcriptomic and proteomic data allows researchers to understand how changes in gene expression are translated into functional proteins. Similarly, linking metabolomic data with genomic information provides insights into how genetic variations influence metabolic pathways, shedding light on disease pathogenesis.

In oncology, multi-omics integration has proven particularly impactful. By combining genomic, transcriptomic, and proteomic data, researchers can identify biomarkers for early cancer detection, classify tumors into molecular subtypes, and predict therapeutic outcomes. This approach has led to the development of personalized cancer therapies that target specific molecular aberrations, improving patient survival rates and quality of life.

Beyond cancer, multi-omics approaches are being applied to a wide range of diseases, including autoimmune disorders, cardiovascular diseases, neurodegenerative conditions, and infectious diseases. For example, integrating microbiomic and metabolomic data has uncovered links between gut microbiota dysbiosis and metabolic disorders such as diabetes and obesity. Similarly, combining genomic and proteomic data has advanced our understanding of neurodegenerative diseases like Alzheimer's, revealing potential biomarkers and therapeutic targets.

The success of multi-omics integration relies heavily on advanced computational tools and machine learning algorithms capable of processing and analyzing large-scale datasets. These technologies enable the identification of hidden patterns and correlations within complex datasets, driving discoveries that were previously unattainable.

In conclusion, omics technologies and their integration represent the cornerstone of precision medicine. By unraveling the molecular intricacies of diseases, they pave the way for highly personalized and effective healthcare solutions. This chapter will explore the methodologies, challenges, and future directions in applying omics data to revolutionize the field of precision medicine.

4.9.1 Opportunities for omics data and AI in advancing precision medicine

4.9.1.1 Deciphering cellular and genomic heterogeneity in diseases

Classical genetics divided diseases into two categories, including Mendelian diseases and complex traits and diseases. Mendelian diseases are attributed to a single gene mutation with a simple pattern of inheritance, while complex traits and diseases are due to multiple gene mutations, alterations, and interactions [193]. These characters directly cause greater challenges for understanding complex traits of diseases. However, as these are far more common than Mendelian diseases, understanding the complex traits of diseases has more socioeconomic impact [194].

Heterogeneity is one of the major obstacles to understanding the aetiologies of complex traits and diseases. For example, cancer is a highly heterogeneous disease with multiple clonal and subclonal populations leading to tumor evolution and drug resistance, which push bulk sequencing technologies to their limits. Currently, single cell and ST technologies and AI advance the knowledge of the heterogeneity and diversity of cell populations. Several tools can be used to estimate copy number aberrations and clonal deconstruction of cancer evolution, such as inferCNV, SCONCE, SCITE, and LACE [195–198]. One of the key questions closely related to therapy is which cells are more likely to metastasize and whether the metastases are seeded early in tumor evolution progression or isolated from later cells. Currently, large omics genomic data provide the possibility of distinguishing phylogenetic patterns between different cell populations. These may help monitor cancer progression and identify combination treatment strategies.

4.9.2 Uncovering drug repurposing strategies in existing omics data

Drug discovery is a time-consuming, laborious, expensive, and high-risk process. It usually takes 10 to 15 years to develop a new drug, with a 2.01% average success rate for developing a new molecular entity [199]. The cost of drug development is increasing year by year. The compound databases have facilitated numerous studies of the similarity of compounds based on various properties. Recently, drug repositioning has been an attractive approach to discovering drugs for diseases. Various data-driven computational or analyzing methods were reported for drug repurposing. One active direction is drug repurposing between cell lines and clinical samples, like TCGA patients. Drug repurposing methods can be divided into two categories, including network-based and structure-based methods. For network-based drug mining, Chyr et al. recently developed an approach named DOTA [200]. It is a novel and robust network-based deep learning approach to repositioning drugs for AD treatment. DOTA considers drug targets, side effects, and associations with other diseases in its predictions, which increases the discovery of mechanistically effective drugs for AD. Structured-based drug mining is also widely used. The core idea of gene signature-based drug mining is that candidate drugs should restore genomic signatures that are altered by the disease compared with the controls. The major types of signatures, upstream biomedical big data resources, and downstream applications are summarized in Fig. 4.2. Several resources and databases play important roles in drug mining, such as the drug-gene interaction database (DGIdb) [201] and the Open Targets database [202–204]. Combined with ML algorithms such as deep learning, these databases could allow the field of drug mining to flourish. An unresolved challenge in small-molecule design is how to represent chemical structures best. There are a plethora of representations ranging from simple circular fingerprints, such as the extended-connectivity fingerprint, to sophisticated symmetry functions. It is still unclear which structure representation is best suited for which small-molecule design problems. Recently, the DeepMind company developed Alphafold2 to predict protein structures with significantly improved accuracy [151]. It will be interesting to see if the rise of ML studies in the field of cheminformatics will provide more guidance on the best choices for future structure representations.

4.9.3 Pharmacokinetics–pharmacodynamics models

Traditional drug development methods are time- and resource-intensive and carry a substantial risk of adverse effects. Hence, a more efficient approach is urgently needed to predict pharmacology and toxicology by applying systems approaches to understanding pharmacokinetics-pharmacodynamics (PKPD). Although there are several topics, including nanoparticle delivery and release in this area, here, we briefly discuss multiscale modeling to predict drug absorption, distribution, metabolism, excretion (ADME), and drug interactions at target sites from the cellular level to system level based on multimodality imaging. One of the most critical challenges is the application of computational modeling for PKPD analysis. Pharmacokinetics models can describe or predict the time course of drug concentration in different body compartments like blood and heart. In contrast, pharmacodynamics models focus on the time course of drug effects at the site of action. Both types of models are playing increasingly important roles in contemporary drug research. However, if they were viewed as two separate processes, they may not fully facilitate drug research. Some pharmaceutical companies like SimCYP and Entelos have been working on a combined PKPD approach to assist drug research and development [205, 206]. They directly use organ-level PKPD data to predict concentration changes for a new drug. However, this traditional modeling approach has limited accuracy because it does not consider the biophysical properties of different tissues and interstitial tumor space. Tang et al. studied how drug release at the tissue level based on the biophysical properties of interstitial tumor space [207]. This is the first work to systematically integrate molecular imaging, biophysical modeling, and signaling pathway modeling to study drug ADME and interactions at target sites.

In short, the heterogeneity of complex traits and diseases leads to diversity in patient data. It requires multi-omics data at the individual level, AI techniques, and intensive analyses to understand the underlying mechanisms and advance knowledge of "precision medicine."

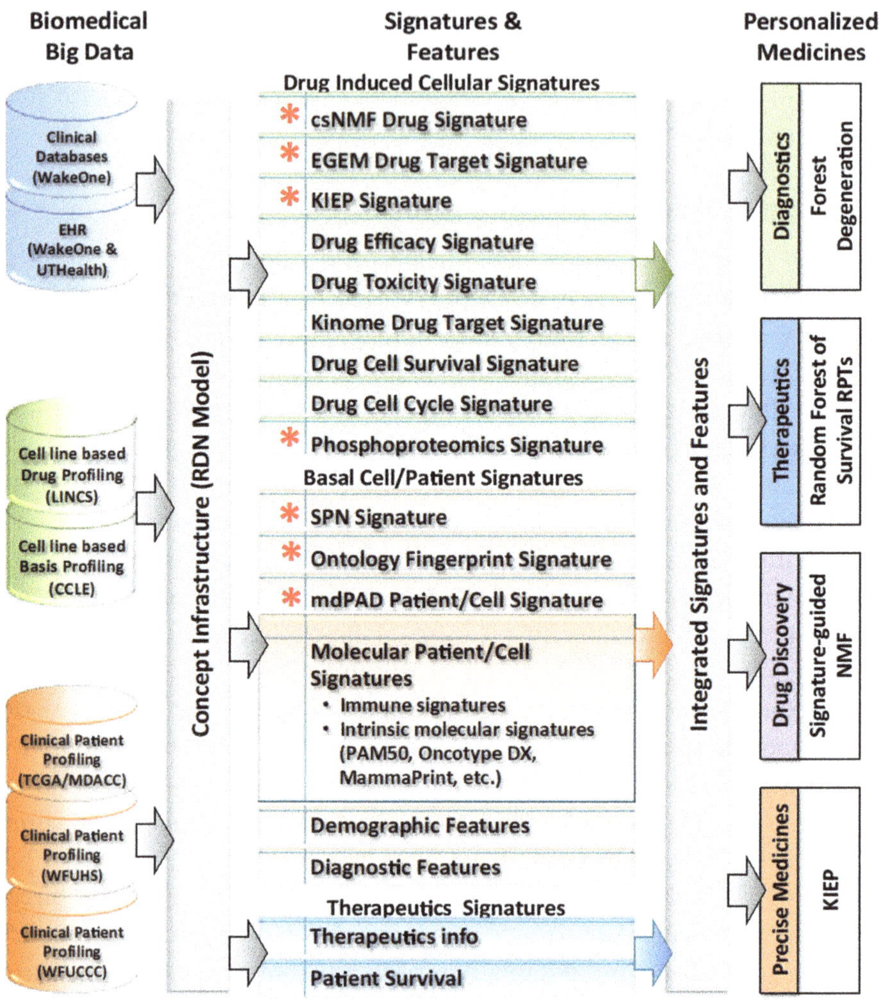

Fig. 4.2 Summary of signatures: Sources, types, and applications. Red star: structured signature.

4.10 Ethical Issues of AI in Immunotherapy

Combining AI with immunotherapy presents unique ethical challenges that must be addressed to harness their full potential responsibly. These issues encompass privacy and data security, bias and fairness, accountability and transparency, autonomy and human oversight, and societal impacts, necessitating comprehensive solutions and regulatory framework [208–212].

4.10.1 Privacy and data security

Immunotherapy relies heavily on detailed personal and genetic data for tailoring treatments. AI applications require large-scale data collection, which increases the risk of breaches and misuse of sensitive health information [213–215]. Moreover, the complex nature of genetic and immune profiling data makes it challenging to anonymize completely, raising the risk of re-identification. To mitigate these concerns, robust encryption, de-identification techniques, and strict access controls are essential. Legal frameworks such as the Health Insurance Portability and Accountability Act (HIPAA) in the United States and the General Data Protection Regulation (GDPR) in the European Union provide safeguards for patient data by emphasizing security standards, explicit consent, and transparency in data usage [216–218].

4.10.2 Bias and fairness

AI models often reflect biases in their training datasets, potentially leading to unequal treatment outcomes in immunotherapy. Variations in immune responses across ethnic, gender, and age groups may result in AI systems favoring certain populations while being less effective for others. Furthermore, the high costs and resource intensity of AI-based immunotherapy raise concerns about accessibility, potentially widening existing healthcare disparities. Addressing these challenges requires curating diverse, representative datasets and applying fairness-aware machine learning techniques. Initiatives such as the European Union's proposed AI Act enforce stringent testing for bias in high-risk AI applications, including healthcare. Regular audits and input from diverse stakeholders can promote fairness and build trust in these technologies [219].

4.10.3 Accountability and transparency

When AI influences critical decisions, such as dosage, drug combination, or patient-specific strategies, clarity is needed around who is responsible if adverse outcomes occur. Clear guidelines must define liability among AI developers, clinicians, and healthcare institutions to prevent ambiguity. An ideal framework would establish shared accountability while ensuring equitable responsibility. Additionally, the complexity of AI models, especially deep learning algorithms, can hinder transparency, making it difficult for clinicians and patients to understand the basis of AI recommendations. This opacity may affect trust and complicate informed consent. Regulations like the EU Medical Device Regulation (MDR) emphasize transparency and safety in AI-powered devices, supporting clearer communication and greater accountability [220, 221].

4.10.4 Autonomy and human oversight

Over-reliance on AI risks undervaluing clinicians' judgment and may lead to overly mechanistic care approaches. Ethical integration of AI requires maintaining human oversight, where clinicians retain final decision-making authority. Guidance from the World Health Organization (WHO) stresses the importance of preserving autonomy through informed consent and transparent communication. Training clinicians to collaborate effectively with AI ensures a balanced approach that combines human expertise with AI-driven insights [222].

4.10.5 Societal impacts

The societal implications of AI in healthcare extend to concerns about job displacement, inequitable access, and healthcare disparities. Advanced AI systems may disproportionately benefit wealthy institutions or regions, leaving others underserved. Policymakers must prioritize equitable access through subsidies, public–private partnerships, and initiatives like the AI for Humanity Initiative, which advocates for responsible and inclusive deployment of AI technologies.

In conclusion, integrating AI with immunotherapy offers great promise for personalized cancer care but requires careful consideration of ethical issues related to data privacy, bias, accountability, transparency, and trust. Developing robust ethical guidelines, transparent AI systems, and inclusive datasets will be essential to ensure that AI-augmented immunotherapy is safe, fair, and accessible to all patients.

References

1. Kussmann, M. *Omics: technologies and translations*. From Physics to Daily Life; 2014: p. 121–152.
2. Sajid, M. S., et al. Introduction to omics technologies. In *Omics approaches in veterinary parasitology*. CRC Press; 2024; p. 22–46.
3. Manzoni, C., et al. Genome, transcriptome and proteome: the rise of omics data and their integration in biomedical sciences. *Brief Bioinform*. 2018; **19**(2): p. 286–302.
4. Marchev, A. S., et al. Metabolomics and health: from nutritional crops and plant-based pharmaceuticals to profiling of human biofluids. *Cell Mol Life Sci*. 2021; **78**(19-20): p. 6487–6503.

5. Robinson, C. J., B. J. Bohannan, and V. B. Young. From structure to function: the ecology of host-associated microbial communities. *Microbiol Mol Biol Rev*. 2010; **74**(3): p. 453–476.
6. McKenna, A., et al. The Genome Analysis Toolkit: a MapReduce framework for analyzing next-generation DNA sequencing data. *Genome Res*. 2010; **20**(9): p. 1297–1303.
7. Genovese, G., et al. BCFtools/liftover: an accurate and comprehensive tool to convert genetic variants across genome assemblies. *Bioinformatics*. 2024; **40**(2): p. btae038.
8. Li, H., et al. The sequence alignment/map format and SAMtools. *Bioinformatics*. 2009; **25**(16): p. 2078–2079.
9. Poplin, R., et al. A universal SNP and small-indel variant caller using deep neural networks. *Nat Biotechnol*. 2018; **36**(10): p. 983–987.
10. Wang, K., M. Li, and H. Hakonarson. ANNOVAR: functional annotation of genetic variants from high-throughput sequencing data. *Nucleic Acids Res*. 2010; **38**(16): p. e164.
11. Cingolani, P. Variant annotation and functional prediction: SnpEff. *Methods Mol Biol*. 2022; **2493**: p. 289–314.
12. Landrum, M. J., et al. ClinVar: improving access to variant interpretations and supporting evidence. *Nucleic Acids Res*. 2018; **46**(D1): p. D1062–D1067.
13. Talevich, E., et al. CNVkit: genome-wide copy number detection and visualization from targeted DNA sequencing. *PLoS Comput Biol*. 2016; **12**(4): p. e1004873.
14. Zafar, H., et al. Monovar: single-nucleotide variant detection in single cells. *Nat Methods*. 2016; **13**(6): p. 505–507.
15. Bohrson, C. L., et al. Linked-read analysis identifies mutations in single-cell DNA-sequencing data. *Nat Genet*. 2019; **51**(4): p. 749–754.
16. Dong, X., et al. Accurate identification of single-nucleotide variants in whole-genome-amplified single cells. *Nat Methods*. 2017; **14**(5): p. 491–493.
17. Wu, C. Y., et al. Integrative single-cell analysis of allele-specific copy number alterations and chromatin accessibility in cancer. *Nat Biotechnol*. 2021; **39**(10): p. 1259–1269.
18. Timothy Tickle, I. T., et al. *inferCNV of the Trinity CTAT Project*. 2019.
19. Muller, S., et al. CONICS integrates scRNA-seq with DNA sequencing to map gene expression to tumor sub-clones. *Bioinformatics*. 2018; **34**(18): p. 3217–3219.
20. Gao, T., et al. Haplotype-aware analysis of somatic copy number variations from single-cell transcriptomes. *Nat Biotechnol*. 2023; **41**(3): p. 417–426.
21. Malikic, S., et al. Integrative inference of subclonal tumour evolution from single-cell and bulk sequencing data. *Nat Commun*. 2019; **10**(1): p. 2750.
22. Satas, G., et al. SCARLET: single-cell tumor phylogeny inference with copy-number constrained mutation losses. *Cell Syst*. 2020; **10**(4): p. 323–332. e8.
23. Chen, X., et al. Manta: rapid detection of structural variants and indels for germline and cancer sequencing applications. *Bioinformatics*. 2016; **32**(8): p. 1220–1222.
24. do Valle, I. F., et al. Optimized pipeline of MuTect and GATK tools to improve the detection of somatic single nucleotide polymorphisms in whole-exome sequencing data. *BMC Bioinform*. 2016; **17**(Suppl 12): p. 341.
25. Mandal, R., et al. Genetic diversity of tumors with mismatch repair deficiency influences anti-PD-1 immunotherapy response. *Science*. 2019; **364**(6439): p. 485–491.

26. Binnewies, M., et al. Understanding the tumor immune microenvironment (TIME) for effective therapy. *Nat Med*. 2018; **24**(5): p. 541–550.
27. Dong, Z. Y., et al. Potential predictive value of TP53 and KRAS mutation status for response to PD-1 blockade immunotherapy in lung adenocarcinoma. *Clin Cancer Res*. 2017; **23**(12): p. 3012–3024.
28. Okamura, R., et al. ARID1A alterations function as a biomarker for longer progression-free survival after anti-PD-1/PD-L1 immunotherapy. *J Immunother Cancer*. 2020; **8**(1): p. e000438.
29. Wu, Y. M., et al. Inactivation of CDK12 delineates a distinct immunogenic class of advanced prostate cancer. *Cell*. 2018; **173**(7): p. 1770–1782. e14.
30. Miao, D., et al. Genomic correlates of response to immune checkpoint therapies in clear cell renal cell carcinoma. *Science*. 2018; **359**(6377): p. 801–806.
31. Tischkowitz, M., et al. Small-cell carcinoma of the ovary, hypercalcemic type-genetics, new treatment targets, and current management guidelines. *Clin Cancer Res*. 2020; **26**(15): p. 3908–3917.
32. Pham, T. V., et al. Role of ultraviolet mutational signature versus tumor mutation burden in predicting response to immunotherapy. *Mol Oncol*. 2020; **14**(8): p. 1680–1694.
33. Shin, D. S., et al. Primary resistance to PD-1 blockade mediated by JAK1/2 mutations. *Cancer Discov*. 2017; **7**(2): p. 188–201.
34. Sun, J. Y., et al. Resistance to PD-1/PD-L1 blockade cancer immunotherapy: mechanisms, predictive factors, and future perspectives. *Biomark Res*. 2020; **8**: p. 35.
35. Skoulidis, F., et al. STK11/LKB1 mutations and PD-1 inhibitor resistance in KRAS-mutant lung adenocarcinoma. *Cancer Discov*. 2018; **8**(7): p. 822–835.
36. Chen, X., et al. Pan-cancer analysis of KEAP1 mutations as biomarkers for immunotherapy outcomes. *Ann Transl Med*. 2020; **8**(4): p. 141.
37. Kato, S., et al. Hyperprogressors after immunotherapy: analysis of genomic alterations associated with accelerated growth rate. *Clin Cancer Res*. 2017; **23**(15): p. 4242–4250.
38. Boichard, A., et al. APOBEC-related mutagenesis and neo-peptide hydrophobicity: implications for response to immunotherapy. *Oncoimmunology*. 2019; **8**(3): p. 1550341.
39. Dentro, S. C., et al. Characterizing genetic intra-tumor heterogeneity across 2,658 human cancer genomes. *Cell*. 2021; **184**(8): p. 2239–2254. e39.
40. Jamal-Hanjani, M., et al. Tracking the evolution of non-small-cell lung cancer. *N Engl J Med*. 2017; **376**(22): p. 2109–2121.
41. Cun, Y., et al. Copy-number analysis and inference of subclonal populations in cancer genomes using Sclust. *Nat Protoc*. 2018; **13**(6): p. 1488–1501.
42. Shen, R., and V. E. Seshan. FACETS: allele-specific copy number and clonal heterogeneity analysis tool for high-throughput DNA sequencing. *Nucleic Acids Res*. 2016; **44**(16): p. e131.
43. Favero, F., et al. Sequenza: allele-specific copy number and mutation profiles from tumor sequencing data. *Ann Oncol*. 2015; **26**(1): p. 64–70.
44. Ha, G., et al. TITAN: inference of copy number architectures in clonal cell populations from tumor whole-genome sequence data. *Genome Res*. 2014; **24**(11): p. 1881–1893.
45. Nik-Zainal, S., et al. The life history of 21 breast cancers. *Cell*. 2012; **149**(5): p. 994–1007.

46. Miller, C. A., et al. SciClone: inferring clonal architecture and tracking the spatial and temporal patterns of tumor evolution. *PLoS Comput Biol.* 2014; **10**(8): p. e1003665.
47. Roth, A., et al. PyClone: statistical inference of clonal population structure in cancer. *Nat Methods*. 2014; **11**(4): p. 396–398.
48. Xiao, Y., et al. FastClone is a probabilistic tool for deconvoluting tumor heterogeneity in bulk-sequencing samples. *Nat Commun*. 2020; **11**(1): p. 4469.
49. Salcedo, A., et al. A community effort to create standards for evaluating tumor subclonal reconstruction. *Nat Biotechnol.* 2020; **38**(1): p. 97–107.
50. Tanner, G., et al. Benchmarking pipelines for subclonal deconvolution of bulk tumour sequencing data. *Nat Commun*. 2021; **12**(1): p. 6396.
51. Martinez-Ruiz, C., et al. Genomic-transcriptomic evolution in lung cancer and metastasis. *Nature*. 2023; **616**(7957): p. 543–552.
52. Lee, T. I., and R. A. Young. Transcriptional regulation and its misregulation in disease. *Cell.* 2013; **152**(6): p. 1237–1251.
53. Fernández-Moya, S. M., J. Ehses, and M. A. Kiebler. The alternative life of RNA—sequencing meets single molecule approaches. *FEBS Lett.* 2017; **591**(11): p. 1455–1470.
54. Wang, R., et al. Splicing defects in rare diseases: transcriptomics and machine learning strategies towards genetic diagnosis. *Brief Bioinform.* 2023; **24**(5): p. bbad284.
55. Rogalska, M. E., et al. Transcriptome-wide splicing network reveals specialized regulatory functions of the core spliceosome. *Science*. 2024; **386**(6721): p. 551–560.
56. Booth, B. J., et al. RNA editing: expanding the potential of RNA therapeutics. *Mol Ther.* 2023; **31**(6): p. 1533–1549.
57. Zhu, L., et al. Non-coding RNAs: the key detectors and regulators in cardiovascular disease. *Genomics*. 2021; **113**(1): p. 1233–1246.
58. Zhou, J., X. Zheng, and H. Shen. Targeting RNA-splicing for SMA treatment. *Mol Cell.* 2012; **33**: p. 223–228.
59. Wu, R., et al. Disruption of nuclear speckle integrity dysregulates RNA splicing in C9ORF72-FTD/ALS. *Neuron*. 2024; **112**(20): p. 3434–3451. e11.
60. Hanson, B., M. J. Wood, and T. C. Roberts. Molecular correction of Duchenne muscular dystrophy by splice modulation and gene editing. *RNA Biol.* 2021; **18**(7): p. 1048–1062.
61. Claborn, M. K., et al. Nusinersen: a treatment for spinal muscular atrophy. *Ann Pharmacother*. 2019; **53**(1): p. 61–69.
62. Kędzierska, H., and A. Piekiełko-Witkowska. Splicing factors of SR and hnRNP families as regulators of apoptosis in cancer. *Cancer Lett.* 2017; **396**: p. 53–65.
63. Tang, J., et al. A critical update on the strategies towards small molecule inhibitors targeting serine/arginine-rich (SR) proteins and serine/arginine-rich proteins related kinases in alternative splicing. *Bioorg Med Chem*. 2022; **70**: p. 116921.
64. Wang, Z., M. Gerstein, and M. Snyder. RNA-seq: a revolutionary tool for transcriptomics. *Nat Rev Genet.* 2009; **10**(1): p. 57–63.
65. Luecken, M. D., and F. J. Theis. Current best practices in single-cell RNA-seq analysis: a tutorial. *Mol Syst Biol.* 2019; **15**(6): p. e8746.
66. Rhoads, A., and K. F. Au. PacBio sequencing and its applications. *Genom, Proteom Bioinform*. 2015; **13**(5): p. 278–289.

67. Nanopore, O. Oxford Nanopore announcement sets sequencing sector abuzz. *Nat Biotechnol.* 2012; **30**(4): p. 295–296.
68. Trapnell, C., L. Pachter, and S. L. Salzberg. TopHat: discovering splice junctions with RNA-seq. *Bioinformatics.* 2009; **25**(9): p. 1105–1111.
69. Dobin, A., et al. STAR: ultrafast universal RNA-seq aligner. *Bioinformatics.* 2013; **29**(1): p. 15–21.
70. Ghosh, S., and C.-K. K. Chan. Analysis of RNA-seq data using TopHat and Cufflinks. *Methods Mol Biol.* 2016; **1374**: p. 339–361.
71. Shen, S., et al. rMATS: robust and flexible detection of differential alternative splicing from replicate RNA-seq data. *Proc Natl Acad Sci.* 2014; **111**(51): p. E5593–E5601.
72. Li, Y. I., et al. Annotation-free quantification of RNA splicing using LeafCutter. *Nat Genet.* 2018; **50**(1): p. 151–158.
73. Kahles, A., et al. SplAdder: identification, quantification and testing of alternative splicing events from RNA-seq data. *Bioinformatics.* 2016; **32**(12): p. 1840–1847.
74. Vaquero-Garcia, J., et al. RNA splicing analysis using heterogeneous and large RNA-seq datasets. *Nat Commun.* 2023; **14**(1): p. 1230.
75. Kim, P., et al. ExonSkipDB: functional annotation of exon skipping event in human. *Nucleic Acids Res.* 2020; **48**(D1): p. D896–D907.
76. Yang, M., et al. ExonSkipAD provides the functional genomic landscape of exon skipping events in Alzheimer's disease. *Brief Bioinform.* 2021; **22**(5): p. bbaa438.
77. Wu, S., et al. ASCancer Atlas: a comprehensive knowledgebase of alternative splicing in human cancers. *Nucleic Acids Res.* 2023; **51**(D1): p. D1196–D1204.
78. Chang, J.-W., H.-S. Yeh, and J. Yong. Alternative polyadenylation in human diseases. *Endocrinol Metab.* 2017; **32**(4): p. 413–421.
79. Zhang, Y., et al. Alternative polyadenylation: methods, mechanism, function, and role in cancer. *J Exp Clin Cancer Res.* 2021; **40**: p. 1–19.
80. Cui, Y., et al. Alternative polyadenylation transcriptome-wide association study identifies APA-linked susceptibility genes in brain disorders. *Nat Commun.* 2023; **14**(1): p. 583.
81. Mostafa, S. M., and C. Moore. Cleavage and polyadenylation factors are potential regulators of adipogenesis. *BMC Res Notes.* 2024; **17**(1): p. 242.
82. Chen, W., et al. Alternative polyadenylation: methods, findings, and impacts. *Genom, Proteom Bioinform.* 2017; **15**(5): p. 287–300.
83. Yu, F., et al. Poly (A)-seq: a method for direct sequencing and analysis of the transcriptomic poly (A)-tails. *PloS One.* 2020; **15**(6): p. e0234696.
84. Welch, J. D., et al. EnD-seq and AppEnD: sequencing 3′ ends to identify nontemplated tails and degradation intermediates. *RNA.* 2015; **21**(7): p. 1375–1389.
85. Weng, X., and T. E. Juenger. A high-throughput 3′-Tag RNA sequencing for large-scale time-series transcriptome studies. *Methods Mol Biol.* 2022; **2398**: p. 151–172.
86. Shiraki, T., et al. Cap analysis gene expression for high-throughput analysis of transcriptional starting point and identification of promoter usage. *Proc Natl Acad Sci.* 2003; **100**(26): p. 15776–15781.
87. Xia, Z., et al. Dynamic analyses of alternative polyadenylation from RNA-seq reveal a 3′-UTR landscape across seven tumour types. *Nat Commun.* 2014; **5**(1): p. 5274.

88. Ye, C., et al. APAtrap: identification and quantification of alternative polyadenylation sites from RNA-seq data. *Bioinformatics*. 2018; **34**(11): p. 1841–1849.
89. Arefeen, A., et al. TAPAS: tool for alternative polyadenylation site analysis. *Bioinformatics*. 2018; **34**(15): p. 2521–2529.
90. Huang, K., et al. CAFuncAPA: a knowledgebase for systematic functional annotations of APA events in human cancers. *NAR Cancer*. 2023; **5**(1): p. zcad004.
91. Hood, J. L., and R. B. Emeson. Editing of neurotransmitter receptor and ion channel RNAs in the nervous system. *Curr Top Microbiol Immunol*. 2012; **353**: p. 61–90.
92. Roth, S. H., et al. Increased RNA editing may provide a source for autoantigens in systemic lupus erythematosus. *Cell Rep*. 2018; **23**(1): p. 50–57.
93. Vlachogiannis, N. I., et al. Increased adenosine-to-inosine RNA editing in rheumatoid arthritis. *J Autoimmun*. 2020; **106**: p. 102329.
94. George, C. X., et al. Adenosine deaminases acting on RNA, RNA editing, and interferon action. *J Interferon & Cytokine Res*. 2011; **31**(1): p. 99–117.
95. Frezza, V., et al. RNA editing in cancer progression. *Cancers*. 2023; **15**(21): p. 5277.
96. Gulei, D., et al. CRISPR-based RNA editing: diagnostic applications and therapeutic options. *Expert Rev Mol Diagn*. 2019; **19**(2): p. 83–88.
97. Lo Giudice, C., et al. Investigating RNA editing in deep transcriptome datasets with REDItools and REDIportal. *Nat Protoc*. 2020; **15**(3): p. 1098–1131.
98. Wu, S., et al. ADeditome provides the genomic landscape of A-to-I RNA editing in Alzheimer's disease. *Brief Bioinform*. 2021; **22**(5): p. bbaa384.
99. Wu, S., et al. Genetic control of RNA editing in neurodegenerative disease. *Brief Bioinform*. 2023; **24**(2): p. bbad007.
100. Wu, S., et al. The integrative studies on the functional A-to-I RNA editing events in human cancers. *Genom, Proteom & Bioinform*. 2023; **21**(3): p. 619–631.
101. Lo Giudice, C., G. Pesole, and E. Picardi. REDIdb 3.0: a comprehensive collection of RNA editing events in plant organellar genomes. *Front Plant Sci*. 2018; **9**: p. 482.
102. Zhu, H., et al. REIA: a database for cancer A-to-I RNA editing with interactive analysis. *Int J Biol Sci*. 2022; **18**(6): p. 2472.
103. Beňačka, R., et al. Non-coding RNAs in human cancer and other diseases: overview of the diagnostic potential. *Int J Mol Sci*. 2023; **24**(22): p. 16213.
104. Adjeroh, D. A., et al. Challenges in LncRNA biology: views and opinions. *Non-coding RNA*. 2024; **10**(4): p. 43.
105. Li, J., et al. Long noncoding RNA XIST: mechanisms for X chromosome inactivation, roles in sex-biased diseases, and therapeutic opportunities. *Genes Dis*. 2022; **9**(6): p. 1478–1492.
106. Hajjari, M., and A. Salavaty. HOTAIR: an oncogenic long non-coding RNA in different cancers. *Cancer Biol Med*. 2015; **12**(1): p. 1.
107. Shaheen, N., et al. MicroRNAs regulation in Parkinson's disease, and their potential role as diagnostic and therapeutic targets. *NPJ Parkinson's Dis*. 2024; **10**(1): p. 186.
108. Riva, P., A. Ratti, and M. Venturin. The long non-coding RNAs in neurodegenerative diseases: novel mechanisms of pathogenesis. *Curr Alzheimer Res*. 2016; **13**(11): p. 1219–1231.

109. Skommer, J., et al. Small molecules, big effects: the role of microRNAs in regulation of cardiomyocyte death. *Cell Death & Dis*. 2014; **5**(7): p. e1325–e1325.
110. Song, Z., R. Gao, and B. Yan. Potential roles of microRNA-1 and microRNA-133 in cardiovascular disease. *Rev Cardiovasc Med*. 2020; **21**(1): p. 57–64.
111. Sebastian-delaCruz, M., et al. The role of lncRNAs in gene expression regulation through mRNA stabilization. *Non-coding RNA*. 2021; **7**(1): p. 3.
112. Biferali, B., et al. Long non-coding RNAs and their role in muscle regeneration. *Curr Top Develop Biol*. 2024; **158**: p. 433–465.
113. Zou, Y., and H. Xu. Involvement of long noncoding RNAs in the pathogenesis of autoimmune diseases. *J Transl Autoimmun*. 2020; **3**: p. 100044.
114. Giraldez, M. D., et al. Comprehensive multi-center assessment of small RNA-seq methods for quantitative miRNA profiling. *Nat Biotechnol*. 2018; **36**(8): p. 746–757.
115. An, J., et al. miRDeep*: an integrated application tool for miRNA identification from RNA sequencing data. *Nucleic Acids Res*. 2013; **41**(2): p. 727–737.
116. Gruber, A. R., S. H. Bernhart, and R. Lorenz. The ViennaRNA web services. *RNA Bioinform*. 2015; **1269**: p. 307–326.
117. Lorenz, R., et al. ViennaRNA Package 2.0. *Algorithms Mol Biol*. 2011; **6**: p. 1–14.
118. Ma, X.-K., et al. CIRCexplorer pipelines for circRNA annotation and quantification from non-polyadenylated RNA-seq datasets. *Methods*. 2021; **196**: p. 3–10.
119. RNAcentral Consortium. RNAcentral 2021: secondary structure integration, improved sequence search and new member databases. *Nucleic Acids Res*. 2021; **49**(D1): p. D212–D220.
120. Zhao, Y., et al. NONCODE 2016: an informative and valuable data source of long non-coding RNAs. *Nucleic Acids Res*. 2016; **44**(D1): p. D203–D208.
121. Zheng, Y., et al. NPInter v5. 0: ncRNA interaction database in a new era. *Nucleic Acids Res*. 2023; **51**(D1): p. D232–D239.
122. Yang, M., et al. lncRNAfunc: a knowledgebase of lncRNA function in human cancer. *Nucleic Acids Res*. 2022; **50**(D1): p. D1295–D1306.
123. Tsimberidou, A. M., et al. Transcriptomics and solid tumors: the next frontier in precision cancer medicine. *Seminars in Cancer Biology*. Vol. 84. Academic Press; 2022.
124. Badowski, C., B. He, and L. X. Garmire. Blood-derived lncRNAs as biomarkers for cancer diagnosis: the Good, the Bad and the Beauty. *NPJ Precis Oncol*. 2022; **6**(1): p. 40.
125. Ren, Z., et al. A group of long non-coding RNAs in blood acts as a specific biomarker of Alzheimer's disease. *Mol Neurobiol*. 2023; **60**(2): p. 566–575.
126. Liu, X., et al. Non-coding RNAs expression in SARS-CoV-2 infection: pathogenesis, clinical significance, and therapeutic targets. *Signal Transduct Target Ther*. 2023; **8**(1): p. 441.
127. Alpuche-Lazcano, S. P., R. J. Scarborough, and A. Gatignol. MicroRNAs and long non-coding RNAs during transcriptional regulation and latency of HIV and HTLV. *Retrovirology*. 2024; **21**(1): p. 5.
128. Li, H.-C., C.-H. Yang, and S.-Y. Lo. Long noncoding RNAs in hepatitis B virus replication and oncogenesis. *World J Gastroenterol*. 2022; **28**(25): p. 2823.
129. Godar, S., et al. Growth-inhibitory and tumor-suppressive functions of p53 depend on its repression of CD44 expression. *Cell*. 2008; **134**(1): p. 62–73.

130. Blauw, H. M., et al. SMN1 gene duplications are associated with sporadic ALS. *Neurology*. 2012; **78**(11): p. 776–780.
131. Suk, T. R., and M. W. Rousseaux. The role of TDP-43 mislocalization in amyotrophic lateral sclerosis. *Mol Neurodegener*. 2020; **15**(1): p. 45.
132. Furquim, S. R., et al. The role of TTN gene variants in dilated cardiomyopathy. *ABC Heart Fail Cardiomyop*. 2024; **4**(3): p. e20240056.
133. Khosravi, H. M., and M. F. Jantsch. Site-directed RNA editing: recent advances and open challenges. *RNA Biol*. 2021; **18**(sup1): p. 41–50.
134. Kung, C.-P., L. B. Maggi Jr, and J. D. Weber. The role of RNA editing in cancer development and metabolic disorders. *Front Endocrinol*. 2018; **9**: p. 762.
135. Gabel, A. M., et al. APC mutations dysregulate alternative polyadenylation in cancer. *Genom Biol*. 2024; **25**(1): p. 255.
136. Cao, Z., Y. Jia, and B. Zhu. BNP and NT-proBNP as diagnostic biomarkers for cardiac dysfunction in both clinical and forensic medicine. *Int J Mol Sci*. 2019; **20**(8): p. 1820.
137. Kaufmann, I., et al. Human Fip1 is a subunit of CPSF that binds to U-rich RNA elements and stimulates poly (A) polymerase. *EMBO J*. 2004; **23**(3): p. 616–626.
138. Hashemi, M., et al. Pre-clinical and clinical importance of miR-21 in human cancers: tumorigenesis, therapy response, delivery approaches and targeting agents. *Pharmacol Res*. 2023; **187**: p. 106568.
139. Bhan, A., and S. S. Mandal. LncRNA HOTAIR: a master regulator of chromatin dynamics and cancer. *Biochim Biophys Acta*. 2015; **1856**(1): p. 151–164.
140. Li, J., et al. Circular RNA MYLK promotes tumour growth and metastasis via modulating miR-513a-5p/VEGFC signalling in renal cell carcinoma. *J Cell Mol Med*. 2020; **24**(12): p. 6609–6621.
141. Cox, J., and M. Mann. MaxQuant enables high peptide identification rates, individualized p.p.b.-range mass accuracies and proteome-wide protein quantification. *Nat Biotechnol*. 2008; **26**(12): p. 1367–1372.
142. Eng, J. K., T. A. Jahan, and M. R. Hoopmann. Comet: an open-source MS/MS sequence database search tool. *Proteomics*. 2013; **13**(1): p. 22–24.
143. Kong, A. T., et al. MSFragger: ultrafast and comprehensive peptide identification in mass spectrometry-based proteomics. *Nat Methods*. 2017; **14**(5): p. 513–520.
144. Cox, J., et al. Accurate proteome-wide label-free quantification by delayed normalization and maximal peptide ratio extraction, termed MaxLFQ. *Mol Cell Proteom*. 2014; **13**(9): p. 2513–2526.
145. Zhong, Q., et al. Protein posttranslational modifications in health and diseases: functions, regulatory mechanisms, and therapeutic implications. *Med Comm*. 2023; **4**(3): p. e261.
146. Anand, S. K., et al. CLUMPS-PTM: spatial clustering of post-translational modifications across cancer types. *Cancer Res*. 2023; **83**(7_Supplement): p. 3132–3132.
147. Lv, H., F. Y. Dao, and H. Lin. DeepKla: an attention mechanism-based deep neural network for protein lysine lactylation site prediction. *Imeta*. 2022; **1**(1): p. e11.
148. Safari-Alighiarloo, N., et al. Protein–protein interaction networks (PPI) and complex diseases. *Gastroenterol Hepatol Bed Bench*. 2014; **7**(1): p. 17.
149. Singh, K. Understanding protein–protein interactions: techniques and applications in drug development. *Multidisciplinary J Mol Biol Biochem*. 2024; **1**(2): p. 17–25.

150. Shannon, P., et al. Cytoscape: a software environment for integrated models of biomolecular interaction networks. *Genom Res*. 2003; **13**(11): p. 2498–2504.
151. Cramer, P. AlphaFold2 and the future of structural biology. *Nat Struct Mol Biol*. 2021; **28**(9): p. 704–705.
152. Alipour, M. Molecular mechanism of helicobacter pylori-induced gastric cancer. *J Gastrointest Cancer*. 2021; **52**(1): p. 23–30.
153. Soto, D., C. Song, and M. E. McLaughlin-Drubin. Epigenetic alterations in human papillomavirus-associated cancers. *Viruses*. 2017; **9**(9): p. 248.
154. Mager, D. Bacteria and cancer: cause, coincidence or cure? A review. J Transl Med. 2006; **4**(1): p. 1–18.
155. Douglas, G. M., R. G. Beiko, and M. G. Langille. Predicting the functional potential of the microbiome from marker genes using PICRUSt. *Methods Mol Biol*. 2018; **1849**: p. 169-177.
156. Aßhauer, K. P., et al. Tax4Fun: predicting functional profiles from metagenomic 16S rRNA data. *Bioinformatics*. 2015; **31**(17): p. 2882–2884.
157. Matchado, M. S., et al. On the limits of 16S rRNA gene-based metagenome prediction and functional profiling. *Microb Genom*. 2024; **10**(2): p. 001203.
158. Mostacci, N., et al. Informed interpretation of metagenomic data by StrainPhlAn enables strain retention analyses of the upper airway microbiome. *mSystems*. 2023; **8**(6): p. e0072423.
159. Lu, J., and S. L. Salzberg. Ultrafast and accurate 16S rRNA microbial community analysis using Kraken 2. *Microbiome*. 2020; **8**(1): p. 124.
160. Truong, D. T., et al. MetaPhlAn2 for enhanced metagenomic taxonomic profiling. *Nat Methods*. 2015; **12**(10): p. 902–903.
161. Namiki, T., et al. MetaVelvet: an extension of Velvet assembler to de novo metagenome assembly from short sequence reads. In *Proceedings of the 2nd ACM conference on bioinformatics, computational biology and biomedicine*. 2011.
162. Davis, N. M., et al. Simple statistical identification and removal of contaminant sequences in marker-gene and metagenomics data. *Microbiome*. 2018; **6**: p. 1–14.
163. Knights, D., et al. Bayesian community-wide culture-independent microbial source tracking. *Nat Methods*. 2011; **8**(9): p. 761–763.
164. Ghaddar, B., M. J. Blaser, and S. De. Denoising sparse microbial signals from single-cell sequencing of mammalian host tissues. *Nat Comput Sci*. 2023; **3**: p. 741–747.
165. Robinson, W., et al. Identification of intracellular bacteria from multiple single-cell RNA-seq platforms using CSI-Microbes. *Sci Adv*. 2024; **10**(27): p. eadj7402.
166. Chong, J., and J. Xia. Computational approaches for integrative analysis of the metabolome and microbiome. *Metabolites*. 2017; **7**(4): p. 62.
167. Guan, S.-W., et al. Weighted gene coexpression network analysis and machine learning reveal oncogenome associated microbiome plays an important role in tumor immunity and prognosis in pan-cancer. *J Transl Med*. 2023; **21**(1): p. 537.
168. Yang, M., et al. A multi-omics machine learning framework in predicting the survival of colorectal cancer patients. *Comput Biol Med*. 2022; **146**: p. 105516.
169. Esteban-Gil, A., et al. ColPortal, an integrative multiomic platform for analysing epigenetic interactions in colorectal cancer. *Sci Data*. 2019; **6**(1): p. 255.

170. Wang, P., et al. microbioTA: an atlas of the microbiome in multiple disease tissues of Homo sapiens and Mus musculus. *Nucleic Acids Res.* 2023; **51**(D1): p. D1345–D1352.
171. Chen, K.-P., et al. BIC: a database for the transcriptional landscape of bacteria in cancer. *Nucleic Acids Res.* 2023; **51**(D1): p. D1205–D1211.
172. Finotello, F., E. Mastrorilli, and B. Di Camillo. Measuring the diversity of the human microbiota with targeted next-generation sequencing. *Brief Bioinform.* 2018; **19**(4): p. 679–692.
173. Gail, M. H., Y. Wan, and J. Shi. Power of microbiome beta-diversity analyses based on standard reference samples. *Am J Epidemiol.* 2021; **190**(3): p. 439–447.
174. Caporaso, J. G., et al. QIIME allows analysis of high-throughput community sequencing data. *Nat Methods.* 2010; **7**(5): p. 335–336.
175. Schloss, P. D., et al. Introducing mothur: open-source, platform-independent, community-supported software for describing and comparing microbial communities. *Appl Environ Microbiol.* 2009; **75**(23): p. 7537–7541.
176. Jari Oksanen, F., et al. *Community ecology package.* R package version 2.3-5; 2016.
177. Elkrief, A., et al. The intimate relationship between gut microbiota and cancer immunotherapy. *Gut Microbes.* 2019; **10**(3): p. 424–428.
178. Sudhakar, P., et al. Computational biology and machine learning approaches to understand mechanistic microbiome-host interactions. *Front Microbiol.* 2021; **12:** p. 618856.
179. Friedman, J., and E. J. Alm. Inferring correlation networks from genomic survey data. *PLoS Comput Biol.* 2012; **8**(9): p. e1002687.
180. Fang, H., et al. CCLasso: correlation inference for compositional data through Lasso. *Bioinformatics.* 2015; **31**(19): p. 3172–3180.
181. Heinken, A., et al. AGORA2: large scale reconstruction of the microbiome highlights widespread drug-metabolising capacities. *bioRxiv.* 2020: p. 2020.11.09.375451.
182. McGregor, K., A. Labbe, and C. M. Greenwood. MDiNE: a model to estimate differential co-occurrence networks in microbiome studies. *Bioinformatics.* 2020; **36**(6): p. 1840–1847.
183. Reiman, D., B. T. Layden, and Y. Dai. MiMeNet: exploring microbiome-metabolome relationships using neural networks. *PLOS Comput Biol.* 2021; **17**(5): p. e1009021.
184. Cheng, J., et al. ViRBase v3.0: a virus and host ncRNA-associated interaction repository with increased coverage and annotation. *Nucleic Acids Res.* 2022; **50**(D1): p. D928–D933.
185. Ma, Q., Y. Tan, and L. Wang. GACNNMDA: a computational model for predicting potential human microbe-drug associations based on graph attention network and CNN-based classifier. *BMC Bioinform.* 2023; **24**(1): p. 35.
186. Algavi, Y. M., and E. Borenstein. A data-driven approach for predicting the impact of drugs on the human microbiome. *Nat Commun.* 2023; **14**(1): p. 3614.
187. Hermida, L. C., E. M. Gertz, and E. Ruppin. Predicting cancer prognosis and drug response from the tumor microbiome. *Nat Commun.* 2022; **13**(1): p. 2896.
188. Borghaei, H., et al. Nivolumab versus docetaxel in advanced nonsquamous non-small-cell lung cancer. *N Engl J Med.* 2015; **373**(17): p. 1627–1639.
189. Mao, J., et al. Gut microbiome is associated with the clinical response to anti-PD-1 based immunotherapy in hepatobiliary cancers. *J Immunother Cancer.* 2021; **9**(12): p. e003334.
190. Hu, M., et al. Gut microbiome for predicting immune checkpoint blockade-associated adverse events. *Genom Med.* 2024; **16**(1): p. 16.

191. Lythgoe, M., et al. Immune modulation and the oral live biotherapeutic product, MRx0518, in treatment-naïve patients with cancer: updated safety data. *Am Soc Clin Oncol.* 2023; **41**: p. 3145.
192. Parra, E., et al. 1024P Baseline biomarkers associated with clinical benefit in patients with solid tumors refractory to immune checkpoint inhibitors (ICIs) treated with live biotherapeutic MRx0518 in combination with pembrolizumab. *Ann Oncol.* 2021; **32**: p. S858–S859.
193. Frazer, K. A., et al. Human genetic variation and its contribution to complex traits. *Nat Rev Genet.* 2009; **10**(4): p. 241–251.
194. Bomprezzi, R., P. Kovanen, and R. Martin. New approaches to investigating heterogeneity in complex traits. *J Med Genet.* 2003; **40**(8): p. 553–559.
195. Durante, M. A., et al. Single-cell analysis reveals new evolutionary complexity in uveal melanoma. *Nat Commun.* 2020; **11**(1): p. 1–10.
196. Hui, S., and R. Nielsen. SCONCE: a method for profiling copy number alterations in cancer evolution using single-cell whole genome sequencing. *Bioinformatics.* 2022; **38**(7): p. 1801–1808.
197. Jahn, K., J. Kuipers, and N. Beerenwinkel. Tree inference for single-cell data. *Genom Biol.* 2016; **17**(1): p. 1–17.
198. Ramazzotti, D., et al. LACE: Inference of cancer evolution models from longitudinal single-cell sequencing data. *J Comput Sci.* 2022; **58**: p. 101523.
199. Xue, H., et al. Review of drug repositioning approaches and resources. *Int J Biol Sci.* 2018; **14**(10): p. 1232–1244.
200. Chyr, J., H. Gong, and X. Zhou. DOTA: deep learning optimal transport approach to advance drug repositioning for Alzheimer's disease. *Biomolecules.* 2022; **12**(2): p. 196.
201. DGIdb. 2020-10-21]. www.dgidb.org.
202. Carvalho-Silva, D., et al. Open targets platform: new developments and updates two years on. *Nucleic Acids Res.* 2019; **47**(D1): p. D1056–D1065.
203. Freshour, S. L., et al. Integration of the Drug–Gene Interaction Database (DGIdb 4.0) with open crowdsource efforts. *Nucleic Acids Res.* 2021; **49**(D1): p. D1144–D1151.
204. *Open Targets.* http://www.opentargets.org.
205. Dickinson, G. L., et al. Incorporating in vitro information on drug metabolism into clinical trial simulations to assess the effect of CYP2D6 polymorphism on pharmacokinetics and pharmacodynamics: dextromethorphan as a model application. *J Clin Pharmacol.* 2007; **47**(2): p. 175–186.
206. Bangs, A. Predictive biosimulation and virtual patients in pharmaceutical RD. *Stud Health Technol Inform.* 2005; **111**: p. 37–42.
207. Tang, L., et al. An integrated multiscale mechanistic model for cancer drug therapy. *Int Sch Res Notices.* 2012; **2012**(1): p. 818492.
208. Johnson, S. L. J. AI, machine learning, and ethics in health care. *J Leg Med.* 2019; **39**(4): p. 427–441.
209. Six Dijkstra, M. W., et al. Ethical considerations of using machine learning for decision support in occupational health: an example involving periodic workers' health assessments. *J Occup Rehabil.* 2020; **30**: p. 343–353.

210. Safdar, N. M., J. D. Banja, and C. C. Meltzer. Ethical considerations in artificial intelligence. *Eur J Radiol*. 2020; **122**: p. 108768.
211. O'Sullivan, S., et al. Legal, regulatory, and ethical frameworks for development of standards in artificial intelligence (AI) and autonomous robotic surgery. *Int J Med Robot*. 2019; **15**(1): p. e1968.
212. Ho, C. W., J. Ali, and K. Caals. Ensuring trustworthy use of artificial intelligence and big data analytics in health insurance. *Bull World Health Organ*. 2020; **98**(4): p. 263.
213. Yadav, N., et al. Data privacy in healthcare: in the era of artificial intelligence. *Indian Dermatol Online J*. 2023; **14**(6): p. 788–792.
214. Karimian, G., et al. The ethical issues of the application of artificial intelligence in healthcare: a systematic scoping review. *AI Ethics*. 2022; **2**(4): p. 539–551.
215. Farhud, D. D., and S. Zokaei. Ethical issues of artificial intelligence in medicine and healthcare. *Iran J Public Health*. 2021; **50**(11): p. i.
216. Philibert, R. A., et al. Methylation array data can simultaneously identify individuals and convey protected health information: an unrecognized ethical concern. *Clin Epigenetics*. 2014; **6**: p. 1–6.
217. Price, W. N., and I. G. Cohen. Privacy in the age of medical big data. *Nat Med*. 2019; **25**(1): p. 37–43.
218. Regulation, General Data Protection. General data protection regulation (GDPR)–official legal text. *Gen Data Prot Regul*. 2016.
219. Quinn, P., and P. De Hert. The Patients' Rights Directive (2011/24/EU)—providing (some) rights to EU residents seeking healthcare in other Member States. *Comput Law Secur Rev*. 2011; **27**(5): p. 497–502.
220. Aboy, M., T. Minssen, and E. Vayena. Navigating the EU AI Act: implications for regulated digital medical products. *NPJ Digit Med*. 2024; **7**(1): p. 237.
221. Niemiec, E. Will the EU Medical Device Regulation help to improve the safety and performance of medical AI devices? *Digit Health*. 2022; **8**: p. 20552076221089079.
222. WHO guidelines on ethical issues in public health surveillance. Geneva: World Health Organization; 2017. Licence: CC BY-NC-SA 3.0 IGO.

CHAPTER 5

Genome-Wide Association Studies and Imaging Genetics

5.1 Definition and Its Evolution

Genome-wide association studies (GWASs) have become a cornerstone of modern genomics, offering a systematic approach to identifying genetic variants, particularly single-nucleotide polymorphisms (SNPs), associated with traits or diseases. By analyzing the entire genome in large cohorts, GWAS has provided unprecedented insights into the genetic architecture of complex conditions. This approach enables researchers to uncover biological pathways, identify potential therapeutic targets, and explore the genetic basis of phenotypic variability.

The emergence of GWAS was made possible by landmark achievements in genomics, including the completion of the Human Genome Project in 2003, which provided a reference map of the human genome, and the International HapMap Project in 2002, which characterized common patterns of genetic variation across populations. These milestones were accompanied by advances in high-throughput genotyping technologies, making it feasible to analyze millions of SNPs across thousands of individuals.

The first successful GWAS, published in 2005, identified variants in the *CFH* gene associated with age-related macular degeneration (AMD) [1]. This discovery demonstrated the potential of GWAS to uncover genetic loci linked to complex diseases and sparked a wave of subsequent studies. For example, the *APOE* gene was identified as a significant risk factor for Alzheimer's disease (AD), while loci in the 9p21 region were linked to coronary artery disease (CAD).

GWAS operates on the principle of linkage disequilibrium (LD), where genetic variants located near each other on the genome are inherited together. By leveraging LD patterns, GWAS identifies SNPs statistically associated with specific traits or diseases, even if the causal variants are not directly genotyped.

Despite its successes, GWAS is not without limitations. One key challenge is the issue of "missing heritability," where the cumulative effects of identified variants explain only a fraction of the genetic contribution to a trait [2, 3]. In addition, population stratification remains a significant concern, as subtle genetic differences between study populations can lead to false associations if not properly accounted for [4, 5]. Addressing these limitations requires larger sample sizes, diverse cohorts, and advanced statistical methods, including fine-mapping and multi-omics integration. One key challenge is the issue of "missing heritability," where the cumulative effects of identified variants explain only a fraction of the genetic contribution to a trait. In addition, GWAS findings have been predominantly derived from individuals of European ancestry, raising concerns about the applicability of results to other populations [2].

Overall, GWAS represents a transformative approach in genetic research, providing a robust framework for identifying genetic contributions to human traits and diseases. With the integration of emerging technologies, such as machine learning and artificial intelligence (AI), GWAS is poised to play an even greater role in advancing precision medicine.

The success of GWAS is rooted in its ability to leverage the principle of LD, where genetic variants located near each other on the genome tend to be inherited together. This principle, extensively described in foundational studies such as those by Gabriel et al. [6] and Reich et al. [7], underpins the statistical power of GWAS to identify loci associated with traits even when causal variants are not directly genotyped. LD patterns allow researchers to map genetic associations across populations efficiently, making them a critical aspect of GWAS methodology. By analyzing common genetic variants across large populations, GWAS identifies statistical associations between SNPs and traits, pinpointing loci that may influence disease risk or other phenotypes.

5.2 Study Design

GWAS relies on well-structured study designs to ensure valid and reproducible results. Two primary designs are commonly used in GWAS: case-control studies and cohort studies. In a case-control study, individuals are divided into two groups—cases (individuals with the disease or trait of interest) and controls (individuals without the disease or trait). Genetic variants are then compared between these two groups to identify differences in allele frequencies associated with the condition. For example, a case–control GWAS examining AD might compare genetic variants in patients diagnosed with the disease to those in healthy individuals of the same age group.

In contrast, cohort studies follow a group of individuals over time, recording phenotypic and genotypic data to identify genetic variants associated with the development of specific diseases or traits. The accuracy of GWAS findings depends heavily on phenotype selection, which refers to the accurate definition and measurement of the trait or disease being studied. Poorly defined phenotypes can lead to misclassification, reducing statistical power and increasing the risk of false-positive findings. For example, knee osteoarthritis (OA) can be measured using static traits, such as X-ray images for Kellgren-Lawrence (KL) scores, or dynamic traits, such as changes in cartilage thickness over time.

Population stratification is another critical consideration in GWAS design. Genetic differences between populations can confound results if not properly accounted for. Methods such as principal component analysis (PCA) and mixed-model approaches are commonly used to correct for population structure. For example, PCA was used in the UK Biobank GWAS to account for subtle population differences across British regions.

Finally, GWASs require large sample sizes to detect variants with small effect sizes. The statistical power of GWAS increases with the number of participants, allowing researchers to identify even modest genetic contributions to complex traits. Projects such as the UK Biobank and the Genetic Components of Knee Osteoarthritis (GeCKO) database exemplify large-scale datasets that have enabled robust genetic discoveries.

5.3 Genotyping and Quality Control

Modern genotyping platforms and SNP array technologies represent the cornerstone of GWAS, enabling large-scale interrogation of genetic variants across diverse populations. Platforms such as Illumina's Infinium Global Screening Array (GSA) and Affymetrix Axiom arrays leverage hybridization-based microarray technology to assay hundreds of thousands of SNPs efficiently and reproducibly. For example, the Illumina GSA can simultaneously genotype over 700,000 SNPs and includes population-specific markers, increasing its utility for ancestrally diverse cohorts. Beyond SNP arrays, whole-genome sequencing (WGS) and whole-exome sequencing (WES) are increasingly employed to identify rare variants and structural variations, offering an unprecedented resolution of the genomic landscape.

However, raw genotyping data is prone to systematic and random errors, necessitating a robust quality control (QC) pipeline. QC ensures data reliability and prevents biases from propagating into downstream analyses. The key QC steps include the following:

- SNP filtering: SNPs with low minor allele frequency (MAF) (e.g., <1%) are removed due to their limited statistical power. Moreover, SNPs with low genotyping call rates (e.g., <95%) are excluded to avoid missingness-driven biases.
- Sample-level QC: Samples with high rates of missing genotypes, extreme heterozygosity rates, or discordant sex information (e.g., from mismatches between self-reported and genetically inferred sexes) are flagged and removed.

- Hardy-Weinberg equilibrium (HWE): SNPs significantly deviating from HWE in control populations (p < 1e-6) suggest potential genotyping errors, population substructure, or natural selection pressures. For example, significant HWE deviations in a GWAS dataset could indicate an artifact stemming from batch effects during genotyping.

Tools such as PLINK are indispensable for QC workflows, offering functionalities for filtering SNPs, identifying duplicate samples, and visualizing heterozygosity and missingness rates. More advanced pipelines, such as Hail, are built for cloud-based, scalable QC workflows in biobank-scale datasets (e.g., UK Biobank). For example, in the GIANT consortium study investigating height-associated loci, extensive QC pipelines filtered SNPs below a 0.05 MAF threshold and excluded samples with >3% missing data, ensuring robust association signals.

Failing to rigorously apply QC measures can lead to an inflation of false positives, compromising the study's integrity and reproducibility, a risk particularly concerning given the sheer scale of GWAS datasets.

5.4 Statistical Analysis

Statistical analysis in GWAS aims to identify genetic variants associated with complex traits or diseases, accounting for millions of hypothesis tests conducted simultaneously. The two most commonly used regression models are as follows:

- Linear regression: Used for continuous traits such as cholesterol levels, BMI, or height.
- Logistic regression: Used for binary phenotypes, such as disease status (case/control).

Both models estimate β coefficients (effect sizes), standard errors, and p-values, assessing the statistical significance of genotype-phenotype associations. For example, in a study on AD, logistic regression was used to identify the APOE ε4 allele as a significant risk factor with an odds ratio (OR) exceeding 3.5.

However, raw regression analysis is often confounded by external factors. Covariate adjustment is essential to account for confounders such as age, sex, population structure (via ancestry principal components [PCs]), and technical batch effects. PCA is typically used to generate ancestry PCs, which are included as covariates in the regression model. For instance, the 1000 Genomes Project demonstrated how failing to adjust for population stratification could lead to spurious associations driven by ancestral differences rather than true genotype–phenotype relationships.

Given the sheer volume of statistical tests in GWAS (often exceeding one million SNPs tested simultaneously), multiple testing correction is critical. The Bonferroni correction applies a stringent threshold for genome-wide significance (e.g., $p < 5 \times 10^{-8}$), effectively reducing type I errors but increasing type II errors (false negatives). Alternatively, the false discovery rate (FDR), often controlled via the Benjamini-Hochberg procedure, offers a more balanced approach for exploratory analyses.

More sophisticated models, such as linear mixed models (LMMs) implemented in tools such as Bayesian Optimization for Linear Mixed Models (BOLT-LMM) and Genome-wide Complex Trait Analysis (GCTA), account for relatedness among individuals and polygenic effects. For example, in the UK Biobank study on obesity traits, BOLT-LMM was used to model both genetic and environmental variance components, enhancing statistical power and reducing population structure-induced confounding.

These statistical frameworks, when combined with appropriate correction measures and computational tools, ensure GWAS findings are reproducible and biologically meaningful.

Following the statistical analysis, the GWAS results are visualized through Manhattan and regional association plots (Fig. 5.1) to identify significant genetic loci associated with the trait of interest. The Manhattan plot displays

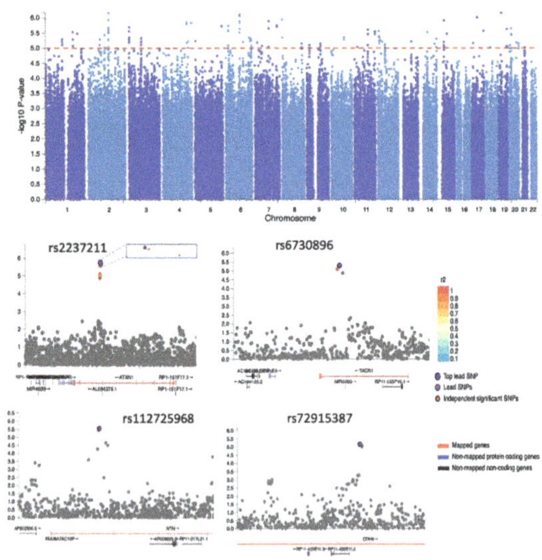

Fig. 5.1 Manhattan plots and regional association and linkage disequilibrium plots for four genome-wide significant loci. The y-axis represents the negative logarithm (base 10) of the SNP *p*-value, and the x-axis represents the position on the chromosome.

the distribution of association signals across the genome, with the x-axis representing chromosomal positions and the y-axis denoting the negative logarithm (base 10) of the p-values. Each point corresponds to an SNP, and the red dashed line marks the genome-wide significance threshold. To further explore these significant loci, regional association and LD plots were generated. These plots provide a detailed view of the genomic regions surrounding the lead SNPs, illustrating the association strength of nearby variants and their LD structure. The lead SNPs are highlighted, with surrounding variants colored based on their LD values, indicating how strongly they are correlated with the lead SNP. Gene annotations below the plots help contextualize the genetic landscape, highlighting nearby protein-coding and noncoding genes. The clustering of significant SNPs in high LD regions suggests potential causal variants or regulatory elements that may contribute to the observed genetic associations.

5.5 Population Stratification and Confounders

Population stratification remains one of the most pervasive confounding factors in GWAS, arising from allele frequency differences between subpopulations due to distinct evolutionary histories and demographic events. For instance, in a GWAS for asthma susceptibility, an SNP might falsely appear associated with the trait if one subpopulation has both a higher frequency of the allele and a higher prevalence of asthma.

PCA is the gold standard approach for detecting and adjusting for population structure. PCA identifies axes of genetic variation (PCs) that capture ancestry-related genetic differences. These PCs are subsequently included as covariates in regression models to mitigate stratification bias. For example, the Wellcome Trust Case-Control Consortium (WTCCC) demonstrated that including the top 10 ancestry PCs effectively removed stratification biases in their GWAS findings.

Beyond PCA, LMMs provide an alternative for addressing both population structure and relatedness. LMMs incorporate a genetic relationship matrix (GRM) to model random effects associated with cryptic relatedness and polygenic background. Tools such as BOLT-LMM and Genome-wide Efficient Mixed-Model Association (GEMMA) have been widely adopted for this purpose. For example, the Neale Lab GWAS analysis of body mass index (BMI) used BOLT-LMM to adjust for both ancestry PCs and subtle relatedness among samples in the UK Biobank dataset.

Environmental and sociocultural confounders further complicate GWAS interpretation. For example, socioeconomic status might correlate with both ancestry and disease risk factors, creating residual confounding if unaccounted for in statistical models. Stratified analyses and meta-analysis approaches can help reduce such biases, especially in large-scale consortium-based studies (e.g., the GIANT consortium and the PGC consortium).

Ultimately, integrating PCA, mixed models, and robust covariate adjustment is essential for reducing stratification-driven false positives, ensuring that reported associations reflect true genetic effects rather than artifacts of population structure or hidden confounders.

5.6 Image GWAS

GWAS has revolutionized our understanding of the genetic underpinnings of complex diseases by linking specific genetic variants to disease phenotypes [8]. However, traditional GWAS approaches often fail to incorporate the rich spatial and functional information provided by medical imaging, which can bridge the gap between genetic variation and disease-related phenotypic changes [9]. Imaging GWAS, a rapidly evolving interdisciplinary field, integrates genetic data with medical imaging modalities to uncover associations between genetic variants and imaging-derived traits, offering a deeper understanding of the molecular mechanisms driving diseases. This chapter provides an overview of imaging GWAS, discussing its methodologies, applications, challenges, and potential in medical research.

Fig. 5.2 provides an overview of the imaging GWAS workflow. Genetic and imaging data are collected from study participants (step 1), and phenotypic features are extracted from imaging data, such as brain regions of interest (step 2). Single-variant association tests identify relationships between specific genetic variants and imaging-derived phenotypes (step 3), where variants such as rs123 show significant genotype–phenotype associations. Notably, genetic variants may not always be causal but can act as proxies for true causal variants due to LD (Step 4). Genome-wide association tests further identify significant signals across the genome, revealing both known and novel associations that surpass the significance threshold (step 5). Follow-up analyses, including fine-mapping, single-cell analysis, and meta-analysis, enable the identification of causal variants and their functional roles (step 6).

Recent advancements in imaging genetics have demonstrated the utility of self-supervised learning algorithms in identifying features from medical images that are associated with genetic variations. For instance, a study by Xie et al. [10] introduced Genetic InfoMax, a framework that employs mutual information maximization to extract informative representations from high-dimensional imaging genetics data, enhancing the discovery of genotype–phenotype associations. Moreover, novel statistical methods have been developed to address the

high-dimensional nature of imaging genetics data. Altmann et al. [11] proposed CLUB-PLS, a partial least squares-based framework that effectively links high-dimensional genetic data to brain imaging data, facilitating the identification of genetic factors influencing brain structure and function.

These advancements underscore the potential of imaging genetics to uncover complex relationships between genetic variations and imaging-derived phenotypes, providing insights into various diseases' biological mechanisms.

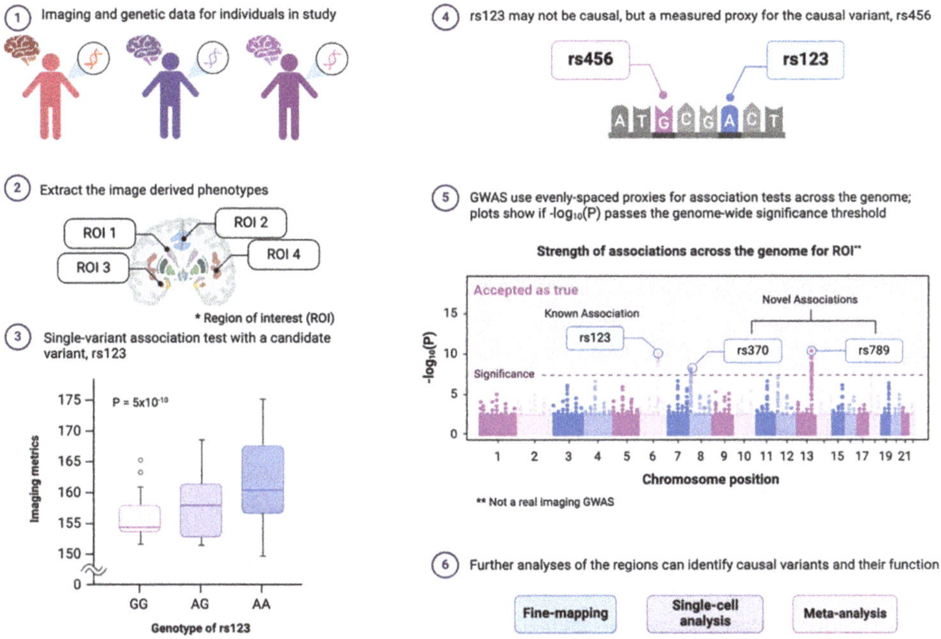

Fig. 5.2 Workflow of imaging genome-wide association studies (imaging GWASs). The figure outlines the stepwise process of imaging GWAS, which integrates genetic data and imaging-derived phenotypes to identify genotype–phenotype associations: (1) Imaging and genetic data collection: genetic and imaging data are gathered from study participants for analysis; (2) image-derived phenotype extraction: brain imaging data are processed to extract quantitative phenotypes from regions of interest (ROIs); (3) single-variant association testing: a candidate genetic variant (e.g., rs123) is tested for association with an imaging phenotype. The boxplot shows significant differences in the phenotype across genotypes (GG, AG, AA) with a *p*-value of 5×10^{-10}; (4) proxy and causal variants: the variant rs123 may not be causal but instead act as a proxy for a causal variant (rs456) due to linkage disequilibrium; (5) genome-wide association mapping: Manhattan plots display association signals across the genome, identifying known associations (e.g., rs123) and novel loci (e.g., rs370 and rs789) that exceed the genome-wide significance threshold; (6) downstream analyses: fine-mapping, single-cell analysis, and meta-analysis are performed to identify causal variants and investigate their biological functions.

5.6.1 Data collection and preprocessing

In imaging GWAS, data collection involves obtaining both genomic and imaging data from participants. Genomic data are typically gathered using genome-wide genotyping platforms such as SNP arrays or WGS. These data undergo rigorous QC to ensure reliability, including filtering for call rate, MAF, and HWE. Imaging data are acquired using standardized protocols across various modalities, such as Magnetic Resonance Imaging (MRI), Positron Emission Tomography (PET), Computed Tomography (CT), or ultrasound, to minimize variability across participants and studies.

For brain imaging, FreeSurfer [12], a widely used software package, can be utilized to process MRI data and extract structural brain features, such as cortical thickness, surface area, and subcortical volumes, based on automated segmentation of brain regions. These features can then be linked to genetic data for GWAS analysis. Atlases such as Destrieux, Automated Anatomical Labeling (AAL), and the Harvard-Oxford atlas further aid in segmenting brain regions and defining phenotypes for genetic association analyses.

For cardiac imaging, tools such as 3D Slicer [13] and Segment software process cardiac MRI and CT images to extract key metrics, including left ventricular ejection fraction, myocardial wall thickness, and ventricular volumes. These quantitative phenotypes are critical for studying genetic influences on cardiac structure and function, particularly in conditions such as hypertrophic cardiomyopathy and heart failure. In addition, CMRtools [14] facilitates automated segmentation and precise quantification of cardiac motion and morphology, enabling high-throughput analyses.

For lung imaging, tools such as Lung-RADS [15] and 3D Lung Analysis process chest CT scans to extract imaging-derived phenotypes such as lung volume, emphysema scores, and nodule characteristics. These metrics are particularly valuable for identifying genetic variants associated with respiratory diseases, including chronic obstructive pulmonary disease (COPD) and pulmonary fibrosis. Furthermore, PET imaging combined with radiotracers, analyzed using tools such as PMOD, enables quantification of lung inflammation and metabolic activity, providing additional insights into disease mechanisms.

For vascular and kidney imaging, tools such as Mimics and Carotid MRI Viewer are used to analyze vessel wall thickness, atherosclerotic plaque volume, and blood flow dynamics from ultrasound, MRI, and CT angiography data. These vascular phenotypes can be linked to genetic variants that influence cardiovascular disease risk, vascular remodeling, and kidney function. In kidney imaging specifically, segmentation tools enable the measurement of renal cortex volume, cyst burden, and related traits, aiding studies of genetic contributions to polycystic kidney disease and chronic kidney disease.

By leveraging a wide array of imaging tools and modalities across multiple organ systems, imaging GWAS enables the extraction of diverse and quantitative phenotypes. These traits, spanning structural, functional, and metabolic indicators, are integrated with genomic data to uncover novel genetic associations. This approach provides invaluable insights into the biological mechanisms underlying complex diseases, paving the way for improved diagnostics and personalized treatments across a wide range of medical disciplines.

5.6.2 Statistical analysis

The statistical analysis in imaging GWAS typically starts with univariate approaches, where each genetic variant is tested for its association with individual imaging traits. This is often done using a linear regression model, expressed as:

$$Y_i = \beta_0 + \beta_1 X_i + \sum_{j=1}^{k} \gamma_j Z_{ij} + \epsilon_i, \tag{5.1}$$

where Y_i represents the imaging phenotype for the ith individual, X_i is the genotype (e.g., the number of minor alleles) for a specific SNP, Z_{ij} are the covariates (e.g., age, sex, and population structure) with corresponding coefficients γ_j, β_0 is the intercept, β_1 is the effect size of the SNP on the phenotype, and ϵ_i is the residual error term.

Logistic regression is used when the imaging trait is binary (e.g., presence or absence of a feature), with the model expressed as

$$\log it(P_i) = \ln\left(\frac{P_i}{1 - P_i}\right) = \beta_0 + \beta_1 X_i + \sum_{j=1}^{k} \gamma_j Z_{ij}, \tag{5.2}$$

where P_i is the probability of the trait being present for the ith individual.

Multivariate approaches extend this framework to simultaneously analyze multiple imaging traits. For example, in canonical correlation analysis (CCA), the relationships between a set of genetic variants X and multiple imaging phenotypes Y are modeled to maximize their mutual correlation:

$$maximize\, corr(a^T X, b^T Y), \tag{5.3}$$

where a and b are weight vectors that define linear combinations of the genetic variants and imaging phenotypes, respectively.

Recent advancements, such as spatial GWAS, incorporate spatial information from imaging data. For example, spatial models use adjacency matrices or kernel functions to capture the relationships between regions of interest (ROIs). A common spatial GWAS model is expressed as:

$$Y = X\beta + WY + \epsilon, \tag{5.4}$$

where Y is the vector of imaging phenotypes, X is the matrix of genotypes, β is the vector of SNP effect sizes, W is the spatial weight matrix encoding ROI interactions, and ϵ is the error term.

These mathematical models enable imaging GWAS to evaluate the influence of genetic variants on imaging traits while accounting for covariates, correlated traits, and spatial dependencies.

Functional genomics tools are often employed to interpret significant genetic associations. For instance, expression quantitative trait loci (eQTL) analyses link genetic variants to gene expression changes, while transcription factor binding studies reveal regulatory mechanisms. Single-cell transcriptomics and epigenetic datasets add cellular-level context, bridging the gap between genetic variants and imaging phenotypes.

5.6.3 Applications of imaging GWAS

Imaging GWAS has emerged as a transformative approach to elucidate the genetic underpinnings of complex diseases across multiple organ systems by linking genetic variants to imaging-derived phenotypes. By leveraging high-throughput imaging data, GWAS provides an opportunity to bridge genetic variation with structural, functional, and molecular changes in tissues and organs, offering unprecedented insights into disease mechanisms, early detection, and potential therapeutic targets.

In neurological and psychiatric disorders, imaging GWAS has significantly advanced our understanding of brain structure and function. For instance, in AD, imaging GWASs have revealed strong associations between APOE ε4 alleles and hippocampal atrophy [16]. Hippocampal atrophy, a hallmark of AD, is an imaging-derived phenotype that correlates with disease progression. By identifying such genetic associations, imaging GWAS provides insight into neurodegeneration pathways, which is crucial for early diagnosis and targeted therapies. Furthermore, schizophrenia studies have linked genetic variants near

ZNF804A to reductions in white matter integrity, as measured using diffusion tensor imaging (DTI) [17]. This highlights the role of genetic variants in altering brain connectivity, advancing our understanding of the neurobiological basis of psychiatric disorders.

In cardiovascular diseases, imaging GWAS has uncovered genetic contributors to cardiovascular traits, enhancing our understanding of disease risk. For example, studies using cardiac MRI identified genetic loci associated with left ventricular ejection fraction and myocardial wall thickness, traits that are predictive of hypertrophic cardiomyopathy and heart failure [18]. The discovery of these loci helps elucidate pathways involved in myocardial remodeling and contractility, enabling targeted investigations into early disease markers and therapeutic interventions. In addition, imaging-derived phenotypes such as carotid artery wall thickness measured using ultrasound have been linked to loci influencing vascular inflammation and atherosclerosis risk, providing a genetic basis for understanding cardiovascular disease progression.

In oncology, imaging GWAS has integrated radiogenomics to predict genetic determinants of tumor traits and treatment responses. For instance, in lung cancer, GWASs have linked genetic variants to tumor metabolic activity and volume using PET imaging [19]. These associations are particularly valuable for identifying genetic markers of tumor aggressiveness and treatment resistance, advancing the personalization of cancer therapies. Imaging phenotypes extracted using tools such as PyRadiomics and 3D Slicer allow the quantification of tumor morphology and metabolic changes, which, when paired with genomic data, help uncover mechanisms driving tumor progression and metastasis.

In ophthalmology, imaging GWAS has been instrumental in identifying genetic determinants of retinal phenotypes. For example, optical coherence tomography (OCT) has been used to measure retinal nerve fiber layer thickness, which is associated with genetic variants influencing the risk of glaucoma [20]. These findings provide insights into the genetic architecture of ocular diseases and enable early identification of individuals at high risk of visual impairment, facilitating timely interventions.

In respiratory diseases, imaging GWAS has linked genetic variants to lung structure and function phenotypes. For instance, genetic loci associated with emphysema severity and lung volume were identified using chest CT scans [21]. These imaging-derived phenotypes are critical for understanding the genetic basis of COPD and lung fibrosis, offering pathways for potential therapeutic development. PET imaging with radiotracers further enhances these analyses by quantifying metabolic and inflammatory processes in the lungs, revealing how genetic variation influences disease activity.

Imaging GWAS also has advanced research in other organ systems. In kidney diseases, renal imaging phenotypes such as cortex volume and cyst burden, measured using MRI, have been linked to genetic determinants of polycystic kidney disease (PKD) [22]. These discoveries have shed light on the mechanisms underlying cyst formation and renal decline, offering new therapeutic targets for PKD. Similarly, in musculoskeletal research, traits such as bone mineral density and trabecular thickness, derived from CT scans, have been associated with genetic loci influencing osteoporosis risk.

By linking genetic variation to imaging-derived phenotypes across organ systems, imaging GWAS not only identifies disease-associated loci but also provides a deeper understanding of the biological mechanisms underlying disease. This integrative approach enhances early detection, informs risk prediction models, and guides the development of personalized treatments, marking a significant advancement in precision medicine.

5.6.4 Challenges in imaging GWAS

Despite its transformative potential, imaging GWAS faces several critical challenges related to computational, statistical, ethical, and biological aspects. Overcoming these challenges is essential to fully realize the potential of imaging GWAS for advancing precision medicine and disease understanding.

One of the most significant challenges is the high dimensionality of imaging datasets. Unlike traditional GWAS, where genetic data are typically limited to millions of SNPs, imaging GWAS must contend with millions of voxel-level features from modalities such as MRI, CT, and PET scans. This results in a feature space that vastly exceeds the number of participants, making it susceptible to overfitting and increased false-positive rates. Traditional dimensionality reduction methods, such as PCA, can reduce this feature space, but these approaches may miss subtle but biologically meaningful variations. More advanced methods, including deep learning–based feature extraction and autoencoder-based embeddings, have shown promise for capturing complex, nonlinear patterns in imaging data while maintaining interpretability.

The challenge of multimodal data integration further complicates the analysis. Imaging modalities such as MRI, PET, CT, and ultrasound capture distinct biological processes at varying spatial resolutions and scales. For example, MRI captures structural features, while PET reveals metabolic activity. Aligning these data types to a common reference space for cross-modal analysis requires sophisticated data fusion techniques, such as multitask learning, CCA, and Bayesian hierarchical models. The variability in data collection protocols and imaging hardware across sites also adds complexity, requiring harmonization tools and batch effect correction algorithms to ensure compatibility of imaging data from different sources. Without such adjustments, heterogeneity in imaging datasets can obscure true genetic signals.

Another key issue is population stratification, a problem common to all GWASs but especially relevant in imaging GWAS. Genetic variation and imaging phenotypes may differ across ethnic and ancestral groups, and without accounting for this variation, spurious associations may arise. This issue is often addressed using PCA or LMMs to adjust for population structure. However, many imaging GWASs are still conducted on European-ancestry populations, which limits the generalizability of results to other populations. There is a growing need for diverse population representation to ensure equitable health outcomes, especially for diseases that present differently across populations. Without diversity, the genetic risk factors uncovered in one population may not apply to others, leading to health disparities in disease prediction and intervention.

Statistical power is another significant challenge in imaging GWAS. Given the high-dimensional nature of imaging phenotypes, there is a need for large sample sizes to achieve sufficient power to detect significant associations while correcting for multiple comparisons. Unlike traditional GWAS, which can access datasets with hundreds of thousands of participants, imaging GWAS is often limited to smaller sample sizes due to the costs and logistical complexities of collecting imaging data. The need for large samples has prompted the creation of large-scale collaborations and consortium-based initiatives, such as the ENIGMA Consortium and the UK Biobank, which provide access to imaging datasets with corresponding genomic information. Collaborative efforts to share and harmonize data are essential to overcome the sample size limitations, but differences in imaging acquisition protocols and data preprocessing pipelines remain obstacles to fully leveraging pooled datasets.

Biological interpretation of imaging GWAS findings poses another major challenge. Unlike clinical phenotypes, such as blood pressure or cholesterol levels, imaging-derived phenotypes such as voxel-based morphometry or DTI metrics are more abstract and require contextualization. Interpreting how genetic variants influence these imaging-derived traits is not straightforward. Moreover, while GWAS can identify genome-wide significant loci, pinpointing the causal variants and the biological mechanisms driving these associations is challenging. Functional validation often requires multi-omics integration, where genomics, transcriptomics, and proteomics are combined to trace the downstream effects of genetic variation on biological pathways. Functional annotation databases such as Genotype-Tissue Expression (GTEx) and ENCODE can help map genetic loci to relevant tissues and pathways, but more sophisticated tools are needed to link these loci to specific imaging phenotypes.

Data privacy and ethical considerations present unique challenges in imaging GWAS. Imaging data, particularly 3D MRI and CT scans, are highly personally identifiable. Advanced image processing techniques can reconstruct facial images from brain MRIs, raising privacy concerns. Moreover, when imaging data are linked to genomic data, the potential for re-identification increases significantly, as genetic information is inherently unique to each individual. Privacy-preserving techniques, such as data anonymization and de-identification protocols, are essential but not foolproof. One emerging solution is the use of federated learning, a machine learning technique that allows researchers to train models on distributed datasets without sharing raw imaging or genomic data. By using federated learning, multiple research institutions can collaborate on large-scale imaging GWAS while maintaining the privacy and security of participant data. Secure multi-party computation (SMPC) and homomorphic encryption are other promising approaches for enabling secure, multi-institutional analysis of sensitive imaging-genomics data. However, these techniques remain computationally expensive and are not widely adopted in current GWASs.

The computational burden of imaging GWAS is another significant challenge. Unlike traditional GWAS, where only SNP data are analyzed, imaging GWAS involves processing massive 3D image datasets with voxel-level granularity. These datasets are not only large in volume but also computationally intensive to analyze. For example, large-scale studies such as those conducted on the UK Biobank involve thousands of MRI, CT, and PET scans for over 100,000 participants, producing petabytes of imaging data. Processing and analyzing these datasets requires access to high-performance computing (HPC) environments and cloud-based infrastructure. Cloud-based platforms, such as AWS, Google Cloud, and DataLad, offer solutions, but the financial costs and regulatory constraints on cloud usage pose barriers for smaller research teams. Emerging solutions such as containerized workflows (e.g., Docker and Singularity) allow for reproducible data analysis pipelines that can be run on cloud platforms or local HPC systems.

Reproducibility and standardization are also essential for the long-term success of imaging GWAS. Different software packages (e.g., FSL, SPM, and FreeSurfer) and image-processing pipelines can produce slightly different imaging-derived phenotypes. These inconsistencies can affect downstream GWAS analyses, leading to differences in the loci identified as genome-wide significance. Standardization efforts, such as harmonization protocols and data curation platforms (such as XNAT), aim to improve consistency in image preprocessing and feature extraction. However, achieving a universal standard for image-derived phenotypes remains an ongoing challenge.

Imaging GWAS faces a range of computational, ethical, and biological challenges. The high dimensionality of imaging phenotypes, multimodal data integration, population stratification, statistical power, biological interpretation, and computational burden all contribute to the complexity of conducting imaging GWAS. To overcome these obstacles, researchers are adopting strategies such as data harmonization, federated learning, multi-site collaborations, and the development of privacy-preserving computation methods. Advances in AI-driven feature extraction, multi-omics integration, and standardization of imaging pipelines will play a critical role in addressing these challenges. As the field evolves, the development of ethically sound, privacy-preserving methods for analyzing large, distributed datasets will become essential to ensure participant confidentiality and regulatory compliance. Overcoming these challenges will enable imaging GWAS to uncover new insights into the genetic architecture of complex diseases, advancing personalized medicine and precision health.

5.6.5 Future directions

Future advancements in imaging GWAS are poised to revolutionize our understanding of complex diseases by broadening the scope of research and integrating cutting-edge technologies. While brain imaging has been the primary focus, there is a growing emphasis on expanding imaging GWAS to explore multi-organ interactions. For instance, combining brain MRI with echocardiographic or vascular imaging data can uncover shared genetic pathways influencing both cardiovascular and neurological systems. Such studies are crucial for investigating systemic diseases, such as hypertension, stroke, and AD, where multi-organ crosstalk plays a significant role. Similarly, integrating imaging phenotypes of organs such as the lungs, kidneys, and liver could provide a more comprehensive understanding of multi-organ dysfunction in diseases such as diabetes and metabolic syndrome.

AI and deep learning techniques represent another major frontier in imaging GWAS. Deep learning models, particularly convolutional neural networks (CNNs) and autoencoders, are increasingly being applied to extract latent features from imaging data that may not be visible through traditional analyses. These latent features can serve as novel quantitative traits for GWAS, enabling researchers to capture subtle patterns of structural or functional changes. AI-driven phenotyping can also automate the processing of large-scale imaging data, addressing challenges related to high dimensionality and heterogeneity. For instance, unsupervised machine learning approaches can identify clusters of phenotypes that correspond to distinct disease subtypes, offering new insights into disease progression and treatment response.

The integration of imaging GWAS with other omics layers, such as transcriptomics, proteomics, and metabolomics, holds great potential for unraveling the molecular mechanisms linking genetic variants to imaging-derived traits. For example, combining genomic data with gene expression profiles can help identify tissue-specific pathways underlying structural or functional changes observed in imaging phenotypes. Proteomic and metabolomic data further enable the characterization of downstream effects, connecting genetic variation to disease phenotypes at the protein and metabolic levels. These multi-omics approaches will facilitate the construction of comprehensive genotype-to-phenotype pathways, offering a systems-level understanding of complex diseases.

Advancements in single-cell imaging and spatial transcriptomics further enhance the future of imaging GWAS. By mapping genetic variation to cellular and spatial patterns within tissues, these technologies can reveal how specific genetic variants influence tissue architecture and function at single-cell resolution. This is particularly relevant in oncology, where tumor heterogeneity can be explored by integrating imaging-derived traits, genetic mutations, and spatial gene expression data to predict disease aggressiveness and treatment outcomes.

Another promising direction is the application of longitudinal imaging GWAS, where imaging and genetic data are collected over time to monitor disease progression and treatment effects. Longitudinal studies can identify dynamic genetic influences on imaging phenotypes, enabling researchers to distinguish causal genetic factors from secondary disease effects. This approach has the potential to enhance early detection of diseases and improve patient stratification for clinical trials.

Findings from imaging GWAS are also expected to play a critical role in precision medicine. By identifying genetic variants associated with imaging biomarkers, researchers can develop predictive models for disease diagnosis, prognosis, and therapeutic response. For example, imaging biomarkers linked to genetic variants could be used to stratify patients based on their risk of disease progression or likelihood of treatment response, enabling personalized interventions tailored to an individual's genetic profile. In addition, these imaging-derived biomarkers can guide the development of novel therapeutics by identifying targetable pathways associated with disease phenotypes.

Finally, the expansion of population-scale biobanks and imaging repositories will accelerate the progress of imaging GWAS. Large-scale initiatives, such as the UK Biobank and other national imaging studies, are providing vast datasets with genomic, imaging, and clinical information. These resources will enable researchers to conduct more robust and generalizable studies, uncovering genetic determinants of imaging traits across diverse populations. Future efforts must prioritize the inclusion of underrepresented populations to ensure the findings of imaging GWAS are equitable and applicable across global populations.

5.7 Post-GWAS Analyses

GWAS provides valuable insights into genetic loci associated with traits or diseases, but identifying the specific causal variants, understanding their biological function, and interpreting their role within cellular and molecular pathways require post-GWAS analyses. These analyses bridge the gap between statistical associations and biological mechanisms, transforming GWAS findings into actionable insights for precision medicine and functional genomics.

5.7.1 Fine mapping and identifying causal variants

Fine mapping aims to pinpoint the causal variant(s) responsible for the observed association signal within a GWAS locus, often narrowing down from hundreds of correlated SNPs in LD to a handful of candidate variants. While GWAS signals typically cover large LD blocks, fine mapping employs statistical and computational approaches to estimate the posterior probability of causality for each SNP in the region.

For example, the CAusal Variants Identification in Associated Regions (CAVIAR) tool estimates the probability that a variant is causal based on summary statistics and LD structure. Similarly, Sum of Single Effects (SuSiE) decomposes association signals into individual causal components, improving precision in multi-SNP loci.

A classic example is the fine mapping of the FTO locus associated with obesity. Initial GWAS findings identified a broad region on chromosome 16 linked to BMI. Through fine mapping, researchers pinpointed a specific variant (rs1421085) that disrupted a conserved motif, leading to altered expression of the IRX3 and IRX5 genes in adipocytes.

Advanced techniques such as colocalization analysis further refine fine mapping by integrating GWAS summary statistics with eQTL data. Tools such as COLOC assess whether the same causal variant drives both the GWAS signal and gene expression variation, providing a more precise understanding of the regulatory mechanisms at play.

5.7.2 Functional annotation of significant SNPs

Significant SNPs from GWAS are often located in noncoding regions, making functional interpretation challenging. Functional annotation involves integrating GWAS signals with epigenomic data, gene expression data, and chromatin interaction maps to infer the biological function of associated variants.

- eQTL: eQTL analysis identifies SNPs that regulate gene expression levels, often using datasets from GTEx or eQTLGen consortia. For example, an eQTL analysis linked the GWAS variant rs12740374 at the SORT1 locus to reduced SORT1 expression in the liver, explaining its association with low-density lipoprotein cholesterol (LDL-C) levels.
- Epigenomic data: Integrating data from assays such as ChIP-seq, ATAC-seq, and DNAse-seq helps identify SNPs located in regulatory elements, such as enhancers or transcription factor binding sites. Tools such as RegulomeDB and HaploReg annotate SNPs with regulatory potential, while datasets from the ENCODE and Roadmap Epigenomics Project provide high-resolution maps of regulatory landscapes.
- Chromatin interaction data: Techniques such as Hi-C and Capture-C reveal physical interactions between distal regulatory regions and gene promoters, enabling researchers to connect noncoding SNPs with their target genes.

5.7.3 Pathway and gene set enrichment analyses

While GWAS identifies individual SNPs or genes associated with traits, biological pathways often involve the concerted action of multiple genes. Pathway and gene set enrichment analyses (PSEA and GSEA) determine whether

GWAS-identified genes are overrepresented in specific biological pathways, molecular functions, or cellular components.

- GSEA: This method tests whether a predefined set of genes (e.g., from Kyoto Encyclopedia of Genes and Genomes (KEGG) or Gene Ontology (GO) databases) is enriched among GWAS-associated genes.
- Multi-marker Analysis of GenoMic Annotation (MAGMA): MAGMA maps GWAS summary statistics to genes and performs pathway-level enrichment testing, identifying broader biological themes underlying genetic associations.
- Data-driven Expression-Prioritized Integration for Complex Traits (DEPICT): DEPICT uses co-expression data to infer the functional relevance of associated genes and highlights enriched pathways.

For instance, pathway analysis of GWAS results for schizophrenia revealed significant enrichment in synaptic signaling pathways, highlighting the role of neuronal function in disease etiology.

Moreover, protein–protein interaction (PPI) networks, constructed using tools such as STRING or Cytoscape, help visualize and prioritize candidate genes based on their connectivity and centrality in biological pathways.

5.7.4 Mendelian randomization

Mendelian randomization (MR) is a statistical approach used to infer causal relationships between an exposure, such as cholesterol levels or BMI, and an outcome, such as CAD or type 2 diabetes, using genetic variants as instrumental variables (IVs). The rationale behind MR lies in overcoming the limitations of traditional observational studies, which are often plagued by confounding variables and reverse causation. Observational studies may suggest associations, but they frequently fail to establish causality due to unmeasured confounding factors, such as lifestyle, environmental influences, or socioeconomic status. MR capitalizes on the principles of Mendel's laws of inheritance, where alleles are randomly allocated during gamete formation and are therefore less susceptible to confounding or reverse causation. This property mimics the random allocation of participants in a randomized controlled trial (RCT), making MR a powerful approach for causal inference.

The methodological framework of MR revolves around the use of genetic variants, typically SNPs, strongly associated with the exposure of interest. For instance, variants in the PCSK9 gene, which lower LDL-C levels, have been widely used to assess the causal effect of cholesterol on CAD. These SNPs serve as proxies or IVs to estimate the causal relationship. The core statistical method in MR relies on the ratio of effect sizes, where the association between the genetic variant and the outcome is divided by its association with the exposure. Various statistical approaches refine this causal estimate, including inverse-variance weighted (IVW) regression, MR-Egger regression, weighted median estimation, and MR-PRESSO (Pleiotropy Residual Sum and Outlier). For example, MR-Egger accounts for horizontal pleiotropy, a scenario where genetic variants influence the outcome through pathways independent of the exposure. Weighted median methods provide reliable causal estimates even if up to half of the instruments are invalid, while MR-PRESSO identifies and adjusts for outliers that may violate MR assumptions.

To ensure valid causal inference, MR relies on three fundamental assumptions. The first is the relevance assumption, which states that the genetic variant must be strongly associated with the exposure. Weak instruments can reduce statistical power and bias the results, often assessed using the F-statistic, where values below 10 indicate weak instrument bias. The second assumption, independence, ensures that the genetic variant is not associated with confounders that might distort the exposure-outcome relationship. This assumption is challenging to test directly, but sensitivity analyses can provide indirect validation. The third assumption, exclusion restriction, posits that the genetic variant influences the outcome exclusively through the exposure and not via any alternative biological pathways. Violations of this assumption, often due to pleiotropy, are addressed using sensitivity analyses such as MR-Egger or MR-PRESSO.

In recent years, MR has evolved beyond simple one-exposure, one-outcome models. Two-sample MR, where genetic associations with exposure and outcome are drawn from separate GWAS datasets, has become standard practice. This approach enables large-scale analysis using publicly available summary statistics, such as those from the UK Biobank or the GWAS Catalog. Multivariable MR (MVMR) extends this framework to assess the causal effects of multiple exposures simultaneously. For example, researchers can evaluate the independent effects of LDL-C, high-density lipoprotein (HDL) cholesterol, and triglycerides on CAD risk within the same model. Bidirectional MR further allows researchers to test the directionality of causation, such as determining whether obesity causes depression or vice versa. Techniques such as Steiger filtering and latent causal variable (LCV) models refine these analyses, distinguishing between genetic correlation and true causal relationships.

The tools supporting MR analyses have also become increasingly sophisticated. Software packages such as TwoSampleMR in R provide streamlined workflows for conducting two-sample MR studies, while MR-Base serves as a centralized database and analysis platform for exploring causal hypotheses. Other tools, such as MR-PRESSO, focus specifically on identifying and correcting pleiotropic outliers, enhancing the robustness of MR findings. Real-world applications of MR have validated therapeutic targets and informed drug development. A landmark MR study using genetic variants in PCSK9 and HMGCR genes confirmed that lowering LDL-C causally reduces CAD risk. These findings directly influenced the development and approval of PCSK9 inhibitors, now widely used to manage cholesterol levels in clinical settings.

Despite its strengths, MR is not without limitations. Horizontal pleiotropy, where genetic variants affect the outcome through multiple pathways, remains a persistent challenge. In addition, population stratification can introduce bias if genetic instruments derived from one population are applied to another with differing allele frequencies or environmental exposures. Weak instrument bias and limited statistical power in rare exposures or outcomes further complicate MR analysis. Careful instrument selection, sensitivity analyses, and robust statistical models are essential to address these limitations.

5.7.5 Polygenic risk scores

Polygenic risk scores (PRSs) are a quantitative measure that aggregates the cumulative effect of multiple genetic variants across the genome to estimate an individual's genetic predisposition to a particular trait or disease. Unlike Mendelian disorders, which are driven by rare high-penetrance variants, most common diseases, such as diabetes, cardiovascular disease, and psychiatric disorders, are influenced by numerous genetic variants, each contributing a small effect size. PRS leverages GWAS summary statistics to combine the effects of these variants into a single score, offering a genome-wide risk prediction metric for complex traits. The rationale behind PRS lies in the polygenic inheritance model, where the collective contribution of hundreds or thousands of common genetic variants influences the trait variance in a population. PRS is now widely recognized as a powerful tool for risk stratification, personalized medicine, and disease prediction.

The methodology behind constructing PRS begins with identifying genetic variants associated with the trait or disease of interest using GWAS data. Typically, SNPs with a specified genome-wide significance threshold (e.g., $p < 5 \times 10^{-8}$) are included, although more inclusive thresholds (e.g., $p < 0.05$) are sometimes used in a "polygenic pruning and thresholding" approach to maximize predictive power. Each selected SNP is assigned a weight (β coefficient) derived from the GWAS effect size, reflecting the strength and direction of its association with the phenotype. The PRS for an individual is then calculated as the sum of the weighted effects of all selected SNPs that they carry, mathematically expressed as

$$\mathrm{PRS} = \sum_{i=1}^{n} \beta_i G_i, \tag{5.5}$$

where β_i represents the effect size of the ith SNP, and G_i denotes the genotype dosage for that SNP (0, 1, or 2 copies of the risk allele). In practice, this computation is often implemented using software tools such as PLINK, PRSice-2, or LDpred, which automate SNP selection, weighting, and aggregation processes.

An essential step in constructing PRS is LD adjustment, as neighboring SNPs are often correlated due to their non-random co-inheritance. Methods such as LDpred and Polygenic Risk Scores using Continuous Shrinkage (PRS-CS) adjust SNP weights based on LD patterns to account for these correlations. LDpred, for example, uses a Bayesian shrinkage approach to optimize SNP effect sizes, while PRS-CS applies a continuous shrinkage before effect size estimates. These adjustments are critical to prevent overfitting and ensure that the PRS captures true polygenic effects rather than statistical noise arising from LD structure.

The validity and utility of PRS depend heavily on the ancestral match between the GWAS discovery dataset and the population in which the PRS is applied. Genetic architecture, allele frequencies, and LD patterns vary across populations, leading to reduced predictive accuracy when PRS derived from European-ancestry GWAS is applied to non-European populations. For instance, PRS derived from European cohorts often show significantly lower predictive power when tested in African or South Asian populations. Addressing these disparities requires more diverse GWAS datasets, trans-ancestry meta-analyses, and population-specific PRS recalibration.

PRS has demonstrated substantial predictive power for various diseases, such as breast cancer, CAD, and schizophrenia. For example, individuals in the top decile of a CAD PRS have been shown to have a threefold higher risk of developing CAD compared to those in the bottom decile. In psychiatry, PRS for schizophrenia has been used to stratify individuals based on their genetic liability and predict outcomes such as treatment response or disease progression. These findings highlight PRS as a promising tool for early risk identification, enabling targeted preventive interventions and personalized healthcare strategies.

Despite its growing utility, PRS is not without limitations. The predictive power of PRS remains modest for many traits, with much of the heritability still unexplained, especially for diseases with strong environmental or lifestyle components. Moreover, PRS does not account for gene-environment interactions, where environmental factors may amplify or mitigate genetic risk. Ethical considerations also emerge, particularly regarding the use of PRS in clinical decision-making, insurance, and employment. Genetic discrimination, data privacy concerns, and unequal access to PRS-based tools in underrepresented populations remain pressing challenges.

In addition to disease prediction, PRS is increasingly being used in genetic discovery and downstream functional analyses. By identifying individuals at extreme ends of the PRS distribution, researchers can study high-risk or low-risk groups in greater detail, uncovering novel biological mechanisms underlying disease susceptibility. Integration with other omics data, such as transcriptomics, proteomics, and metabolomics, is further enhancing the interpretability and utility of PRS.

In conclusion, PRSs represent a significant advancement in the field of genetic epidemiology, offering a scalable and cost-effective approach to quantifying genetic susceptibility to complex diseases. While methodological refinements, increased diversity in GWAS datasets, and ethical considerations remain areas of ongoing development, PRS holds immense potential for transforming disease risk prediction, enabling precision medicine, and advancing our understanding of complex trait genetics. As PRS becomes more refined and widely adopted, it is likely to play an integral role in both clinical and research settings, paving the way for more personalized and equitable healthcare solutions.

5.8 Conclusion

This chapter explores the advancements, methodologies, and applications of GWAS and imaging genetics, emphasizing their transformative role in unraveling the genetic architecture of complex diseases. GWAS has significantly contributed to identifying genetic variants, particularly SNPs, associated with a wide array of traits and diseases. By leveraging large cohorts and high-throughput genotyping technologies, GWAS has unveiled critical insights into biological pathways, highlighted potential therapeutic targets, and advanced precision medicine initiatives. Imaging genetics further bridges the gap between genetic variations and phenotypic manifestations by integrating genomic data with imaging-derived traits. This integration has elucidated the genetic underpinnings of structural and functional changes in various organs, notably in brain and cardiovascular research. For instance, imaging GWAS has revealed associations between APOE ε4 and hippocampal atrophy in AD and between genetic loci and cardiac structure traits predictive of heart disease. This chapter also discusses post-GWAS analyses, including fine mapping, functional annotation, pathway enrichment, MR, and PRS, all of which are essential for translating GWAS findings into biological understanding and clinical applications. These analyses provide deeper insights into causal genetic variants and their functional roles in disease pathophysiology.

However, the field faces challenges such as missing heritability, population stratification, data harmonization, and computational complexities associated with high-dimensional imaging data. Addressing these issues requires larger, more diverse cohorts, advanced statistical models, and standardized imaging protocols. Incorporating AI, multi-omics integration, and longitudinal data analyses are promising strategies for overcoming these challenges and enhancing discovery power. Future directions highlight the importance of expanding imaging GWAS across diverse populations and organ systems, integrating cutting-edge technologies such as AI and deep learning for feature extraction, and adopting ethical, privacy-preserving data-sharing practices. These advancements will not only improve the understanding of genetic contributions to disease but also pave the way for early detection, risk prediction, and personalized therapeutic interventions.

In conclusion, the application of GWAS and imaging genetics holds tremendous potential for advancing biomedical research and precision medicine. Continued interdisciplinary collaboration and technological innovation are crucial to unlocking novel insights into complex diseases and improving health outcomes globally.

References

1. Klein, R.J., et al. Complement factor H polymorphism in age-related macular degeneration. *Science*. 2005; **308**(5720): p. 385–389.
2. Manolio, T.A., F.S. Collins, and N.J. Cox. Finding the missing heritability of complex diseases. *Nature*. 2009; **461**(7265): p. 747–753.
3. Eichler, E.E., J. Flint, and G. Gibson. Missing heritability and strategies for finding the underlying causes of complex disease. *Nat Rev Genet*. 2010; **11**(6): p. 446–450.

4. Price, A.L., N.J. Patterson, and R.M. Plenge. Principal components analysis corrects for stratification in genome-wide association studies. *Nat Genet.* 2006; **38**(8): p. 904–909.
5. Marchini, J., B. Howie, and S. Myers. A new multipoint method for genome-wide association studies by imputation of genotypes. *Nat Genet.* 2007; **39**(7): p. 906–913.
6. Gabriel, S.B., S.F. Schaffner, and H. Nguyen. The structure of haplotype blocks in the human genome. *Science.* 2002; **296**(5576): p. 2225–2229.
7. Reich, D.E., M. Cargill, and S. Bolk. Linkage disequilibrium in the human genome. *Nature.* 2001; **411**(6834): p. 199–204.
8. Visscher, P.M., et al. Five years of GWAS discovery. *Am J Hum Genet.* 2012; **90**(1): p. 7–24.
9. Cen, X., et al. Towards interpretable imaging genomics analysis: methodological developments and applications. *Inf Fusion.* 2024; **102**(102032): p. 102032.
10. Xie, Y., et al. Genetic InfoMax: exploring mutual information maximization in high-dimensional imaging genetics studies. *ArXiv.* 2023.
11. Altmann, A., et al. Tackling the dimensions in imaging genetics with CLUB-PLS. *arXiv [q-bio.GN]*. 2023.
12. Khadhraoui, E., et al. Automated brain segmentation and volumetry in dementia diagnostics: A narrative review with emphasis on FreeSurfer. *Front Aging Neurosci.* 2024; **16**: p. 1459652.
13. Fedorov, A., et al. 3D Slicer as an image computing platform for the Quantitative Imaging Network. *Magn Reson Imaging.* 2012; **30**(9): p. 1323–1341.
14. Luo, C., et al. Assessing the accuracy of CMRtools software for diagnosing liver iron overload in thalassemia patients: Influencing factors and optimisation strategies. *Front Med (Lausanne).* 2024; **11**: p. 1424294.
15. Martin, M.D., et al. Lung-RADS: Pushing the limits. *Radiographics.* 2017; **37**(7): p. 1975–1993.
16. Jack, C.R., Jr., et al. Hypothetical model of dynamic biomarkers of the Alzheimer's pathological cascade. *Lancet Neurol.* 2010; **9**(1): p. 119–128.
17. Hibar, D.P., et al. Novel genetic loci associated with hippocampal volume. *Nat. Commun.* 2017; **8**(1): p. 13624.
18. Levin, M.G., et al. Multi-ancestry multivariate genome-wide analysis highlights the role of common genetic variation in cardiac structure, function, and heart failure-related traits. *medRxiv,* 2021: p. 2021.08.03.21261508.
19. Aerts, H.J.W.L., et al. Decoding tumour phenotype by noninvasive imaging using a quantitative radiomics approach. *Nat Commun.* 2014; **5**(1): p. 4006.
20. Craig, J.E., et al. Multitrait analysis of glaucoma identifies new risk loci and enables polygenic prediction of disease susceptibility and progression. *Nat Genet.* 2020; **52**(2): p. 160–166.
21. Sandford, A.J., L. Joos, and P.D. Paré. Genetic risk factors for chronic obstructive pulmonary disease. *Curr Opin Pulm Med.* 2002; **8**(2): p. 87–94.
22. Lu, Y.-Q. and Y. Wang. Multi-omic analysis reveals genetic determinants and therapeutic targets of chronic kidney disease and kidney function. *Int J Mol Sci.* 2024; **25**(11): p. 6033.

CHAPTER 6

Single-Cell Sequencing Data Analysis: Basic Analysis

Single-cell sequencing represents a groundbreaking advancement in molecular biology and genomics, allowing researchers to analyze the genome, transcriptome, epigenome, and proteome at the resolution of individual cells. Unlike traditional bulk sequencing methods that average signals across heterogeneous cell populations, single-cell sequencing captures the unique molecular profiles of individual cells, offering unparalleled insights into cellular heterogeneity. However, the analysis of single-cell sequencing data presents unique computational challenges, requiring specialized methods to extract meaningful biological insights from these complex, high-dimensional datasets.

In this chapter, we will cover the fundamental preprocessing methods for single-cell RNA sequencing (scRNA-seq), single-cell assay for transposase-accessible chromatin sequencing (scATAC-seq), and cellular indexing of transcriptomes and epitopes by sequencing (CITE-seq) data. These methods encompass essential steps including matrix generation, quality control (QC), and normalization. Preprocessing techniques specific to single-cell spatial transcriptomics and proteomics will be discussed in Chapters 8 and 9, respectively. Beyond preprocessing, this chapter will also introduce key analytical approaches commonly applied to single-cell data. These include feature selection and dimensionality reduction for simplifying complex datasets, clustering for identifying cell subpopulations, differential analysis to uncover gene- or region-specific variations, and advanced visualization techniques for interpreting and presenting results effectively.

6.1 Introduction to Single-Cell Sequencing

Single-cell sequencing encompasses a diverse array of technologies designed to profile various molecular modalities, including RNA, DNA, chromatin accessibility, and proteins at single-cell resolution. Early applications, such as scRNA-seq, have focused on capturing gene expression patterns and have since become foundational tools in transcriptomics. Building on this, technologies such as scATAC-seq enable the study of chromatin accessibility to uncover regulatory elements and transcription factor activity. Moreover, advancements in multi-omic approaches now enable the simultaneous measurement of multiple layers of molecular information, such as CITE-seq, which integrates transcriptomic and proteomic profiling, allowing simultaneous measurement of RNA and surface protein expression within individual cells. These technologies provide unprecedented insights into cellular heterogeneity and regulatory mechanisms, advancing our understanding of complex biological systems.

These technologies leverage microfluidics, droplet-based platforms, and combinatorial indexing methods to isolate and process thousands of individual cells in parallel. The continuous refinement of these platforms has improved their throughput, sensitivity, and resolution, making single-cell sequencing accessible to a broad range of research applications.

6.1.1 Single-Cell RNA Sequencing

scRNA-seq involves a series of steps to analyze gene expression at the individual cell level. As shown in Fig. 6.1, the current scRNA-seq process typically includes the following steps: tissue dissection, cellular dissociation, single-cell isolation, RNA extraction and fragmentation, complementary DNA (cDNA) synthesis, library preparation, sequencing, genome mapping, generation of single-cell expression profiles, and clustering and annotation to identify distinct cell types or states [1, 2]. The process begins with tissue dissection, where the tissue of interest is isolated and prepared. This is followed by cellular dissociation, in which the tissue is enzymatically or mechanically dissociated into a suspension of individual cells. Next, single-cell isolation is performed, where individual cells are separated using techniques such as microfluidics or droplet-based methods. Once isolated, the RNA molecules within each cell are extracted and fragmented into smaller segments. These RNA fragments are reverse-transcribed into cDNA [3, 4]. The cDNA is then amplified, and sequencing adapters are added to prepare for sequencing. In the next stage, the prepared cDNA libraries are subjected to high-throughput sequencing, producing raw sequence data. These sequences are then aligned to a reference genome (genome mapping), enabling the identification of genes and their expression levels. The mapped data is processed to generate single-cell expression profiles, where each cell's gene expression is quantified and represented in a matrix. Finally, computational methods are used for single-cell analysis, such as clustering and cell-type annotation, where cells with similar gene expression patterns are grouped and annotated to identify distinct cell types or states.

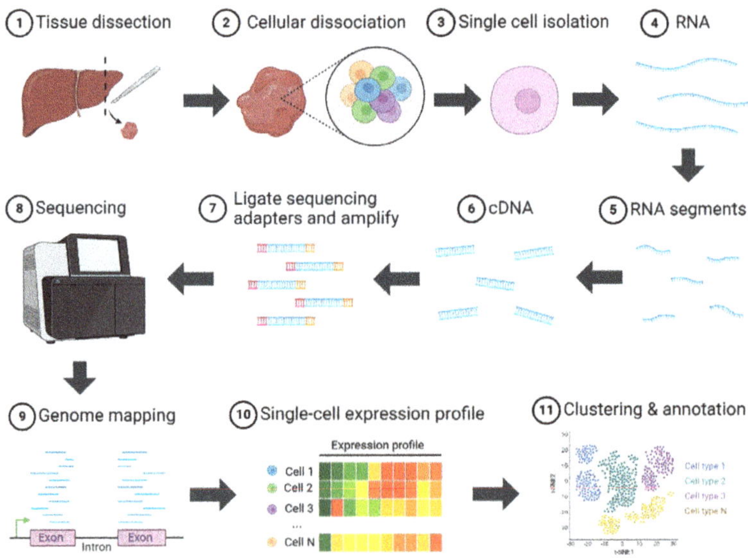

Fig. 6.1 Workflow of single-cell RNA sequencing Created with BioRender.com.

Modern scRNA-seq technologies can be broadly categorized into three main approaches: plate-based methods, droplet-based methods, and combinatorial indexing-based methods, each with unique advantages and limitations tailored to specific research needs [5].

Plate-based scRNA-seq technologies were among the first to be developed and include methods such as SMART-seq [6] and SMART-seq2 [7]. These technologies excel in capturing full-length transcripts, making them ideal for applications requiring detailed analysis of transcript isoforms or splice variants. Plate-based methods are highly sensitive and accurate but are limited in scalability due to the labor-intensive and costly process of isolating cells into individual wells. As a result, these methods are typically used for studies focusing on a smaller number of cells where high-resolution transcriptomic data is required.

Droplet-based scRNA-seq technologies, such as Drop-seq [4] and 10x Genomics, represent a significant leap in scalability and cost-efficiency. These methods encapsulate individual cells in microfluidic droplets, where RNA is reverse-transcribed into cDNA and tagged with unique molecular identifiers (UMIs) and cell barcodes. This design enables the simultaneous processing of thousands to millions of cells in a single experiment, making droplet-based technologies ideal for studies requiring high-throughput profiling. However, droplet-based methods typically capture only the 3' or 5' ends of transcripts, which limits their ability to detect full-length isoforms but provides sufficient information for gene expression analysis and cell-type identification.

Combinatorial indexing-based scRNA-seq technologies, such as SPLiT-seq [8] and sci-RNA-seq [9], enable transcriptome profiling without requiring physical cell isolation. Instead, these methods use iterative rounds of molecular barcoding to uniquely label transcripts from individual cells. This approach offers a highly scalable solution for profiling large numbers of cells at a reduced cost. While combinatorial indexing is less sensitive than plate-based methods and lacks the resolution for full-length transcript analysis, it is well-suited for studies requiring population-level insights across diverse cell types.

Each scRNA-seq technology has its strengths and limitations, making the choice of platform highly dependent on the specific research goals. Plate-based methods are optimal for in-depth studies of individual cells, droplet-based approaches are favored for large-scale cell population analysis, and combinatorial indexing provides a cost-effective solution for massive-scale studies. The continued advancement of these technologies is expanding their applications and providing researchers with powerful tools to unravel the complexity of cellular systems.

6.1.2 Single-Cell ATAC Sequencing

The scATAC-seq workflow, as illustrated in Fig. 6.2, is designed to profile chromatin accessibility at the single-cell level [10]. The process begins with the binding of Tn5 transposases to open chromatin, where the hyperactive Tn5 enzyme preferentially targets accessible regions of the genome. This step is crucial as it selectively marks open chromatin while avoiding regions of tightly packed, closed chromatin.

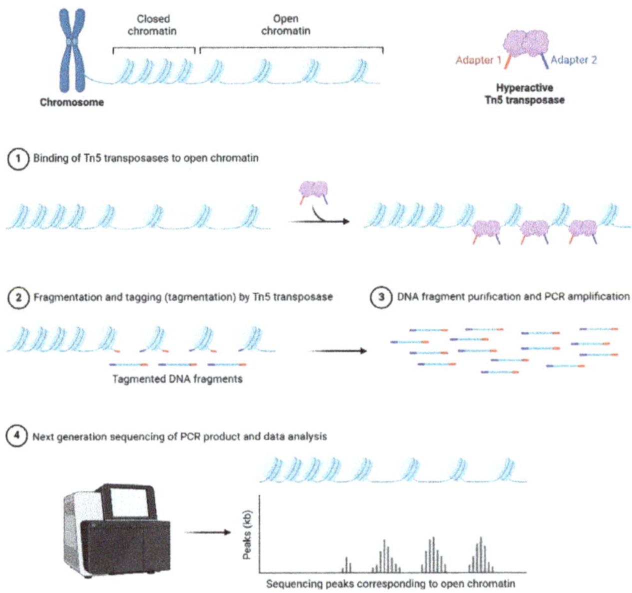

Fig. 6.2 Workflow of single-cell ATAC sequencing Created with BioRender.com.

Once bound, the Tn5 transposase performs fragmentation and tagging (tagmentation), simultaneously cutting the accessible DNA regions and integrating sequencing adapters into the fragmented DNA. This step produces small, adapter-tagged DNA fragments that are ready for further processing.

Following tagmentation, the DNA fragments undergo purification and polymerase chain reaction (PCR) amplification, enriching the library for sequencing. This step ensures that only relevant chromatin-accessible regions are amplified while removing unwanted background DNA. The amplified DNA library is then subjected to next-generation sequencing (NGS), where the sequences corresponding to open chromatin regions are read and analyzed.

The final step involves data processing and analysis, where sequencing reads are mapped to the reference genome, and peaks corresponding to accessible chromatin regions are identified. These peaks indicate transcriptionally active regulatory elements, such as promoters and enhancers, providing insights into gene regulatory mechanisms at the single-cell level.

scATAC-seq [10] is the first technique to enable chromatin accessibility profiling at single-cell resolution, laying the foundation for subsequent advancements. Another widely adopted platform is 10x Genomics Chromium Single Cell ATAC, which leverages microfluidic technology to partition individual cells into droplets along with barcoded gel beads. The Tn5 transposase is then used to tag accessible chromatin regions, enabling the simultaneous analysis of thousands of cells. This method is highly scalable and has become a standard for high-throughput scATAC-seq studies.

For researchers seeking alternatives to microfluidics, sci-ATAC-seq [11] offers a combinatorial indexing approach. This technique uses multiple rounds of barcoding to label individual cells, eliminating the need for specialized equipment and enabling the profiling of tens of thousands of cells in a single experiment. Another notable method is droplet-based scATAC-seq (dscATAC-seq) [12], which combines droplet microfluidics with Tn5 transposase to achieve high sensitivity and throughput. This approach is particularly useful for studying rare cell populations due to its ability to capture subtle chromatin accessibility differences.

In summary, current scATAC-seq technologies have become essential for studying chromatin accessibility at single-cell resolution, providing insights into gene regulatory mechanisms and cell-type-specific epigenetic landscapes.

6.1.3 CITE-seq

The CITE-seq workflow, as illustrated in Fig. 6.3, integrates scRNA-seq with protein expression profiling using antibody-derived tags (ADTs) [13]. The process begins with ADT binding to the cell surface proteins of interest, where antibodies conjugated with unique nucleotide barcodes (ADT) specifically recognize and label cell surface markers.

Next, tagged microbeads and cells with bound ADTs are encapsulated within droplets, utilizing a microfluidic system that co-encapsulates single cells and barcoded beads in an oil-based droplet. This step ensures that both messenger ribonucleic acid (mRNA) and surface protein information from individual cells are retained for simultaneous analysis. Following encapsulation, cells are lysed within the droplet to release mRNA, allowing both RNA and ADTs to become available for hybridization with the microbead tags.

In the subsequent step, mRNA and ADTs hybridize into the microbead tags, where polyadenylated mRNA and uniquely barcoded antibody tags are captured onto bead-linked oligonucleotides. The captured material is then processed for cDNA synthesis and library preparation, where cDNA products derived from mRNA (~300 bp) and ADT oligo products (<180 bp) are separated based on size for efficient sequencing.

Finally, libraries are pooled and sequenced, allowing for both transcriptomic (gene expression) and proteomic (antibody-tag) data to be analyzed in parallel. By leveraging cell barcodes, scRNA-seq and ADT data can be integrated, providing a comprehensive multi-modal dataset for clustering and visualization. CIE-seq enables researchers to simultaneously investigate transcriptomic and protein-level information at single-cell resolution.

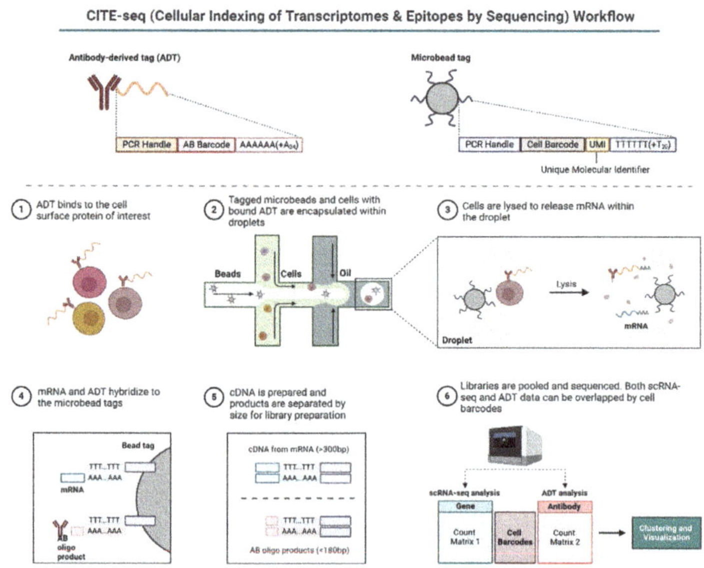

Fig. 6.3 Workflow of the cellular indexing of transcriptomes and epitopes by sequencing Created with BioRender.com.

Single-cell sequencing has proven indispensable for answering questions that require a granular understanding of cellular diversity and dynamics. In cancer biology, it has been instrumental in dissecting intratumoral heterogeneity, identifying therapy-resistant clones, and uncovering the mechanisms underlying metastasis [14, 15]. In immunology, single-cell analyses have facilitated the characterization of immune cell subsets and their functional states in response to infections or autoimmune diseases [16, 17]. Developmental biology has also greatly benefited, with single-cell sequencing enabling the reconstruction of cell differentiation trajectories and uncovering lineage-specific gene regulatory networks [18, 19].

Beyond basic research, single-cell analysis is becoming a cornerstone in precision medicine. It enables the identification of biomarkers, elucidation of disease mechanisms, and development of targeted therapeutic strategies by focusing on the molecular signatures of individual cells [20, 21]. These advancements highlight the transformative potential of single-cell technologies in both research and clinical applications.

6.2 scRNA-seq Data Preprocessing

The preprocessing of scRNA-seq data typically involves several key steps [22]: (1) processing raw FASTQ data to generate a gene (or transcript) expression count matrix via alignment and quantification; (2) performing QC to remove low-quality cells and minimize interference with downstream analyses; (3) normalizing the expression count matrix to eliminate biases across genes and cells; and (4) removing batch effects of systematic differences introduced by experimental variations across different batches, improving the integration of datasets for downstream analysis. Each of these steps will be discussed in detail in the following subsections.

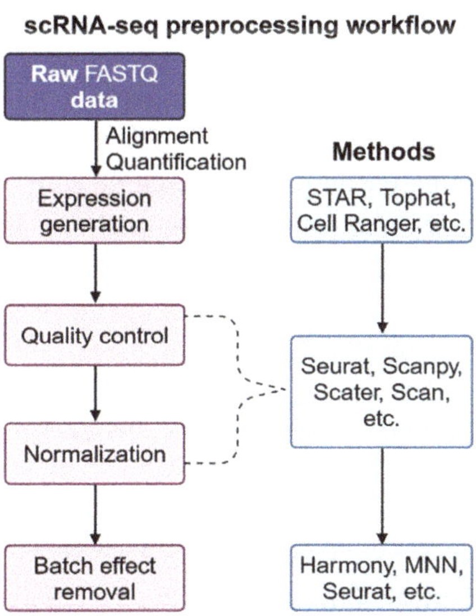

Fig. 6.4 Overview of single-cell RNA sequencing data preprocessing.

6.2.1 Generate Expression Count Matrix

For the raw sequencing data of single-cell transcriptomes, typically provided in FASTQ format, an appropriate alignment method is employed to process the data. Tools such as STAR [23] or TopHat [24] are commonly used for mapping high-quality sequencing reads to a reference genome with high accuracy. These methods align reads to the genome and quantify them into expression count matrices, where rows correspond to genes and columns represent individual cells. The resulting matrix serves as the foundational input for downstream scRNA-seq analysis.

In addition to general alignment tools, specialized pipelines have been developed to streamline the preprocessing of single-cell data. Among these, Cell Ranger is a widely used tool specifically designed for generating expression count matrices from scRNA-seq data. Developed by 10x Genomics, Cell Ranger [25] performs read alignment, transcript quantification, and quality filtering in an integrated workflow, producing a high-quality count matrix ready for further analysis. The robustness and ease of use of tools such as Cell Ranger have made them standard in scRNA-seq workflows, ensuring consistency and reliability in data preprocessing.

6.2.2 Quality Control

scRNA-seq technology, while highly transformative, is not without its technical limitations. Issues such as cell damage during dissociation, failures in library preparation, multiple cells being captured in the same droplet, or low transcript coverage and capture rates can lead to missing or unreliable gene expression data. These limitations introduce noise and artifacts into the data, potentially impacting downstream analyses.

To address these challenges, QC is an essential step in scRNA-seq preprocessing. QC aims to identify and filter out low-quality cells that could otherwise skew analytical results. This process typically involves evaluating cells based on specific metrics and setting thresholds to exclude those that fail to meet quality standards. The commonly used QC criteria include

1. Cell counting depth: Cells with extremely low or abnormally high sequencing depths may indicate insufficient RNA capture or potential artifacts, such as doublets (two cells mistakenly sequenced as one).
2. Number of genes detected per cell: Cells with very few detected genes often represent poor-quality samples, such as damaged cells or droplets

with ambient RNA. Conversely, cells with an unusually high number of detected genes may signal the presence of doublets or multiplets.

3. Proportion of mitochondrial gene counts: An elevated percentage of mitochondrial gene expression often indicates stressed or dying cells, as such cells tend to release more mitochondrial RNA.

By setting appropriate thresholds for these variables, researchers can systematically exclude problematic cells, improving the overall reliability of the data. Modern scRNA-seq analysis pipelines often integrate tools such as Seurat [26], Scanpy [27], Scater [28], or Scran [29], which automate QC processes and provide visualization methods to help fine-tune filtering criteria.

6.2.3 Data Normalization

The goal of data normalization in scRNA-seq is to correct for systematic differences in sequencing coverage across cells, ensuring that these differences do not bias downstream analyses, such as clustering, trajectory inference, or differential gene expression analysis. By standardizing gene expression measurements, normalization enables meaningful comparisons of transcriptomic profiles between cells. The most widely used normalization approach for scRNA-seq data involves scaling gene expression counts based on a size factor, which accounts for the variability in sequencing depth across cells. This size factor is typically estimated using the total expression count for each cell, representing the relative deviation caused by differences in library size. Normalized expression values are obtained by dividing the raw counts for each cell by its size factor, followed by a log transformation. The log transformation serves to compress the dynamic range of expression values, mitigating the impact of extreme counts and making the data more suitable for downstream analyses, such as *LogNormalize()* function in Seurat [26] and *normalize_total()* function in Scanpy [27].

In addition to this basic normalization strategy, more advanced methods have been developed to address specific challenges in single-cell data, such as zero inflation (due to dropout events) and highly variable sequencing depth. For example, normalization methods may incorporate gene-specific scaling or regularization to enhance the robustness of the results. One widely used method is SCTransform [30], which employs a regularized negative binomial regression–based approach to model gene expression while accounting for sequencing depth and technical noise. SCTransform also corrects for overdispersion and zero inflation, making it particularly effective for single-cell datasets. Another popular tool is Scran [29], which uses a deconvolution approach to estimate size factors for pools of cells, improving accuracy for small datasets or datasets with high variability in cell sizes.

Proper normalization is a critical preprocessing step, as it directly impacts the quality and interpretability of downstream analyses. Careful selection of normalization techniques and tools is essential to ensure that systematic biases are minimized while preserving biologically meaningful variations in the data.

6.2.4 Batch Effect Removal

scRNA-seq technology is advancing rapidly, with numerous platforms now available for generating single-cell transcriptomic data [31]. These platforms allow researchers to capture gene expression at single-cell resolution, providing unprecedented insights into cellular heterogeneity. However, data generated from different platforms often exhibit significant batch effects due to technical variations such as sequencing chemistry, instrument calibration, and sample handling. These batch effects can obscure true biological signals and pose a major challenge when integrating multi-platform scRNA-seq datasets.

Addressing batch effects is critical for ensuring the reliability and accuracy of downstream analyses, including clustering, trajectory inference, and differential expression (DE) studies. However, correcting batch effects is particularly challenging due to their often nonlinear and complex nature, making it difficult to harmonize datasets while preserving biologically meaningful variations. Effective batch correction strategies must therefore strike a balance between removing unwanted technical artifacts and maintaining essential biological differences.

To overcome these challenges, researchers have developed a variety of algorithms for integrating scRNA-seq data across different platforms. For example, Harmony [32] applies principal component analysis (PCA) for dimensionality reduction, clustering similar cells across batches while ensuring batch diversity within each cluster. These methods enable effective alignment of datasets while preserving biologically relevant differences. Seurat [26], a popular tool for scRNA-seq data analysis, uses canonical correlation analysis (CCA) to identify shared sources of variation between datasets (batches) and then aligns the datasets in a low-dimensional space using mutual nearest neighbors (MNN), which identifies pairs of cells that are nearest neighbors across batches to mitigate technical variations while preserving biological distinctions. In addition, Seurat's anchor-based integration method selects biologically similar cell pairs ("anchors") across batches and utilizes them to harmonize datasets effectively. These approaches ensure that integrated datasets retain meaningful biological insights while reducing unwanted batch effects.

6.3 scATAC-seq Data Preprocessing

scATAC-seq data preprocessing is a crucial step in analyzing chromatin accessibility at the single-cell level. The goal is to extract high-quality chromatin accessibility information from raw sequencing data while minimizing noise and technical artifacts. A well-structured preprocessing pipeline ensures accurate downstream analysis, such as clustering, differential accessibility analysis, and regulatory network inference. The preprocessing workflow typically includes several steps, as shown in Fig. 6.5.

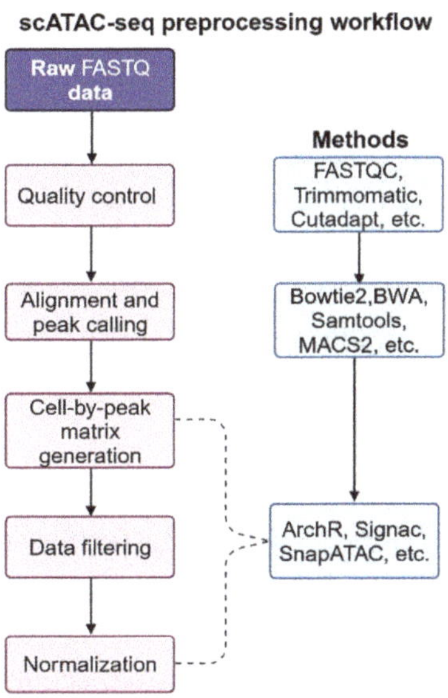

Fig. 6.5 Overview of single-cell ATAC sequencing data preprocessing.

6.3.1 Raw Data Quality Control and Preprocessing

The preprocessing pipeline begins with raw sequencing data, typically in FASTQ format. Before alignment, it is essential to assess data quality using tools such as FastQC [33], which provides key quality metrics, including base quality scores, guanine-cytosine (GC) content distribution, and adapter contamination. To ensure high-quality reads, preprocessing tools such as Trimmomatic [34] or Cutadapt [35] are used to trim low-quality bases, remove adapter sequences, and filter out contaminants. This step generates clean FASTQ files, reducing sequencing errors and improving mapping efficiency in the subsequent steps.

6.3.2 Read Alignment and Peak Calling

High-quality reads are then aligned to a reference genome using specialized alignment tools such as Bowtie2 [36] or BWA [37], optimized for handling short-read sequencing data. To mitigate technical biases introduced by PCR amplification, duplicate reads are identified and removed using tools such as Picard [38] or samtools [39]. The next crucial step is peak calling, which identifies regions of open chromatin indicative of regulatory elements such as enhancers and promoters. This is commonly performed using MACS2 [40], which generates peak files in BED format, providing a genome-wide map of accessible chromatin regions.

6.3.3 Generation of the Cell-by-Peak Matrix

Following peak calling, the next step is to construct a sparse cell-by-peak matrix, where each row represents a single cell and each column represents a genomic region (peak). This matrix is generated by assigning reads to peak regions for each individual cell. Tools such as SnapATAC [41] and ArchR [42] facilitate this process by associating sequencing reads with corresponding cellular barcodes. This step is fundamental for downstream analysis, as it enables quantification of chromatin accessibility at the single-cell level.

6.3.4 Data Quality Control and Filtering

To ensure robust analysis, rigorous QC is performed on both cells and peaks. Low-quality cells are removed based on criteria such as an excessively low number of reads, high levels of mitochondrial reads, or excessive read mapping to genomic blacklist regions (known artifacts in sequencing data). In addition, peaks that are detected in only a few cells are filtered out to reduce noise. These QC steps refine the cell-by-peak matrix, ensuring that only high-confidence signals are retained for further analysis.

6.3.5 Data Normalization and Bias Correction

Since scATAC-seq data exhibits inherent biases, including variability in sequencing depth across cells, normalization is essential to ensure comparability. Term frequency-inverse document frequency (TF-IDF) normalization is widely used to adjust for differences in read coverage across peaks and cells. Moreover, log-normalization and batch correction methods, implemented in tools such as Signac [43] and ArchR [42], help correct for technical variation, facilitating accurate downstream analyses such as clustering, trajectory inference, and regulatory network reconstruction.

Preprocessing scATAC-seq data is a complex but essential step in obtaining high-quality chromatin accessibility profiles at the single-cell level. The inherent sparsity and technical noise of scATAC-seq data require a carefully designed preprocessing pipeline to ensure accurate and reproducible results. By implementing rigorous QC measures, such as removing low-quality cells and filtering unreliable peaks, researchers can enhance data reliability and mitigate sequencing artifacts. Efficient alignment and peak calling strategies further refine chromatin accessibility signals, enabling the identification of regulatory elements with high confidence.

6.4 CITE-seq Data Preprocessing

CITE-seq is a multimodal single-cell sequencing technology that simultaneously measures transcriptomic and surface protein expression in individual cells. This dual-modality profiling enhances the resolution of cellular states and enables a more comprehensive characterization of cell types. However, the preprocessing of CITE-seq data is more complex than standard scRNA-seq due to the integration of RNA and ADT data. A well-defined preprocessing pipeline is crucial for ensuring the accuracy and reliability of downstream analyses. The preprocessing workflow typically consists of the following steps (Fig. 6.6).

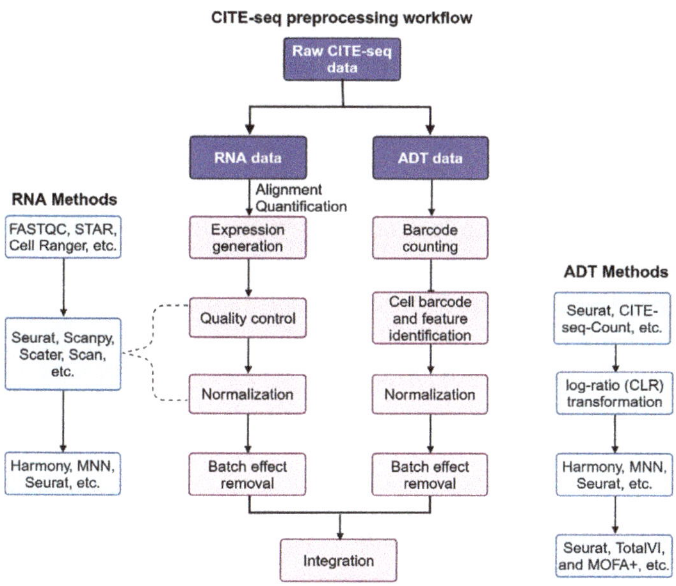

Fig. 6.6 Overview of CITE-seq data preprocessing.

6.4.1 Raw Data Processing and Quality Control

CITE-seq generates two distinct data types: gene expression data from RNA and protein abundance data from ADTs. The raw sequencing data, typically in FASTQ format, undergoes initial QC using tools such as FastQC [33] to assess sequencing quality, adapter contamination, and base composition. Adapter sequences, low-quality reads, and PCR duplicates are removed using preprocessing tools such as Trimmomatic [34] or Cutadapt [35].

6.4.2 Alignment and UMI Processing

For RNA sequencing, reads are aligned to a reference transcriptome using alignment tools such as STAR or Cell Ranger, which also collapse UMIs to reduce PCR amplification biases. The protein-derived ADT reads, which contain barcode sequences rather than standard genomic sequences, do not undergo traditional alignment but are instead directly processed to count the occurrence of each antibody-derived barcode per cell.

6.4.3 Cell Barcode and Feature Identification

Cell barcodes are filtered to remove empty droplets and low-quality cells with abnormally low UMI counts. Simultaneously, protein feature counts are extracted and linked to corresponding cells. Tools such as Seurat [26] and CITE-seq-Count [44] facilitate this step, ensuring that each cell has both transcriptomic and protein expression profiles correctly assigned.

6.4.4 Data Normalization and Batch Correction

Since RNA and protein expression data are measured using different methodologies, they require independent normalization strategies. RNA data is typically normalized using log-transformation or SCTransform, while ADT data is often normalized using centered log-ratio (CLR) transformation to account for differences in sequencing depth and amplification efficiency. Moreover, batch correction techniques, such as Harmony or MNN correction, may be applied to remove technical variations arising from different experimental conditions or antibody batches.

6.4.5 Quality Control and Filtering

Further QC is performed to identify and remove cells with excessive mitochondrial RNA content, unusually low or high feature counts, or abnormal ADT distributions. High background noise in the ADT signal, often due to nonspecific antibody binding, is corrected by thresholding or denoising methods such as isotype control-based correction or machine learning–based noise removal.

6.4.6 Integration and Downstream Compatibility

The final preprocessed dataset consists of a combined RNA and protein expression matrix, enabling multimodal analyses such as joint clustering, cell-type annotation, and multimodal trajectory inference. Tools such as Seurat [26], TotalVI [45], and MOFA+ [46] allow seamless integration of RNA and protein features, facilitating advanced analyses such as multimodal DE and regulatory network inference.

By implementing a robust preprocessing pipeline, researchers can ensure the accuracy and interpretability of CITE-seq data. Proper preprocessing minimizes noise, corrects technical biases, and enhances the biological insights gained from multimodal single-cell profiling, making CITE-seq a powerful tool for dissecting cellular heterogeneity and functional states.

6.5 Feature Selection and Dimensionality Reduction

Feature selection and dimensionality reduction are critical preprocessing steps in single-cell data analysis, ensuring that the data is both biologically informative and computationally manageable. These steps streamline downstream analyses such as clustering, trajectory inference, and DE analysis by focusing on the most relevant features while mitigating noise and redundancy.

6.5.1 Feature Selection

Feature selection aims to identify genes that are biologically meaningful and contribute to the variation in the dataset while discarding low-quality genes dominated by random noise. By narrowing the focus to informative features, feature selection enhances the signal-to-noise ratio and improves the efficiency and interpretability of subsequent analyses.

In scRNA-seq, feature selection typically focuses on identifying genes that exhibit high variability across cells. The rationale behind this approach is that genes with variable expression are more likely to be involved in driving cellular diversity and biological processes. Methods such as variance-based filtering, where genes with low variance across cells are discarded, are commonly used. In addition, methods such as the "mean-variance model" (e.g., the method implemented in Seurat [26]) assess genes based on their mean expression levels and variance, allowing for the selection of genes that are not only variable but also biologically significant. Another widely used approach is the identification of highly expressed or differentially expressed genes (DEGs) across cell clusters or conditions. Gene set enrichment analysis (GSEA) [47] can also be employed to select genes that belong to specific biological pathways or functional categories, further enhancing the interpretability of the analysis.

For scATAC-seq data, feature selection involves identifying accessible chromatin regions (peaks) that contribute to cellular diversity. Since the goal of scATAC-seq is to understand chromatin accessibility and its relationship to gene regulation, peaks that show variability in accessibility across cells are prioritized. Common approaches for feature selection in scATAC-seq data include filtering for regions with high variability in accessibility between cells, which may be indicative of regulatory activity in different cell populations. Moreover, integrating scATAC-seq data with scRNA-seq data can enhance feature selection by linking accessible chromatin regions to gene expression. For example, open chromatin regions near the promoters or enhancers of highly variable genes are often selected for further analysis, as they are likely to play a role in regulating gene expression. Methods such as SnapATAC [41] and ArchR [42] can help in feature selection by efficiently mapping chromatin accessibility data to specific genomic regions, filtering out regions with little or no variability, and prioritizing regulatory regions likely to affect gene expression or cellular behavior. The selection of peaks associated with key transcription factors or regulatory networks further enhances the interpretability of the results.

In CITE-seq, feature selection involves identifying both RNA and protein markers that provide complementary information about cellular identity and function. CITE-seq data combines scRNA-seq with surface protein expression data, enabling the simultaneous analysis of transcriptomic and proteomic features. Feature selection for CITE-seq typically includes the identification of highly expressed genes (from scRNA-seq) and cell surface proteins (from CITE-seq) that are both variable across cells and biologically relevant to the study's context. For scRNA-seq features in CITE-seq, the same variance-based methods used for single-cell RNA data are applied, with a focus on genes that exhibit cell-type-specific expression patterns or are known to be involved in key biological processes. For protein features, methods such as marker-based filtering or clustering approaches are used to identify surface proteins that distinguish cell types or states. The integration of both RNA and protein features can be achieved using advanced algorithms that combine data from both modalities, enhancing the resolution of cell-type identification and providing a more comprehensive view of cellular behavior.

6.5.2 Dimensionality Reduction

Dimensionality reduction is a crucial step in scRNA-seq analysis, enabling researchers to simplify high-dimensional data while retaining essential biological information. Single-cell datasets, often containing thousands of genes measured across thousands to millions of cells, are difficult to interpret directly, making dimensionality reduction essential for visualization and downstream analysis. There are two main approaches: linear and nonlinear methods. PCA is the most commonly used linear method in single-cell analysis. PCA reduces dimensionality by identifying the principal components, linear combinations of genes, that capture the largest variance in the data. By projecting the data onto these components, PCA retains the most informative features while significantly reducing the number of dimensions. PCA is often used as an intermediate step before applying clustering or other advanced analyses.

However, for capturing complex cellular relationships, nonlinear methods such as t-distributed stochastic neighbor embedding (t-SNE) and uniform manifold approximation and projection (UMAP) are more effective. t-SNE is widely used for visualizing high-dimensional data in two or three dimensions, particularly for distinguishing discrete cell populations. It excels at preserving local relationships, making it highly effective for identifying distinct clusters of cell types. However, t-SNE can distort global structures, meaning that relative distances between clusters may not always reflect true biological relationships. In addition, t-SNE is computationally intensive, especially for large datasets, and requires careful tuning of parameters such as perplexity to achieve optimal results. UMAP is a more recent technique that balances local and global structure preservation, making it ideal for both visualization and analysis of cellular hierarchies and transitions. Unlike t-SNE, UMAP can better capture continuous trajectories, such as differentiation pathways, and is more robust to parameter choices. It is also computationally efficient, scaling well to large single-cell datasets while producing clearer and more interpretable embeddings. Due to these advantages, UMAP has become a preferred method for single-cell research, particularly in applications requiring both cluster separation and an accurate representation of underlying biological gradients.

We have provided a detailed introduction to the algorithms behind PCA, t-SNE, and UMAP in Chapter 1. Readers may refer to Chapter 1 for an in-depth explanation. A comparison of PCA, t-SNE, and UMAP is presented in Table 6.1.

Table 6.1 Comparison and applications of PCA, t-SNE, and UMAP.

Method	Type	Preserves local structure	Preserves global structure	Speed	Common use cases
PCA	Linear	No	Yes	Fast	Preprocessing, clustering
t-SNE	Nonlinear	Yes	No	Slow	Visualization of clusters
UMAP	Nonlinear	Yes	Yes	Fast	Visualization, trajectory analysis

Both feature selection and dimensionality reduction are foundational to single-cell analysis workflows, enabling researchers to distill complex, high-dimensional data into manageable and interpretable forms while preserving key biological insights. These steps ensure that analyses remain both computationally efficient and biologically meaningful, facilitating discoveries about cellular heterogeneity, differentiation, and function.

6.6 Clustering and Cell-Type Identification

Clustering and cell-type identification are essential steps in scRNA-seq analysis. These steps enable researchers to group cells with similar transcriptomic profiles into clusters, often representing distinct cell types or states, and to annotate these clusters to uncover the underlying biology. This section introduces the clustering methods commonly used in single-cell analysis, tools for implementing these methods, and strategies for cell-type annotation.

6.6.1 Clustering Methods

Clustering methods play a crucial role in single-cell sequencing analysis by identifying groups of cells with similar gene expression patterns. Given the high-dimensional nature of single-cell data, these methods must account for sparsity, noise, and complex relationships between cells. To address these challenges, clustering approaches can be broadly categorized into graph-based and density-based methods, each designed to capture meaningful biological variation while maintaining computational efficiency.

Graph-based clustering is one of the most widely used techniques in scRNA-seq analysis. These methods construct a k-nearest neighbor (k-NN) graph, where nodes represent individual cells and edges are weighted based on transcriptional similarity. The resulting graph is then partitioned into distinct communities using clustering algorithms such as Louvain and Leiden algorithms. The Louvain algorithm [48] is a modularity-based clustering method that identifies densely connected subgraphs within the k-NN graph. It works iteratively in two phases: first, each node (cell) is assigned to its own cluster, and then nodes are merged into communities if doing so increases modularity, a measure of clustering quality. In the second phase, the detected communities are aggregated into "supernodes," and the process repeats until modularity gain stabilizes. The Louvain method is computationally efficient and scalable, making it a common choice in single-cell analysis tools such as Seurat [26] and Scanpy [27]. However, it has limitations, such as the potential for generating poorly connected clusters, especially in large datasets.

To address these limitations, the Leiden algorithm [49] was developed as an improved version of Louvain. Leiden clustering refines community detection in three key ways: (1) it ensures better initialization to prevent disconnected subclusters, (2) it applies a refined partitioning strategy to improve the consistency of clustering results, and (3) it introduces an additional refinement step to further optimize modularity. Due to these improvements, Leiden clustering has become the preferred approach for large-scale single-cell studies, ensuring more reliable and biologically meaningful clusters.

Unlike graph-based approaches, density-based clustering methods identify clusters by locating regions of high cell density separated by regions of lower density. These methods are particularly useful for identifying rare cell populations and irregularly shaped clusters. One of the most popular density-based clustering methods is Density-Based Spatial Clustering of Applications with Noise (DBSCAN) [50]. DBSCAN operates by defining clusters based on a minimum number of points (*MinPts*) that must be found within a specified radius ε. The algorithm follows these steps: (1) it selects a random cell and determines whether it has at least *MinPts* neighbors within distance ε, (2) if this condition is satisfied, the cell is designated as a core point, and a cluster is formed by expanding outward to include all connected core points, and (3) if a cell does not meet the *MinPts* requirement, it is considered noise. DBSCAN is highly effective for discovering heterogeneous cell populations and handling outliers but is sensitive to parameter tuning, which can impact cluster detection quality.

Several bioinformatic tools integrate clustering methods into single-cell analysis workflows, providing users with efficient computational frameworks and visualization options. Seurat [26] is an R-based toolkit widely used for scRNA-seq analysis. It supports graph-based clustering using Louvain or Leiden algorithms and provides intuitive visualization tools such as UMAP and t-SNE to explore clustering results. Scanpy [27] is a Python-based framework optimized for analyzing large single-cell datasets. It implements efficient graph-based clustering methods and offers seamless integration with dimensionality reduction and downstream analyses. Each clustering approach has distinct advantages, and the choice depends on dataset characteristics, computational requirements, and biological objectives. Combining multiple clustering strategies can improve the robustness of cell population identification, leading to more accurate biological interpretations in single-cell research.

6.6.2 Cell-Type Annotation

After clustering cells based on their gene expression profiles, the next critical step is cell-type annotation, which involves assigning biological identities to each cluster. This process translates computationally derived groups into meaningful biological entities, such as specific cell types or functional states. Annotation requires a combination of biological expertise and computational methods, including manual identification using marker genes, automated annotation tools, and reference databases.

Manual cell-type annotation relies on known marker genes, genes uniquely or highly expressed in specific cell types. By comparing the expression profiles of clusters to established markers, researchers can infer the identity of each cell population. For example, CD4 and CD8A are well-known markers distinguishing helper and cytotoxic T cells [51], respectively, while SOX2 and NANOG are key markers of pluripotent stem cells [52].

Reference databases provide curated gene expression signatures for known cell types, serving as valuable resources for both manual and automated annotations. These databases help standardize cell-type identification and improve the consistency of annotations across studies. CellMarker [53, 54] is a comprehensive database containing cell-type-specific marker genes across multiple species and tissues. Researchers can query this database to find validated markers for manual annotation or integrate it with computational tools for automated classification. PanglaoDB [55] is a widely used resource for human and mouse single-cell data, offering extensive coverage of known cell-type markers. It supports both exploratory marker searches and direct integration with automated annotation pipelines.

With the increasing scale and complexity of single-cell datasets, automated annotation tools have become essential for efficient and reproducible cell-type identification. These tools leverage machine learning and reference datasets to systematically assign cell identities. For example, SingleR [56] compares the transcriptome of each individual cell to curated reference datasets. It computes similarity scores between the input single-cell data and reference profiles, assigning the most likely cell type based on transcriptomic similarity. SingleR is widely used for its ability to work with multiple reference datasets and its straightforward application. CellTypist [57] is a machine learning–based annotation tool that utilizes curated cell-type reference models. CellTypist employs logistic regression classifiers trained on thousands of annotated single-cell profiles, providing rapid and scalable cell-type predictions. It also allows users to update models with custom reference data for enhanced accuracy.

The remarkable success of large language models (LLMs) in natural language processing (NLP) and computer vision has opened new avenues for tackling cell-type annotation in scRNA-seq data. Recently, several computational tools have emerged that leverage LLMs for automated and accurate cell-type identification [58–63]. One such tool, scBERT [62], enables unbiased, data-driven discovery of gene expression patterns and long-range dependencies for cell-type annotation. It achieves this by utilizing multi-head attention mechanisms to extract information from diverse representation subspaces. Unlike the traditional transformer encoder used in Bidirectional Encoder Representations from Transformers (BERT), single-cell Bidirectional Encoder Representations from Transformers (scBERT) employs Performer [64], a scalable variant of transformers designed to handle large datasets efficiently. In the fine-tuning step, scBERT applies a classifier to the pretrained Performer model, enabling high-precision cell-type annotation. Another LLM-based tool, single-cell Generative Pretrained Transformer (scGPT) [65], is built upon a generative pretrained foundation model to learn cell representations from diverse single-cell datasets. During pretraining, scGPT utilizes stacked transformer layers with multi-head attention, allowing it to simultaneously learn cell and gene embeddings. The model is trained in an autoregressive manner using zero-shot learning, where it first generates gene expression values from cell embeddings and gradually improves by leveraging prior knowledge of gene expression relationships. In the fine-tuning step, scGPT incorporates a supervised model to annotate unknown cells by introducing a multilayer perceptron (MLP)-based classifier into the generative pretrained framework. These advancements highlight the potential of LLMs in revolutionizing cell-type annotation, offering scalable, high-throughput, and context-aware methods for analyzing complex single-cell datasets.

Clustering and cell-type annotation are central to extracting biological insights from scRNA-seq data. By leveraging graph-based or density-based clustering methods, researchers group cells with similar expression patterns, while marker-based and automated annotation approaches refine these clusters into biologically meaningful identities. The integration of reference databases and machine learning models continues to enhance the accuracy and scalability of annotation. Ultimately, these steps enable researchers to investigate cellular diversity, developmental trajectories, and disease mechanisms, providing a deeper understanding of the biological systems at play.

6.7 Differential Expression Analysis

DE analysis is a cornerstone of scRNA-seq studies, enabling the identification of genes that vary significantly between groups of cells, such as cell clusters, conditions, or time points. These genes often serve as markers for cell types, states, or responses to perturbations. This section outlines the key methods, statistical models, and tools used in DE analysis for single-cell data.

6.7.1 Methods for Identifying Marker Genes or Differentially Expressed Genes

Marker genes or DEGs are identified by analyzing gene expression differences between groups of cells. This process helps in understanding cellular heterogeneity, identifying distinct cell types or states, and uncovering molecular mechanisms underlying biological processes. The identification of DEGs is essential for interpreting scRNA-seq data, as it enables researchers to link gene expression patterns to functional characteristics of cells. The specific comparisons made to identify DEGs depend on the study design and research objectives.

One common approach is to compare gene expression between different cell clusters. In this context, clusters represent groups of cells with similar transcriptomic profiles, typically corresponding to specific cell types or functional states. By identifying genes that distinguish one cluster from another, researchers can infer key molecular signatures associated with each cell type or state. These marker genes provide insights into lineage identity, functional specialization, and potential transitions between cellular states.

Another important comparison involves analyzing DEGs between experimental conditions, such as healthy versus diseased cells. This approach helps in uncovering molecular alterations associated with disease progression, therapeutic responses, or environmental influences. By identifying genes that are upregulated or downregulated in diseased cells, researchers can gain insights into pathophysiological mechanisms, potential biomarkers, and therapeutic targets.

Temporal and spatial comparisons also play a crucial role in DEG analysis. Temporal comparisons track gene expression changes over time, providing insights into dynamic processes such as cell differentiation, response to stimuli, or developmental transitions. Spatial comparisons, on the other hand, analyze gene expression variations across different regions of a tissue or organ, revealing spatially regulated genes that define tissue architecture, microenvironment interactions, or spatial gradients of signaling molecules.

To perform DE analysis in scRNA-seq data, two common approaches are employed: cluster-level comparisons and single-cell comparisons. In cluster-level comparisons, cells within a cluster are aggregated, and average expression values are compared across clusters. This approach helps mitigate technical noise and biological variability inherent to single-cell data. However, it may mask cellular heterogeneity within clusters, potentially overlooking rare subpopulations or intermediate cell states.

Single-cell comparisons, in contrast, analyze expression values at the level of individual cells while accounting for variability. This method provides higher resolution and sensitivity, enabling the detection of subtle transcriptional differences and rare cell populations. However, it is computationally intensive and requires sophisticated statistical models to handle sparsity, technical noise, and dropout effects in scRNA-seq data. Despite these challenges, single-cell comparisons are particularly valuable for studying dynamic cellular processes, lineage transitions, and plasticity in heterogeneous cell populations.

By integrating these approaches, researchers can achieve a comprehensive understanding of gene expression dynamics in single-cell data, ultimately contributing to advancements in developmental biology, disease research, and precision medicine.

6.7.2 Statistical Models for Single-Cell Data

The statistical analysis of scRNA-seq data is a complex task due to the unique characteristics of the data, including sparsity, high variability, and zero inflation. These features pose challenges to traditional statistical methods, making it essential to employ specialized models that can accurately capture the underlying biological signals while accounting for technical artifacts and noise. To achieve robust and reliable DE analysis, researchers rely on a variety of statistical models designed to handle these complexities.

One of the commonly used methods in scRNA-seq analysis is the Wilcoxon rank-sum test, a nonparametric test employed in tools such as Seurat [26] and Scanpy [27]. This test compares the ranks of gene expression values between two groups without making any assumptions about the distribution of the data. This is particularly advantageous in single-cell analyses where gene expression distributions can be highly skewed and standard parametric tests might not be suitable. The Wilcoxon rank-sum test is widely used in scRNA-seq because of its simplicity and robustness, making it a reliable choice for detecting DEGs between two distinct conditions or cell populations.

Negative binomial models are another class of statistical models frequently used in scRNA-seq analysis. These models are particularly effective at dealing with overdispersed count data, a common feature of single-cell transcriptomics. Overdispersion occurs when the variance in the data exceeds the mean, which is often observed in scRNA-seq datasets due to biological heterogeneity and technical variability. The negative binomial distribution is flexible enough to model this extra variability, making it well-suited for both bulk RNA-seq and single-cell RNA-seq applications. Tools such as DESeq2 [66] and edgeR [67] utilize negative binomial models to account for variability across cells and provide more accurate DE analysis by considering both the mean and variance of gene expression counts. These models help researchers identify genes whose expression significantly differs between conditions while also adjusting for the inherent overdispersion in the data.

Zero-inflated models are particularly important for handling the dropout events commonly encountered in scRNA-seq. Dropout refers to the phenomenon where some genes are not detected in certain cells, resulting in an excess of zeros in the data. Zero-inflated models address this issue by modeling two separate processes: one that determines whether a gene is expressed in a cell (i.e., whether the gene's count is non-zero) and another that models the level of gene expression given that the gene is detected. This dual modeling approach allows for more accurate interpretation of the data, particularly in the context of sparse datasets. Tools such as Model-based Analysis of Single-cell Transcriptomics (MAST) [68] use zero-inflated models to improve sensitivity and account for both technical and biological covariates that may influence gene expression. By incorporating these covariates, MAST enhances the detection of DEGs and allows for more nuanced analyses of gene expression patterns across different cell types or conditions.

Each of these statistical approaches, Wilcoxon rank-sum, negative binomial, and zero-inflated models, offers distinct advantages in the analysis of scRNA-seq data. The choice of model depends on the specific characteristics of the dataset, the research question at hand, and the computational resources available. By leveraging these specialized statistical methods, researchers can overcome the challenges posed by the complexities of scRNA-seq data, enabling more accurate and meaningful insights into gene expression dynamics at the single-cell level.

6.8 Basic Visualization Techniques

Visualization plays a critical role in scRNA-seq data analysis, enabling researchers to explore datasets, assess data quality, and interpret biological findings. Effective visualizations help uncover patterns, highlight differences between cell populations, and identify important genes or clusters. This section outlines fundamental visualization techniques commonly used in scRNA-seq analyses, focusing on QC metrics, dimensionality reduction results, and gene expression patterns.

6.8.1 Quality Control Metrics Visualization

Before conducting downstream analyses, it is essential to evaluate the quality of scRNA-seq data using various visualization techniques. Violin plots and scatter plots are commonly employed to examine QC metrics such as the number of detected genes per cell, total UMI counts, and the proportion of mitochondrial gene expression. Violin plots provide an overview of the distribution of these metrics across all cells, helping to identify low-quality cells that should be removed. Scatter plots allow researchers to assess relationships between QC metrics, such as the correlation between total RNA counts and the number of detected genes, which can indicate the presence of cell doublets or low-complexity cells. These visualizations ensure that only high-quality cells are retained for further analysis.

6.8.2 Dimensionality Reduction Visualization

Since scRNA-seq datasets contain thousands of genes per cell, dimensionality reduction techniques are used to summarize and visualize the data in a lower-dimensional space. PCA is a linear method that reduces data complexity while preserving variance, often used as an initial step for downstream analyses. However, nonlinear techniques such as UMAP and t-SNE are preferred for visualizing cell clusters. UMAP effectively preserves both global and local structure, making it a powerful tool for cluster separation, while t-SNE excels at capturing local relationships between cells. These techniques allow researchers to identify distinct cell populations, track cellular transitions, and assess batch effects or technical artifacts.

6.8.3 Gene Expression Visualization

To interpret biological insights from scRNA-seq data, gene expression visualization techniques such as heatmaps and dot plots are commonly used. Heatmaps display the expression levels of selected genes across different clusters or conditions, revealing marker genes that define specific cell types. They provide a high-resolution view of expression patterns across thousands of cells, aiding in cluster annotation and DE analysis. Dot plots, on the other hand, summarize gene expression across clusters by representing both the percentage of cells expressing a gene and its average expression level. These plots are particularly useful for comparing marker gene expression across multiple cell populations. Moreover, feature plots (e.g., scatter plots overlaying gene expression on UMAP or t-SNE projections) allow for the spatial visualization of gene expression patterns in the reduced dimensional space, helping to pinpoint genes associated with specific cell types or states.

By integrating these visualization techniques, researchers can gain deeper insights into scRNA-seq datasets, ensuring data quality, identifying meaningful biological patterns, and effectively communicating their findings. These methods serve as essential tools for interpreting complex single-cell data and guiding downstream analyses.

6.9 Conclusion

Single-cell sequencing technologies have revolutionized our ability to explore cellular heterogeneity, uncover regulatory mechanisms, and understand dynamic biological processes at an unprecedented resolution. This chapter has provided an overview of the fundamental steps involved in single-cell sequencing data analysis, covering preprocessing, feature selection, clustering, DE analysis, and visualization. By implementing rigorous computational approaches, researchers can extract meaningful insights from scRNA-seq, scATAC-seq, and CITE-seq data, leading to a deeper understanding of cellular identity, gene regulation, and disease mechanisms.

The preprocessing of single-cell sequencing data is a crucial step in ensuring high-quality downstream analysis. For scRNA-seq, preprocessing involves generating expression count matrices, QC, normalization, and batch effect removal, all of which contribute to reducing technical noise and improving data reliability. Similarly, scATAC-seq preprocessing focuses on QC, read alignment, peak calling, and normalization, which are essential for accurately mapping chromatin accessibility. CITE-seq preprocessing integrates RNA and protein expression data, requiring careful handling of multimodal data through QC, normalization, and batch correction to ensure seamless integration and meaningful interpretation.

Beyond preprocessing, feature selection and dimensionality reduction techniques enable efficient data exploration by identifying key genes or chromatin regions that drive cellular variation. Clustering methods facilitate the discovery of distinct cell populations, while cell-type annotation provides biological context by leveraging known marker genes. DE analysis allows researchers to identify genes associated with specific cell states, developmental transitions, or disease conditions, offering insights into molecular mechanisms underlying cellular behavior.

Visualization plays a pivotal role in interpreting single-cell sequencing data, providing intuitive representations of QC metrics, dimensionality reduction results, and gene expression patterns. Techniques such as violin plots, UMAP, t-SNE, and heatmaps help researchers assess data quality, explore cellular diversity, and communicate findings effectively. These visual tools are essential for validating computational analyses and guiding biological interpretations.

In conclusion, single-cell sequencing data analysis is a multifaceted process that requires careful preprocessing, robust computational methods, and effective visualization strategies. As single-cell technologies continue to evolve, integrating multi-omic data, improving computational scalability, and developing more sophisticated analysis frameworks will be critical for advancing our understanding of cellular complexity. The insights gained from single-cell studies hold immense potential for biomedical research, driving discoveries in developmental biology, immunology, cancer research, and regenerative medicine.

References

1. Hawrylycz, M.J., et al. An anatomically comprehensive atlas of the adult human brain transcriptome. *Nature*. 2012; **489**(7416): p. 391–399.
2. Ziegenhain, C., et al. Comparative analysis of single-cell RNA sequencing methods. *Mol Cell*. 2017; **65**(4): p. 631–643 e4.

3. Klein, A.M., et al. Droplet barcoding for single-cell transcriptomics applied to embryonic stem cells. *Cell*. 2015; **161**(5): p. 1187–1201.
4. Macosko, E.Z., et al. Highly parallel genome-wide expression profiling of individual cells using nanoliter droplets. *Cell*. 2015; **161**(5): p. 1202–1214.
5. Sant, P., K. Rippe, and J.P. Mallm. Approaches for single-cell RNA sequencing across tissues and cell types. *Transcription*. 2023; **14**(3–5): p. 127–145.
6. Ramskold, D., et al. Full-length mRNA-Seq from single-cell levels of RNA and individual circulating tumor cells. *Nat Biotechnol*. 2012; **30**(8): p. 777–782.
7. Picelli, S., et al. Full-length RNA-seq from single cells using Smart-seq2. *Nat Protoc*. 2014; **9**(1): p. 171–181.
8. Rosenberg, A.B., et al. Single-cell profiling of the developing mouse brain and spinal cord with split-pool barcoding. *Science*. 2018; **360**(6385): p. 176–182.
9. Martin, B.K., et al. Optimized single-nucleus transcriptional profiling by combinatorial indexing. *Nat Protoc*. 2023; **18**(1): p. 188–207.
10. Buenrostro, J.D., et al. Single-cell chromatin accessibility reveals principles of regulatory variation. *Nature*. 2015; **523**(7561): p. 486–490.
11. Cusanovich, D.A., et al. A single-cell atlas of in vivo mammalian chromatin accessibility. *Cell*. 2018; **174**(5): p. 1309–1324 e18.
12. Lareau, C.A., et al. Droplet-based combinatorial indexing for massive-scale single-cell chromatin accessibility. *Nat Biotechnol*. 2019; **37**(8): p. 916–924.
13. Stoeckius, M., et al. Simultaneous epitope and transcriptome measurement in single cells. *Nat Methods*. 2017; **14**(9): p. 865–868.
14. Douglas, G.M., et al. PICRUSt2 for prediction of metagenome functions. *Nat Biotechnol*. 2020; **38**(6): p. 685–688.
15. Ghandi, M., et al. Next-generation characterization of the cancer cell line encyclopedia. *Nature*. 2019; **569**(7757): p. 503–508.
16. Li, X. and Y. Zhang. A structural view of salicylic acid perception. *Nat Plants*. 2020; **6**(10): p. 1197–1198.
17. Bhatt, A.S. Making the microbiome personal. *Nat Med*. 2019; **25**(5): p. 708.
18. Nieuwdorp, M. Looking for the start of metabolic disease in the gut. *Nat Med*. 2019; **25**(5): p. 709.
19. Butler, A., et al. Integrating single-cell transcriptomic data across different conditions, technologies, and species. *Nat Biotechnol*. 2018; **36**(5): p. 411–420.
20. York, A. A short and long distance relationship. *Nat Rev Microbiol*. 2020; **18**(11): p. 603–603.
21. Sakata, R.C., et al. Base editors for simultaneous introduction of C-to-T and A-to-G mutations. *Nat Biotechnol*. 2020; **38**(7): p. 865–869.
22. Amezquita, R.A., et al. Orchestrating single-cell analysis with bioconductor. *Nat Methods*. 2020; **17**(2): p. 137–145.
23. Dobin, A., et al. STAR: ultrafast universal RNA-seq aligner. *Bioinformatics*. 2013; **29**(1): p. 15–21.
24. Kim, D., et al. TopHat2: accurate alignment of transcriptomes in the presence of insertions, deletions and gene fusions. *Genome Biol*. 2013; **14**(4): p. 1–13.
25. Zheng, G.X., et al. Massively parallel digital transcriptional profiling of single cells. *Nat Commun*. 2017; **8**: p. 14049.

26. Stuart, T., et al. Comprehensive integration of single-cell data. *Cell.* 2019. **177**(7): p. 1888–1902 e21.
27. Wolf, F.A., P. Angerer, and F.J. Theis. SCANPY: large-scale single-cell gene expression data analysis. *Genome Biol.* 2018; **19**(1): p. 1–5.
28. McCarthy, D.J., et al. Scater: pre-processing, quality control, normalization and visualization of single-cell RNA-seq data in R. *Bioinformatics.* 2017. **33**(8): p. 1179–1186.
29. Lun, A.T., D.J. McCarthy, and J.C. Marioni. A step-by-step workflow for low-level analysis of single-cell RNA-seq data with bioconductor. *F1000Res.* 2016; **5**: p. 2122.
30. Hafemeister, C. and R. Satija. Normalization and variance stabilization of single-cell RNA-seq data using regularized negative binomial regression. *Genome Biol.* 2019; **20**(1): p. 296.
31. Mereu, E., et al. Benchmarking single-cell RNA-sequencing protocols for cell atlas projects. *Nat Biotechnol.* 2020; **38**(6): p. 747–755.
32. Korsunsky, I., et al. Fast, sensitive and accurate integration of single-cell data with Harmony. *Nat Methods.* 2019; **16**(12): p. 1289–1296.
33. Andrews, S. FastQC: a quality control tool for high throughput sequence data. 2010. Available online at: http://www.bioinformatics.babraham.ac.uk/projects/fastqc.
34. Bolger, A.M., M. Lohse, and B. Usadel. Trimmomatic: a flexible trimmer for Illumina sequence data. *Bioinformatics.* 2014; **30**(15): p. 2114–2120.
35. Martin, M. Cutadapt removes adapter sequences from high-throughput sequencing reads. *EMBnet. journal.* 2011; **17**(1): p. 10–12.
36. Langmead, B. and S.L. Salzberg. Fast gapped-read alignment with Bowtie 2. *Nat Methods.* 2012. **9**(4): p. 357–359.
37. Li, H. and R. Durbin. Fast and accurate short read alignment with Burrows-Wheeler transform. *Bioinformatics.* 2009; **25**(14): p. 1754–1760.
38. Broad Institute. Picard toolkit. Broad Institute, GitHub repository. 2018. Available from: http://broadinstitute.github.io/picard/
39. Li, H., et al. The sequence alignment/Map format and SAMtools. *Bioinformatics*; 2009; **25**(16): p. 2078–2079.
40. Zhang, Y., et al. Model-based analysis of ChIP-Seq (MACS). *Genome Biol.* 2008; **9**(9): p. 1–9.
41. Fang, R., et al. Comprehensive analysis of single cell ATAC-seq data with SnapATAC. *Nat Commun.* 2021; **12**(1): p. 1337.
42. Granja, J.M., et al. ArchR is a scalable software package for integrative single-cell chromatin accessibility analysis. *Nat Genet.* 2021; **53**(3): p. 403–411.
43. Stuart, T., et al. Single-cell chromatin state analysis with Signac. *Nat Methods.* 2021; **18**(11): p. 1333–1341.
44. Roelli, P., et al. Hoohm/CITE-seq-Count: 1.4.2 (1.4.2). *Zenodo.* 2019. https://doi.org/10.5281/zenodo.2590196.
45. Gayoso, A., et al. Joint probabilistic modeling of single-cell multi-omic data with totalVI. *Nat Methods.* 2021; **18**(3): p. 272–282.
46. Argelaguet, R., et al. MOFA+: a statistical framework for comprehensive integration of multi-modal single-cell data. *Genome Biol*, 2020; **21**(1): p. 1–17.
47. Subramanian, A., et al. Gene set enrichment analysis: a knowledge-based approach for interpreting genome-wide expression profiles. *Proc Natl Acad Sci U S A.* 2005; **102**(43): p. 15545–15550.

48. Blondel, V.D., et al. Fast unfolding of communities in large networks. *J Stat Mech: Theory Exp*. 2008; **2008**(10): p. P10008.
49. Traag, V.A., L. Waltman. and N.J. van Eck, From Louvain to Leiden: guaranteeing well-connected communities. *Sci Rep*. 2019; **9**(1): p. 1–12.
50. Ester, M., et al. A density-based algorithm for discovering clusters in large spatial databases with noise. *In KDD*. 1996; **96**(34): p. 226–231.
51. Davis, M.M. and P.J. Bjorkman, T-cell antigen receptor genes and T-cell recognition. *Nature*. 1988; **334**(6181): p. 395–402.
52. Chambers, I. and A. Smith, Self-renewal of teratocarcinoma and embryonic stem cells. *Oncogene*. 2004; **23**(43): p. 7150–7160.
53. Hu, C., et al. CellMarker 2.0: an updated database of manually curated cell markers in human/mouse and web tools based on scRNA-seq data. *Nucleic Acids Res*. 2023; **51**(D1): p. D870–D876.
54. Zhang, X., et al. CellMarker: a manually curated resource of cell markers in human and mouse. *Nucleic Acids Res*. 2019; **47**(D1): p. D721–D728.
55. Franzen, O., L.M. Gan, and J.L.M. Bjorkegren. PanglaoDB: a web server for exploration of mouse and human single-cell RNA sequencing data. *Database* (*Oxford*). 2019. **2019**: p. 1–9.
56. Aran, D., et al. Reference-based analysis of lung single-cell sequencing reveals a transitional profibrotic macrophage. *Nat Immunol*. 2019; **20**(2): p. 163–172.
57. Dominguez Conde, C., et al. Cross-tissue immune cell analysis reveals tissue-specific features in humans. *Science*. 2022; **376**(6594): p. eabl5197.
58. Xu, J., et al. CIForm as a transformer-based model for cell-type annotation of large-scale single-cell RNA-seq data. *Brief Bioinform*. 2023; **24**(4): p. 1–11.
59. Chen, J., et al. Transformer for one stop interpretable cell type annotation. *Nat Commun*. 2023; **14**(1): p. 223.
60. Jiao, L., et al. scTransSort: transformers for intelligent annotation of cell types by gene embeddings. *Biomolecules*. 2023; **13**(4).
61. Song, T., et al. TransCluster: a cell-type identification method for single-cell RNA-Seq data using deep learning based on transformer. *Front Genet*. 2022; **13**: p. 1038919.
62. Yang, F., et al. scBERT as a large-scale pretrained deep language model for cell type annotation of single-cell RNA-seq data. *Nat Mach Intell*. 2022; **4**(10): p. 852–866.
63. Cui, H., et al. scGPT: toward building a foundation model for single-cell multi-omics using generative AI. *Nat Methods*. 2024; **21**(8): p. 1470–1480.
64. Choromanski, K., et al. Rethinking attention with performers. *arXiv preprint arXiv:2009.14794*. 2020.
65. Cui, H., et al. scGPT: towards building a foundation model for single-cell multi-omics using generative AI. *bioRxiv*. 2023: p. 2023.04. 30.538439.
66. Love, M.I., W. Huber, and S. Anders. Moderated estimation of fold change and dispersion for RNA-seq data with DESeq2. *Genome Biol*. 2014; **15**(12): p. 1–21.
67. Robinson, M.D., D.J. McCarthy, and G.K. Smyth. edgeR: a bioconductor package for differential expression analysis of digital gene expression data. *Bioinformatics*. 2010; **26**(1): p. 139–140.
68. Finak, G., et al. MAST: a flexible statistical framework for assessing transcriptional changes and characterizing heterogeneity in single-cell RNA sequencing data. *Genome Biol*. 2015; **16**: p. 1–13.

CHAPTER 7

Single-Cell Sequencing Data Analysis: Advanced Analysis

Advanced single-cell sequencing data analysis enables a deeper understanding of cellular heterogeneity, dynamic transitions, and regulatory mechanisms by leveraging sophisticated computational approaches. This chapter explores key methodologies that extend beyond basic clustering and differential expression analysis, addressing challenges such as data imputation, gene regulatory network (GRN) inference, trajectory and pseudotime analysis, cell cycle dynamics, and cell–cell communication.

7.1 Introduction

The fast advance of single-cell technologies allows for dissecting cellular heterogeneity at different omics layers with an unprecedented resolution, which boosts our understanding of complex biological systems or processes, including cancer, the immune system, and chronic diseases, and thus provides valuable insights for clinical and translational studies. Many protocols have been developed to quantify single-cell transcriptomes, and many computational methods have been developed to study single-cell omics data. In this chapter, we will focus on reviewing machine learning methods in single-cell data analysis. We will start with the preprocessing of single-cell RNA sequencing (scRNA-seq) data, such as data imputation and cross-platform batch effects removal, and then cell cycle and cell type identification. Following that, we will summarize the advanced data analysis, such as CNV estimate, single-cell pseudo-time trajectory analysis, phylogenetic tree inference, cell-cell interaction (CCI), and regulatory network inference. Finally, we will present the latest challenges, such as multi-omics integration and integrated analysis of scRNA-seq and spatial transcriptome.

7.1.1 Overview of Advanced Challenges in Single-Cell Analysis

Single-cell analysis has transformed the field of biology by enabling researchers to uncover cellular heterogeneity and complex biological processes at unprecedented resolution. However, the field faces several advanced challenges that require innovative solutions. One of the primary issues is the high dimensionality and sparsity of single-cell datasets. These datasets often contain thousands of genes across millions of cells, resulting in large, sparse matrices with many genes exhibiting zero expression in individual cells, which complicates downstream analyses.

Another major challenge is the presence of batch effects and technical variability introduced by differences in experimental protocols, sequencing platforms, or sample preparation methods. Disentangling these technical artifacts from true biological signals is critical to ensure meaningful insights. Additionally, scalability and computational complexity pose significant hurdles as the size of datasets continues to grow. Efficient algorithms and high-performance computing resources are necessary to process these data effectively.

Data integration across multiple omics layers, such as transcriptomics, epigenomics, and proteomics, represents another complex challenge. Aligning and analyzing multi-omics data in a biologically meaningful way requires advanced computational frameworks. Similarly, accurate cell type annotation, especially for novel or rare cell types, is hindered by the limited availability of comprehensive reference datasets and ambiguities in marker gene expression. Furthermore, understanding dynamic and temporal processes like cell differentiation or signaling pathways often relies on trajectory inference (TI), which is inherently difficult due to the snapshot nature of single-cell measurements. Finally, while advanced machine learning approaches, including deep learning, are increasingly applied to single-cell data, their lack of interpretability limits their utility in deriving biologically actionable insights.

7.1.2 Importance and Significance of Deeper Insights from Single-Cell Analysis

The ability to extract deeper insights from single-cell analysis is pivotal for advancing our understanding of biology and enabling new therapeutic strategies. Single-cell analysis allows for a detailed characterization of cellular heterogeneity within tissues, revealing rare cell types and distinct cellular states. Such insights are essential for understanding complex phenomena like disease progression, immune system dynamics, and tissue regeneration.

In developmental biology, single-cell analysis helps decode lineage trajectories and cell fate decisions, providing a comprehensive view of developmental pathways. This knowledge has direct implications for regenerative medicine and stem cell therapy. In the context of diseases, single-cell techniques uncover cell-specific alterations in gene expression, epigenetic regulation, and signaling pathways, elucidating the mechanisms underlying conditions such as cancer, neurodegenerative disorders, and autoimmune diseases. These findings can lead to the discovery of novel biomarkers and therapeutic targets.

Single-cell analysis also plays a critical role in precision medicine by enabling the identification of patient-specific cellular signatures and responses to treatment. This approach facilitates the development of tailored therapies, improving patient outcomes. Furthermore, the integration of multi-omics data provides a holistic view of cellular functions by exploring the interplay between transcription, chromatin accessibility, and protein expression. Such integration deepens our understanding of complex biological systems. We will explore in greater depth in Chapter 10.

Finally, addressing the challenges of single-cell analysis drives the development of advanced computational tools and methods, which improve scalability, accuracy, and interpretability. By overcoming these challenges and leveraging the deeper insights offered by single-cell analysis, researchers can unlock the complexities of cellular systems, ultimately transforming basic biology and translational medicine.

7.2 Single-Cell RNA Sequencing Data Imputation

scRNA-seq revolutionizes the traditional profiling of transcriptomes, making it able to fully characterize the transcriptomes of individual cells at unprecedented throughput. A major problem for scRNA-seq is the sparsity of the expression matrix with a tremendous number of zero values. Most of these zero or near-zero values are artificially caused by technical defects called dropouts [1]. A pressing need in scRNA-seq data analysis remains identifying and handling the dropout events that, otherwise, will severely hinder downstream analysis and attenuate the power of scRNA-seq on a wide range of biological and medical applications. Therefore, applying computational approaches to address problems of missingness and noise is very important and timely, particularly in light of the increasingly popular and large amount of scRNA-seq data.

7.2.1 Computational Methods for scRNA-seq Data Imputation

Several methods have been recently proposed to address the challenges resulting from excessive zero values in scRNA-seq. MAGIC [1] imputes missing expression values by sharing information across similar cells, based on the idea of heat diffusion. ScImpute [2] learns each gene's dropout probability in each cell and then imputes the dropout values, borrowing information from other similar cells selected based on the genes unlikely affected by dropout events. SAVER [3] borrows information across genes using a Bayesian approach to estimate unobserved true expression levels of genes. DrImpute [4] imputes dropouts by simply averaging the expression values of similar cells defined by clustering. VIPER [5] borrows information from a sparse set of local neighborhood cells of similar expression pattern to impute the expression measurements in the cell of interest based on nonnegative sparse regression models. Meanwhile, some other methods aim at the same goal by denoising the scRNA-seq data. DCA [6] uses a deep count autoencoder network to denoise scRNA-seq datasets by learning the count distribution, overdispersion, and sparsity of the data. ENHANCE [7] recovers denoised expression values based on principal component analysis (PCA) on raw scRNA-seq data. Another imputation method, DeepImpute [8], uses a deep neural network with dropout layers and loss functions to learn patterns in the data, allowing for scRNA-seq imputation.

While existing studies have adopted varying approaches for dropout imputation and yielded promising results, they either borrow information from similar cells or aggregate (co-expressed or similar) genes of the observed data, which will lead to over smoothing (e.g., MAGIC) and remove natural cell-to-cell stochasticity in gene expression (e.g., scImpute). Moreover, the imputation performance will be significantly reduced for rare cells, which have limited information and are common for many scRNA-seq studies. Alternatively, SCRABBLE [9] attempts to leverage bulk data as a constraint on matrix regularization to impute dropout events. However, most scRNA-seq studies often lack matched bulk RNA-seq data and thus limit its practicality. Additionally, due to the nontrivial distinction between true and false zero counts, imputation and denoising need to account for both the intra-cell-type dependence and inter-cell-type specificity. In view of the above concerns, a deep generative model would be a better choice to learn the true data distribution and then generate new data points with some variations, which are then independently used to impute the missing values and avoid overfitting.

Deep generative models have been widely used for missing value imputation in different fields [10–12]. Among deep generative models, generative adversarial networks (GANs) have evoked increasing interest in the computer vision community since its first introduction in 2014 [13]. GANs have become an active area of research with multiple variants developed [14–18] and hold promising in data imputation [19] because of its capability of learning and mimicking any distribution of data. Given the great success of GANs in inpainting, we hypothesize that similar deep neural net architectures could be used to impute dropouts in scRNA-seq data. In the following, we introduce scIGANs, a GAN-based framework designed for single-cell RNA-seq imputation.

7.2.2 Generative Adversarial Networks and Improved Wasserstein GANs

We here show that the GANs can be applied to scIGANs. The GANs training strategy is to define a game between two competing networks. The generator network maps a source of noise to the input space. The discriminator network receives either a generated sample or a true data sample and must distinguish between the two. The generator is trained to fool the discriminator. Formally, the game between the generator G and discriminator D is the minimax objective:

$$\min_G \max_D \mathop{\mathbb{E}}_{x\sim\mathbb{P}_r}[\log(D(x))] - \mathop{\mathbb{E}}_{\tilde{x}\sim\mathbb{P}_g}[\log(1-D(\tilde{x}))], \tag{7.1}$$

where D is the discriminator that can be any network, $\mathbb{P}_r$ is the real data distribution and $\mathbb{P}_g$ is the model distribution implicitly defined by $\tilde{x} = G(z)$, $z \sim p(z)$; G is the generator, which can be any network, z can be sampled from any noise distribution p, such as the uniform distribution or a spherical Gaussian distribution.

It is difficult to train the original GANs model since minimizing the objective function corresponds to minimizing the Jensen–Shannon divergence between $\mathbb{P}_r$ and $\mathbb{P}_g$, which is not continuous with respect to the generator's parameters. Earth-mover (Wasserstein-1) distance $W(q,p)$ is used to deal with such difficulty [20]. Such a model is called Wasserstein GANs (WGANs), in which the objective function is constructed as

$$\min_G \max_{D\in\mathcal{D}} \mathop{\mathbb{E}}_{x\sim\mathbb{P}_r}[D(x)] - \mathop{\mathbb{E}}_{\tilde{x}\sim\mathbb{P}_g}[D(\tilde{x})], \tag{7.2}$$

where $\mathcal{D}$ is the set of 1-Lipschitz function, the definition of other symbols are the same as the original GANs model. To enforce the Lipschitz constraint on the critic, one can clip the weights of the critic to lie within a compact space $[-c,c]$. The set of functions satisfying this constraint is a subset of the k-Lipschitz functions for some k which depends on c and the critic architecture. Researchers introduced an alternative way to enforce the Lipschitz constraint, usually called improved WGANs (IWGANs), which are widely used in training GAN models [21]. The objective is

$$\min_G \max_{D\in\mathcal{D}} \mathop{\mathbb{E}}_{\tilde{x}\sim\mathbb{P}_g}[D(\tilde{x})] - \mathop{\mathbb{E}}_{x\sim\mathbb{P}_r}[D(x)] + \lambda \mathop{\mathbb{E}}_{\hat{x}\sim\mathbb{P}_{\hat{x}}}[(\|\nabla_{\hat{x}}D(\hat{x})\|_2 - 1)^2], \tag{7.3}$$

where $\hat{x}$ is sampled from the straight lines between pairs of points sampled from the real data distribution and the generator distribution. λ is a predefined parameter. This method is used to train scIGANs [22], GANs for scRNA-seq data imputation.

7.2.3 The Generative Adversarial Networks for scRNA-seq Imputation

The basic idea is that scIGANs can learn the nonlinear gene–gene dependencies from complex, multicell-type samples and train a generative model to generate realistic expression profiles of defined cell types [23]. To train scIGANs, the real single-cell expression profiles are first reshaped to images and fed to GANs, wherein each cell corresponds to an image with the normalized gene expression representing the pixel (Fig. 7.1). The generator generates fake images by transforming a 100-dimensional latent variable into single-cell gene expression profiles. The discriminator evaluates whether the images are authentic or generated. These two networks are trained concurrently while competing against one another to improve the performance of both (Fig. 7.1).

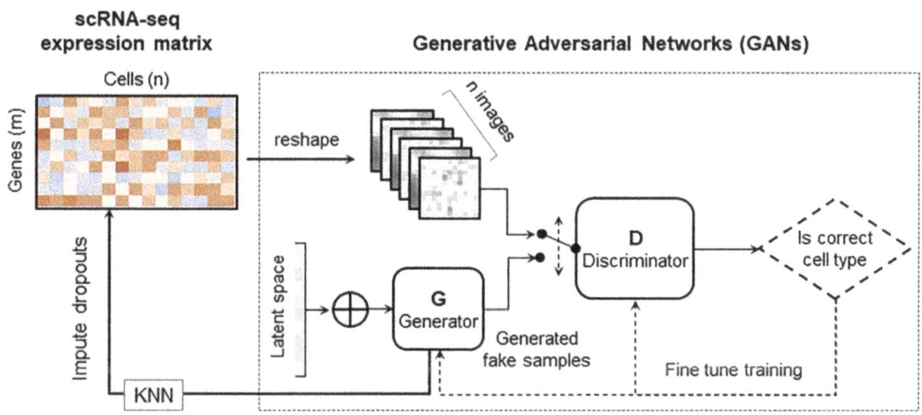

Fig. 7.1 Overview of scIGANs approach.

Once trained, the generative model is used to generate scRNA-seq data of defined cell types. And then we propose to infer the true expression of dropouts from the generated realistic cells. The most important benefit of using generated cells instead of the real cells for scIGANs is to avoid overfitting for the cell type of big population but insufficient power for rare cells. The generator can produce a set of cells of any number with the expression profiles faithfully characterizing the demand cell type; then the k-nearest neighbors (KNN) approach is used to impute the dropouts of the same cell type in the real scRNA-seq data.

The generator network of scIGANs is defined as $G(z, La_z; \theta)$. The inputs of the generator are: $z \sim \text{norm}(0,1)$, and label $La_z \sim U(1,9)$. Denote θ as the parameters need to be learned. The generator is defined in the following steps (Fig. 7.2):

1. Do transposed convolution on z with channels = 32*8, kernel size = 8, strides = 2, padding = 0, denote the result by DConv1_1.
2. Expand the dimension of La_z to (9,1,1), then do transposed convolution with channels = 32*8, kernel size = 8, strides = 2, padding = 0, denote the result by DConv1_2.
3. Concatenate z_n and La_n as Concat1, then do transposed convolution with channels = 32*4, kernel size = 4, strides = 2, padding = 1, denote the result by DConv2.
4. Do transposed convolution on DConv2 with channels = 1, kernel size = 4, strides = 2, padding = 1, and denote the result by Dconv3, which is the output of Generator.

The discriminator network is defined as $D(x, La_x; w)$. The inputs of the discriminator are samples of real data $x \sim \mathbb{P}_r$(or $\tilde{x} \sim \mathbb{P}_g$) representing the expression profile of an individual cell, and a label of x(or $\tilde{x}$) denoted by La_x representing the cell type or subpopulation. Denote w as the parameters need to be learned. The discriminator is defined in the following steps (Fig. 7.2):

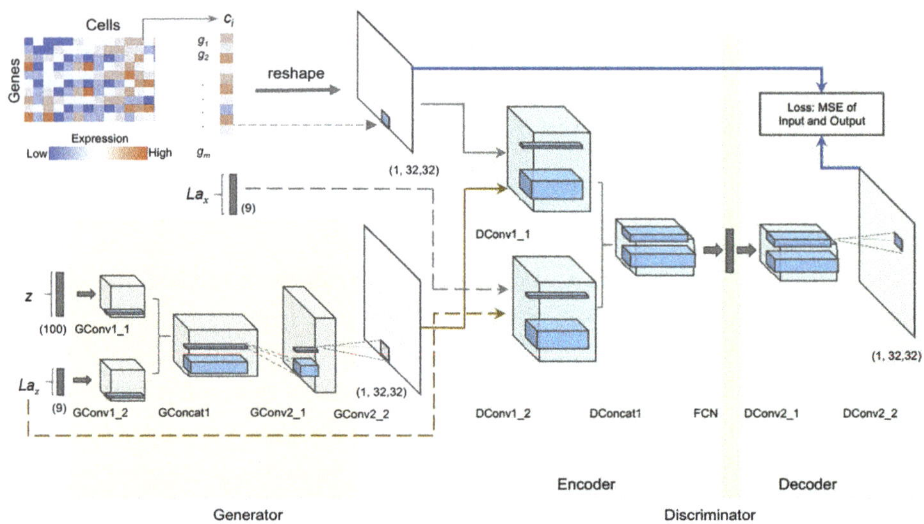

Fig. 7.2 The architecture of generative adversarial networks trained using the single-cell RNA-seq expression matrix.

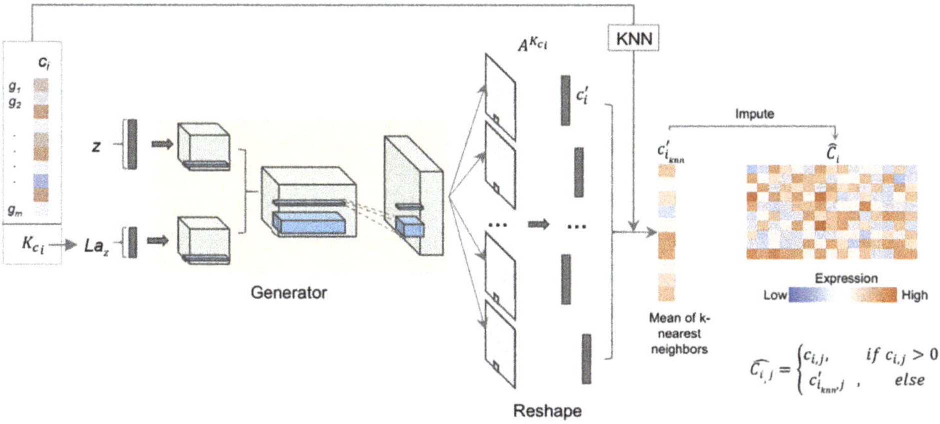

Fig. 7.3 Impute the scRNA-seq expression based on the generated expression matrix using the mean of k-nearest neighbors.

1. Do convolution on x or $\tilde{x}$ with channels = 32, kernel size = 8, strides = 4, padding = 2, denote the result by Conv1_1.
2. Do convolution on Conv1_1 with channels = 32*2, kernel size = 4, strides = 2, padding = 1, denote the result by Conv1_2.
3. Expand the dimension of La_x to (9,1,1), then repeat it to dimension (9,4,4), denote the result by Tile1.
4. Concatenate Conv1_2 and Tile1 as Concat1, then do convolution with channels = 1, kernel size = 4, strides = 2, padding = 0, denote the result by Conv2, which is the output of discriminator.

With a well-trained GANs model, for a given cell c_i which belongs to sub-population K_{ci}, we generate a candidate set $A^{K_{ci}}$ with n_{can} expression profiles. Denote c'_{iknn} as the k nearest neighbors using Euclidian distance in set $A^{K_{ci}}$. We then use the following equation to impute jth gene in cell c_i (Fig. 7.3):

$$\hat{c}_{i,j} = \begin{cases} c_{i,j}, & if\ c_{i,j} > 0 \\ c'_{iknn,j}, & else \end{cases}.$$

7.3 Gene Regulatory Network Inference

Studies have revealed that genes cannot work alone [24]. Instead, they constantly influence one another, which can be called epistasis [25], involving the interaction between two or more genes. These interplays are important for molecular regulation, signal transduction, biological networks, and lots of functional pathways [26]; therefore, network modeling of genes is significantly helpful for our knowledge about key regulators related to biochemical pathways. GRN [27] is a set of genes, or parts of genes, that interact with each other to control a specific cell function and has been extensively utilized based on graph model for functional analysis in recent years. GRN is essential to revealing questions of cellular identity [28] and has been demonstrated to play important roles in searching for disease-related biomarkers and drug position targets [29] and extensively utilized as an important tool for analyzing genomics data. Network modeling has long been employed as a powerful tool to understand and interpret complex biological systems [30–35]. Boolean models focus on discovering potential combinations of transcription factors (TFs) needed to transit from one state to another, including single-cell network synthesis (SCNS) toolkit [30] and BoolTraineR [31]. SCODE [32] is a network construction tool that relies on an ordinary differential equation (ODE) model to map differentiation in single-cell transcriptome data. Regression-based network modeling algorithms hold the premise that the expression of all genes can be attributed to a simple weighted linear equation. GENIE3 [33] adopts random forests to break down the prediction of a regulatory network using certain genes into the same different regression problems. In each regression problem, the expression count of one of the genes is predicted based on the expression patterns of all the other genes. GRNBoost2 [34] adopts a self-tuning algorithm, which uses gradient boosting rather than a global estimate for the decision trees in the boosting ensembles, based on GENIE3 architecture. SCINET [35] is a recently published computational method that detects direct regulatory interactions between TFs and targets by combining co-expression with motif enrichment analysis. The author revealed that cell types affect the interactions between disease-related genes, and marker genes show there exists higher cell-type-specific enrichment than those in the global network. SCENIC [36] is a correlation-based method that couples gene co-expression with TF-binding motif analyses of modules of co-expressed genes to identify GRN modules, predict TF regulators, and identify the single-cell level activity of putative TF targets. SINGE adopted kernel-based Granger Causality regression, which can be used to improve the irregular pseudotimes and dropout values using ordered single cells [37].

Boolean models explore genes relationship by binarizing the gene expression values, which have the benefit of fewer parameters and thus avoid overfitting. SCODE provides a command-line interface, which is convenient to conduct an analysis. GENIE3 was designed for bulk RNA-seq data, but it has been used for scRNA analysis, and it considers the interaction of a certain number of genes in one calculation and can estimate the nonlinear dependencies between genes. Compared to other tools, SCENIC provides instructions to select genes in the analysis. SINGE has better performance over other GRN inference methods to evaluate ChIP-seq and ChIP-chip data. Network modeling of scRNA-seq data represents one of the key tools to develop a new perspective on cellular biology and molecular mechanisms underlying pathophysiology and guide precision medicine, but different types of regulatory networks provide different information, and the key point for us is to extract conclusions among these types of networks from reasonable biological aspects.

7.4 Trajectory Inference and Pseudotime Analysis

Inferring the position of each individual cell on the lineage trajectory based on the scRNA-seq profiles is one of the promising applications of scRNA-seq. Dynamic processes such as cell cycle, cell differentiation, and cell activation [38, 39] can be modeled computationally using TI methods. The inferred trajectories can be cyclic, linear, bifurcating, tree-shaped, or disconnected graphs.

Numbers of TI methods have come out over the past few years. Most of them focus on correctly ordering the cells along a fixed trajectory. Monocle 2 [40] is a well-known tool developed by Trapnell *et al.* It uses DDRTree [41], a scalable reversed graph embedding algorithm, to find the projections between gene expression profiles level and lower-dimensional spaces, while simultaneously learning a principal tree on a population of single cells in this reduced space. Single-cell Trajectories Reconstruction, Exploration, and Mapping (STREAM) can accurately reconstruct complex trajectories and provide informative and intuitive visualizations of highlighted important genes along the inferred trajectories [42]. New cells can be mapped to the inferred trajectories without reconstruction as well.

Several other methods have also been proposed to infer the topology of scRNA-seq data [43–45]. These methods identify the order of cells along branches and obtain topological connections between these branches. Partitioned approximate graph abstraction (PAGA) was proposed to construct a KNN graph on cells and then perform a Louvain clustering algorithm to identify the membership of cells. Both continuous and disconnected structures are preserved in data at multiple resolutions [45]. One of the major updates in Monocle 3 [43] is the use of PAGA to automatically partition cells to learn disjoint or parallel trajectories. In addition, RNA velocity [46] is a concept for inferring cell RNA dynamics from the deviation of the observed ratio of spliced and unspliced mRNA from an inferred steady state.

There are several future challenges in TI that need to be taken into account. First, compared with the actual cell number of transcriptome analysis, many existing methods only allow the measurement of a very limited number of cells. Second, it is necessary to define the features to be used in constructing trajectories. Features with the same expression patterns usually retain important information of cells that belong to the same lineage. Third, there should be a definite evaluation method to compare the performance of different TI algorithms as previously described by Saelens *et al.*

7.4.1 Monocle

Monocle [40] is one of the most widely used computational frameworks for TI and pseudotime analysis in scRNA-seq. It reconstructs cellular differentiation pathways by ordering cells along a continuous trajectory based on their transcriptional profiles. This enables researchers to infer developmental processes, lineage relationships, and cell fate transitions from heterogeneous single-cell data.

Mathematical formulation of Monocle

The core idea of Monocle is to learn a low-dimensional representation of cells and arrange them in an optimal order to reflect their progression through a biological process. The TI pipeline in Monocle involves three main steps: (1) Monocle initially reduces the high-dimensional gene expression matrix X (cells × genes) into a lower-dimensional space using PCA or independent component analysis (ICA):

$$X' = W^T X, \tag{7.4}$$

where X' is the transformed matrix, and W^T is the projection matrix that captures the most variance in gene expression. (2) After dimensionality reduction, Monocle constructs a cell trajectory by modeling differentiation as a tree structure. It uses a minimum spanning tree (MST) on a KNN graph:

$$min \sum_{(i,j)\in E} d(i,j), \tag{7.5}$$

where $d(i,j)$ represents the distance between cells i and j, and the goal is to connect all cells with minimal total distance while maintaining a biologically plausible trajectory. (3) Monocle assigns each cell a pseudotime value, indicating its position along the trajectory. This is done by defining a root cell (often the most undifferentiated state) and computing the shortest path distance t_i from the root:

$$t_i = \min_{p \in P} \sum_{(j,k) \in P} d(j,k), \tag{7.6}$$

where P represents all possible paths in the MST from the root to cell i. This results in a continuous ordering of cells along the inferred trajectory.

7.4.2 RNA Velocity

RNA velocity is a powerful computational approach for inferring the future transcriptional state of a cell based on scRNA-seq data. Unlike traditional pseudotime methods, which only order cells based on their current gene expression profiles, RNA velocity predicts the direction and speed of cell state transitions by analyzing spliced and unspliced RNA transcripts. This method provides insights into cell fate decisions, differentiation trajectories, and dynamic regulatory processes, making it a valuable tool for understanding developmental biology, disease progression, and cellular plasticity.

RNA velocity relies on modeling the transcriptional process by examining the balance between newly transcribed unspliced pre-mRNA and mature spliced mRNA. The method assumes that gene expression follows a kinetic model governed by three key parameters: the transcription rate (α), the splicing rate (β), and the degradation rate (γ). The relationship between spliced (s) and unspliced (u) transcripts over time is described by the following differential equations:

$$\frac{ds}{dt} = \alpha - \beta s \tag{7.7}$$

$$\frac{du}{dt} = -\alpha u + \gamma s, \tag{7.8}$$

where s represents the spliced mRNA abundance, u represents the unspliced mRNA abundance, and the rates α, β, and γ govern the transcriptional dynamics of a given gene. By solving these equations, RNA velocity estimates whether a gene's expression is increasing or decreasing in a particular cell, ultimately predicting its future state.

The computational workflow for RNA velocity consists of three main steps. First, the method quantifies spliced and unspliced transcripts by classifying sequencing reads as intronic (unspliced) or exonic (spliced) using specialized tools like Velocyto. This produces two count matrices: U (unspliced) and S

(spliced), which serve as input for velocity estimation. Next, RNA velocity for each gene is calculated by assessing the steady-state ratio between unspliced and spliced transcripts using a linear regression model, where genes with an excess of unspliced transcripts are inferred to be upregulated, and those with reduced unspliced transcripts are inferred to be downregulated. Finally, RNA velocity vectors are projected onto a low-dimensional embedding, such as PCA, UMAP, or t-SNE, to visualize the predicted future trajectories of individual cells. The velocity vector for each cell (v_c) is computed as:

$$v_c = \sum_g w_g v_{g,c}, \tag{7.9}$$

where w_g is the weight of gene g, and $v_{g,c}$ is the velocity of gene g in cell c. This results in a vector field representation, where arrows indicate the predicted movement of cells within the transcriptional landscape.

By leveraging the dynamic nature of mRNA processing, RNA velocity provides a quantitative and predictive framework for studying cellular transitions, offering deeper insights into how gene expression evolves over time.

7.5 Cell Cycle Analysis

Cell cycle is a key component in the biological processes, which drive the transcriptional heterogeneity in cell differentiation [47], cell state, and oncogenesis [48, 49]. Studying the assignment of cell cycle phases is also of great significance to the occurrence, development, and treatment of tumors. Consequently, accurately identifying the cell cycle phases is the key to understand different biological processes [50].

At present, most research uses cell cycle information obtained from experiments, such as utilizing chemical induction [51], counterflow centrifugation elutriation [52], and DNA content [53] to investigate the enrichment of cells in different cell cycle phases. However, these methods have the disadvantages of complicated operation, low sensitivity, long experiment period, and may introduce biological batch effects.

7.5.1 Computational Methods to Predict Cell Cycle Phases

Several computational methods have been developed to analyze cyclic or circadian processes on the single-cell resolution, which can be roughly divided into continuous and discrete predictions of cell cycle phases. Continuous prediction gives the order of cells continuously distributed within each phase [54], suggesting that they could also be used to locate single cells along the circular cell cycle trajectory, which we called pseudo time in the cell cycle. Continuous assignment methods include Cyclum [55], Cyclops [56], Peco [57], and Oscope [58]. Cyclum and Cyclops use autoencoder, an unsupervised learning technique, to analyze a cell-gene expression matrix. Cyclum projects a single cell with high resolution onto a nonlinear periodic trajectory in low dimension, where the pseudo times are represented by radians in the range $[0, 2\pi]$ [55], by removing linear effects in the scRNA-seq data. Unlike Cyclum, Cyclops uses linear projection to project data onto a closed elliptical curve in low-dimensional space [56]. It employs square root and division in the autoencoder model, which makes optimization more complex. Peco is a supervised approach that uses the data generated from FUCCI fluorescence images and scRNA-seq to train the "naive Bayes" predictor for predicting the continuous cell cycle phase [57]. Oscope identifies oscillating genes and uses them to order single cells at different cell cycle phases [58]. Beyond the continuously quantitative description of cell cycle progress, classification of cells into discrete states is also applied in the study of cell cycle identification. Cyclone selects pairs of genes whose expression changes with the cell cycle phase and classifies single cells into specific cell cycle stages according to the selected marker gene pairs [50]. Liu *et al.* introduced cell cycle time-series as a consensus traveling salesman problem (TSP) to recover cell cycle pseudo time (ReCAT) from unsynchronized single-cell data. In their study, a hidden Markov model (HMM) based on Bayes scores and mean scores was designed to segment the pseudo times into G1, S, and G2/M [59].

7.5.2 Cell Cycle Pseudotime Estimation Algorithm

Overview of CCPE approach

scRNA-seq data is a cell-specific gene expression matrix with high dimensionality and sparsity. Traditional clustering methods have low efficiency for computing high-dimensional and sparse matrices. Therefore, it is necessary to introduce dimension reduction in the model. CCPE [60] is a novel cell cycle pseudotime estimation method to characterize cell cycle timing from scRNA-seq data. CCPE maps high-dimensional scRNA-seq data onto a helix in 3D space, where 2D space is used to capture the cycle information in scRNA-seq data and one dimension to predict the chronological orders of cells along the cell cycle, which we call cell cycle pseudotime. scRNA-seq data is repeatedly transformed from high-dimensional to low-dimensional and then mapped back to high-dimensional. At the same time, CCPE iteratively optimizes the discriminative dimensionality reduction via learning a helix until convergence (Fig. 7.4). CCPE is applied to several analyses and applications to demonstrate its ability to accurately estimate the cell cycle pseudotime and stages.

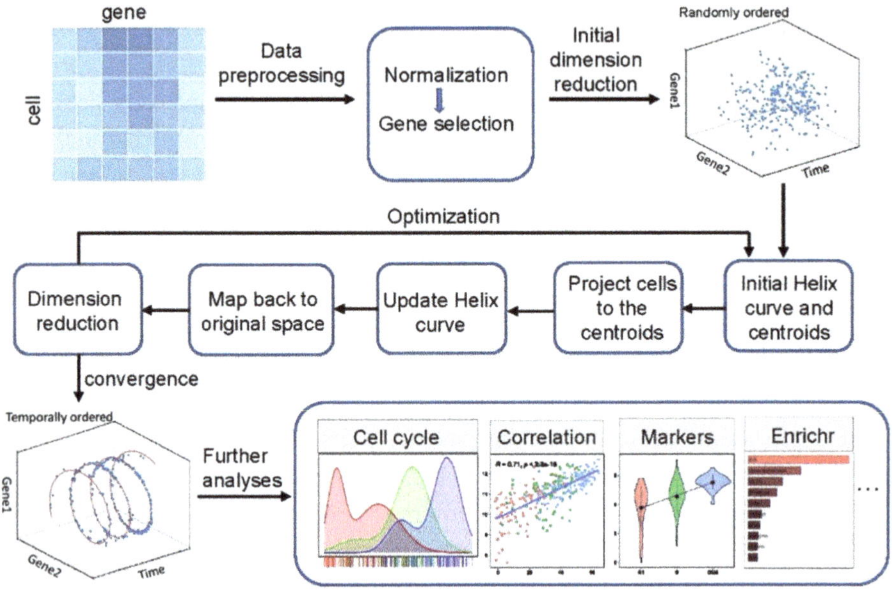

Fig. 7.4 Overview of the CCPE approach.

Construction of the CCPE model

Learning a helix in the reduced dimension. We suppose scRNA-seq data X with N cells and D genes lie in the high dimension. CCPE considers the linear projection $Z = f(X) = W^T X$ to infer the embedded expression profiles $Z_{d\times N}$ ($d \ll D$) from X and the reversed linear projection is $X = f^{-1}(Z) = WZ$, where d is the reduced dimension and $W \in R^{D\times d}$ with $W^T W = I$. We construct a circular helix $\widehat{Z} = f(x, y, z)$ in 3D dimension to get the best fit of Z. A circular helix of radius a and slope v/a (or pitch $2\pi v$) is described as follows for cell i:

$$\widehat{Z}_i = f_{Helix}(\widehat{x}_i, \widehat{y}_i, \widehat{z}_i) = \begin{cases} \widehat{x}_i = v\theta_i \\ \widehat{y}_i = a\sin\theta_i \\ \widehat{z}_i = a\cos\theta_i \end{cases}. \tag{7.10}$$

Then CCPE formulate the following object function to obtain the reduced dimension via learning a helix:

$$\min_{W,Z,\hat{Z}} \sum_{i=1}^{N} \|X_i - WZ_i\|^2 + \sum_{i=1}^{N} \|Z_i - \hat{Z}_i\|^2 \tag{7.11}$$

$$s.t.$$

$$W^T W = I$$

$$\hat{Z}_i = f_{Helix}(\hat{x}_i, \hat{y}_i, \hat{z}_i) = \begin{cases} \hat{x}_i = v\theta_i \\ \hat{y}_i = a \sin\theta_i \\ \hat{z}_i = a \cos\theta_i \end{cases}$$

$$a > 0,\ v > 0,$$

where $X = [X_1, X_2, \ldots, X_N] \in R^{D\times N}$ is the scRNA-seq data, $W = [W_1, W_2, \ldots, W_d] \in R^{D\times d}$ is an orthogonal set of d linear basis vectors $W_i \in R^D$, $Z = [Z_1, Z_2, \ldots, Z_N] \in R^{d\times N}$ is represented by the embedded expression profiles of X in low-dimension R^d. $\hat{Z}$contains fitted points of Z on a circular helix with the same dimension as Z and $\hat{x}_i$, $\hat{y}_i$, $\hat{z}_i$ is the coordinates of cell i projected on the circular helix.

Furthermore, cells in the same cell cycle phase should cluster together on the helix, so CCPE considers the clustering objective into the optimization problem as below:

$$\begin{aligned} \min_{W, Z, \hat{Z}, Y} & \sum_{i=1}^{N} \|X_i - WZ_i\|^2 + \lambda \sum_{i=1}^{N} \|Z_i - \hat{Z}_i\|^2 \\ & + \gamma \left[\sum_{k=1}^{K} \sum_{i=1}^{N_k} r_{i,k} \|Z_i - Y_k\|^2 + \sigma\Omega(R) \right] \end{aligned} \tag{7.12}$$

$$s.\ t.\ \ W^T W = I,\ \hat{Z}_i = f_{Helix},\ \sum_{k=1}^{K} r_{i,k} = 1,\ \ r_{i,k} \geq 0,\ \forall i,\ \forall k,$$

where N_k is the number of cells in cluster k. $r_{i,k}$ is the weight of soft clustering based on the assumption of K clusters. Y_k represents the centroid of cluster k. $\Omega(R)$ is the negative entropy regularization and $\sigma > 0$ is the regularization parameter for $\Omega(R)$. $\lambda > 0$, $\gamma > 0$ are parameters that indicate the importance of each component of the objective function. The solution of $r_{i,k}$ in terms of $\sum_{k=1}^{K} r_{i,k} = 1$ and formula of $\Omega(R)$ are described in Cun et al. [41] as the following

$$r_{i,k} = \exp\left(-\frac{\|Z_i - Y_k\|^2}{\sigma}\right) \Big/ \sum_{k=1}^{K} \exp\left(-\frac{\|Z_i - Y_k\|^2}{\sigma}\right) \tag{7.13}$$

$$\Omega(R) = \sum_{i=1}^{N} \sum_{k=1}^{K} r_{i,k} \log r_{i,k}. \tag{7.14}$$

Optimization of CCPE's objective function. The object function (7.12) is optimized using alternating structure optimization, which has been successfully applied to several optimization problems [61]. The parameters are divided into two parts $\{W, Z, Y\}$ and $\{\hat{Z}\}$ to be optimized, and CCPE solves one group by fixing the other group alternatively until convergence.

First, CPPE optimizes $\{W, Z, Y\}$ by fixing $\{\hat{Z}\}$. Given a known helix, CCPE can see $\hat{Z}$ as a constant matrix $C \in R^{d \times N}$. After simple matrix manipulation, function (7.12) with respect to $\{W, Z, Y\}$ can be rewritten as the following optimization problem:

$$\begin{aligned} \min_{W,Z,Y} tr[XX^T - 2W^T XZ^T + (1 + \lambda + \gamma)ZZ^T - \\ 2\lambda CZ^T + \lambda CC^T - 2\gamma R^T Z^T Y + \gamma Y\Gamma Y^T], \end{aligned} \tag{7.15}$$

where $\Gamma = diag(1^T R)$ and R is the weight matrix of soft clustering. Set L equals formula (7.15) and the first derivative of L with respect to Y to zero:

$$\frac{\partial L}{\partial Y} = -2\gamma ZR + 2\gamma Y\Gamma = 0. \tag{7.16}$$

Then we get the optimization of Y as:

$$Y = ZR\Gamma^{-1}. \tag{7.17}$$

Substituting Y into L and set the first derivative of L with respect to Z to zero, we can get the optimization of Z as:

$$Z = (W^T X + \lambda C)[(1 + \lambda + \gamma)I - \gamma R\Gamma^{-1} R^T]^{-1} = (W^T X + \lambda C)Q, \tag{7.18}$$

where $Q = [(1+\lambda+\gamma)I - \gamma R\Gamma^{-1}R^T]^{-1}$ and the inverse of $[(1+\lambda+\gamma)I - \gamma R\Gamma^{-1}R^T]$ exists. Similarly, substituting Z into L, the objective function becomes the following optimization problem:

$$L = \max_{W} tr\left(CQX^T W\right):\ W^T W = I. \tag{7.19}$$

This is the constrained quadratic problem, which has the closed-form solution [62] of W as follows:

$$W = VI_{D\times d}U^T, \tag{7.20}$$

where V is an $D \times D$ unitary matrix, U is an $d \times d$ unitary matrix, and $U\Sigma V^T$ is the singular value decomposition of matrix CQX^T, Σ is an $d \times D$ rectangular diagonal matrix with non-negative real numbers on the diagonal.

Second, given W, Z, and Y, we can obtain $\widehat{Z}$ easily by solving the following curve-fitting problem:

$$\min_{a,v,\theta} \sum_{i=1}^{N} \left\| Z_i - \widehat{Z}_i \right\|^2 :\ \widehat{Z}_i = f_{Helix}. \tag{7.21}$$

Overall, the optimization process of the problem (7.12) is given in Algorithm 1.

Algorithm 1

1. **Input:** scRNA-seq data X, parameters λ, σ and γ, number of clusters K.
2. Initialize Z and Y
3. **Repeat**
4. Obtain $\widehat{Z}$ by solving (7.21) via Helix fitting
5. Compute R with each element as (7.13)
6. $\Gamma = diag(1^{\mathrm{T}} R)$
7. $Q = [(1+\lambda+\gamma)I - \gamma R\Gamma^{-1}R^T]^{-1}$
8. $A = \widehat{Z}QX^T$
9. Perform SVD on A such that $A = U\Sigma V^T$

(Continued)

10. $W = VI_{D\times d}U^T$
11. $Z = (W^T X + \lambda \hat{Z})Q$
12. $Y = ZR\Gamma^{-1}$
13. **Until** Convergence

7.5.3 Strategy Development to Remove Cell Cycle Effects from scRNA-seq Data

scRNA sequencing made it possible to study heterogeneity in gene expression in high resolution. Such heterogeneity exists due to technical noise and different biological factors [63]. The cell cycle is a major source of bias, which introduces greater within-cell-type heterogeneity, causing quite different expression profiles between cell types [64]. Three methods have been developed to remove the cell cycle effect from their expression levels. The most popular ones are Seurat [65] and ccRemover [64]; both of them rely on known cell cycle genes. Seurat calculated cell cycle phase scores based on S and G2M cell cycle markers. Cells that do not express these markers may be in the G1 phase. Cell cycle effects will be removed during data normalization [65]. ccRemover is a PCA-based method that identifies components related to cell cycle with a larger component score by comparing with the control group. Thus, it can retain other factors while thoroughly removing cell cycle effect [64]. Cyclum is another method to be developed for removing the cell cycle effect. Cyclum removes the cell cycle effect by subtracting the reconstructed matrix of nonlinear components in the hidden layer from the expression level [55].

7.6 Cell-Cell Communication

Cell-cell interaction, also known as cell-cell communication, governs the functional activities of cells and coordinates multiple-cell actions [66]. The dynamic network formed through interaction and cooperation between cells with adjacent or remote partners plays crucial roles in numerous biological processes [67]. Gene expression data in many tissues and organs, such as brain [68, 69], heart [70], and lungs [71, 72], has revealed CCI plays an important role in organ function. Studying CCI within disease/tumors and their surrounding microenvironments can bring to light how cells communicate with their surroundings and help guide the development of effective treatment strategies [73]. CCI leverages diverse molecules, including ions fluxes, metabolites, secreted vesicles, and so on. A majority part of the interaction is mediated by secreted cell surface ligand and receptor [74]. The increasing development of single-cell sequencing technologies provides an unparalleled opportunity to infer the ligand-receptor (LR) interactions at a high-resolution cell state map.

Many statistical tools have been developed to perform such inferences [75–79]. iTALK [75] adopts a product score to deduce LR pairs within the highly expressed genes using public LR databases. The input of iTALK is a count expression matrix, with known cell type information. It has the ability to process multiple datasets and can take the batch effects and variability in LR expression across datasets. CellPhoneDB [76] predicts enriched LR interactions between two cell types based on the expression, taking subunit architecture for both ligands and receptors into consideration. The author generated a null distribution for each LR pair between cell types and obtained the likelihood of cell-type enrichment of each LR to predict enriched signaling interactions. NATMI [77] also uses the transcriptome profiling of every ligand and receptor in a single-cell dataset with known cell labels, and then predicts edges between any pair of cell types from a ligand sending cell to a receptor, and finally generates a cell connectivity summary network matrix. SingleSignalR [78] is an R package designed to perform LR interaction inference and visualization. The tool first integrates existing LR pairs databases as LRdb and computes the average gene signature expression across all the cells to get a regularized score of each LR pair, which depends on external specific information to the cell types. CellChat is also an R package developed to infer, visualize, and analyze CCI using scRNA-seq data [79]. It not only contains the multi-subunit structure of LR, but also includes stimulatory and inhibitory cofactors, resulting a more comprehensive database compared to other popular tools. A mass action-based model is applied to infer the probability of interaction between a LR, which takes the percentage of cells in each group among all the cells into consideration.

Each method has its own advantage over others. iTALK can be applied to several types of datasets, including time series data and different platform data, but it does not propose any cutoff for the scores of the LR interactions. CellPhoneDB has the outweighing that it provides online analysis and considers the heteromeric composition of the ligands and receptors. NATMI is a network-based tool used to estimate which cell type pairs are most likely to communicate. The regularized score of SingleSignalR has better performance in controlling false positives over other tools. Beyond the purely fundamental research enterprise of interpreting the cell–cell biological messages, CellChat can also be used to compare communication networks in different states of an organ. With the increasing studies to explore potential mechanism and usage of CCI, we are looking forward to more CCI-related occurrences and progress of a disease, and also drug treatment and resistance will be discovered and verified.

7.7 Conclusion

Advancements in single-cell sequencing data analysis have revolutionized our ability to explore cellular heterogeneity, gene regulation, and dynamic biological processes at an unprecedented resolution. This chapter covered several key aspects of advanced single-cell analysis, including data imputation, GRN inference, TI, cell cycle analysis, and cell-cell communication. These methods go beyond basic clustering and differential expression analysis, enabling deeper insights into cellular states, transitions, and regulatory mechanisms.

One of the fundamental challenges in scRNA-seq analysis is the presence of technical noise and dropout events, which obscure true biological signals. To address this, computational methods for data imputation, including GANs and IWGANs, have been developed. The introduction of scIGANs, a specialized GAN-based framework for scIGANs, has significantly improved the accuracy of gene expression recovery, thereby enhancing downstream analyses such as clustering and TI.

Understanding the regulatory mechanisms governing cellular function requires GRN inference. Advanced computational approaches leverage single-cell multi-omics data to reconstruct transcriptional regulatory interactions, offering insights into how gene expression is controlled at the single-cell level. These inferred networks are crucial for identifying key transcription factors and regulatory circuits involved in cellular differentiation and disease progression.

TI and pseudotime analysis provide a means to reconstruct dynamic biological processes, such as cellular differentiation, immune activation, and reprogramming. By modeling continuous gene expression changes, TI algorithms allow researchers to infer lineage relationships and identify key transition points between cellular states. These methods have been instrumental in developmental biology and cancer research, shedding light on how cells progress through differentiation or acquire malignant properties.

The cell cycle plays a central role in cell fate decisions and proliferation dynamics. Computational approaches for cell cycle analysis facilitate the identification of cell cycle phases, estimation of pseudotime along the cycle, and removal of cell cycle effects from gene expression data to prevent confounding biological interpretations. The development of the CCPE algorithm further enhances our ability to analyze cell cycle progression in single-cell datasets, aiding in the study of proliferation-associated processes in development and disease.

Finally, cell-cell communication analysis is essential for understanding how cells interact within complex tissues and microenvironments. Recent advancements in computational frameworks have enabled the prediction of LR interactions at the single-cell level, providing insights into intercellular signaling networks. These analyses are particularly valuable for studying immune responses, tumor microenvironments, and tissue development.

In conclusion, the field of single-cell analysis continues to evolve rapidly, driven by methodological innovations that address technical challenges and expand our understanding of cellular systems. As computational approaches become more sophisticated and integrative, they will unlock even deeper insights into cellular behavior, gene regulation, and tissue dynamics, ultimately contributing to breakthroughs in developmental biology, immunology, and precision medicine.

References

1. van Dijk, D., et al. Recovering gene interactions from single-cell data using data diffusion. *Cell*. 2018; **174**(3): p. 716–729 e27.
2. Li, W.V. and J.J. Li. An accurate and robust imputation method scImpute for single-cell RNA-seq data. *Nat Commun*. 2018; **9**(1): p. 997.
3. Huang, M., et al. SAVER: gene expression recovery for single-cell RNA sequencing. *Nat Methods*. 2018; **15**(7): p. 539–542.
4. Gong, W., et al. DrImpute: imputing dropout events in single cell RNA sequencing data. *BMC Bioinformatics*. 2018; **19**(1): p. 220.
5. Chen, M. and X. Zhou. VIPER: variability-preserving imputation for accurate gene expression recovery in single-cell RNA sequencing studies. *Genome Biol*. 2018; **19**(1): p. 196.
6. Eraslan, G., et al. Single-cell RNA-seq denoising using a deep count autoencoder. *Nat Commun*. 2019; **10**(1): p. 390.
7. Wagner, F., D. Barkley. and I. Yanai. Accurate denoising of single-cell RNA-Seq data using unbiased principal component analysis. *bioRxiv*. 2019: p. 655365.
8. Arisdakessian, C., et al. DeepImpute: an accurate, fast, and scalable deep neural network method to impute single-cell RNA-seq data. *Genome Biol*. 2019; **20**(1): p. 211.
9. Peng, T., et al. SCRABBLE: single-cell RNA-seq imputation constrained by bulk RNA-seq data. *Genome Biol*. 2019; **20**(1): p. 88.
10. Mattei, P.-A. and J. Frellsen. missIWAE: deep generative modelling and imputation of incomplete data. In Third workshop on Bayesian deep learning (NeurIPS 2018). 2018.

11. Zhang, H., P. Xie. and E. Xing. Missing value imputation based on deep generative models. arXiv preprint arXiv:1808.01684. 2018.
12. Mattei, P.-A. and J. Frellsen. MIWAE: deep generative modelling and imputation of incomplete data sets. in International Conference on Machine Learning. 2019. PMLR.
13. Goodfellow, I., et al. Advances in neural information processing systems. *Curran Associates Inc*. 2014; **27**: p. 2672–2680.
14. Gulrajani, I., et al. Improved training of Wasserstein GANs. *Adv Neural Inf Process Syst*. 2017; **30**: p. 5767–5777.
15. Ghahramani, A., F.M. Watt. and N.M. Luscombe. Generative adversarial networks simulate gene expression and predict perturbations in single cells. *bioRxiv*. 2018: p. 262501.
16. Miyato, T., et al. Spectral normalization for generative adversarial networks. arXiv preprint arXiv:1802.05957. 2018.
17. Sutskever, I., O. Vinyals, and Q. V. Le. Sequence to sequence learning with neural networks. *Adv Neural Inf Process Syst*. 2014; **27**.
18. Radford, A. Unsupervised representation learning with deep convolutional generative adversarial networks. arXiv preprint arXiv:1511.06434. 2015.
19. Yoon, J., J. Jordon. and M. Schaar. Gain: missing data imputation using generative adversarial nets. In International conference on machine learning. 2018. PMLR.
20. Yang, Q., et al. Low-dose CT image denoising using a generative adversarial network with Wasserstein distance and perceptual loss. *IEEE Trans Med Imaging*. 2018; **37**(6): p. 1348–1357.
21. Gulrajani, I., et al. Improved training of Wasserstein GANs. arXiv. 2017.
22. Xu, Y., et al. scIGANs: single-cell RNA-seq imputation using generative adversarial networks. *Nucleic Acids Res*. 2020; **48**(15): p. e85.
23. Marouf, M., et al. Realistic in silico generation and augmentation of single-cell RNA-seq data using generative adversarial networks. *Nat Commun*. 2020; **11**(1): p. 166.
24. Cordell, H.J. Detecting gene–gene interactions that underlie human diseases. *Nat Rev Genet*. 2009; **10**(6): p. 392–404.
25. Phillips, P.C. Epistasis—the essential role of gene interactions in the structure and evolution of genetic systems. *Nat Rev Genet*. 2008; **9**(11): p. 855–867.
26. Harvey, K.F., X. Zhang. and D.M. Thomas. The Hippo pathway and human cancer. *Nat Rev Cancer*. 2013; **13**(4): p. 246–257.
27. Fiers, M., et al. Mapping gene regulatory networks from single-cell omics data. *Brief Funct Genomics*. 2018; **17**(4): p. 246–254.
28. Han, L., et al. Single cell transcriptomics identifies a signaling network coordinating endoderm and mesoderm diversification during foregut organogenesis. *Nat Commun*. 2020; **11**(1): p. 4158.
29. Cha, J. and I. Lee. Single-cell network biology for resolving cellular heterogeneity in human diseases. *Exp Mol Med*. 2020; **52**(11): p. 1798–1808.
30. Woodhouse, S., et al. SCNS: a graphical tool for reconstructing executable regulatory networks from single-cell genomic data. *BMC Syst Biol*. 2018; **12**(1): p. 59.
31. Lim, C.Y., et al. BTR: training asynchronous Boolean models using single-cell expression data. *BMC Bioinformatics*. 2016; **17**(1): p. 355.
32. Matsumoto, H., et al. SCODE: an efficient regulatory network inference algorithm from single-cell RNA-Seq during differentiation. *Bioinformatics*. 2017; **33**(15): p. 2314–2321.

33. Huynh-Thu, V.A., et al. Inferring regulatory networks from expression data using tree-based methods. *PLOS One*. 2010; **5**(9).
34. Moerman, T., et al. GRNBoost2 and Arboreto: efficient and scalable inference of gene regulatory networks. *Bioinformatics*. 2019; **35**(12): p. 2159–2161.
35. Mohammadi, S., J. Davila-Velderrain. and M. Kellis. Reconstruction of cell-type-specific interactomes at single-cell resolution. *Cell Syst*. 2019; **9**(6): p. 559–568 e4.
36. Aibar, S., et al. SCENIC: single-cell regulatory network inference and clustering. *Nat Methods*. 2017; **14**(11): p. 1083–1086.
37. Deshpande, A., et al. Network inference with granger causality ensembles on single-cell transcriptomic data. *bioRxiv*. 2021.
38. Etzrodt, M., M. Endele. and T. Schroeder. Quantitative single-cell approaches to stem cell research. *Cell Stem Cell*. 2014; **15**(5): p. 546–558.
39. Tanay, A. and A. Regev. Scaling single-cell genomics from phenomenology to mechanism. *Nature*. 2017; **541**(7637): p. 331–338.
40. Qiu, X., et al. Reversed graph embedding resolves complex single-cell trajectories. *Nat Methods*. 2017; **14**(10): p. 979–982.
41. Mao, Q., et al. Dimensionality reduction via graph structure learning. In Proceedings of the 21th ACM SIGKDD International Conference on Knowledge Discovery and Data Mining. 2015.
42. Chen, H., et al. Single-cell trajectories reconstruction, exploration and mapping of omics data with STREAM. *Nat Commun*. 2019; **10**(1): p. 1903.
43. Cao, J., et al. The single-cell transcriptional landscape of mammalian organogenesis. *Nature*. 2019; **566**(7745): p. 496–502.
44. Street, K., et al. Slingshot: cell lineage and pseudotime inference for single-cell transcriptomics. *BMC Genomics*. 2018; **19**(1): p. 477.
45. Wolf, F.A., et al. PAGA: Graph abstraction reconciles clustering with trajectory inference through a topology preserving map of single cells. *Genome Biol*. 2019; **20**(1): p. 59.
46. La Manno, G., et al. RNA velocity of single cells. *Nature*. 2018; **560**(7719): p. 494–498.
47. Pauklin, S. and L. Vallier. The cell-cycle state of stem cells determines cell fate propensity. *Cell*. 2013; **155**(1): p. 135–147.
48. Bar-Joseph, Z., et al. Genome-wide transcriptional analysis of the human cell cycle identifies genes differentially regulated in normal and cancer cells. *Proc Natl Acad Sci U S A*. 2008; **105**(3): p. 955–960.
49. Kastan, M.B. and J. Bartek. Cell-cycle checkpoints and cancer. *Nature*. 2004; **432**(7015): p. 316–323.
50. Scialdone, A., et al. Computational assignment of cell-cycle stage from single-cell transcriptome data. *Methods*. 2015; **85**: p. 54–61.
51. Vassilev, L.T. Cell cycle synchronization at the G2/M phase border by reversible inhibition of CDK1. *Cell Cycle*. 2006; **5**(22): p. 2555–2666.
52. Ly, T., et al. A proteomic chronology of gene expression through the cell cycle in human myeloid leukemia cells. *Elife*. 2014; **3**: p. e01630.
53. Sasagawa, Y., et al. Quartz-Seq: a highly reproducible and sensitive single-cell RNA sequencing method, reveals non-genetic gene-expression heterogeneity. *Genome Biol*. 2013; **14**(4): p. R31.

54. Sakaue-Sawano, A., et al. Visualizing spatiotemporal dynamics of multicellular cell-cycle progression. *Cell*. 2008; **132**(3): p. 487–498.
55. Liang, S., et al. Latent periodic process inference from single-cell RNA-seq data. *Nat Commun*. 2020; **11**(1): p. 1441.
56. Anafi, R.C., et al. CYCLOPS reveals human transcriptional rhythms in health and disease. *Proc Natl Acad Sci U S A*. 2017; **114**(20): p. 5312–5317.
57. Hsiao, C.J., et al. Characterizing and inferring quantitative cell cycle phase in single-cell RNA-seq data analysis. *Genome Res*. 2020; **30**(4): p. 611–621.
58. Leng, N., et al. Oscope identifies oscillatory genes in unsynchronized single-cell RNA-seq experiments. *Nat Methods*. 2015; **12**(10): p. 947–950.
59. Liu, Z., et al. Reconstructing cell cycle pseudo time-series via single-cell transcriptome data. *Nat Commun*. 2017; **8**(1): p. 22.
60. Liu, J., Yang, M., Zhao, W., Zhou, X. CCPE: cell cycle pseudotime estimation for single cell RNA-seq data. *Nucleic Acids Res*. 2022; **50**(2): pp. 704–716.
61. Ando, R. K., T. Zhang, and P. Bartlett. A framework for learning predictive structures from multiple tasks and unlabeled data. *J Mach Learn Res*. 2005; **6**(11): p. 1817–1853.
62. Lai, R. and S. Osher. A splitting method for orthogonality constrained problems. *J Sc Comput*. 2014; **58**(2): p. 431–449.
63. Buettner, F., et al. f-scLVM: scalable and versatile factor analysis for single-cell RNA-seq. *Genome Biol*. 2017; **18**(1): p. 212.
64. Barron, M. and J. Li. Identifying and removing the cell-cycle effect from single-cell RNA-sequencing data. *Sci Rep*. 2016; **6**: p. 33892.
65. Butler, A., et al. Integrating single-cell transcriptomic data across different conditions, technologies, and species. *Nat Biotechnol*. 2018; **36**(5): p. 411–420.
66. SINGER, S.J., Intercellular communication and cell–cell adhesion. *Science*. 1992; **225**(5052): p. 1671–1677.
67. Camp, J.G., et al. Multilineage communication regulates human liver bud development from pluripotency. *Nature*. 2017; **546**(7659): p. 533–538.
68. Yuzwa, S.A., et al. Proneurogenic ligands defined by modeling developing cortex growth factor communication networks. *Neuron*. 2016; **91**(5): p. 988–1004.
69. Sheikh, B.N., et al. Systematic identification of cell–cell communication networks in the developing brain. *iScience*. 2019; **21**: p. 273–287.
70. Skelly, D.A., et al. Single-cell transcriptional profiling reveals cellular diversity and intercommunication in the mouse heart. *Cell Rep*. 2018. **22**(3): p. 600–610.
71. Zepp, J.A., et al. Distinct mesenchymal lineages and niches promote epithelial self-renewal and myofibrogenesis in the lung. *Cell*. 2017; **170**(6): p. 1134–1148 e10.
72. Cohen, M., et al. Lung single-cell signaling interaction map reveals basophil role in macrophage imprinting. *Cell*. 2018; **175**(4): p. 1031–1044 e18.
73. Kumar, M.P., et al. Analysis of single-cell RNA-seq identifies cell–cell communication associated with tumor characteristics. *Cell Rep*. 2018; **25**(6): p. 1458–1468 e4.
74. Armingol, E., et al. Deciphering cell–cell interactions and communication from gene expression. *Nat Rev Genet*. 2021; **22**(2): p. 71–88.
75. Wang, Y., et al. iTALK: an R Package to characterize and illustrate intercellular communication. *bioRxiv*, 2019.

76. Efremova, M., et al. CellPhoneDB: inferring cell–cell communication from combined expression of multi-subunit ligand-receptor complexes. *Nat Protoc*. 2020; **15**(4): p. 1484–1506.
77. Hou, R., et al. Predicting cell-to-cell communication networks using NATMI. *Nat Commun*. 2020; **11**(1): p. 5011.
78. Cabello-Aguilar, S., et al. SingleCellSignalR: inference of intercellular networks from single-cell transcriptomics. *Nucleic Acids Res*. 2020; **48**(10): p. e55.
79. Jin, S., et al. Inference and analysis of cell–cell communication using CellChat. *Nat Commun*. 2021; **12**(1): p. 1088.

CHAPTER 8

Spatial Transcriptomics Data Analysis

8.1 Introduction to Spatial Transcriptomics

8.1.1 Background on Spatial Transcriptomics

The ability to capture gene expression patterns at single-cell resolution has transformed our understanding of cellular diversity, functional states, and developmental processes. However, traditional transcriptomics approaches, such as bulk RNA sequencing and single-cell RNA sequencing (scRNA-seq), typically lack spatial information, isolating cells from their native microenvironments [1–4]. This spatial context is critical in understanding how cells function within their tissue or organ structures. Spatial transcriptomics (ST) addresses this gap by enabling high-throughput, genome-wide expression profiling while preserving the spatial coordinates of transcripts in situ. By combining spatial resolution with transcriptomic data, ST provides a multidimensional view of cellular function and organization that aligns molecular profiles with histological structures [5, 6].

Since its introduction, ST has rapidly evolved with the development of diverse methodologies and platforms. High-resolution techniques, such as single-molecule fluorescence in situ hybridization (smFISH) and multiplexed error-robust FISH (MERFISH), allow visualization of specific gene expressions within tissues, while sequencing-based approaches, like Slide-seq and Visium, map global gene expression with spatial coordinates. These technologies vary in resolution, target specificity, and scalability, catering to different experimental needs in the research landscape.

The spatial context of gene expression is essential for understanding the functional dynamics within multicellular systems. Cells are not isolated entities; they interact with neighboring cells and respond to microenvironmental cues, which collectively drive tissue-specific functions, cellular differentiation, and tissue remodeling. By integrating spatial context, ST enables researchers to study cell-cell communication, tissue architecture, and local microenvironments in unprecedented detail [7–10]. This is particularly crucial for identifying rare cell types or unique subpopulations that might be overlooked by traditional methods, as well as for understanding how these cells influence their surroundings.

8.1.2 Applications and Significance in Biological Research

The applications of ST span various fields, including developmental biology, oncology, neuroscience, and immunology, each benefiting from the ability to map cellular heterogeneity in spatially structured tissues. In cancer research, ST enables the exploration of tumor microenvironments, allowing researchers to dissect cellular interactions between tumor cells and immune cells, track the spatial spread of cancerous cells, and study drug responses in a tissue-specific manner [11–14]. The ability to map spatial gene expression also aids in understanding tumor heterogeneity, a major factor in cancer progression and therapy resistance.

In developmental biology, ST can reveal the orchestrated gene expression patterns underlying tissue morphogenesis and organogenesis, which are inherently spatial processes. Similarly, in disease contexts, the spatial distribution of transcripts can reveal cellular heterogeneity within tumors, map immune cell infiltration, and track cellular interactions across pathological and non-pathological states. These insights not only advance our understanding of disease mechanisms but also aid in identifying potential therapeutic targets based on spatially restricted gene expression patterns [4, 15, 16].

In neuroscience, ST has been instrumental in unraveling the organization of complex brain tissues. The spatial mapping of gene expression in the brain helps define cell types, neuronal circuits, and region-specific molecular profiles critical for functions such as memory, cognition, and sensory processing [17–21]. Additionally, ST in brain tissue studies facilitates research into neurodegenerative diseases by providing insights into how cellular populations and gene expression patterns vary spatially in response to disease progression.

Moreover, ST is a valuable tool in studying immune responses, as it allows researchers to observe immune cell localization and organization within tissues during infection, inflammation, or autoimmune diseases. By examining how immune cells are spatially distributed and interact with other cells, scientists can gain a deeper understanding of immune system dynamics, tissue-specific immune responses, and immune dysregulation in diseases.

We constructed a new ST database, named SPASCER (*https://ccsm.uth.edu/SPASCER*), designed to help understand the heterogeneity of tissue organizations, region-specific microenvironments, and intercellular interactions across tissue architectures at multiple levels [22]. SPASCER contains datasets from 43 studies, including 1,082 sub-datasets from 16 organ types across four species (Fig. 8.1), and provides systematic annotations of ST, including spatially patterned genes, spatially patterned pathways, gene regulatory networks, cell–cell interactions (CCIs), and ST deconvolution and interactions, which could be used to systematically study the spatial heterogeneity of tissue organizations and their related biological processes at the unicellular level.

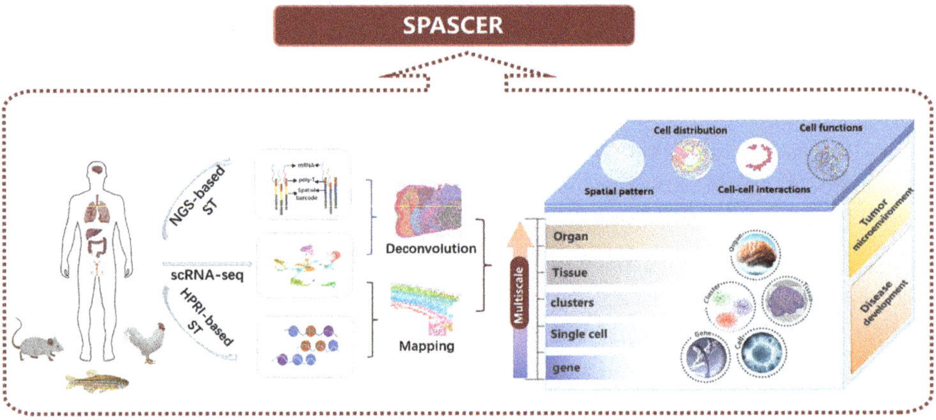

Fig. 8.1 The main workflow of SPASCER.

Overall, ST is transforming biological research by offering a high-resolution, spatially resolved view of gene expression. This technology is unlocking new avenues in basic and clinical research, enabling discoveries that were previously obscured by traditional transcriptomics methods. As ST technology continues to evolve, it holds the potential to become an integral tool in precision medicine, providing spatially informed insights that could drive personalized therapeutic strategies based on the unique spatial profiles of individual patients' tissues.

8.2 Overview of Spatial Transcriptomics Technologies

ST has seen rapid technological advancements, each contributing unique benefits and limitations suited to different research needs. These methods generally fall into three broad categories: imaging-based techniques, sequencing-based approaches, and emerging technologies that combine or extend the capabilities of the existing platforms. Below, we discuss the features, advantages, and limitations of each approach, as well as their applicability to various types of biological research.

Table 8.1 Summary of the existing spatial omics technologies.

Technology	Method	Resolution	Single Cell	Number of Cells	Number of Genes
Imaging-based technologies	ISS	–	Yes	450	39
	MERFISH	0.1 μm	Yes	100–40,000	130–10,000
	STARmap	–	Yes	~30,000	160–1,020
	seqFISH	–	Yes	~3,000	~10,000
	EEL FISH	–	Yes	~120,000	~900
	Xenium	0.05 μm	Yes	~10,000	266–480
Sequencing-based technologies	Tomo-seq	–	No	–	12,000
	Visium	55 μm	No	20,000–40,000	~20,000
	Geo-seq	–	Yes	5–40	~8,000
	Slide-seq V2	10 μm	No	–	550 UMI
	HDST	2 μm	No	~60,000	44 UMI
	Stereo-seq	0.5 μm	Yes	~100,000	~1000
	Visium HD	2 μm	Yes	~20,000	~7,600

8.2.1 Imaging-Based Techniques

Imaging-based approaches use fluorescent tagging of mRNA molecules in situ and high-resolution fluorescence microscopy to detect and differentiate between single mRNA transcripts, which enable visualization of specific gene expression patterns within tissue sections at a cellular or even subcellular level. Imaging-based methods can be broadly subdivided into in situ hybridization (ISH) or in situ sequencing (ISS) methods. Some of the most prominent ISH techniques include smFISH, STARmap, and MERFISH (Table 8.1).

MERFISH allows simultaneous visualization of thousands of mRNA transcripts in a single tissue section through a barcoding strategy and error-correcting codes. This technique can capture comprehensive gene expression profiles at subcellular resolution, making it well-suited for detailed mapping of gene expression within complex tissues, such as the brain. However, MERFISH requires sophisticated imaging equipment and complex data analysis, potentially limiting its accessibility.

Spatially Resolved Transcript Amplicon Readout Mapping (STARmap) is an innovative ST technique designed to profile gene expression at high spatial resolution within intact tissues. It employs a unique strategy based on hydrogel-tissue chemistry to preserve the spatial organization of the tissue while enabling in situ RNA sequencing. Through targeted gene barcoding and fluorescence microscopy, STARmap can measure the expression of hundreds of genes simultaneously with subcellular precision. This method is particularly powerful for exploring spatial patterns of gene expression in complex tissues, such as the brain, where the spatial arrangement of cells is crucial for understanding function.

Xenium is a cutting-edge ST platform developed by 10x Genomics that enables high-resolution gene expression profiling directly within tissue sections. This technology combines advanced ISH chemistry with automated imaging to detect hundreds of RNA transcripts simultaneously at subcellular resolution. Xenium is designed to preserve the spatial architecture of tissues, making it ideal for studying complex biological systems, such as tumor microenvironments, neural circuits, and organ development. The platform provides robust sensitivity and specificity for transcript detection, allowing researchers to map gene expression patterns, identify cell types, and uncover spatially resolved molecular interactions within intact tissues. By integrating seamlessly with other 10x Genomics tools, Xenium facilitates multimodal analyses, enabling deeper insights into spatial biology.

ISH methods provide high spatial resolution and are excellent for studies focusing on specific genes or pathways with limited targets. However, the high-resolution imaging required for these techniques can be resource-intensive and may not scale well for high-throughput, whole-transcriptome analyses.

8.2.2 Sequencing-Based Approaches

Sequencing-based ST technologies use spatially encoded barcodes or molecular tags to record gene expression in situ, which are then sequenced to generate genome-wide profiles. Major platforms in this category include ST from 10x Genomics Visium, Slide-seq, Stereo-seq, Light-seq, and SeqFISH.

10x Genomics Visium Spatial Gene Expression technology, often referred to as ST, is based on spatial barcoding of fixed tissue sections placed on specialized slides containing spatially organized capture spots. Each spot captures mRNA within its immediate vicinity, and the captured RNA is sequenced, generating spatially resolved transcriptomic data. Visium is widely used due to its ease of use and commercial availability, although its resolution is limited by the size of the capture spots, making it more suitable for studies of heterogeneous tissues with relatively large cellular structures. This method can be applied to fixed or frozen tissue slices and has been widely used in various fields, including immunology, cancer immunology, and neuroscience.

Slide-seq is a relatively recent technology that uses beads coated with spatially barcoded oligonucleotides to capture transcripts from tissue sections placed on top of the beads. Each bead's unique barcode captures spatial information, and transcripts captured by each bead are subsequently sequenced. This technology provides high spatial resolution and can capture whole-transcriptome profiles across entire tissue sections. However, due to its reliance on bead-based capture, Slide-seq requires optimized protocols for bead positioning and tissue sectioning. Moreover, the enhanced version of Slide-seq, known as Slide-seqV2, has achieved significantly higher spatial resolution (approaching single-cell resolution) and improved sensitivity (exceeding 1,000 transcripts per bead). This advancement enables the identification of underlying genetic programs that were inadequately captured by the original Slide-seq.

Sequential fluorescence in situ hybridization (SeqFISH) combines sequencing and ISH techniques by applying sequential rounds of hybridization to detect thousands of genes in situ. This technique uses a set of probes that target different transcripts in each round, with a unique fluorescent signal generated for each target gene. SeqFISH achieves high-resolution, whole-transcriptome profiling and can capture single-cell gene expression. However, the sequential imaging required for SeqFISH can be time-consuming and resource-intensive, limiting its use for large sample sizes.

Spatio-temporal Enhanced Resolution Omics Sequencing (Stereo-seq) is a recently developed next-generation sequencing (NGS)-based spatial profiling technology that leverages randomly barcoded DNA nanoballs arranged in arrays to achieve nanoscale resolution, approaching single-cell precision. This method provides an extensive field of view (FOV) at the centimeter scale while maintaining cellular-level resolution. Stereo-seq has been successfully applied in diverse research areas, including mouse organogenesis, plant biology, neuroscience, and developmental biology, showcasing its versatility in studying complex biological systems.

Sequencing-based approaches provide comprehensive, genome-wide transcriptomic profiles and are well-suited for analyzing complex tissue samples with heterogeneous cell populations. However, they tend to have lower resolution than ISH techniques and are often limited by the physical constraints of the spatial barcoding or capture mechanisms.

8.2.3 Emerging Technologies

In addition to established ISH and sequencing-based methods, several emerging technologies are advancing ST by combining multiple modalities or enhancing throughput and resolution. Notable examples include ISS, spatial barcoding innovations, and hybrid platforms.

ISS technologies, such as STARmap, perform sequencing directly within tissue sections, preserving spatial information while capturing gene expression profiles. ISS can achieve single-cell resolution and offers scalability for high-throughput studies, although it often requires complex protocols and specialized sequencing instruments.

Spatial barcoding innovations have led to the development of technologies like Stereo-seq, which uses nanodots and micro-patterned chips to capture transcriptomic data with ultra-high resolution over large tissue areas. This approach aims to overcome limitations in resolution and throughput that exist in traditional sequencing-based techniques, making it a promising technology for large-scale studies. Hybrid platforms are also emerging, integrating multiple spatial omics modalities to generate a more comprehensive molecular profile of tissues. For example, recent methods combine proteomics, epigenomics, and transcriptomics in a spatially resolved format, offering multilayered insights into gene expression, protein abundance, and epigenetic modifications within the same tissue sections.

Emerging technologies provide higher resolution, greater scalability, and, in some cases, multi-omic capabilities. However, these approaches often require complex instrumentation and high computational capacity, and they may still be in experimental stages with limited availability.

8.3 Data Processing and Computational Methods

The analysis of ST data requires a suite of computational tools to process raw data, map gene expression profiles to spatial coordinates, and ensure data quality and reliability. From image preprocessing to spatial mapping and quality control (QC), each step requires careful consideration to generate accurate and reproducible results. This section outlines key computational processes, including image processing, alignment and spatial mapping, normalization, and QC in ST.

8.3.1 Image Processing and Preprocessing

Image processing is essential for ST methods that rely on microscopy, such as ISH-based techniques (e.g., smFISH, MERFISH) and ISS approaches. Accurate image processing ensures precise localization of gene expression signals and helps delineate cell boundaries within tissue sections. Software like CellProfiler [23] and ilastik [24] enables automated cell segmentation and feature extraction. Segmentation algorithms are applied to identify individual cells, nuclei, or tissue regions in microscopic images. Tissue segmentation can be challenging due to variability in cell morphology, tissue density, and image quality. Machine learning techniques, particularly deep learning models like convolutional neural networks (CNNs), are commonly used for automated, high-throughput segmentation, although manual curation may still be needed in complex tissues. Deep learning-based models, such as U-Net, offer advanced segmentation capabilities for complex tissue structures.

However, traditional image-based segmentation methods are limited in their ability to fully leverage the rich information provided by ST profiling. To address this situation, researchers have developed a variety of cell segmentation methods for spatial transcriptomes. Petukhov et al. have developed subcellular spatial transcriptomics cell segmentation (SCS) [25], a method that integrates imaging and sequencing data to enhance the accuracy of cell segmentation. Utilizing a transformer neural network, SCS adaptively learns the spatial relationship of each spot to the cell center for precise spot-to-cell assignments.

To identify nuclei, the paired staining image and sequencing section are first aligned, ensuring that image pixels correspond to spot coordinates. The nuclei are then segmented from the aligned staining image using the watershed algorithm implemented in Spateo, resulting in masks for individual nuclei. The model comprises two main components: an encoder and a classifier. The encoder generates a hidden representation, z, for each spot. To address the typically sparse gene expression detected per spot, the model incorporates neighborhood gene expression data. Specifically, the expression profile of each spot is augmented by integrating the profiles of its 50 nearest neighbors, providing a more comprehensive representation for subsequent processing. The expression profiles x, and the distance vectors s, for the center spot and its neighboring spots are processed through two distinct fully connected layers. These layers project the inputs into 64-dimensional vectors, which are subsequently combined through summation. The classifier takes as input the center spot's representation from the final encoder layer and predicts the gradient direction from a spot to its cell center and determines whether the spot is inside or outside a cell. Before this transformation, layer normalization is applied. Through this method, each cell could be divided into two parts: a cytoplasm part and a nucleus part.

Tested on two novel subcellular ST platforms, SCS outperformed traditional image-based methods by delivering higher accuracy, detecting more cells, and producing more realistic estimates of cell size. Additionally, SCS enables detailed RNA localization analysis through its spot assignments, further validating the segmentation results and providing deeper insights into subcellular RNA distribution.

Besides, ST data can be noisy, particularly in fluorescence-based methods where autofluorescence or nonspecific binding may introduce background signals. Computational approaches, such as background subtraction and denoising algorithms, are applied to enhance signal clarity. These preprocessing steps help improve the accuracy of downstream analysis by ensuring that only true signals are retained.

8.3.2 Alignment and Spatial Mapping

Spatially resolved genomic technologies have enabled the study of the physical organization of cells and tissues, offering insights into local cellular interactions. However, precisely aligning spatial observations across different slices, samples, scales, individuals, and technologies remains challenging. In spatial genomics, five main approaches have been developed for aligning spatial coordinates across samples. Probabilistic alignment of spatial transcriptomic experiments (PASTE) [5] was designed to align neighboring tissue slices in ST data. PASTE employs an optimal transport framework to map spatial locations between adjacent slices. Its objective function balances transcriptional similarity with the spatial proximity of locations. While PASTE is efficient and reliable, it is limited to linear alignments, which often lack the flexibility needed to handle complex distortions in data.

Although alignment is not the primary aim of Splotch, it incorporates a linear autocorrelation model to adjust ST slices and align specific regions of tissue. In addition, two landmark-based methods for spatial alignment have been introduced. The effortless generic Gaussian process (GP) landmark transfer (Eggplant) approach uses GP regression to align slices. Eggplant maps gene expression values from each misaligned slice onto a common coordinate system (CCS). This method requires users to identify corresponding landmarks in both the misaligned and template samples. Eggplant performs these template transfers independently for each slice and gene, overlooking correlations between them. The ST imaging framework (STIM) leverages computer vision techniques to align ST data onto a CCS. However, for both Eggplant and STIM, identifying shared landmark locations can be particularly challenging in tissues without well-defined structures, such as tumors.

Gaussian process spatial alignment (GPSA) [26] is a two-layer GP model. The first layer maps the spatial locations of observed samples onto a CCS, while the second layer transforms the CCS into the observed readouts. This method facilitates advanced spatially aware analyses that would be challenging or imprecise with unaligned data. GPSA enables applications such as variance analysis, the construction of a dense three-dimensional (3D) atlas from sparse two-dimensional (2D) slices, and performing association tests across multiple data modalities. These capabilities make GPSA a powerful tool for handling complex spatial data.

The alignment of ST is essential to compare and integrate spatial data, enabling researchers to study how cells and tissues are organized and interact within their microenvironment. Alignment helps overcome challenges such as distortions, variations in sample preparation, and differences in spatial scales, facilitating downstream analyses like constructing 3D tissue atlases, identifying spatial patterns of gene expression, and integrating multimodal data for a comprehensive understanding of biological systems.

8.3.3 Normalization and Quality Control

Normalization and QC are essential steps to ensure the reliability and comparability of ST data. These steps help mitigate technical variability and enable meaningful comparisons across cells, tissue regions, or samples. Several bioinformatics packages, including SCTransform (for normalization) and Scanpy (for QC), provide robust options for handling ST data. Integrated workflows are available within packages such as Giotto and SpaCy, offering end-to-end pipelines for data processing and analysis.

Normalization corrects for variations in sequencing depth, probe hybridization efficiency, and capture efficiency across spots or cells. Common normalization methods include scaling gene expression data to account for total expression counts, log transformation to reduce the impact of highly expressed genes, and batch correction to mitigate inter-sample variability. More advanced methods, such as those based on probabilistic modeling (e.g., SCTransform), can adjust for cell-specific and spatial biases, enabling more accurate representation of relative gene expression levels across the tissue.

QC metrics are used to assess data quality at multiple stages, from initial image acquisition to final expression matrices. Metrics such as signal-to-noise ratio, alignment accuracy, and dropout rates (i.e., missing values due to low capture efficiency) help evaluate the reliability of ST data. In sequencing-based methods, additional QC metrics include sequencing depth, mapping rate, and the percentage of mitochondrial or ribosomal RNA reads, which can indicate sample quality. Low-quality spots or cells are typically excluded to improve overall data quality and reduce noise in downstream analysis.

ST data often contain outliers due to technical artifacts, such as doublet cells, contamination, or spatial misalignments. Computational methods, such as clustering-based filtering and outlier detection algorithms, are applied to identify and remove outliers. In complex tissues, spatial statistics (e.g., Moran's *I*) are used to assess spatial autocorrelation and detect cells or regions that deviate significantly from expected patterns, ensuring that only biologically relevant signals are retained.

8.4 Data Analysis Techniques in Spatial Transcriptomics

Data analysis in ST focuses on extracting meaningful biological insights from spatially resolved gene expression patterns within complex tissue architectures. Advanced analysis techniques help in identifying spatially variable genes (SVGs), clustering cells based on their spatial context, constructing CCI networks, and visualizing spatial expression data. This section discusses essential techniques, including spatial gene expression analysis, cell clustering, spatial pattern detection, and CCI inference.

8.4.1 Spatially Variable Gene Analysis

Identifying genes with spatially distinct expression patterns is a fundamental step in ST analysis. SVGs display significant variation across different regions or cell types within a tissue, revealing genes that are critical to tissue structure and function.

To identify SVGs, spatial statistics are applied to quantify gene expression variability across spatial coordinates. Common methods include Moran's *I* and Geary's *C*, which measure spatial autocorrelation, or neighborhood-based methods that evaluate expression differences between neighboring cells or spots. Statistical tests, such as spatial differential expression (DE) analysis, are used to determine the significance of spatial gene variation, allowing researchers to pinpoint genes that contribute to tissue-specific functions or regional characteristics. In tissues with clear spatial structures, such as the liver or brain, detecting zonation (i.e., distinct expression domains or zones) helps define functionally distinct regions. Trendsceek [27] treats ST data as marked point processes and evaluates the significant association between spatial distributions and gene expression levels for point pairs. SpatialDE [28], on the other hand, employs GP regression to partition gene expression variability into spatial and independent noise components, determining significance by comparing it against a null model without spatial variation. SPARK [29] builds upon SpatialDE by incorporating GP regression with ten spatial kernels to capture diverse spatial expression patterns, thereby boosting statistical sensitivity. SPARK-X advances this concept by leveraging a covariance test framework to assess the dependency between spatial locations and gene expression.

To tackle SVG identification, scGCO [30] integrates graph cuts—a method borrowed from computer vision—with hidden Markov random fields to pinpoint spatially dependent candidate regions for individual genes, which are subsequently tested under a complete spatial randomness framework. Squidpy identifies SVGs using Moran's *I*, with significance assessed through *p*-values derived from standard normal approximations based on randomized permutations.

Here we take SPARK as an example for a specific explanation. SPARK is built upon a generalized linear spatial model (GLSM) framework and incorporates various spatial kernels to effectively handle count data generated by smFISH or sequencing-based ST technologies. Utilizing a newly developed penalized quasi-likelihood (PQL) algorithm, SPARK achieves scalability, enabling the analysis of tens of thousands of genes across extensive spatial locations. A defining feature of SPARK is its reliance on a mixture of χ^2 distributions, which serve as the foundation for its test statistic distribution. Additionally, it leverages a recently developed Cauchy combination rule to integrate information across multiple spatial kernels, enabling accurate *p*-value calibration. This ensures stringent control of type I error rates at the transcriptome-wide scale while enhancing the power to detect spatially expressed (SE) genes, outperforming existing methodologies in sensitivity and precision.

$$Y_i \sim \text{Poisson}\left(N_i \lambda_i\right) \tag{8.1}$$

$$\log\left(\lambda_i\right) = x_i^T \beta + b_i + \epsilon_i \tag{8.2}$$

$$b = \left(b_1, \ldots, b_n\right)^T \sim MVN(0, \tau_1 K) \tag{8.3}$$

$$\epsilon = (\epsilon_1, \ldots, \epsilon_n)^T \sim MVN(0, \tau_2 \mathrm{I}). \tag{8.4}$$

In the above Eq. (8.1–8.4), for the gene of interest, let $y_i(s_i)$ represent the gene expression count observed for the *i*-th sample at location s_i. Define $x_i(s_i)$

as a k-dimensional vector of covariates, which includes a constant term (set to one) for the intercept and $k - 1$ additional explanatory variables for the i-th sample. These explanatory variables may include batch effects, cell-cycle information, or other relevant factors that need to be accounted for during the analysis. Additionally, $N_i(s_i)$ denotes the normalization factor for the i-th sample. In this work, $N_i(s_i)$ is calculated as the total count of gene expression across all genes in the sample, as the primary focus is on relative gene expression levels. Alternatively, $N_i(s_i)$ can be set to one if the analysis is aimed at absolute gene expression levels.

To model the observed gene expression count data, we assume an overdispersed Poisson distribution, where $\lambda_i(s_i)$ represents the Poisson rate parameter. This parameter captures the underlying gene expression level for the i-th sample and, in a spatial context, can also be interpreted as an unobserved spatial random process at location s_i. The logarithm of the latent variable $\lambda_i(s_i)$ is modeled as a linear combination of three components: β, a, k-dimensional vector of coefficients, where the intercept reflects the average log-expression of the gene across spatial locations, and the remaining $k - 1$ coefficients correspond to the effects of the explanatory variables. ϵ_i, the residual error term, assumed to be independently and identically distributed as $N(0,\tau^2)$ with variance τ^2. $b_i(s_i)$, a zero-mean stationary GP that captures the spatial correlation structure among locations.

SPARK introduces a data-generative model and employs a model-based hypothesis-testing framework to identify spatial patterns, setting it apart from earlier tools for exploring spatial data. Unlike methods that rely on variograms or semi-variograms to visualize spatial autocorrelation, SPARK leverages a generative approach to capture underlying spatial structures. Additionally, its hypothesis-testing framework differentiates it from simpler spatial test statistics such as Moran's I and Geary's C, which are commonly used to assess spatial autocorrelation.

8.4.2 Cell Clustering and Spatial Subtyping

ST provides high-dimensional gene expression data tied to spatial coordinates, enabling the identification of distinct cell types and their spatial organization within tissues. Two critical analytical steps in this domain are cell clustering and spatial subtyping, both of which are essential for uncovering tissue architecture and cellular interactions.

Cell clustering involves grouping cells based on their gene expression profiles to identify distinct cell types or states. Common clustering methods leverage techniques from machine learning, such as principal component analysis (PCA) or non-negative matrix factorization (NMF), to reduce dimensionality, followed by clustering algorithms like k-means, Louvain, or Leiden. For ST data, methods such as Seurat or Scanpy often combine transcriptomic and spatial information for clustering (Table 8.2).

Table 8.2 Summary of the clustering methods for spatial transcriptomics.

Tools	Task	Method	Software
GraphST	Clustering	Graph neural network, contrastive learning	Python
SpatialPCA	Clustering	Spatial probabilistic PCA	R
STAGATE	Clustering	Adaptive graph attention auto-encoder	Python
conST	Clustering	Masked autoencoder, contrastive learning	Python
SpaGCN	Clustering	Graph convolutional network	Python
BayesSpace	Clustering	Markov chain Monte Carlo random field	R
PRECAST	Clustering	Gaussian mixture model	R
BASS	Clustering	Bayesian analysis	R
DeepST	Clustering	Variational graph autoencoder	Python

An additional consideration for clustering in ST is the incorporation of spatial relationships between cells or spots. Tools like SpaGCN [31] and STAGATE [9] integrate graph neural networks (GNNs) and spatial proximity into the clustering process, enabling the identification of spatially coherent cell clusters. SpaGCN is a graph convolutional network-based method designed to analyze ST data by integrating gene expression, spatial coordinates, and histological information. By leveraging graph convolution, SpaGCN incorporates the gene expression profiles of neighboring spots to refine the representation of each spot. This approach facilitates the identification of spatial domains that exhibit consistent gene expression patterns and align with histological features. Following this, SpaGCN performs domain-guided DE analysis to identify genes that are specifically enriched within the detected spatial domains, providing insights into localized biological processes and tissue structures.

Spatial subtyping extends clustering by further refining groups of cells based on their spatial arrangement and local microenvironment. This process accounts for spatial patterns of gene expression, such as spatially restricted gene modules or gradients across tissue regions. Spatial subtyping often employs spatial statistics, such as Moran's *I* or Geary's *C*, to detect autocorrelations in gene expression and identify spatially distinct regions. Advanced spatial subtyping methods, like BayesSpace, use Bayesian modeling to assign cells or spots to spatial subtypes while accounting for spatial dependency and variability. These methods are particularly effective in characterizing tissue substructures, such as tumor niches, gradients of differentiation, or inflammation zones. Recent advances integrate cell clustering with spatial subtyping to provide a holistic view of tissue organization. For example, frameworks like SpatialPCA [32] and GIANT decompose ST data into principal components or modules, which can be used to both cluster cells and identify spatially distinct subtypes. This integration allows researchers to investigate not only cell type composition but also their spatial context, uncovering patterns such as cellular neighborhoods or zonation within tissues.

The combination of cell clustering and spatial subtyping in ST has been transformative in studying tissues with complex spatial organization, such as tumors, developing organs, or inflamed tissues. However, challenges remain, including batch effects, noise in the data, and computational scalability for large datasets. Developing robust methods that integrate spatial, transcriptomic, and multimodal data will be critical for advancing this field further.

8.4.3 Cell–Cell Interaction and Microenvironment Analysis

A variety of computational tools have been developed to infer CCIs activities from scRNA-seq data using curated signaling databases. These approaches primarily focus on quantifying interactions between ligand and receptor pairs, often relying on explicitly defined mathematical functions. For instance, methods may compute the product of ligand and receptor expression levels or apply nonlinear models such as Hill functions.

Different tools emphasize specific aspects of CCI. For example, CellPhoneDB [33], ICELLNET [34], and CellChat [35] account for the multi-subunit nature of protein complexes in signaling interactions. Meanwhile, SoptSC [36] and NicheNet [37] integrate downstream intracellular signaling and gene–gene interactions to capture more comprehensive communication dynamics. Additionally, scTensor [38] explores higher-order CCI patterns using hypergraphs to model complex signaling networks.

While many tools originally developed for nonspatial transcriptomic data have been adapted for spatial applications, only a limited number of methods have been specifically tailored to spatial datasets. Giotto [39], for instance, creates a spatial proximity graph to infer interactions mediated by ligand-receptor pairs located on cell membranes. CellPhoneDB v3 focuses on interactions within spatially defined microenvironments by restricting analysis to cell clusters. stLearn associates the co-expression of ligand and receptor genes with the spatial heterogeneity of cell types. In addition, probabilistic approaches like SVCA [40] and machine learning models such as MISTy [41] investigate spatially restricted intercellular gene–gene interactions, while NCEM establishes functional relationships between cell type, spatial context, and gene expression patterns. Despite these advancements, most existing methods focus on local CCI, analyzing interactions at the level of individual cell pairs or within small neighborhoods. This localized approach neglects broader, collective aspects of CCI, such as competitive dynamics or large-scale communication patterns among cells, limiting the understanding of global spatial interactions.

Optimal transport theory has recently been adapted for analyzing transcriptomic data, including CCI inference. Researchers have extended optimal transport methods to a collective framework, enabling the preservation of distribution comparability while ensuring that the total signal remains within the constraints of ligand and receptor availability. This approach also incorporates spatial range limits for signaling and accommodates multiple competing species. Unlike traditional optimal transport techniques, the collective optimal transport method simultaneously optimizes the total transported mass and the coupling between ligands and receptors.

To achieve this, entropy regularization is introduced, which enforces marginal distribution inequalities. This reformulation positions collective optimal transport as a specific case within the broader unbalanced optimal transport framework. Building on this concept, COMMunication analysis by Optimal Transport (COMMOT) [42] was developed as a tool to infer CCI. It can analyze multiple ligand-receptor pairs concurrently for ST datasets or spatially annotated scRNA-seq data, where spatial distances are derived from paired imaging data. COMMOT provides insights into spatial signaling directions, evaluates downstream gene expression effects of CCC through ensemble tree models, and offers robust visualization tools to aid in the interpretation of these analyses.

Here we take COMMOT as an example for a specific explanation. For an ST dataset containing n_s spots, n_l ligand types, and n_r receptor types, the collective optimal transport framework identifies an optimal coupling $P^* \in R_+^{n_l \times n_r \times n_s \times n_s}$, where $P^*_{i,j,k,l}$ quantifies the signaling intensity from a sender cell k to a receiver cell l via ligand III and receptor j. This is formulated as a minimization task, $\min_{P \in \Gamma} \sum_{(i,j) \in I} \alpha_{(i,j)} {<P_{i,j,..,}, C_{(i,j)}>}_F$, where $\Gamma = \{P \in R_{+}^{n_l \times n_r \times n_s \times n_s} : P_{i,j,.,.} = 0 \text{ for } (i,j) \notin I, \sum_{j,l} P_{i,j,k,l} \leq X_{i,k}, \sum_{i,k} P_{i,j,k,l} \leq X_{j,l}\}$ the objective is to minimize the weighted cost associated with each ligand–receptor pair. The index set I defines the ligand and receptor species pairs that can interact, while $X_{i,k}$ represents the expression level of gene i at spot k. The cost matrix $C(i,j)$, which is species-specific, is a modified distance matrix reflecting the spatial proximity between spots, where distances greater than the spatial range of ligand III are set to infinity. To account for competition between molecules and cells, it is assumed that each receptor species or cell has a limited capacity for interactions. As a result, a stronger inferred interaction between a receptor and one ligand species or cell diminishes the potential for interaction with other ligand species or cells.

8.5 Integrating Spatial Transcriptomics with Multi-Omics Data

Integrating ST with other omics layers, such as genomics, epigenomics, proteomics, and metabolomics, enables researchers to study complex biological processes from multiple perspectives. This multi-omics approach enhances our understanding of how spatial gene expression patterns are regulated and how they relate to other molecular features within tissues. By combining spatial information with multi-omics data, researchers can explore relationships between gene expression, protein activity, metabolic states, and genomic features within a precise spatial context.

The integration of multi-omics data with ST enables a more comprehensive view of cellular states and tissue organization. Each omics layer provides unique insights that, when combined, offer a synergistic perspective on cellular function. Transcriptomics alone provides information on gene expression patterns, but other omics layers reveal how these patterns are regulated and how they translate to cellular functions. For example, combining ST with epigenomic data (such as chromatin accessibility) identifies regulatory elements that drive spatially distinct expression patterns, while integration with proteomics uncovers how gene expression translates into protein abundance within specific regions.

Multi-omics integration is particularly valuable in disease studies, where understanding the interplay between genomic mutations, transcriptomic changes, protein dysregulation, and metabolic alterations can elucidate disease mechanisms. In cancer, for instance, integrating ST with genomics helps identify tumor-specific mutations and their spatial distribution within the tumor microenvironment, providing insights into tumor heterogeneity and progression.

ST combined with metabolomics reveals spatial patterns of metabolic activity, which is crucial in tissues like the liver and brain, where spatial metabolic gradients drive tissue function. Similarly, combining transcriptomics with proteomics can help identify signaling pathways active within specific regions or cell types, allowing researchers to map localized cellular interactions and pathways within the tissue.

8.5.1 Computational Tools for scRNA-seq Integration in Spatial Transcriptomics

NGS-based ST techniques, such as ST, 10x Visium, and Slide-seq, are capable of profiling RNA expression across the entire transcriptome from spatially organized spots. However, these spots often encompass multiple cells, which limits their spatial resolution. This constraint prevents these methods from achieving whole-transcriptome data at single-cell spatial precision. To overcome these challenges, bioinformaticians have developed a range of integrative approaches that combine ST with scRNA-seq data (Table 8.3).

Table 8.3 Summary of the methods for scRNA-seq integration in ST.

Tools	Task	Method	Software
Tangram	Clustering	Deep learning	Python
gimVI	Clustering	Probabilistic model	Python
SpaGE	Clustering	Joint embedding-based	Python
Seurat	Clustering	Joint embedding-based	R
SpaOTsc	Clustering	Optimal transport	Python
novospaRC	Clustering	Optimal transport	Python
LIGER	Clustering	Joint embedding-based	R
stPlus	Clustering	Joint embedding-based	Python
RCTD	Deconvolution	Probabilistic model	R
SpatialDWLS	Deconvolution	Weighted least squares	R
PASTE2	Integration	Partial fused Gromov–Wasserstein optimal transport	Python
PRECAST	Integration	Gaussian mixture model, discrete hidden Markov random field	R
DeepST	Integration	Variational graph autoencoder	Python
BASS	Integration	Bayesian analysis, multi-sample analysis	R
Cell2location	Integration	Probabilistic model	Python
DSTG	Deconvolution	Deep learning	Python

For instance, SpaGE [43] leverages the domain adaptation algorithm alongside k-nearest-neighbor regression to estimate these missing spatial expression patterns. gimVI employs a deep generative model to infer spatial distributions of transcripts that are otherwise undetected. Tangram utilizes non-convex optimization within a deep learning framework to align scRNA-seq data to spatial coordinates. Both novoSpaRc and SpaOTsc employ optimal transport principles to construct spatial relationships between cells based on scRNA-seq profiles. Seurat v3 applies canonical correlation analysis (CCA) to integrate spatial and single-cell datasets into a shared latent space, enabling the projection of single-cell transcriptomic data onto spatial spots. LIGER [44] combines integrative NMF with shared factor neighborhood graphs to predict spatial gene expression. Finally, stPlus integrates auto-encoders with weighted k-nearest-neighbor methods to model spatial expression patterns. By merging ST with scRNA-seq, these computational methods allow researchers to predict and reconstruct the spatial organization of gene expression, effectively addressing the limitations of current ST technologies.

Additionally, several other methods focus on estimating cell type compositions within spatial spots. For instance, RCTD [45] uses cell type profiles obtained from scRNA-seq, combined with supervised learning, to decompose mixtures of cell types. SpatialDWLS [46] applies a weighted least-squares framework to infer cell type proportions, while Stereoscope employs a probabilistic model-based approach to deconvolve cell mixtures using scRNA-seq data. Cell2location [47] leverages gene expression signatures of cell subpopulations from scRNA-seq data to estimate the abundance of each cell type at individual spots. STRIDE [48] relies on topic profiles derived from scRNA-seq data to infer cell type compositions in spatial mixtures, whereas DestVI [49] employs variational inference and latent variable models to estimate proportions of cell types. Collectively, these methods provide researchers with robust tools for dissecting the cellular composition of ST spots, offering a deeper understanding of tissue organization and cellular heterogeneity.

8.5.2 Computational Tools for Other Multi-Omics Integration in Spatial Transcriptomics

Integrating ST and histologic data has emerged as a powerful computational challenge aimed at enhancing our understanding of tissue organization by combining gene expression information with histological context. This integration leverages the strengths of both data types: ST provides detailed molecular profiles, while histological data offers high-resolution structural and morphological insights. Computational approaches to achieve this integration vary in methodology, but they share the goal of linking spatial gene expression patterns with histological features to create a unified map of tissues.

One major class of computational methods involves image-based alignment and feature extraction. These approaches begin by identifying spatial locations in histological images that correspond to ST data spots or cells. Tools such as SpaHist and ST-Net use CNNs to extract histological features from tissue images and align them with ST coordinates. These extracted features can be combined with gene expression data to uncover relationships between molecular signatures and tissue morphology. By integrating these datasets, researchers can identify SVGs associated with histological structures, such as blood vessels, tumor margins, or specific cell types.

Another key direction is the use of multimodal deep learning models to jointly analyze transcriptomic and histological data. Methods such as Tangram and HE2RNA employ deep learning frameworks to predict ST patterns from histological images. These models use non-convex optimization or transformer architectures to map histological image features directly onto gene expression profiles, enabling the reconstruction of molecular landscapes in regions where ST data may be sparse. By leveraging the rich visual data in histological images, these tools improve the resolution of spatial transcriptomic datasets and enhance their interpretability. Additionally, Starfysh offers a comprehensive platform for multimodal analysis of spatial transcriptomic datasets. Utilizing reference-free deep generative modeling, archetypes, and shared histological features, Starfysh jointly models transcriptomic and histological data to infer proportions of fine-grained, context-specific cellular states while generating cell type-specific gene expression profiles for downstream applications.

The integration of multi-omics data with ST can be accomplished through several strategies, each suited to different experimental setups and research questions. One approach to multi-omics integration involves spatial co-registration, where data from different omics layers are mapped to the same spatial coordinates within a tissue section. For example, ST data can be combined with mass spectrometry imaging (MSI) data to link gene expression with protein or metabolite abundances in the same tissue region. Co-registration is often facilitated by image alignment and registration algorithms that map the same tissue section or adjacent sections with high spatial precision.

Recent technological advances have enabled multi-omics profiling within the same tissue section. Techniques such as spatial-CITE-seq (which combines

ST with protein profiling) and spatial ATAC-seq (which combines ST with chromatin accessibility data) allow researchers to measure multiple omics layers simultaneously. These approaches reduce sample preparation variability and enable direct spatial correlation between different molecular layers, enhancing the reliability of multi-omics analysis.

In cases where different omics data layers are collected from separate tissue sections or samples, computational methods are used to integrate these datasets. Data fusion algorithms align molecular profiles by matching patterns across datasets, often using probabilistic models, graph-based methods, or machine learning techniques. For example, tools like multi-omics factor analysis (MOFA) and Seurat's multi-omics integration pipelines use dimensionality reduction and matrix factorization to find shared features across omics layers, while machine learning models like neural networks can align ST with MSI data. Various computational tools have been developed to facilitate the integration of ST with other omics data. These tools typically provide functionalities for data alignment, normalization, and feature correlation.

Harmony is an integration tool designed to correct for batch effects and align multi-omics data in a spatial context. MOFA+ (an extension of MOFA) is another tool for multi-omics factor analysis that integrates various data types while preserving spatial structure. These tools are particularly effective in linking ST data with genomics or epigenomics data, as they identify shared patterns while accounting for spatial differences.

Deep learning models, such as variational autoencoders (VAEs) and GNNs, are increasingly used for complex multi-omics integration. These models are adept at handling high-dimensional data and can be trained to recognize complex relationships between omics layers. GNNs, in particular, are well-suited for integrating ST with spatial proteomics or metabolomics data, as they can incorporate spatial neighborhood information and reveal localized interactions in tissues.

8.5.3 Challenges in Integrating Multi-Omics Data with Spatial Transcriptomics

While multi-omics integration provides powerful insights, it also presents several technical and computational challenges. Addressing these challenges is essential for ensuring reliable and biologically meaningful results.

Different omics techniques vary in their spatial resolution. For instance, ST methods like Visium provide expression data at a multicellular resolution, while techniques like MSI typically offer lower spatial resolution for protein and metabolite data. Aligning datasets with mismatched resolution requires advanced image registration techniques and data transformation approaches to aggregate or interpolate data at different scales. Each omics layer may have inherent data sparsity or missing values. ST data often contain dropout events (where some transcripts are undetected), while proteomics or metabolomics data may have incomplete coverage due to detection limits. Computational methods like imputation algorithms and matrix factorization are often employed to handle these issues, but they can introduce noise or bias, making careful validation essential.

Multi-omics datasets often come from different experimental protocols, leading to batch effects that complicate data integration. For example, variations in sample preparation, sequencing depth, or imaging conditions across omics layers can introduce artifacts. Normalization techniques, such as scaling and batch correction algorithms, help address these issues, but may not fully eliminate them, especially when integrating spatial data collected across different tissue sections.

Integrating multiple omics layers generates large, complex datasets that require sophisticated computational models to reveal underlying relationships. Interpreting these complex multi-omics interactions requires a combination of bioinformatics expertise, machine learning, and domain knowledge. For instance, distinguishing biologically relevant signals from technical artifacts or spurious correlations is challenging, especially in heterogeneous tissues with intricate spatial patterns.

8.6 Current Challenges and Future Directions

While ST has rapidly advanced and provided transformative insights across biological and medical fields, several challenges remain. Addressing these limitations and developing new techniques will be crucial for maximizing the impact of ST on scientific discovery and clinical applications. This section explores key challenges, such as technical limitations, data analysis difficulties, and translational barriers, as well as emerging directions that could shape the future of ST research.

8.6.1 Technical Challenges and Limitations

ST platforms vary significantly in their spatial resolution and sensitivity. High-resolution methods, such as multiplexed error-robust fluorescence in situ hybridization (MERFISH) and seqFISH, can achieve single-cell or even subcellular resolution but are typically limited to a pre-selected panel of genes due to technical constraints in multiplexing. Conversely, methods like Visium offer broader transcriptome coverage but are limited to capturing gene expression at a multicellular level due to spot size and diffusion effects. Achieving both high spatial resolution and whole-transcriptome coverage remains a significant challenge, as current technologies require researchers to prioritize either spatial precision or breadth of detection.

Maintaining tissue integrity during sample preparation is critical for accurate ST analysis. Traditional fixation and sectioning methods, such as formalin fixation and paraffin embedding (FFPE), often degrade RNA, limiting the quality of data obtained. Although advances like cryosectioning improve RNA integrity, they require specialized equipment and expertise. Moreover, methods must address challenges in processing highly heterogeneous tissues, which can vary in thickness, cellular density, and structural complexity. Ensuring reliable tissue preservation, particularly for clinical samples, is essential for expanding the applications of ST.

Current ST approaches primarily focus on RNA detection, which offers critical insights into gene expression but provides limited information about protein abundance or genomic alterations. Expanding spatial methods to include other biomolecules, such as proteins and DNA, remains challenging due to technical issues in multiplexing and signal interference. While recent advances, such as spatial proteomics and multi-omics platforms (e.g., spatial-CITE-seq), offer promising solutions, integrating these data types into a cohesive spatial framework with RNA remains an area of active development.

8.6.2 Data Analysis and Computational challenges

ST experiments generate massive datasets, often with thousands of genes measured across tens of thousands of spatial locations. Analyzing these high-dimensional data while accounting for spatial structure introduces computational complexity. Data storage and processing requirements strain traditional computational resources, especially for high-resolution datasets. Developing efficient algorithms and workflows that streamline data processing, feature extraction, and interpretation is critical to enabling broader adoption and scaling of ST analyses.

Integrating ST data with other omics layers (e.g., proteomics, metabolomics) or across tissue samples from different individuals or time points is challenging due to alignment difficulties. Spatial registration methods that accurately map different datasets to a CCS are essential for comparative analyses but are limited by differences in resolution, section thickness, and tissue deformation. Advances in image processing and machine learning algorithms that facilitate accurate spatial alignment and integration of multi-modal data will be crucial to enhance multi-omics studies.

ST data are prone to batch effects and technical noise, resulting from variations in sample preparation, sequencing depth, or signal intensity. QC and normalization methods developed for scRNA-seq often fail to capture spatial-specific biases, making it challenging to compare datasets across samples. Developing spatially aware normalization techniques and QC protocols is essential to improving data reliability and interpretability. Improved validation methods that integrate biological context with data-driven quality assessments are necessary for achieving robust, reproducible analyses.

8.6.3 Biological Interpretation and Translational Challenges

As ST data contain rich information on gene expression, cellular interactions, and spatial organization, extracting biologically meaningful insights remains a challenge. Unlike bulk or scRNA-seq data, ST data require new analysis frameworks that account for spatial dependencies and tissue architecture. Advanced computational models, such as graph-based methods and spatially informed machine learning algorithms, are needed to capture complex spatial patterns, interactions, and gradients in gene expression. Developing user-friendly tools and visualization methods that facilitate interpretation by non-computational biologists is crucial for bridging the gap between data analysis and biological insight.

While ST has enormous potential in clinical research, translating findings into clinical settings faces several challenges. First, ST techniques require costly equipment, specialized reagents, and technical expertise, limiting their accessibility in clinical labs. Second, the clinical relevance of ST findings often depends on validation in larger patient cohorts, which can be difficult to achieve due to the high cost and time-intensive nature of spatial data acquisition. Establishing standards and best practices for ST in clinical research, as well as developing cost-effective workflows, will be essential for advancing precision medicine applications.

ST, especially when applied to human samples, presents ethical and data privacy considerations. Patient-derived tissue samples often contain unique molecular signatures that could be used to identify individuals, raising concerns around data sharing and privacy. In clinical applications, ensuring compliance with privacy regulations (e.g., HIPAA, GDPR) and obtaining proper consent for the use of spatial data are necessary to protect patient rights. Establishing guidelines and secure frameworks for data sharing and storage, especially in collaborative research settings, will be critical for advancing ST in clinical contexts.

8.6.4 Future Directions and Emerging Innovations

Future innovations in ST will likely focus on achieving higher spatial resolution while simultaneously expanding the range of detectable biomolecules. Integrating transcriptomics with proteomics, epigenomics, and metabolomics in a spatially resolved manner will offer a more comprehensive view of cellular states and interactions. For example, emerging techniques such as spatial metabolomics and spatial ATAC-seq will complement RNA profiling, allowing researchers to analyze metabolic activity and chromatin accessibility within specific tissue regions. Continued advancements in molecular imaging, nanoscale detection methods, and multiplexing will be critical for developing comprehensive spatial multi-omics platforms.

Automated image analysis and machine learning are set to play a transformative role in ST. Deep learning models, such as CNNs, are already being used for image segmentation, cell type classification, and spatial pattern recognition in transcriptomics data. Additionally, GNNs hold potential for analyzing ST data by incorporating spatial relationships and neighborhood information. Automated tools that combine machine learning with ST data processing will streamline analyses, reduce user bias, and increase accessibility for researchers without computational expertise.

One of the most promising directions for ST is the development of real-time and in situ profiling techniques, which would enable researchers to study gene expression dynamics in live tissues or organoid cultures. Advances in biosensors and live imaging techniques could enable real-time tracking of gene expression changes in response to stimuli, enhancing our understanding of dynamic processes like wound healing, immune responses, and tumor progression. Real-time spatial profiling would allow researchers to capture temporal changes in spatial gene expression, providing insights into how cellular states and interactions evolve within their native context.

As ST technologies mature, there is increasing potential to create personalized spatial maps of tissue organization for individual patients. Such maps would allow clinicians to capture the unique spatial landscape of a patient's tissue, providing insights into disease heterogeneity, progression, and response to treatment. For example, in oncology, personalized spatial maps of the tumor microenvironment could guide targeted therapies and identify regions that may benefit from different treatment approaches. The development of high-throughput, cost-effective workflows that generate personalized spatial maps will be essential for incorporating ST into precision medicine and personalized treatment plans.

ST is poised to transform our understanding of tissue biology and disease, yet challenges in technology, data analysis, and clinical translation must be addressed to realize its full potential. As higher-resolution, multi-omics, and real-time techniques emerge, ST will become more powerful and versatile, enabling deeper insights into cellular states, interactions, and disease mechanisms. By addressing current limitations and embracing future innovations, ST has the potential to become a cornerstone of systems biology and precision medicine.

References

1. Bressan, D., G. Battistoni, and G.J. Hannon. The dawn of spatial omics. *Science*. 2023; **381**(6657): p. eabq4964.
2. Rao, A., et al. Exploring tissue architecture using spatial transcriptomics. *Nat*. 2021; **596**(7871): p. 211–220.
3. Chen, T.Y., et al. Spatial transcriptomic technologies. *Cells*. 2023; **12**(16).
4. Liu, L., et al. Spatiotemporal omics for biology and medicine. *Cell*. 2024; **187**(17): p. 4488–4519.
5. Zeira, R., et al. Alignment and integration of spatial transcriptomics data. *Nat Methods*. 2022; **19**(5): p. 567–575.
6. Li, B., et al. Benchmarking spatial and single-cell transcriptomics integration methods for transcript distribution prediction and cell type deconvolution. *Nat Methods*. 2022; **19**(6): p. 662–670.
7. Zhang, H., et al. BayesTME: an end-to-end method for multiscale spatial transcriptional profiling of the tissue microenvironment. *Cell Syst*. 2023; **14**(7): p. 605–619.
8. Yang, W., et al. Deciphering cell–cell communication at single-cell resolution for spatial transcriptomics with subgraph-based graph attention network. *Nat Commun*. 2024; **15**(1): p. 7101.
9. Dong, K., S. Zhang. Deciphering spatial domains from spatially resolved transcriptomics with an adaptive graph attention auto-encoder. *Nat Commun*. 2022; **13**(1).
10. Kuppe, C., et al. Decoding myofibroblast origins in human kidney fibrosis. *Nat*. 2021; **589**(7841): p. 281–286.
11. Lindskrog, S.V., et al. An integrated multi-omics analysis identifies prognostic molecular subtypes of non-muscle-invasive bladder cancer. *Nat Commun*. 2021; **12**(1): p. 2301.
12. Andersson, A., et al. Spatial deconvolution of HER2-positive breast cancer delineates tumor-associated cell type interactions. *Nat Commun*. 2021; **12**(1): p. 6012.

13. Berglund, E., et al. Spatial maps of prostate cancer transcriptomes reveal an unexplored landscape of heterogeneity. *Nat Commun*. 2018; **9**(1): p. 2419.
14. Gong, D., et al. Spatial oncology: translating contextual biology to the clinic. *Cancer Cell*. 2024; **42**(10): p. 1653–1675.
15. Asp, M., et al. A spatiotemporal organ-wide gene expression and cell atlas of the developing human heart. *Cell*. 2019; **179**(7): p. 1647–1660.
16. Mantri, M., et al. Spatiotemporal single-cell RNA sequencing of developing chicken hearts identifies interplay between cellular differentiation and morphogenesis. *Nat Commun*. 2021; **12**(1): p. 1771.
17. La Manno, G., et al. Molecular architecture of the developing mouse brain. *Nat*. 2021; **596**(7870): p. 92–96.
18. Network, B.I.C.C. A multimodal cell census and atlas of the mammalian primary motor cortex. *Nat*. 2021; **598**(7879): p. 86–102.
19. Akeret, K., et al. Spatial transcriptome data from coronal mouse brain sections after striatal injection of heme and heme-hemopexin. *Data Brief*. 2022; **41**: p. 107866.
20. Ortiz, C., M. Carlen, and K. Meletis. Spatial transcriptomics: molecular maps of the mammalian brain. *Annu Rev Neurosci*. 2021; **44**: p. 547–562.
21. Joglekar, A., et al. A spatially resolved brain region- and cell type-specific isoform atlas of the postnatal mouse brain. *Nat Commun*. 2021; **12**(1): p. 463.
22. Fan, Z., et al. SPASCER: spatial transcriptomics annotation at single-cell resolution. *Nucleic Acids Res*. 2023; **51**(D1): p. D1138–D1149.
23. Carpenter, A.E., et al. CellProfiler: image analysis software for identifying and quantifying cell phenotypes. *Genome Biol*. 2006; **7**(10): p. R100.
24. Berg, S., et al. ilastik: interactive machine learning for (bio)image analysis. *Nat Methods*. 2019; **16**(12): p. 1226–1232.
25. Chen, H., D. Li, and Z. Bar-Joseph. SCS: cell segmentation for high-resolution spatial transcriptomics. *Nat Methods*. 2023; **20**(8): p. 1237–1243.
26. Jones, A., et al. Alignment of spatial genomics data using deep Gaussian processes. *Nat Methods*. 2023; **20**(9): p. 1379–1387.
27. Edsgard, D., P. Johnsson, and R. Sandberg. Identification of spatial expression trends in single-cell gene expression data. *Nat Methods*. 2018; **15**(5): p. 339–342.
28. Svensson, V., S.A. Teichmann, and O. Stegle. SpatialDE: identification of spatially variable genes. *Nat Methods*. 2018; **15**(5): p. 343–346.
29. Sun, S., J. Zhu, and X. Zhou. Statistical analysis of spatial expression patterns for spatially resolved transcriptomic studies. *Nat Methods*. 2020; **17**(2): p. 193–200.
30. Zhang, K., W. Feng, and P. Wang. Identification of spatially variable genes with graph cuts. *Nat Commun*. 2022; **13**(1): p. 5488.
31. Hu, J., et al. SpaGCN: integrating gene expression, spatial location and histology to identify spatial domains and spatially variable genes by graph convolutional network. *Nat Methods*. 2021; **18**(11): p. 1342–1351.
32. Shang, L., X. Zhou. Spatially aware dimension reduction for spatial transcriptomics. *Nat Commun*. 2022; **13**(1): p. 7203.
33. Efremova, M., et al. CellPhoneDB: inferring cell–cell communication from combined expression of multi-subunit ligand–receptor complexes. *Nat Protoc*. 2020; **15**(4): p. 1484–1506.

34. Noël, F., et al. Dissection of intercellular communication using the transcriptome-based framework ICELLNET. *Nat Commun.* 2021; **12**(1).
35. Jin, S., et al. Inference and analysis of cell-–cell communication using CellChat. *Nat Commun.* 2021; **12**(1): p. 1088.
36. Wang, S., et al. Cell lineage and communication network inference via optimization for single-cell transcriptomics. *Nucleic Acids Res.* 2019; **47**(11): p. e66.
37. Browaeys, R., W. Saelens, and Y. Saeys. NicheNet: modeling intercellular communication by linking ligands to target genes. *Nat Methods.* 2020; **17**(2): p. 159–162.
38. Tsuyuzaki, K., M. Ishii, and I. Nikaido. Sctensor detects many-to-many cell–cell interactions from single cell RNA-sequencing data. *BMC Bioinformatics.* 2023; **24**(1): p. 420.
39. Dries, R., et al. Giotto: a toolbox for integrative analysis and visualization of spatial expression data. *Genome Biol.* 2021; **22**(1): p. 78.
40. Arnol, D., et al. Modeling cell–cell interactions from spatial molecular data with spatial variance component analysis. *Cell Rep.* 2019; **29**(1): p. 202–211.
41. Tanevski, J., et al. Explainable multiview framework for dissecting spatial relationships from highly multiplexed data. *Genome Biol.* 2022; **23**(1): p. 97.
42. Cang, Z., et al. Screening cell–cell communication in spatial transcriptomics via collective optimal transport. *Nat Methods.* 2023; **20**(2): p. 218–228.
43. Abdelaal, T., et al. SpaGE: spatial gene enhancement using scRNA-seq. *Nucleic Acids Res.* 2020; **48**(18): p. e107.
44. Welch, J.D., et al. Single-cell multi-omic integration compares and contrasts features of brain cell identity. *Cell.* 2019; **177**(7): p. 1873–1887.
45. Cable, D.M., et al. Robust decomposition of cell type mixtures in spatial transcriptomics. *Nat Biotechnol.* 2022; **40**(4): p. 517–526.
46. Dong, R., G.C. Yuan. SpatialDWLS: accurate deconvolution of spatial transcriptomic data. *Genome Biol.* 2021; **22**(1): p. 145.
47. Kleshchevnikov, V., et al. Cell2location maps fine-grained cell types in spatial transcriptomics. *Nat Biotechnol.* 2022; **40**(5): p. 661–671.
48. Sun, D., et al. STRIDE: accurately decomposing and integrating spatial transcriptomics using single-cell RNA sequencing. *Nucleic Acids Res.* 2022; **50**(7): p. e42.
49. Lopez, R., et al. DestVI identifies continuums of cell types in spatial transcriptomics data. *Nat Biotechnol.* 2022; **40**(9): p. 1360–1369.

CHAPTER 9

Spatial Proteomics Data Analysis

The subcellular localization of proteins is often tightly regulated by their functions, making the capture of protein spatial localization critically important for a comprehensive understanding of cell biology. Spatial proteomics, as a technique, leverages advances in microscopy imaging technology and mass spectrometry to enable detailed studies of protein spatial distribution characteristics and protein interaction networks. Recently, spatial proteomics has emerged as an essential method for studying protein and cellular spatial localization and has been applied to clinical research on various diseases. Meanwhile, with the rapid development of spatial proteomics, corresponding analytical workflows and computational methods have also seen significant expansion. Therefore, this section will provide a detailed introduction to the imaging principles, data acquisition, data analysis workflows, multimodal omics integration methods in spatial proteomics, and its applications in disease research. Finally, a summary of spatial proteomics and prospects for its future development will be presented.

9.1 Introduction to Spatial Proteomics

9.1.1 Spatial Proteomics Experimental Workflow

Spatial proteomics data can generally be categorized based on imaging principles into the following types: fluorescence-based antibody spatial proteomics methods and mass spectrometry-based spatial proteomics imaging methods (Fig. 9.1 and Table 9.1). Here, we will provide a detailed introduction to the currently available spatial proteomics technologies according to these principles.

Fluorescence-based antibody spatial proteomics methods

Fluorescence-based antibody spatial proteomics methods utilize specific antibodies or fluorescent probes to label target proteins, allowing for precise protein localization. These methods use microscopy to observe and record the spatial distribution of labeled proteins within tissues and cells, generating high-resolution spatial protein maps. Immunohistochemistry (IHC) represents one of the earliest techniques in this category, which relies on specific antibodies for protein localization. However, its limitation lies in the fact that only one protein can be localized per experiment.

Subsequently, multiplex immunofluorescence techniques, such as cyclic immunofluorescence (CycIF) [1], were developed to address this limitation. These methods use multiple fluorescent probes to increase the number of proteins that can be labeled simultaneously, enabling the detection of up to 30 proteins in a single experiment. However, spectral overlap between different fluorescent signals poses a challenge, limiting the capability for accurate quantification of multiple proteins in complex samples.

Currently, one of the most representative fluorescence imaging methods is co-detection by indexing (CODEX) [2]. This technique employs antibodies conjugated with DNA barcodes and performs multiple rounds of staining and imaging, enabling high-throughput detection of up to 50 proteins in a single sample. This method is particularly well-suited for detecting cell markers and conducting spatial analyses at the single-cell level. It has made significant contributions, especially in exploring the tumor microenvironment.

Mass spectrometry–based spatial proteomics imaging methods

Apart from fluorescence-based antibody spatial proteomics methods, mass spectrometry is another mainstream technique for spatial protein imaging. It uses mass spectrometers to separate protein peptides, enabling protein identification. By analyzing mass spectrometry signals and using image analysis software, it can quantitatively assess protein expression. Mass spectrometry-based spatial proteomics methods are divided into two major categories: spatial untargeted proteomics and targeted proteomics.

Mass spectrometry imaging (MSI) [3] is a spatial untargeted mass spectrometry method. It combines mass spectrometry analysis with imaging technology and requires only a small amount of samples to obtain protein-related data in a label-free manner. MSI can measure the spatial distribution of hundreds of analytes across various tissue sections.

Spatial targeted mass spectrometry methods also play a crucial role in research. The two most representative techniques in this category are imaging mass cytometry (IMC) [4] and multiplexed ion beam imaging (MIBI) [5]. IMC uses metal-conjugated antibodies to label target proteins in a sample. A laser ablates the sample point by point, releasing metal ions. The mass spectrometer detects signals based on the mass-to-charge ratio (m/z) of the metal ions, and image analysis software reconstructs the spatial distribution of proteins. IMC enables high-resolution localization and quantitative analysis of multiple protein markers. It is suitable for studies requiring high-dimensional spatial resolution data, such as tumor microenvironment analysis, where it has delivered excellent results. MIBI employs an ion beam to bombard the sample surface, exciting secondary ion signals. The mass spectrometer analyzes the m/z of these ions to distinguish metal-conjugated antibodies labeling different proteins. MIBI achieves high-resolution and highly multiplexed spatial analysis, capable of detecting up to 100 markers simultaneously, making it ideal for multidimensional studies of complex biological samples. Deep visual proteomics (DVP) [6] is currently one of the most promising spatial proteomics methods.

Table 9.1 Spatial proteomics technologies.

Methods	Imaging Categories	Imaging Samples
Co-detection by indexing (CODEX)	Fluorescence antibody based	FFPE/FF
Multiplex immunofluorescence (MIF)	Fluorescence antibody based	FFPE/FF
Cyclic immunofluorescence (CyCIF)	Fluorescence antibody based	FFPE/FF
Mass spectrometry imaging (MSI)	Mass spectrometry imaging	FFPE/FF
Imaging mass cytometry (IMC)	Mass spectrometry imaging	FFPE/FF
Multiplexed ion beam imaging (MIBI)	Mass spectrometry imaging	FFPE/TMA
Deep visual proteomics (DVP)	Mass spectrometry imaging	FFPE/FF

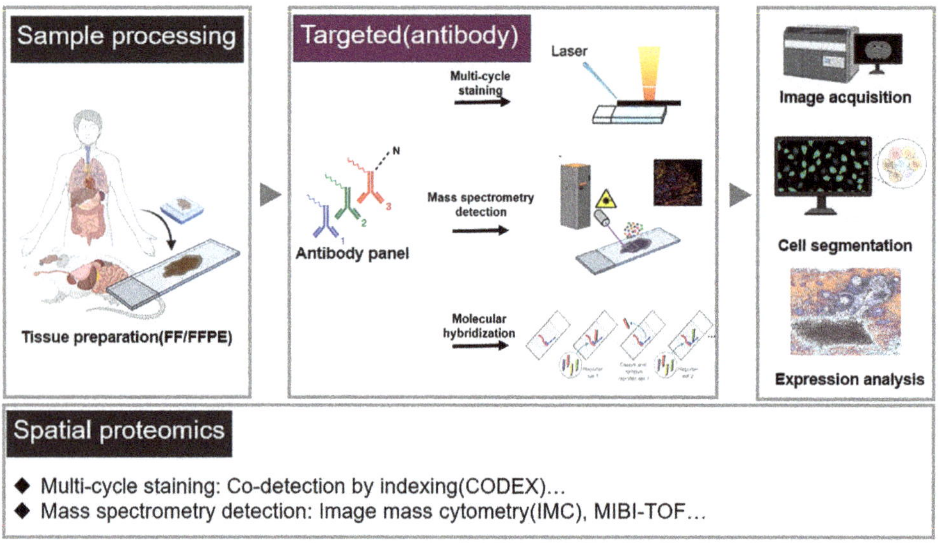

Fig. 9.1 Spatial proteomics.

By utilizing biology image analysis software (BIAS) to coordinate scanning microscopes and laser microdissection microscopes, DVP seamlessly integrates high-information-content imaging, deep learning-based cell segmentation, and machine learning-based cell-type and state identification. This enables in-depth exploration of protein characteristics while retaining complete spatial information, providing molecular insights into proteome changes at the phenotypic level. This approach not only enhances our understanding of single-cell heterogeneity and tissue complexity but also offers a novel analytical perspective for research into health and disease.

9.1.2 Spatial Proteomics Preprocess: Image Segmentation

Once spatial protein images are obtained, the next challenge is how to quickly extract cellular states and corresponding information from these images, which is critical for pathological research and diagnosis. Currently, cell images are primarily segmented manually by pathologists. However, relying solely on visual annotation and identification is highly labor-intensive. Therefore, there is an urgent need for automated segmentation methods to reduce labor and resource consumption. Utilizing computer vision to segment and count cell images can significantly alleviate the workload on medical personnel.

In recent years, numerous image analysis technologies have been applied to tasks such as automated cell segmentation, cell counting, tumor cell identification, and pathology image detection. Cell segmentation forms the foundation of cell image analysis. It divides cell images into nonoverlapping regions based on features such as grayscale, geometry, color, and texture.

Building upon this, advancements in deep learning and neural networks, such as convolutional neural networks (CNNs) and U-Net architectures, have significantly improved the accuracy and efficiency of automated cell segmentation. These models can adapt to complex cellular morphologies and heterogeneous tissue environments, enabling precise delineation of cell boundaries even in noisy or low-contrast images.

Moreover, integrating these segmentation technologies with spatial proteomics data allows researchers to analyze cellular states and protein distributions in a spatial context, providing critical insights into tissue microenvironments and disease mechanisms. Tools leveraging artificial intelligence (AI) can also perform multiscale analyses, identifying both individual cells and larger structural patterns within tissues.

These innovations not only reduce the workload for pathologists but also enhance the reproducibility and scalability of image analysis in clinical and research settings. As these technologies continue to evolve, they will play an increasingly essential role in enabling high-throughput analysis and supporting precision diagnostics in pathology and related fields.

For example, in biological cell image segmentation:

- Early tools such as CellProfiler [7] utilized thresholding methods to distinguish between the background and content, followed by morphological processing and watershed algorithms to address cell clustering. It also extracted features such as geometry, texture, and signal intensity for downstream analysis.
- Ilastik [8] introduced interactive segmentation, where users marked foreground and background regions, and the software used machine learning algorithms such as Random Forest to classify pixels, generate probability maps, and extract target regions with post-processing.
- Deep learning algorithms have since further advanced segmentation accuracy, especially for fluorescence microscopy images, by automatically learning features from large datasets and achieving superior performance in complex scenarios.

Here, we take Mesmer [9] as an example to explain the process of transforming a CODEX image from tissue sections into a spatial protein expression matrix (Fig. 9.2):

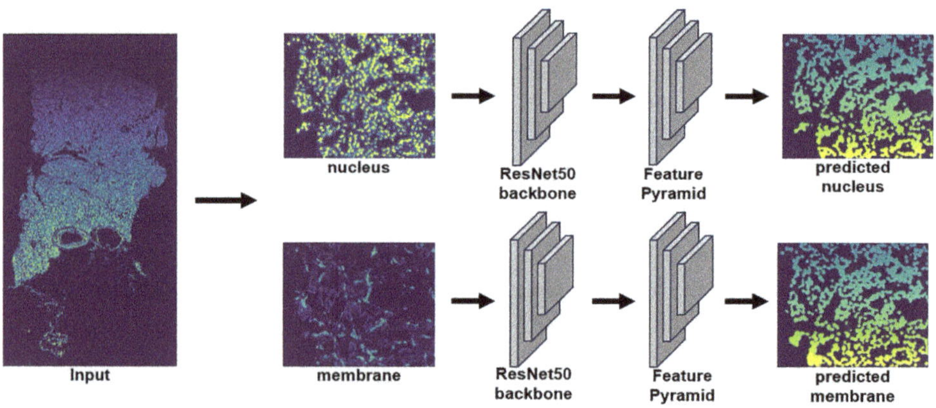

Fig. 9.2 Mesmer model overview.

Mesmer is a fully automated cell segmentation tool developed by DeepCell, designed specifically for cell segmentation tasks in complex biological samples, such as CODEX, MIBI, and other multichannel imaging data. It utilizes a dataset containing 1 million paired whole-cell and nuclear annotations, encompassing various imaging modalities. By training on this extensive dataset, Mesmer generates pretrained parameters that users can leverage to perform detailed segmentation of both nuclei and whole cells across different imaging techniques. This enables highly accurate and efficient cell segmentation for a wide range of applications.

First, tissue sections are labeled using antibodies conjugated with DNA barcodes, where each antibody binds to a specific protein marker. Subsequently, multiple rounds of cyclic detection are performed. In each detection cycle, specific fluorescent probes are paired with the DNA barcodes, and the distribution of the target proteins marked by the antibodies is captured using fluorescence microscopy. After imaging, chemical elution is used to remove the probes and markers, allowing the sample to be reused for the detection of the next antibody's target protein. Each imaging cycle produces high-resolution images of protein distributions, recording the protein signals from each cell or tissue region. After multiple cycles, the combined imaging data can cover up to 50 different proteins, resulting in a multichannel image where each channel corresponds to a specific protein marker. The resulting multichannel image is then input into Mesmer for further analysis.

The Mesmer model is composed of a ResNet50 [10] backbone network and a feature pyramid network (FPN). At the top of the FPN, there are four prediction heads: two dedicated to nuclear segmentation and two for whole-cell segmentation. This architecture enables the model to efficiently extract multiscale features and perform precise segmentation tasks for both nuclei and entire cells.

The inputs for Mesmer consist of two types of protein images: one composed of markers such as 4′,6-diamidino-2-phenylindole (DAPI), which define cell nuclei, and another composed of markers such as CD45 or E-cadherin, which define cell membranes. These two sets of multichannel images are preprocessed through concatenation, normalization, and cropping into fixed-size patches (allowing arbitrary initial sizes). The processed patches are then fed into the ResNet [11] backbone to generate feature representations.

The outputs of the deep learning model are used to generate predictions for nuclei and membranes, categorized into two types:

1. Pixel-wise transform: This output represents the classification probability of each pixel, indicating whether a pixel belongs to the cell's interior region, boundary, or background.
2. Inner distance transform: This output predicts the normalized distance of each pixel to the center of the cell, allowing the precise localization of the cell's centroid.

9.1.3 Spatial Proteomics Preprocess: Expression Matrix Generation

Using these predictions, a marker-based watershed algorithm converts the continuous predictions into discrete label images. In these images, each cell region's pixels are assigned a unique integer label, representing a single cell. With predictions from Mesmer, we can accurately locate each cell within a tissue section. Then, using quantification methods such as the regionprops function in the Python package skimage [12], we can quantify the expression values of all proteins for the identified cells based on signal intensities in the image. This results in a matrix composed of cell positions and the expression values of all marker proteins corresponding to each position. This matrix can then be input into downstream bioinformatics algorithms for further spatial analysis.

9.2 Exploring Cellular Distribution in Tissue: Computational Analysis of Spatial Proteomics

9.2.1 Large Spatial Proteomics Database Resources

Currently, there are multiple publicly available large-scale spatial proteomics datasets and corresponding integrated databases. This section provides a detailed introduction to these resources (Fig. 9.3 and Table 9.2).

The Human BioMolecular Atlas Program (HuBMAP) [13] is a large-scale, multimodal, and multiscale single-cell resolution spatial atlas for the healthy human body. HuBMAP spans various sequencing and imaging technologies, integrating modalities such as single-cell RNA sequencing (scRNA-seq), assay for transposase-accessible chromatin using sequencing (ATAC-seq), spatial transcriptomics (ST), and spatial proteomics. It encompasses data from 23 human tissues, 19 techniques, thousands of tissue sections, and detailed metadata. HuBMAP offers an outstanding interactive visualization platform to assist users in conducting their research, and the dataset is continuously updated. Within HuBMAP, spatial proteomics includes data generated from CODEX, IMC, and MIBI. Both protein expression matrices and raw images are publicly available for download.

Aquila [14] is a database that integrates data from all spatial omics technologies. It hosts a vast collection of spatial omics studies, including spatial proteomics data from CODEX and MIBI. Users can freely filter and download data of interest from the website and view the corresponding original publications. Aquila provides an online platform for basic spatial omics analyses and visualization. All foundational analysis results are visualized within this platform, with database modules including, but not limited to, basic data statistics, replay statistics, neighborhood networks, spatial autocorrelation, and spatial entropy.

SODB [15] also compiles spatial proteomics datasets. Similar to Aquila, SODB aggregates a large number of spatial omics datasets, including spatial proteomics data from MIBI, CODEX, IMC, t-CyCIF, 4i, and 3D-IMC. It offers detailed annotations for all spatial omics data. Users can explore spatial data at multiple scales through interactive visualization tools on SODB's website, including expression view, SOview, annotation view, and comparison view, to perform specific comparisons of gene and cell information. Moreover, all datasets are indexed to research publications and are available for public download.

scProAtlas [16] is a database specifically designed for spatial proteomics, integrating multiple spatial proteomics technologies, including CODEX, MIBI, IMC, CyCIF, and Chip Cytometry [17]. This database collects a vast amount of spatial proteomics data and provides open access to protein expression matrices derived from image segmentation. In addition, scProAtlas summarizes results ranging from single-method spatial proteomics analyses to multimodal integration analyses between spatial proteomics and single-cell technologies, offering users detailed spatial atlases of human tissues. The analytical modules provided by scProAtlas include, but are not limited to, neighborhood analysis,

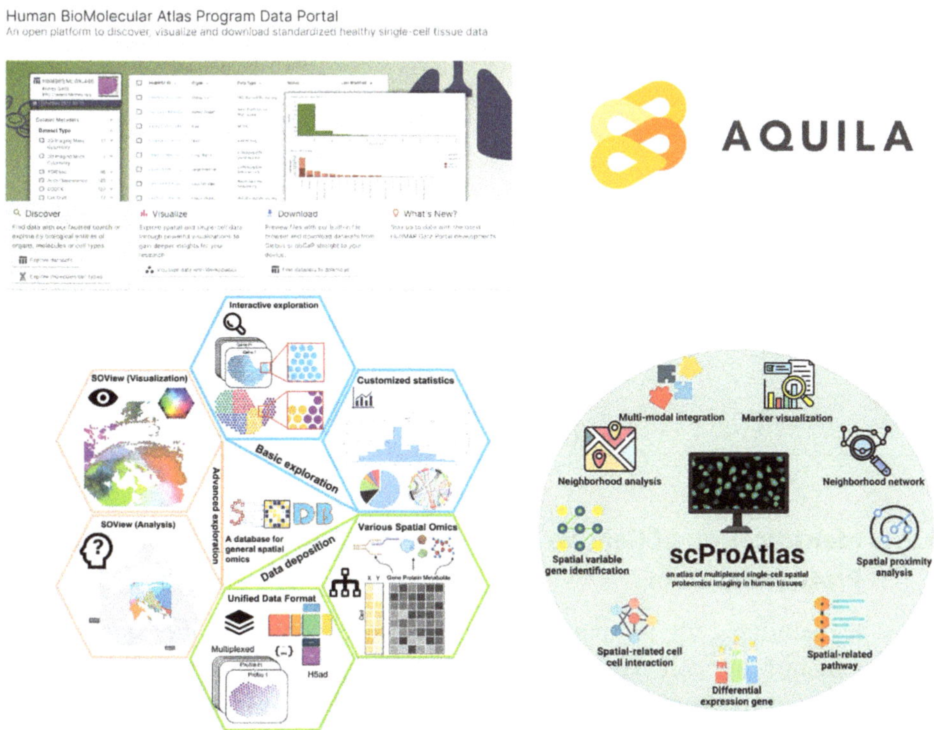

Fig. 9.3 Database resources of HuBMAP, Aquila, SODB, and scProAtlas.

neighborhood network, proximity analysis, spatial variable gene identification for multimodal integration, and cell-cell interaction analysis for multimodal integration.

Table 9.2 Spatial proteomics database.

Database/Data Resource	Data	Analysis
HuBMAP	• CODEX • LC-MS • IMC • MIBI	NA
HTAN	• CODEX • CyCIF • IMC • LC-MS • MIBI • Mass Spectrometry	NA
scProAtlas	• CODEX • IMC • MIBI • CyCIF • Chip Cytometry • MELC	• Cell-type annotation • Neighborhood identification • Neighborhood network • Spatial enrichment • Multimodal integration • Spatial gene identification • Cell–cell interaction
Aquila	• CODEX • CyCIF • IMC • MIBI	• Cell statistics • Spatial statistics • Neighborhood network • Spatial community • Cell–cell interaction • Spatial variable gene
SODB	• CODEX • MIBI • IMC • CyCIF	• Cell annotation • Spatial visualization • Multi-view visualization

9.2.2 Identification of Cellular Phenotypes Within Tissues Using Spatial Proteomics Data

After segmenting cells in spatial proteomics images, one of the most critical steps is identifying cell types. Protein channels in spatial proteomics typically select key cell markers to stain the target cell types of interest within tissues. Therefore, designing appropriate machine learning or deep learning algorithms to identify these cell types accurately becomes essential. Cell-type annotation is one of the most critical issues in transcriptomics. From single-cell transcriptomics to ST, numerous cell-type annotation models have been developed. Similarly, spatial proteomics also includes several annotation algorithms. These models can be categorized based on their annotation approaches, ranging from clustering-based cell-type annotation to multimodal integration for annotating cell types. In addition to clustering-based methods, supervised learning algorithms have been widely adopted in spatial proteomics to classify cell types using pre-labeled datasets, providing robust and efficient annotation for new data. Semi-supervised methods also leverage prior biological knowledge or partially labeled datasets, enabling annotation in cases with limited labeled data. Furthermore, multimodal integration approaches combine spatial proteomics data with complementary omics layers, such as transcriptomics or metabolomics, to enhance the resolution and accuracy of cell-type annotation. These diverse methods not only help accurately identify cell types but also uncover spatial patterns of tissue organization and intercellular interactions. This capability is essential for advancing our understanding of complex biological systems, investigating disease mechanisms, and discovering new biomarkers for precision medicine. By combining computational innovations with spatial proteomics, researchers can push the boundaries of tissue-level biology and translational research.

Here, we will highlight some mainstream models based on different cell-type annotation methods. CELESTA [18] is an unsupervised machine learning algorithm that identifies the cell type of each cell individually using its marker expression profile and spatial information. It avoids manual gating or clustering and minimizes reliance on user-defined hyperparameters by leveraging marker expression and spatial data. CELESTA employs an optimized framework that uses prior knowledge to transparently assign single cells to the most probable cell types (anchor cells). Through iterative processes, it assigns all cells to the initially identified anchor cells.

AnnoSpat [19] is another annotation model developed for spatial proteomics. It utilizes neural networks and point process algorithms to automatically identify cell types and quantify spatial pattern relationships between cells. AnnoSpat integrates both semi-supervised and supervised learning methods, enabling automated cell-type annotation of spatial proteomics data without requiring manually labeled cells for training. In addition, its point process-based algorithm quantifies spatial relationships between multiple cell types, providing insights into cellular interactions and organization.

Astir [20] is another probabilistic model that assigns cells to cell types by integrating prior knowledge of marker proteins. Astir uses a deep recognition neural network for rapid inference and can annotate millions of cells without requiring a previously annotated reference. Astir takes the measured expression data and a predefined set of marker proteins for all expected cell types as input and applies a statistical machine learning model to assign cell-type probabilities to each cell. This enables Astir to rapidly generate cell-type annotation results for large-scale datasets, processing millions of cells within a short period.

ACDC [21] is an automated annotation algorithm designed for high-throughput flow cytometry. It leverages the biological knowledge accumulated by humans and combines it with machine learning algorithms to achieve automated annotation of mass cytometry data. The ACDC method consists of two steps. First, ACDC converts a user-specified table of markers and cell labels into landmark points in high-dimensional space that represent the fingerprints of specific cell types. Then, ACDC performs semi-supervised classification using random walks, gathering information from all the landmark points to classify single-cell events. In ACDC, prior knowledge of canonical cell types is explicitly encoded through a user-specified table, transformed into landmark points, and ultimately fed into a semi-supervised learning algorithm. Finally, the classification results are integrated, and visualization tools are provided to support data interpretation and further analysis.

9.2.3 Identifying Tissue Structure via Neighborhood Analysis

After identifying cell types, performing neighborhood analysis becomes another critical step in spatial proteomics. Cell-type annotation provides information about individual cells, while neighborhood analysis focuses on the spatial aggregation patterns of cells across a tissue section. By calculating the proportions of cell types in local regions, neighborhood analysis allows for the division of the entire tissue into simplified local region distributions. Neighborhood analysis helps uncover critical patterns, such as immune cell clustering in inflamed regions, tumor microenvironment composition, or stromal cell contributions to epithelial organization. It also highlights cell-cell interactions and their influence on processes such as immune responses, tissue repair, or disease progression. For example, proximity analysis can identify interactions between immune and tumor cells, revealing potential regulatory or inhibitory mechanisms within the microenvironment.

By integrating these spatial insights with computational tools, neighborhood analysis provides a more profound understanding of tissue organization and function, offering new avenues for discovering spatially defined biomarkers, exploring disease mechanisms, and informing precision medicine strategies.

One of the most commonly used methods for neighborhood analysis is a clustering-based approach [22] (Fig. 9.4). In brief, this method consists of four steps: (1) Define a fixed-size window centered on each cell across the entire sample. The window size is a crucial hyperparameter, typically set to include cells within a radius equivalent to 10–30 cells in most studies. (2) For each window, calculate the proportion of all cell types within it, generating a window vector for each cell that represents the local cell-type composition. (3) Use a clustering algorithm (e.g., k-nearest neighbors [KNN]) to cluster the window vectors based on their similarity. (4) Manually define the meaning of each cluster based on the average cell-type composition within it. This approach allows researchers to identify the local distribution patterns of cell types around each cell and cluster regions with similar distributions, revealing regions with analogous spatial structures. However, this method often requires prior knowledge to inform the calculations, particularly in determining the window size.

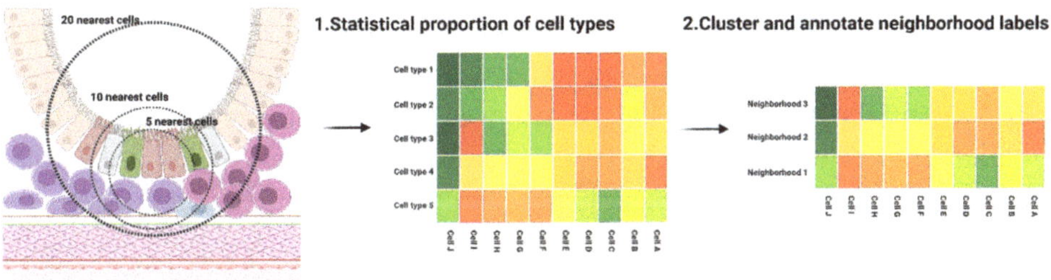

Fig. 9.4 Clustering method of neighborhood analysis.

Another method for identifying cellular neighborhood (CN) information in tissues is CytoCommunity [23], a deep learning model based on graph neural networks (GNNs). Its input includes one-hot encoded vectors representing cell types and spatial coordinates of each cell, which are used to construct an attributed graph representing the tissue structure. Depending on the downstream task, the model can be divided into two components: an unsupervised clustering component and a supervised learning component with differentiable pooling. The graph convolution layer transforms the input graph information into node embeddings, while a softmax layer outputs a soft assignment vector. For unsupervised clustering, the model uses a MinCut loss function to optimize the soft assignment vector, ultimately producing the best clustering results in the sample. For supervised classification tasks, the differentiable pooling module aggregates all node embeddings into a whole-graph embedding vector, enabling multi-sample supervised classification tasks, such as classifying different disease subtypes.

CellSNAP [24] is an information fusion using deep learning model that integrates spatial proteomics data with its original image information. It is applicable to all imaging-based spatial omics data and takes three types of inputs: an expression matrix, spatial location information, and raw images. CellSNAP processes these three inputs by first using a CNN encoder to extract features from the local tissue images surrounding each cell. It then constructs two parallel GNN models: one to process the local image features extracted by the CNN and another to handle the cell's expression profiles. Finally, the information from the two GNNs is combined through an Multilayer perceptron (MLP), enabling the unified representation to be used for downstream tasks such as unsupervised clustering and other analyses.

9.3 Multimodal Integration in Spatial Proteomics

Earlier, we introduced some analysis workflows that can be performed using spatial proteomics alone. In this section, we will provide a detailed introduction to the potential outcomes that can be achieved through the integration of spatial proteomics with other omics data. For example, combining spatial proteomics with transcriptomics links protein expression to gene regulation, while integration with metabolomics reveals how proteins influence localized metabolic pathways. Merging spatial proteomics with genomics enables mapping genetic variations to protein-level effects within tissues, offering valuable insights into complex diseases. Such approaches not only enhance biological understanding but also drive advancements in precision medicine, such as identifying new therapeutic targets and improving disease stratification. By integrating spatial proteomics with other omics layers, researchers can gain a more profound understanding of how molecular processes are organized within tissues. For instance, linking transcriptomics to spatial proteomics allows researchers to correlate gene expression patterns with spatial protein distributions, revealing tissue-specific regulatory mechanisms. Similarly, integration with metabolomics helps map how proteins and metabolites interact locally to drive physiological processes or contribute to disease progression.

Furthermore, combining spatial proteomics with genomics provides the ability to trace how genetic mutations or structural variants influence protein localization and function in a spatial context. This is particularly important in studying diseases such as cancer, where genetic and spatial factors converge to shape the tumor microenvironment. These integrated approaches offer not only enhanced biological insights but also tools to identify novel biomarkers, refine patient stratification, and develop targeted therapeutic strategies.

9.3.1 scRNA-seq and Spatial Proteomics Data Integration

Like ST, the most common integration approach is the combination of scRNA-seq and spatial proteomics. scRNA-seq allows us to study cells from various perspectives; however, its main limitation lies in the lack of spatial context. Integrating spatial proteomics with scRNA-seq naturally becomes an excellent choice, as it compensates for this limitation and provides more diverse results for subsequent analyses.

A commonly used tool is MaxFuse [25], which has provided valuable insights into cell-type annotation in numerous studies. MaxFuse takes labeled reference scRNA-seq data and targeted spatial proteomics data as inputs and applies fuzzy smoothing to both datasets. By calculating canonical correlation based on all features of the cross-modal matched cell pairs, MaxFuse captures shared information between the two modalities, establishing a connection between them. It learns a linear joint embedding that projects cells from different modalities into a shared space. Finally, based on the fuzzy-smoothed joint embedding coordinates, MaxFuse calculates pairwise distances between cross-modal cell pairs and updates the cross-modal cell matching using a linear assignment algorithm. Through iterative optimization, MaxFuse refines the final matching results.

MARIO [26], such as MaxFuse, is a cell-matching algorithm that matches partially overlapping features while fully accounting for the similarities and differences between modalities. MARIO uses singular value decomposition (SVD) on shared features and constructs a cross-modal distance matrix based on the Pearson correlation of the reduced matrices. The matching process is optimized by solving a linear assignment problem, minimizing the distance between cells from the two modalities. Finally, the matching results are refined through multiple rounds of filtering and correction, producing the final matching matrix between scRNA-seq and spatial proteomics data. This enables the transfer of information between the two modalities effectively.

Single-cell multi-omics data alignment (scMODAL) [27] is a deep learning framework capable of integrating scRNA-seq, spatial proteomics, and scATAC-seq. Designed specifically for unpaired datasets, scMODAL projects different single-cell datasets into a shared low-dimensional latent space and aligns cell embeddings using generative adversarial networks (GANs). To accurately identify correspondences among cell populations across datasets, scMODAL leverages prior information from known linked features to identify anchor cell pairs that guide the integration process while preserving the topology of all input features.

9.3.2 Spatial Transcriptomics and Spatial Proteomics Data Integration

In addition to scRNA-seq and spatial proteomics, the integration of ST and spatial proteomics can also be used for label transfer.

STELLAR [28] is a method based on geometric deep learning for discovering and identifying cell types in spatial data at single-cell resolution. STELLAR can match cells in the target dataset with cells in the reference dataset, automatically assign and identify cell types, and discover novel cell types that differ from those in the reference. First, STELLAR uses Graph Convolutional Networks (GCNs) to learn latent cell embeddings for cells in the target and reference datasets. This captures both spatial and molecular similarities among cells, embedding spatially proximate cells with similar gene or protein expression levels into nearby positions in the latent space. Furthermore, in an unannotated target dataset, STELLAR can automatically assign cells to known cell types from the reference dataset and identify cells with unique features as novel cell types not included in the reference. The loss function in STELLAR is divided into two components to support these functionalities. One component separates known cell types in the reference dataset by controlling their intraclass variance using the adaptive margin mechanism. The other generates auxiliary labels for unannotated data to discover novel cell types, with these labels based on the nearest neighbors of cells in the embedding space.

9.4 Spatial Proteomics Toolkits

Currently, there are several general-purpose toolkits that provide comprehensive frameworks for analyzing and visualizing spatial data. These toolkits integrate a broad array of commonly used analytical algorithms, enabling researchers to perform various tasks, such as data preprocessing, statistical analysis, and spatial visualization, all within a unified platform. By combining tools for spatial enrichment analysis, cell-cell interaction studies, and spatial clustering, these toolkits significantly simplify the analysis pipeline. In addition, many of these frameworks offer modular and extensible designs, allowing users to customize workflows and integrate domain-specific methods.

Some toolkits also include advanced features such as machine learning-based modeling, deep learning-driven image segmentation, and high-dimensional data integration, which are crucial for handling the increasing complexity of spatial datasets. With interactive visualization modules, researchers can intuitively explore spatial patterns, co-localization relationships, and tissue architecture, enhancing data interpretation and discovery. These toolkits cater to both novice users and advanced developers by providing comprehensive documentation, tutorials, and examples, fostering a collaborative and accessible research environment. As spatial omics data continues to grow in scale and complexity, such toolkits play an essential role in driving innovation and enabling breakthroughs in the field.

Squidpy [29] is a Python framework that integrates tools from omics and image analysis, enabling scalable characterization of spatial molecular data such as transcriptomes or multivariate proteins. Squidpy offers an efficient infrastructure and various analytical methods to effectively store, manipulate, and interactively visualize spatial omics data. Squidpy is an integrated framework that combines spatial neighborhood graphs and images, equipped with interactive visualization modules. Within the framework provided by Squidpy, various spatial omics analysis methods are integrated, allowing the construction of graphs such as Delaunay, radius, KNN, square, and hexagonal grids based on spatial information. Using permutation tests, Squidpy can compute spatial enrichment and cell–cell interactions within the data. Squidpy can also calculate spatial co-localization and centrality scores to illustrate relationships between clusters or cell types. By leveraging statistical methods such as Ripley's statistics and spatial autocorrelation, Squidpy can analyze spatial statistical patterns of features or cells. For image data, Squidpy offers visualization and processing tools, including image segmentation and feature calculation. In addition, it includes numerous spatial plotting functions to provide effective visualizations for image data.

Vitessce [30] is an interactive web-based visualization framework designed for exploring multimodal and spatially resolved single-cell data. Single-cell and spatially resolved multiomics technologies have made it possible to measure thousands of features across millions of cells. However, visual analysis of high-dimensional data types, such as transcriptomics, proteomics, genome-mapped data, and imaging, remains a significant challenge. Existing large-scale visualization tools are primarily focused on transcriptomics data and lack support for spatial data types (e.g., cell segmentations) or genome-mapped modalities (e.g., chromatin accessibility profiles). Vitessce addresses these limitations by providing a scalable web-based data visualization framework that supports simultaneous exploration of data from transcriptomics (e.g., scRNA-seq, scATAC-seq, CITE-seq, and ST), proteomics (e.g., CODEX, IMC, and MIBI), genome-mapped, and imaging modalities. A notable advantage of Vitessce is its ability to deploy visualizations of spatial and single-cell data across a wide range of environments and computing infrastructures, including static websites, web applications, data portals, Jupyter Notebooks, RStudio, and R Shiny applications.

We have summarized most of the current algorithms and toolkits related to spatial proteomics based on their input types, principles, and applications in Table 9.3:

Table 9.3 Spatial proteomics algorithms.

Models	Methods	Modality	Input	Applications
CELESTA	Anchor matching with prior knowledge	Spatial proteomics	Protein expression and marker list	• Cell-type annotation
AnnoSpat	Semi-supervised clustering with point process	Spatial proteomics	Protein expression and marker list	• Cell-type annotation
Astir	Deep recognition neural network	Spatial proteomics	Protein expression and marker list	• Cell-type annotation
ACDC	Random walk for semi-supervised clustering	Spatial proteomics	Protein expression	• Cell-type annotation
KNN	Clustering cell-type composition in each cell	Spatial proteomics	Cell-type label	• Spatial clustering
CytoCommunity	GNN with MinCut pooling and differentiable pooling	Spatial transcriptomics/ spatial proteomics	Cell-type label	• Spatial clustering

Models	Methods	Modality	Input	Applications
CellSNAP	Fusion deep learning model	Spatial proteomics	Protein expression, spatial location, raw image	• Spatial clustering
MaxFuse	Linear assignment	Spatial proteomics, scRNA-seq/ scATAC-seq	Protein expression, single-cell expression	• Multimodal integration • Cell-type annotation
MARIO	Linear assignment	Spatial proteomics, scRNA-seq/ scATAC-seq	Protein expression, single-cell expression	• Multimodal integration • Cell-type annotation
scMODAL	Generative adversarial networks (GANs)	Spatial proteomics, scRNA-seq/ scATAC-seq	Protein expression, single-cell expression	• Multimodal integration • Cell-type annotation
STELLAR	GCN	Spatial proteomics, scRNA-seq/ spatial proteomics	Protein expression, single-cell expression	• Multimodal integration • Cell-type annotation
Squidpy	Statistics methods	Spatial proteomics/ spatial transcriptomics	Protein/ single-cell expression, spatial location	• Spatial visualization • Spatial autocorrelation • Spatial enrichment • Spatial co-localization • Spatial interaction
Vitessce		Spatial proteomics/ spatial transcriptomics	Protein expression data, cell-type label, spatial location	• Spatial visualization • Single-cell visualization

9.5 Progress in Disease Research Through Spatial Proteomics

Cancer has long been one of the most challenging diseases globally, with rising incidence and mortality rates imposing a significant burden on healthcare systems worldwide. Therefore, exploring cancer mechanisms from various perspectives and developing effective treatment strategies are critical issues in medical research. As an advanced technology, spatial proteomics provides a more comprehensive perspective and novel approaches to studying cancer mechanisms from a spatial viewpoint.

Numerous studies have utilized spatial proteomics to conduct detailed cancer research. For instance, Schürch et al. [22] investigated whether the spatial structure of tissue neighborhoods, such as tertiary lymphoid structures (TLSs), is functionally relevant in the colorectal cancer (CRC) tissue microenvironment. They selected Crohn's-like reaction (CLR), characterized by abundant newly formed TLSs, and diffuse inflammatory infiltration (DII), which lacks TLSs, as the control group. Using CODEX, they constructed tissue microarrays (TMAs) that specifically represent the CRC invasive front tumor immune microenvironment (iTME). In this study, the researchers conducted an in-depth investigation of CNs in CRC samples. By coupling cell types with CNs, they sought to understand the dynamic processes within the iTME. They also analyzed the relationship between CNs and CRC survival rates under different conditions, identifying CNs that are specific to TLS. Subsequently, they built an inter-CN communication network to explore potential spatial communications between CNs and their impact on the iTME.

Liver cancer, as the third leading cause of cancer-related deaths worldwide, is becoming a severe global health issue, with mortality rates continuing to rise. Qiu et al. [31] also utilized CODEX technology to quantify the spatial status of 36 biomarkers in tumor nodules from 401 Hepatocellular Carcinoma (HCC) patients. By integrating clinical characteristics and prognosis information, they constructed a spatial single-cell resolution atlas of HCC and further explored the spatial heterogeneity between HCC samples and the complex cell-cell interaction modules. Moreover, they revealed how the spatial interactions between vimentin-high-expressing macrophages and regulatory T cells (Tregs) promote tumor progression and enhance the response to immunotherapy. In this study, the researchers conducted a detailed analysis of CNs within tissues. Using spatial proteomics data, they identified both cell types and CNs for all cells. Subsequently, based on the identified CN compositions, they clustered all HCC patients and categorized them into five tumor patterns: high levels of immune and stroma-infiltrated niches, low levels of immune and stromal-infiltrated niches, high levels of endothelial cells, higher levels of tumor–immune–mixed niches, and proliferating cell-dominated niches with low levels of immune and stromal cells. By training a random forest model, they determined the molecular importance for each pattern and identified the specific role of vimentin expression in defining tumor expression patterns. Through their analysis, they confirmed the interaction between vimentin-high-expressing macrophages and Tregs in the tumor microenvironment.

To study the spatial interactions between cancer and non-cancer cells, Mo et al. [32] conducted extensive CODEX imaging across six types of cancer and also generated a large number of ST and scRNA-seq samples. To describe tumor structures and niches, they defined tumor microregions as spatially distinct cancer cell clusters separated by stromal components and found that these microregions exhibited differences in size and density across cancer types, with the largest microregions observed in metastatic samples. They further identified spatial subclones within these microregions and observed subclonal structures in numerous sections, discovering significant variations in the functions and distributions of various cell types across different regions. Subsequently, they used RCTD to integrate ST and CODEX data, identifying immune-active (hot) and immune-cold (cold) neighborhoods, as well as enhanced immune exhaustion markers surrounding the three-dimensional (3D) subclones. These findings provide crucial insights into the spatial evolution of tumors through interactions with the local microenvironment in both two-dimensional and 3D spaces.

9.6 Conclusion

Spatial proteomics is an emerging research technology, and the demand for it has been growing rapidly recently, making it an indispensable part of spatial omics studies. By leveraging high-throughput imaging and mass spectrometry technologies, combined with advanced computational methods, spatial proteomics enables precise analysis of protein expression and spatial distribution in tissue or cell samples. By mapping proteins in their spatial context, spatial proteomics provides critical insights into the complex interplay between proteins, cellular architecture, and microenvironment. It has become a powerful tool for studying tissue heterogeneity, identifying cellular niches, and understanding disease progression and therapeutic responses at unprecedented resolution. Moreover, spatial proteomics offers significant applications in areas such as biomarker discovery, drug target identification, and the exploration of protein-protein interactions in both normal and pathological conditions. As technology evolves, innovations such as single-cell resolution imaging, multiplexed protein quantification, and integration with other omics data are further expanding the capabilities of spatial proteomics. The use of machine learning and AI to analyze high-dimensional spatial proteomics data is also paving the way for more automated, scalable, and interpretable analyses. These advancements are transforming spatial proteomics into an essential tool for addressing some of the most pressing questions in biology and medicine, driving forward research in areas such as oncology, neuroscience, and immunology.

Currently, the computational methods used in spatial proteomics rely on multimodal data integration and efficient analytical tools to perform comprehensive analyses. This chapter introduces various methods designed for diverse applications, such as image segmentation algorithms based on machine learning and deep learning (e.g., Mesmer), cell-type annotation models (e.g., CELESTA and AnnoSpat), neighborhood analysis models (e.g., window clustering methods and CytoCommunity), and cross-modal data integration algorithms (e.g., MaxFuse and scMODAL). Moreover, general-purpose toolkits such as Squidpy and Vitessce provide researchers with significant convenience for analyzing biological data. These tools integrate a variety of functionalities, enabling efficient handling, visualization, and interpretation of complex datasets, thus accelerating the discovery process in biological research. These methods not only enable the decomposition and visualization of high-dimensional protein expression data but also reconstruct the spatial structures and interactions of cells within the tissue microenvironment.

In terms of biological applications, spatial proteomics has shown great potential in cancer, neurological diseases, and immunological research. For instance, in cancer studies, spatial proteomics can reveal tumor progression mechanisms and immune evasion by analyzing CNs and their communication networks within the iTME. Moreover, by examining molecular importance across different neighborhoods and employing innovative analytical techniques, researchers can thoroughly investigate critical spatial mechanisms that may exist in localized regions of tissue. In addition, the integration of spatial proteomics with ST and single-cell omics provides a new perspective for multimodal analysis, enabling a more comprehensive understanding of cellular functions.

Looking ahead, spatial proteomics is poised to become even more efficient, precise, and versatile with continuous advancements in imaging technologies and computational algorithms. Innovations such as ultra-high-resolution imaging, automated sample preparation, and real-time data acquisition and analysis tools will significantly enhance the accuracy and throughput of spatial proteomics. The development of large-scale multimodal databases, such as HuBMAP, will provide a robust foundation for integrating spatial proteomics with other omics data, enabling more comprehensive analysis of tissue architecture and disease progression.

Moreover, the future of spatial proteomics lies in its deep integration with AI and machine learning. These technologies will empower researchers to extract meaningful insights from the vast, complex datasets generated by spatial proteomics, automating pattern recognition, biomarker discovery, and predictive modeling. The combination of spatial proteomics with multimodal omics approaches, such as transcriptomics, metabolomics, and epigenomics, will create a holistic view of cellular functions and their spatial context, further unraveling the complexity of biological systems.

These advances will also have transformative implications for precision medicine. Spatial proteomics will enable highly targeted exploration of disease mechanisms, revealing new therapeutic targets and biomarkers with unparalleled spatial and molecular resolution. By integrating patient-specific spatial data, it will drive the development of personalized diagnostics, treatment strategies, and monitoring systems tailored to individual molecular profiles.

In the long term, spatial proteomics has the potential to revolutionize fields such as oncology, neuroscience, and immunology. It will enable researchers to identify critical microenvironments, study tumor-immune interactions, and uncover the spatial organization of neural networks. Ultimately, these capabilities will lead to groundbreaking discoveries, not only in understanding complex diseases but also in advancing preventative healthcare and early intervention, shaping the future of human health and medicine.

References

1. Lin, J.-R., M. Fallahi-Sichani, and P.K. Sorger. Highly multiplexed imaging of single cells using a high-throughput cyclic immunofluorescence method. *Nat Commun*. 2015; **6**(1): p. 8390.
2. Black, S., et al. CODEX multiplexed tissue imaging with DNA-conjugated antibodies. *Nat Protoc*. 2021; **16**(8): p. 3802–3835.
3. Shimma, S. Mass spectrometry imaging. *Mass Spectrom (Tokyo)*. 2022; **11**(1): p. A0102.
4. Chang, Q., et al. Imaging mass cytometry. *Cytometry Part A*. 2017; **91**(2): p. 160–169.
5. Liu, C.C., et al. Reproducible, high-dimensional imaging in archival human tissue by multiplexed ion beam imaging by time-of-flight (MIBI-TOF). *Lab Invest*. 2022; **102**(7): p. 762–770.
6. Mund, A., et al. Deep visual proteomics defines single-cell identity and heterogeneity. *Nat Biotechnol*. 2022; **40**(8): p. 1231–1240.
7. Carpenter, A.E., et al. CellProfiler: image analysis software for identifying and quantifying cell phenotypes. *Genome Biol*. 2006; **7**: p. 1–11.
8. Berg, S., et al. Ilastik: interactive machine learning for (bio) image analysis. *Nat Methods*. 2019; **16**(12): p. 1226–1232.

9. Greenwald, N.F., et al. Whole-cell segmentation of tissue images with human-level performance using large-scale data annotation and deep learning. *Nat Biotechnol.* 2022; **40**(4): p. 555–565.
10. Koonce, B. ResNet 50. In: Koonce, B. *Convolutional Neural Networks with Swift for Tensorflow: Image Recognition and Dataset Categorization.* Apress; 2021: p. 63–72.
11. Targ, S., D. Almeida, and K. Lyman. *ResNet in ResNet: Generalizing Residual Architectures.* Cornell University; 2016.
12. Van der Walt, S., et al. scikit-image: image processing in Python. *PeerJ.* 2014; **2**: p. e453.
13. Jain, S., et al. Advances and prospects for the Human BioMolecular Atlas Program (HuBMAP). *Nat Cell Biol.* 2023; **25**(8): p. 1089–1100.
14. Zheng, Y., et al. Aquila: a spatial omics database and analysis platform. *Nucleic Acids Res.* 2023; **51**(D1): p. D827–D834.
15. Yuan, Z., et al. SODB facilitates comprehensive exploration of spatial omics data. *Nat Methods.* 2023; **20**(3): p. 387–399.
16. Wang, T., et al. scProAtlas: an atlas of multiplexed single-cell spatial proteomics imaging in human tissues. *Nucleic Acids Res.* 2025; **53**(D1): p. D582–D594.
17. Hennig, C., N. Adams, and G. Hansen. A versatile platform for comprehensive chip-based explorative cytometry. *Cytometry A.* 2009; **75**(4): p. 362–370.
18. Zhang, W., et al. Identification of cell types in multiplexed in situ images by combining protein expression and spatial information using CELESTA. *Nat Methods.* 2022; **19**(6): p. 759–769.
19. Mongia, A., et al. AnnoSpat annotates cell types and quantifies cellular arrangements from spatial proteomics. *Nat Commun.* 2024; **15**(1): p. 3744.
20. Geuenich, M. J., et al. Automated assignment of cell identity from single-cell multiplexed imaging and proteomic data. *Cell Syst.* 2021; **12**(12): p. 1173–1186. e5.
21. Lee, H.C., et al. Automated cell type discovery and classification through knowledge transfer. *Bioinformatics.* 2017; **33**(11): p. 1689–1695.
22. Schürch, C.M., et al. Coordinated cellular neighborhoods orchestrate antitumoral immunity at the colorectal cancer invasive front. *Cell.* 2020; **182**(5): p. 1341–1359. e19.
23. Hu, Y., et al. Unsupervised and supervised discovery of tissue cellular neighborhoods from cell phenotypes. *Nat Methods.* 2024; **21**(2): p. 267–278.
24. Zhu, B., et al. Cross-domain information fusion for enhanced cell population delineation in single-cell spatial-omics data. *bioRxiv* [Preprint]. 2024.
25. Chen, S., et al. Integration of spatial and single-cell data across modalities with weakly linked features. *Nat Biotechnol.* 2024; **42**(7): p. 1096–1106.
26. Zhu, B., et al. Robust single-cell matching and multimodal analysis using shared and distinct features. *Nat Methods.* 2023; **20**(2): p. 304–315.
27. Wang, G., et al. scMODAL: a general deep learning framework for comprehensive single-cell multi-omics data alignment with feature links. 2024: p. 2024.10. 01.616142.
28. Brbić, M., et al. Annotation of spatially resolved single-cell data with STELLAR. *Nat Methods.* 2022; **19**(11): p. 1411–1418.
29. Palla, G., et al. Squidpy: a scalable framework for spatial omics analysis. *Nat Methods.* 2022; **19**(2): p. 171–178.

30. Keller, M.S., et al. Vitessce: integrative visualization of multimodal and spatially resolved single-cell data. *Nat Methods*. 2025; **22**(1): p. 63–67.
31. Qiu, X., et al. Spatial single-cell protein landscape reveals vimentin high macrophages as immune-suppressive in the microenvironment of hepatocellular carcinoma. *Nat Cancer*. 2024; **5**: p. 1557–1578.
32. Mo, C.-K., et al. Tumour evolution and microenvironment interactions in 2D and 3D space. 2024; **634**(8036): p. 1178–1186.

CHAPTER 10

Single-Cell Multimodal Integration

Advances in single-cell technologies have enabled high-throughput profiling at single-cell resolution [1–4], providing a profound understanding of cell heterogeneity and functions [3, 5–7]. The emergence of single-cell multi-omics technologies offers many exciting biological opportunities from multiple perspectives, such as genomics, transcriptomics, epigenomics, and proteomics [8–11]. Integrating diverse omics datasets will contribute novel insights into complex cellular microenvironments and biological processes. In this chapter, we will provide a comprehensive overview of the multifaceted domain of single-cell multi-omics integration. This includes an in-depth discussion of its significance in advancing our understanding of complex biological systems and cellular heterogeneity. We will explore the challenges inherent in integrating diverse omics data, such as differences in data modalities, technical noise, and batch effects. In addition, we will classify and elaborate on the different types of integration, including intra-modality, paired inter-modality, and unpaired inter-modality approaches, each addressing specific scenarios in single-cell research. Finally, we will introduce innovative methodologies and state-of-the-art tools that have been developed to overcome these challenges, paving the way for a more unified and holistic view of cellular functions and interactions.

10.1 Introduction

10.1.1 Introduction to Single-Cell Multi-Omics Data

Single-cell multi-omics data provide an unprecedented opportunity to study the diverse molecular layers of individual cells, offering insights into the interplay between various biological processes such as gene expression, chromatin accessibility, DNA methylation, and protein activity. Unlike single-omics approaches, which capture only one dimension of cellular biology, multi-omics methods integrate data from multiple layers, enabling a comprehensive understanding of cellular states, functions, and heterogeneity. Figure 10.1 illustrates the diversity of single-cell multi-omics technologies, organized into four primary categories: genomics, transcriptomics, epigenomics, and proteomics.

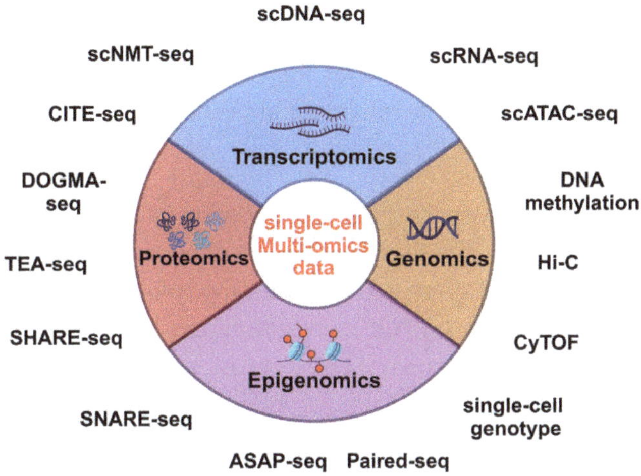

Fig. 10.1 Diversity of single-cell multi-omics technologies.

Single-cell genomics

Single-cell genomics focuses on analyzing the DNA content of individual cells to explore genetic variations, such as mutations, copy number variations, and structural alterations. This omics layer is essential for studying genetic heterogeneity within tissues, identifying clonal evolution in cancer, and uncovering rare genetic alterations that may drive disease. Single-cell genomics techniques, such as whole-genome amplification followed by sequencing, provide high-resolution insights into genetic diversity at the cellular level.

Single-cell transcriptomics

Single-cell transcriptomics, often performed using single-cell RNA sequencing (scRNA-seq), examines the gene expression profiles of individual cells. This technique enables researchers to study cellular heterogeneity, identify rare cell types, and explore dynamic processes such as differentiation and response to stimuli. Transcriptomics serves as a crucial layer for understanding how genetic information is translated into functional molecules, shedding light on the regulatory mechanisms that govern cellular behavior.

Single-cell epigenomics

Epigenomics explores the chromatin landscape and epigenetic modifications, such as DNA methylation and histone modifications, at single-cell resolution. Techniques such as single-cell assay for transposase-accessible chromatin sequencing (scATAC-seq) and single-cell bisulfite sequencing enable researchers to investigate how chromatin accessibility and epigenetic marks regulate gene expression. By integrating epigenomic data with transcriptomic and genomic data, researchers can unravel the complex regulatory networks that drive cell-type specificity and developmental processes.

Single-cell proteomics

Single-cell proteomics measures the protein content of individual cells, offering a direct assessment of the functional molecules that mediate cellular processes. Mass spectrometry–based approaches and antibody-based techniques, such as cytometry by time of flight (CyTOF), are commonly used for proteomics studies. Since proteins are the primary effectors of cellular function, integrating proteomic data with other omics layers provides a more complete understanding of cellular phenotypes and their regulatory mechanisms.

In summary, single-cell multi-omics technologies provide a powerful and versatile toolkit for exploring the molecular complexity of individual cells, enabling researchers to unravel the interplay of various biological layers. This integrative approach holds immense promise for advancing our understanding of cellular mechanisms, disease pathogenesis, and therapeutic interventions.

10.1.2 Significance of Multi-Omics Integration

Single-cell multi-omics technologies have been increasingly used to provide profound insights into the complex cellular microenvironment and biological processes. Integrative analysis of single-cell multi-omics data has offered many exciting biological opportunities and revealed the molecular determinants of human diseases [12–14]. By integrating transcriptomics and epigenomics, scientists can unravel how chromatin accessibility and epigenetic modifications influence gene expression, providing a dynamic understanding of gene regulation. Similarly, the combination of genomics and proteomics links genetic mutations to functional changes at the protein level, bridging the gap between genotype and phenotype. These multi-layered analyses uncover complex regulatory networks and provide context for how cellular functions are orchestrated and disrupted in different states. For example, Stuart et al. [15] integrated scRNA-seq and DNA methylation profiles to reveal the putative mechanisms of cell-type-specific epigenomic regulation within their defined mouse cortical cell types. Nativio et al. [16] identified molecular pathways and epigenetic alterations underlying late-onset Alzheimer's disease by integrating transcriptomic, proteomic, and epigenomic profiling of human brains. Bian et al. [17] reconstructed genetic lineages and traced the epigenomic and transcriptomic dynamics through single-cell multi-omics. Granja et al. [18] identified both patient-shared malignant signatures and patient-specific regulatory features such as RUNX1-linked regulatory elements via integrating single-cell transcriptomic and chromatin-accessibility profiles in acute leukemia analysis. These studies highlight the significance of single-cell multi-omics integration in accelerating the explorations of cell-type identification, gene regulatory, and disease states, as well as illuminating the causes and underlying mechanisms of human diseases.

By capturing the complexity of biological systems at single-cell resolution, multi-omics approaches are poised to revolutionize our understanding of cellular processes and their implications for health and disease. They provide a framework for exploring previously uncharted territories in biology, from deciphering the mechanisms of cellular differentiation and immune response to unraveling the origins of developmental disorders. As computational and experimental techniques continue to advance, single-cell multi-omics will become an even more powerful tool for addressing fundamental questions in biology and translating discoveries into clinical applications.

10.1.3 Challenges in Single-Cell Multi-Omics Data Analysis

Despite significant advancements in computational methods for the integrative analysis of multi-omics data, numerous challenges persist in developing robust and reliable approaches. One major hurdle stems from the variability in datasets, which are often generated from different sources, platforms, species, and experimental conditions. These discrepancies introduce considerable complexity in harmonizing and integrating diverse data types into a unified framework.

Single-cell sequencing data, in particular, are prone to high technical noise and dropout events due to the inherently low amounts of mRNA in individual cells and the inefficiency of signal capture. This technical variability complicates the analysis, often masking subtle biological signals. In addition, data generated at different times or in different laboratories may exhibit significant batch effects, further exacerbating the challenges of integration.

Storage formats for multi-omics data vary widely depending on the platform used, which adds another layer of complexity. Tools for multi-omics integrative analysis typically require data to be in specific formats, necessitating extensive preprocessing. This includes tasks such as data imputation to address missing values, systematic normalization to ensure comparability across datasets, removal of batch effects to minimize technical variability, and stringent quality control checks to maintain data reliability.

The goal of integrative analysis is to address critical biological questions by applying appropriate methodologies. Achieving this requires accounting for the inherent biological differences between tissues, species, and molecular layers (e.g., RNA-seq versus ATAC-seq). These biological nuances are crucial for accurately interpreting the integrated data and extracting meaningful insights. For instance, analyzing multi-omics data derived from different tissue contexts, disease states, and molecular modalities can uncover complex cellular and disease mechanisms that would remain hidden in single-omics studies.

To fully harness the potential of single-cell multi-omics data, there is an urgent need to develop advanced machine learning and statistical approaches. These methods must be capable of integrating heterogeneous datasets while preserving their biological relevance. By addressing these challenges, researchers can unlock the full potential of single-cell multi-omics, enabling a more profound understanding of cellular systems and answering a broad range of biological questions.

10.2 Types of Single-Cell Multi-Omics Integration

The integration of single-cell multi-omics data can be categorized into three main types: intra-modality integration, paired inter-modality integration, and unpaired inter-modality integration [19]. Figure 10.2 shows the integration types using scRNA-seq and scATAC-seq as examples.

10.2.1 Intra-Modality Integration

Advances in single-cell sequencing technology have revolutionized our ability to study cellular heterogeneity, developmental dynamics, and cell-to-cell interactions within diverse biological systems. These advancements have enabled the generation of high-throughput single-cell gene expression data using various

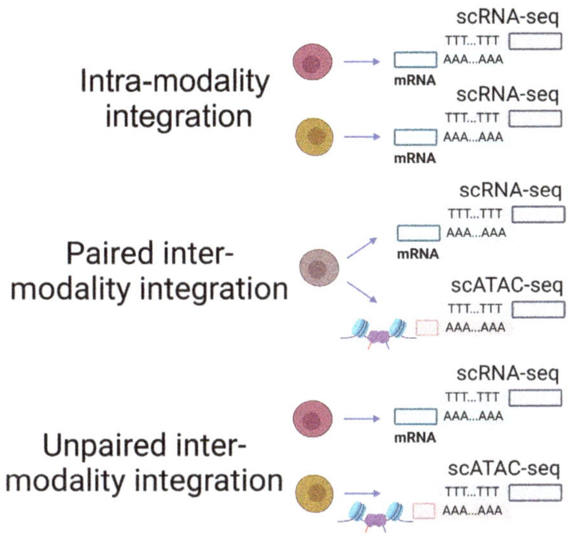

Fig. 10.2 Types of integration between scRNA-seq and scATAC-seq Created with BioRender.com.

sequencing methods [20], offering unprecedented resolution for understanding cellular function. While scRNA-seq data allows researchers to perform detailed cell-type-specific studies, many analyses, such as constructing comprehensive catalogs of healthy and diseased cells, require the integration of datasets from multiple donors and technological platforms.

However, significant differences often exist between scRNA-seq datasets generated from different donors and platforms. These discrepancies introduce batch effects during the data integration process, which can obscure true biological signals and impact downstream analyses reliant on the integrated data. Addressing these challenges is critical for producing reliable and biologically meaningful insights.

Intra-modality integration focuses on harmonizing and analyzing data within a single-omics layer, such as transcriptomics, across various datasets or experimental conditions. This type of integration is essential for mitigating batch effects, reducing technical noise, and addressing inconsistencies arising from variations in data generation protocols. For instance, combining scRNA-seq datasets from multiple studies can help create a more comprehensive atlas of cell types, encompassing broader biological diversity and enhancing the statistical power of downstream analyses. By harmonizing these datasets, intra-modality integration provides a more accurate and holistic understanding of cellular functions and their variations across conditions or populations.

10.2.2 Paired Inter-Modality Integration

scRNA-seq technology provides valuable transcriptomic-level insights at single-cell resolution. However, it is limited to capturing only RNA expression profiles. The advent of single-cell multi-omics technologies has expanded this scope, enabling the measurement of other molecular layers within individual cells, for example, studying chromatin accessibility, chromatin structural organization, histone modification, and DNA methylation status in single cells. Among them, chromatin accessibility is the data generated by single-cell sequencing assay for transposase-accessible chromatin technology (scATAC-seq), which is an epigenomic analysis technology used to measure chromatin accessibility to discover cell-type-specific regulatory mechanisms [21, 22]. Recent advancements in single-cell technologies have enabled the simultaneous measurement of RNA and ATAC sequencing data within the same cell. Methods such as SHARE-seq [23], sci-CAR [24], SNARE-seq [25], and paired-seq [26] generate paired datasets, wherein both transcriptomic and epigenomic information are collected from the same cell. This paired inter-modality integration allows for the direct analysis of multiple molecular layers within individual cells, providing a unique opportunity to study interdependence. For example, paired datasets enable researchers to investigate how chromatin accessibility (scATAC-seq) influences gene expression (scRNA-seq) within the same cellular context. By ensuring direct correspondence between molecular modalities, paired datasets facilitate the construction of more accurate regulatory networks and the discovery of mechanisms driving cellular processes. This integration enhances our understanding of complex biological systems, offering new insights into how molecular layers interact to regulate cell function and behavior.

10.2.3 Unpaired Inter-Modality Integration

Unpaired inter-modality integration addresses the challenge of analyzing multi-omics datasets where different modalities are measured in separate cell populations rather than the same cells. Although these datasets lack direct correspondences between molecular layers, they are often the only available data for many biological systems. Overcoming this limitation requires sophisticated computational methods to infer relationships and align unpaired modalities. Unpaired data not only have different features but also frequently exhibit significant variations in cell count, thus giving rise to a distinct category of integration methods for unpaired data.

10.3 Methods for Single-Cell Multi-Omics Integration

Corresponding integration methods have been developed to address different types of data integration. Intra-modality integration methods handle data within the same modality, while paired inter-modality integration methods, designed for data such as scRNA-seq and scATAC-seq measured from the same cells, leverage techniques like matrix factorization [27], the weighted nearest neighbor (WNN) algorithm [28], and neural networks [29, 30] to derive a unified representation of cellular states. These methods are specifically tailored for paired data, which limits their applicability to unpaired datasets. Unpaired inter-modality data present greater challenges, as they differ not only in feature space but often also in cell number. As a result, a distinct set of integration methods has emerged to address unpaired data. These approaches focus on manifold learning and cell alignment in a shared embedding space, frequently employing neural networks [31–33]. In addition, a non-neural network approach has been proposed that utilizes nonnegative matrix factorization and an online learning algorithm to incorporate new data incrementally without requiring a complete re-computation [34]. In this section, we will describe these integration strategies in detail.

10.3.1 Intra-Modality Integration Methods

scRNA-seq technology is advancing at an extraordinary pace, with more than a dozen platforms now available for generating single-cell transcriptomic data [35]. These platforms enable researchers to capture gene expression at the resolution of individual cells, opening new frontiers in understanding cellular heterogeneity. However, data generated across different experimental platforms often exhibits pronounced batch effects, stemming from technical variations such as sequencing chemistry, instrument calibration, and sample handling. These batch effects can obscure underlying biological signals and present a significant challenge when integrating multi-platform scRNA-seq data.

Eliminating the impact of these batch effects is critical for ensuring the reliability and accuracy of downstream analyses, such as clustering, trajectory inference, and differential expression studies. However, addressing batch effects is particularly challenging due to their often nonlinear and complex nature, which complicates the task of harmonizing datasets while preserving biologically relevant variations.

To tackle these challenges, researchers have developed various algorithms aimed at integrating cross-platform scRNA-seq data. Methods for intra-modality integration employ dimensionality reduction algorithms to reduce the complexity of the data and identify the common biological signal across datasets to align cells or cell populations to integrate scRNA-seq datasets from different sources or experiments [36–39]. Haghverdi et al. introduced a mutual nearest neighbors (MNN) algorithm that identifies corresponding cells between datasets and reconstructs the data in a shared space [40]. While effective, this approach requires computing neighbor relationships in high-dimensional gene expression space, leading to high computational demands in terms of memory and processing time. To overcome this limitation, the fastMNN method was developed, applying the MNN algorithm to a low-dimensional space after principal component analysis (PCA). This innovation significantly reduces runtime while maintaining high accuracy, making it suitable for large-scale datasets.

Another notable approach is Harmony [38], which also employs PCA for dimensionality reduction. Harmony identifies similar cells across different batches in the reduced space and clusters them while ensuring batch diversity within each cluster. This method effectively aligns datasets while preserving critical biological differences, making it a powerful tool for cross-platform data integration.

Recently, deep learning techniques have also been applied to address batch effects. Shaham et al. developed a batch correction method using residual neural networks trained to minimize the maximum mean discrepancy (MMD) between the source and target data distributions [41]. This approach learns a mapping from the source to the target dataset during training, achieving effective batch correction. However, its performance can be limited on smaller datasets due to insufficient training data.

The development of these advanced algorithms underscores the importance of addressing batch effects in scRNA-seq integration. By harmonizing datasets across platforms while retaining meaningful biological signals, these methods facilitate more robust and comprehensive analyses of single-cell data, enabling deeper insights into complex biological systems.

10.3.2 Paired Inter-Modality Integration Methods

Methods for paired inter-modality integration apply matrix factorization [27], WNN algorithm [28], and neural networks [29, 30] to integrate different omics data measured within a cell and to obtain a joint definition of cellular state (Fig. 10.3). Each method addresses specific challenges of multi-omics data integration and offers unique strengths suited to different analytical requirements. These methods are specially designed for paired data, making their application to other unpaired data challenging. The insights gained from paired inter-modality integration are invaluable for revealing coordinated regulatory processes, such as how transcription factors interact with open chromatin regions to modulate gene expression.

Matrix decomposition approaches aim to simplify the data by representing each cell as the product of two components: a vector describing each omics element (e.g., gene, epigenetic locus, or protein) and a corresponding common feature vector. For instance, Multi-Omics Factor Analysis v2 (MOFA+) is an enhanced version of the MOFA [42] framework that leverages factor analysis and sparsity-driven Bayesian modeling to integrate single-cell data from different modalities and experimental conditions [43]. Another method, single-cell aggregation and integration (scAI) [27], specifically targets the integration of epigenetic data, such as chromatin accessibility and DNA methylation, with transcriptomic data. By learning similarities between cells, scAI groups similar cells together, effectively addressing sparsity issues inherent in epigenetic datasets. Although matrix decomposition methods are straightforward and interpretable, their ability to model complex nonlinear relationships is limited, which can restrict their utility in highly intricate datasets.

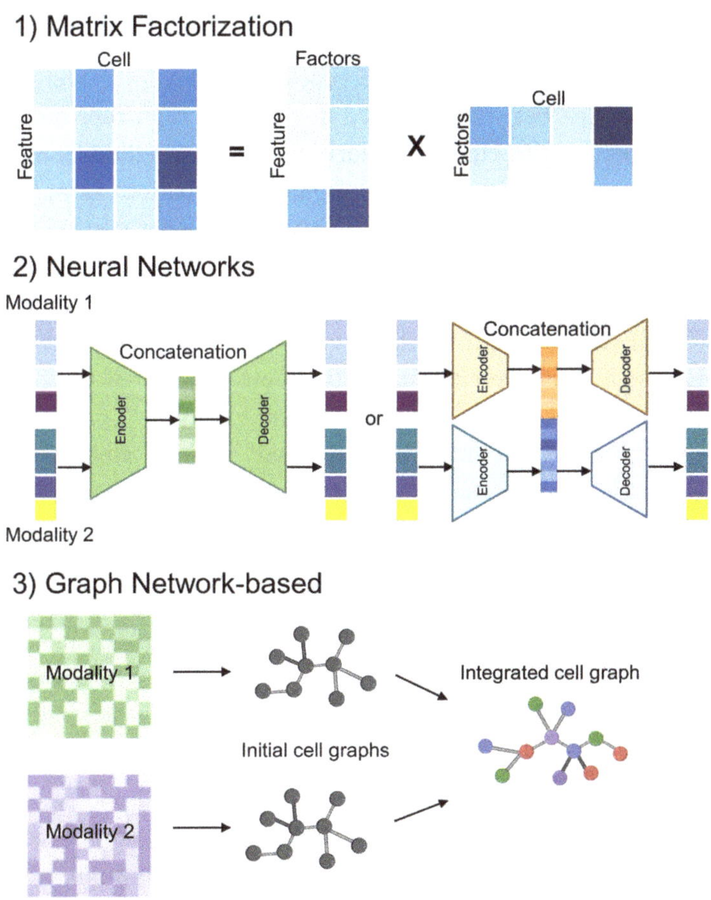

Fig. 10.3 Overview of methods for paired inter-modality integration.

Neural network–based methods, particularly those involving autoencoders, have shown great promise for multi-omics data integration. These methods use unsupervised learning to map high-dimensional data into a shared low-dimensional space, capturing complex relationships across modalities. For example, single-cell multimodal variational autoencoder (scMVAE) [30] utilizes multimodal encoders and random optimization to integrate omics data with similar cellular and feature characteristics. It approximates the joint latent feature distribution using a Gaussian mixture model (GMM) and reconstructs expression profiles through a decoder. Similarly, single-cell Multi-View Profiler (scMVP) [29] focuses on integrating gene expression and chromatin accessibility data from the same cell, employing a multi-view variational autoencoder (VAE) with cluster consistency constraints to project and integrate multi-omics data. Despite their strengths in handling nonlinear relationships, neural network–based methods often require extensive hyperparameter tuning, substantial computational resources, and careful optimization to achieve reliable results.

Network-based approaches, on the other hand, leverage weighted graphs to model relationships between cells. Nodes in the graph represent individual cells, while edges capture their similarities. These methods emphasize the neighborhood structure and often aggregate information from neighboring nodes to improve robustness to noise. For instance, Seurat v4 [28] uses a WNN graph to represent input data. In constructing the WNN graph, it learns specific weights for each cell, reflecting the relative importance of different omics data modalities. While this method has been effective for integrating transcriptomics and epigenomics data, its applicability to other types of paired single-cell multi-omics datasets remains an open question.

In summary, matrix decomposition methods provide simplicity and interpretability, neural network-based approaches excel in modeling nonlinear relationships, and network-based techniques enhance noise robustness through neighborhood aggregation. Together, these methods offer a comprehensive toolkit for the integration of multimodal single-cell data, each tailored to address specific challenges in uncovering complex biological insights. As computational tools continue to evolve, these approaches will play a critical role in advancing our understanding of cellular systems and their inter-omics dependencies.

10.3.3 Unpaired Inter-Modality Integration Methods

Compared to multi-omics data measured in the same cell, single-omics data are easier to obtain experimentally. These unpaired single-cell multi-omics data are typically collected from similar but distinct cells within the same tissue. The primary challenge in integrating such data lies in aligning the cells across different omics datasets. To address this, various machine learning and statistical learning methods have been developed for unpaired multi-omics data integration. These methods can be broadly categorized into Bayesian approaches, manifold learning techniques, and neural network–based models, as illustrated in Fig. 10.4.

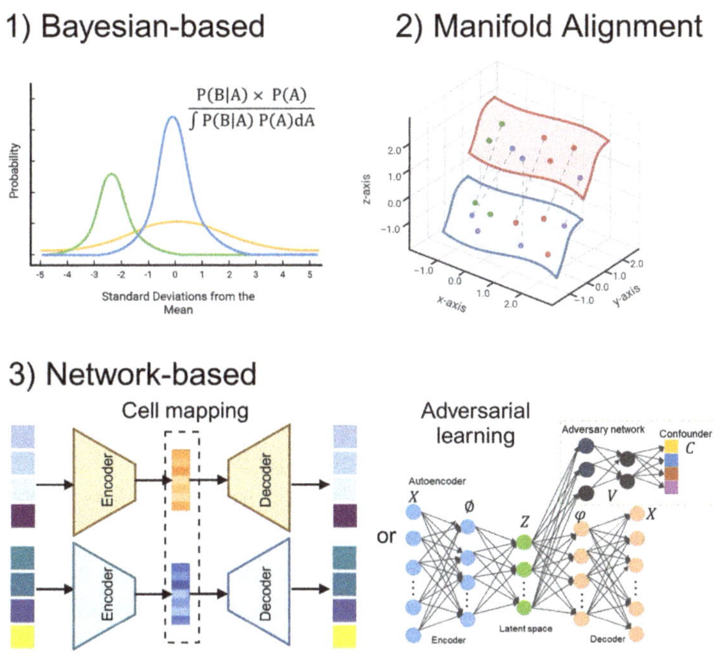

Fig. 10.4 Overview of methods for unpaired inter-modality integration.

Bayesian methods, such as Clonealign [44] and MUSIC [45], represent notable approaches in this category. Clonealign aligns cells measured by scRNA-seq with clones from single-cell DNA sequencing data using a Bayesian latent variable model, thereby integrating heterogeneous scRNA-seq and single-cell DNA sequencing data. MUSIC, on the other hand, links gene expression profiles to specific biological functions by modeling the perturbation effect. It leverages Bayesian outputs to estimate the impact of individual gene perturbations on cellular phenotypes, particularly in single-cell CRISPR data. However, both methods are limited in scope: Clonealign is restricted to integrating single-cell RNA and DNA data, while MUSIC focuses solely on combining single-cell

RNA data with CRISPR screening data. This specificity significantly narrows their applicability to broader single-cell omics integration scenarios.

Manifold learning-based methods aim to predict a shared low-dimensional representation of multi-omics data, facilitating alignment in a reduced-dimensional space. Pamona exemplifies this approach by using manifold learning to derive a low-dimensional representation and aligning datasets via the optimal transport model of partial Gromov-Wasserstein (GW) distance [46]. After alignment, Pamona employs Laplacian eigenmaps to project multi-omics data into a unified low-dimensional space for integration. While effective in certain contexts, Pamona's reliance on memory-intensive distance matrix storage and its diminished performance with larger sample sizes pose significant limitations.

Neural network-based approaches include methods utilizing graph neural networks and adversarial networks. For instance, single-cell graph convolutional network (scGCN) [47] leverages the topological structure between cells to learn high-order cellular representations and relationships, enhancing data integration performance. It enables label transfer across datasets through knowledge graphs, facilitating the application of insights gained from previous datasets to new ones. Other approaches, such as autoencoders, also demonstrate strong capabilities in transfer learning. However, the effectiveness of scGCN depends heavily on the quality of the knowledge graph, and like many artificial intelligence (AI) models, it suffers from limited interpretability due to its complex internal structure and "black box" nature.

Despite the variety of algorithms available for single-cell multi-omics data integration, most are designed to address specific types of omics data and lack the flexibility to solve all three major integration challenges simultaneously. Moreover, effective integration requires careful consideration of the inherent differences between omics features. Developing a universal model capable of integrating diverse single-cell omics data while addressing their unique limitations and data sparsity remains an open and critical area of research for the field.

10.4 Example Algorithm for Single-Cell Multi-Omics Data Integration: Cell Cycle-Aware Network

10.4.1 Background and Significance of Cell Cycle-Aware Network

The emergence of single-cell multi-omics technologies has enabled insights into complex cellular microenvironments and biological processes, offering many exciting biological opportunities from perspectives apart from transcriptomics, such as genomics, epigenomics, proteomics, metabolomics, and spatial transcriptomics [8–10]. In particular, scATAC-seq is an epigenomic profiling technology for studying the chromatin accessibility of individual cells [48–50]. It provides the ability to examine the openness of chromatin regions in the nucleus at the single-cell level, which is unavailable in scRNA-seq data [48]. This enhances our comprehension of the epigenetic state, cell-type heterogeneity, and cell state [51, 52]. However, scATAC-seq data has more extreme sparsity than scRNA-seq data, which also increases the difficulty of analysis based on scATAC-seq data [49, 53, 54]. Meanwhile, cell-type annotation in scATAC-seq data is challenging due to the lack of specifically designed tools and the use of unintuitive cis- and trans-regulatory elements in scATAC-seq data [55]. In recent years, advanced technologies have made it possible to simultaneously characterize gene expression and chromatin accessibility in the same cell, which we call the generated data paired data [23–26]. These techniques provide tools for the integrated analysis of scRNA-seq and scATAC-seq data, which can apply the information obtained from the large amount of annotated scRNA-seq data for the cell-type annotation of scATAC-seq data [56]. However, single-omics data are more readily available than paired scRNA-seq and scATAC-seq data. That is, we can generate scRNA-seq and scATAC-seq data of different single-cell experimental samples separately, but they are from the same organ or tissue, which we refer to as unpaired data in this study [2]. Therefore, in cases where paired data are not abundant, integrating these unpaired data is a better option for researchers to conduct broader studies. However, unpaired data frequently introduce complexity to subsequent analyses due to differences in features and sparsity level, so it is important to develop novel data integration methods that can be applied to unpaired scRNA-seq and scATAC-seq data.

Cell cycle-aware network (CCAN) [57] is a robust method designed to extract intrinsic biological signals masked by context-specific patterns (i.e., cell-type-specific heterogeneity) and confounding factors (i.e., cell cycle effects, batch effects, and noise). This versatile method offers a comprehensive solution capable of addressing all three types of integration issues outlined earlier. CCAN can integrate single-cell multi-omics data and remove cell cycle effects from the integrated data while maintaining heterogeneity between cell types. Its adaptability makes it suitable for a range of tasks, including single-cell multi-omics data integration, batch effect removal, cell cycle effects removal, label transferring, and cell-type prediction.

10.4.2 Basic Structure of Domain Separation Network in CCAN

Domain separation network (DSN) is a specific type of neural network architecture used for domain adaptation; it was designed based on labeled source-domain data and unlabeled target-domain data [58, 59]. In CCAN, source-domain data is gene expression profiles of scRNA-seq data, and target-domain data can be the same or different single-cell omics data, such as gene expression of scRNA-seq data from different protocols and chromatin accessibility of scATAC-seq data. In the basic structure of DSN, it contains one shared encoder and two domain-specific private encoders to extract the common and private representations from input data. The common and private components of the same domain should be totally split to ensure the independence of these parts. A shared decoder reconstructs the input domain by cascading the shared and private embeddings. Given a labeled dataset in a source domain and an unlabeled dataset in a target domain, we mainly use cell type classification as the cross-domain task, training on the shared embedding from the source domain that generalizes to the target domain, which requires the high invariance between shared embeddings of source and target domains. To achieve this goal, alignment is considered in the embedding space to eliminate differences between domains. Objectively, DSN is a model that produces a shared representation that is similar for both domains and a private representation that is different and transfers the classification label from the source domain to the target domain.

10.4.3 Cell Cycle-Aware Module in CCAN

The cell cycle has been recognized as a confounding factor in the analysis of cell-type-dependent processes. In CCAN, to achieve better cell-type transfer from the source domain to the target domain, it is necessary not only to account for differences in the shared embeddings between different domains but also to remove cell cycle effects from the shared representations. Considering the mutual independence between shared and private components in DSN, we introduce a cell cycle-aware module in the private part of DSN to characterize the dynamic process of the cell cycle. The high degree of differences between shared and private components can make shared components out of the influence of private components, that is, the prediction of cell types based on shared components can eliminate the effect of cell cycle modeling based on private components. Taking the source data as X_s and the target data as X_t, the objective of the cell cycle-aware module is to infer the cell cycle pseudotime $Z_{\cdot p}$ for cells from their corresponding profiles X_s or X_t. DSN is an autoencoder-based network; in the encoder, CCAN uses a standard multi-layer perceptron with hyperbolic tangent activation functions in the private part (also as a circular part) and Scaled Exponential Linear Unit (SELU) activation functions in the shared part (also as an acyclic part). The encoder of source data X_s is as follows:

$$Z_s = \begin{bmatrix} Z_{sp}^{(circular)} \\ Z_{ss}^{(acyclic)} \end{bmatrix} = \begin{bmatrix} W_3^{(circular)} \tanh\left(W_2^{(circular)} \tanh\left(W_1^{(circular)} X_s + b_1 \right) + b_2 \right) \\ W_3^{(acyclic)} \operatorname{selu}\left(W_2^{(acyclic)} \operatorname{selu}\left(W_1^{(acyclic)} X_s + b_1 \right) + b_2 \right) \end{bmatrix}, \tag{10.1}$$

where W's and b's are the weight matrices and bias vectors of the encoder; Z_{sp} and Z_{ss} are private embedding and shared embedding of the source domain; and Z_s represents the cascade of Z_{sp} and Z_{ss} in the hidden layer. In the decoder, CCAN uses cosine and sine as the activation functions in the first layer, followed by two layers performing linear transformations. The reconstruction of source data $\widehat{X_S}$ can be represented mathematically as

$$\widehat{X}_s = \left[V^{(circular)} V^{(acyclic)} \right] \begin{bmatrix} \sin Z_{sp}^{(circular)} \\ \cos Z_{sp}^{(circular)} \\ Z_{ss}^{(acyclic)} \end{bmatrix} = V Z_s, \tag{10.2}$$

where V's are the weight matrices of the decoder and $\widehat{X_S}$ is the reconstructed matrix of the source domain generated by the decoder. The target domain has the same autoencoder structure of the cell cycle-aware module in the private part as follows.

$$Z_t = \begin{bmatrix} Z_{tp}^{(circular)} \\ Z_{ts}^{(acyclic)} \end{bmatrix} = \begin{bmatrix} W_6^{(circular)} \tanh\left(W_5^{(circular)} \tanh\left(W_4^{(circular)} X_t + b_1 \right) + b_2 \right) \\ W_3^{(acyclic)} selu\left(W_2^{(acyclic)} \operatorname{selu}\left(W_1^{(acyclic)} X_t + b_1 \right) + b_2 \right) \end{bmatrix} \tag{10.3}$$

$$\widehat{X}_t = \left[V^{(circular)} V^{(acyclic)} \right] \begin{bmatrix} \sin Z_{tp}^{(circular)} \\ \cos Z_{tp}^{(circular)} \\ Z_{ts}^{(acyclic)} \end{bmatrix} = V Z_t, \tag{10.4}$$

where Z_{tp} and Z_{ts} are private embedding and shared embedding of the target domain; Z_t represents the cascade of Z_{tp} and Z_{ts} in the hidden layer; and $\widehat{X}_t$ is the reconstructed matrix of the target domain.

10.4.4 Algorithm Design and Training of CCAN

As illustrated in Fig. 10.5, CCAN uses a DSN to integrate data from source and target domains and transfer the annotations from source domain to target domain. Shared encoders and private encoders in DSN are three-layer perceptrons to learn circular and acyclic embeddings of both modalities separately. The shared embedding function projects a high-dimensional profile of each cell to a low-dimensional vector, which distinguishes biologically meaningful signals from circular confounding factors (private embedding) and transforms the embeddings of cells from different domains into a similar distribution. In the decoder, we use sine and cosine as the activation functions specific for private embeddings, followed by a two-layer perceptron performing noncircular transformations mapping the embedded data to the original space.

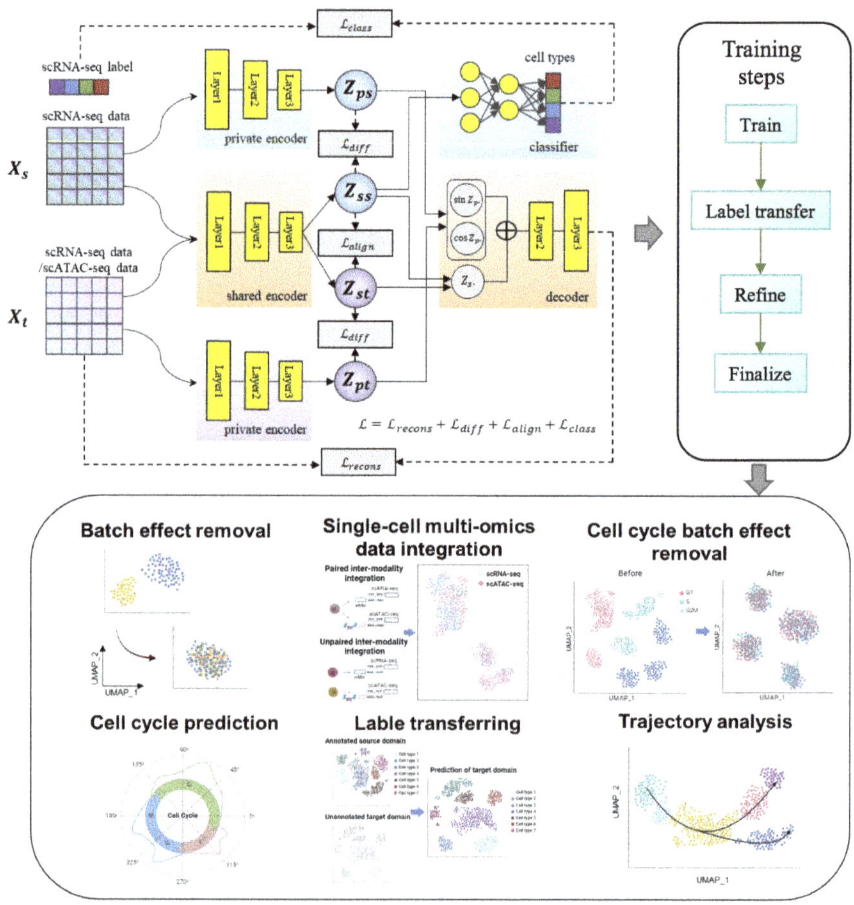

Fig. 10.5 Overview of the CCAN approach.

CCAN uses the labeled transcriptomic profile of scRNA-seq data (source domain) and the unlabeled profile from the same or different omics data (target domain) as input. Taking scRNA-seq data as source domain and scATAC-seq data as target domain as an example, the training of CCAN has four main steps:

Step 1: Pretraining of the cell cycle-aware DSN. CCAN used a cell cycle-aware DSN to perform joint embedding and modality alignment in a common embedding space through a multi-objective loss $\mathcal{L}$. *First*, CCAN trains the source and target autoencoders by minimizing data reconstruction error.

$$\mathcal{L}_{recons} = \frac{1}{N_s}\sum_{i=1}^{N_s}\left\|X_s^{(i)} - \widehat{X_s^{(i)}}\right\|^2 + \frac{1}{N_t}\sum_{i=1}^{N_t}\left\|X_t^{(i)} - \widehat{X_t^{(i)}}\right\|^2, \tag{10.5}$$

where N_s and N_t are the number of cells in source and target data, X_s and X_t are input source and target data, and $\widehat{X}_s$ and $\widehat{X}_t$ are the decoder-reconstructed data. *Second*, it learns shared signals between the scRNA-seq data and scATAC-seq data as well as private signals that are unique to the scRNA-seq data and scATAC-seq data. CCAN applies an orthogonal constraint $\mathcal{L}_{diff}$ to push the features of the shared and the private embeddings apart from each other

$$\mathcal{L}_{diff} = \left\|Z_{ss}^T Z_{sp}\right\|^2 + \left\|Z_{ts}^T Z_{tp}\right\|^2, \tag{10.6}$$

where Z_{ss} and Z_{ts} are the embedded source and target data from the shared encoder, and Z_{sp} and Z_{tp} are the embedded data from domain-specific private encoders. The rationale is to disentangle biological signals specific for cell-type heterogeneity from cell cycle confounders. *Third*, CCAN regularizes the embeddings of scRNA-seq and scATAC-seq data to make their distributions similar. Aligning the distributions across cells from different domains in the shared embedding space can alleviate the out-of-distribution problem between different modalities. CCAN uses MMD loss $\mathcal{L}_{MMD}$ as the alignment loss $\mathcal{L}_{align}$ to align the distribution of the scRNA-seq data and scATAC-seq data

$$\mathcal{L}_{align} = \mathcal{L}_{MMD} = \left\|\frac{1}{N_s}\sum_{i=1}^{N_s} k(Z_{ss}^{(i)}) - \frac{1}{N_t}\sum_{i=1}^{N_t} k(Z_{ts}^{(i)})\right\|^2, \tag{10.7}$$

where k represents the kernel function. *Finally*, after the unsupervised pre-training of the autoencoder, a supervised cell-type classification model can be trained from the aligned shared embedding using the labeled scRNA-seq data, referred to as a cross-entropy classification loss $\mathscr{L}_{class}$

$$\mathscr{L}_{class} = \sum_{i=1}^{K} y_i \log(p_i), \tag{10.8}$$

where y_i is the ground-truth cell-type label of scRNA-seq data and p_i is the predicted probability generated by the cell-type classifier. In summary, the total loss in the pretraining of CCAN is

$$\mathscr{L} = \mathscr{L}_{class} + \mathscr{L}_{recons} + \mathscr{L}_{diff} + \mathscr{L}_{align}. \tag{10.9}$$

Multiple losses will be given different weights during training.

Step 2: Label transferring from source domain to target domain. After pretraining of CCAN, the trained cell-type classification network is applied to the unlabeled scATAC-seq data, transferring the cell-type information of scRNA-seq data to annotate the scATAC-seq data; we regard it as label transferring. CCAN makes it possible to remove circular confounders while performing accurate annotations for scATAC-seq data. After the label transferring in step 2, we will obtain a pseudo-label of the scATAC-seq data.

Step 3: Refining CCAN by introducing a cluster alignment loss. CCAN will improve the joint embedding and label prediction performance using the ground truth of scRNA-seq data and the pseudo-label of the scATAC-seq data. CCAN adds cluster alignment loss $\mathscr{L}_{ca}$ to refine the neural networks.

$$\mathscr{L}_{ca} = \frac{1}{N_s N_t} \sum_{i=1}^{N_s} \sum_{j=1}^{N_t} [\delta d(z_{ss}^i, z_{ts}^j) + (1-\delta)\max(0, m - d(z_{ss}^i, z_{ts}^j))] + \frac{1}{K}\sum_{k=1}^{K} \| \lambda_s - \lambda_t \|^2, \tag{10.10}$$

where m is the distance threshold, λ_s and λ_t are centroids of embedded source and target data, and K is the number of cell types. $\mathscr{L}_{ca}$ is a class-conditional loss that forces the features from the same class to concentrate together and the features from different classes to be separated. In addition, it introduces a conditional feature matching loss to improve the alignment between two domains that aligns the clusters that correspond to the same class but come from different domains. Thus, the updated alignment loss is

$$\mathscr{L}_{align} = \mathscr{L}_{MMD} + \mathscr{L}_{ca}. \tag{10.11}$$

Step 4: Finalization of CCAN model. The last step generates the joint profile of two modalities and finalizes the annotation of scATAC-seq data. Besides, more downstream analyses can be operated on the joint embedding profile of scRNA-seq and scATAC-seq data when cascading the shared embedding of scRNA-seq and scATAC-seq.

10.5 Evaluating Multi-Omics Integration

10.5.1 Metrics for Integration Quality

k-Nearest neighbor batch effect test

The k-nearest neighbor batch effect test (kBET) [60] is a statistical tool designed to assess the presence and magnitude of batch effects in high-dimensional datasets, particularly in single-cell data. kBET evaluates whether cells from different batches are well-mixed in a shared embedding space after integration or normalization. The method begins by constructing neighborhoods for each cell using the k-nearest neighbors (kNN) approach. The kNN is based on a chosen distance metric, such as Euclidean or cosine distance, in a reduced-dimensionality space (e.g., PCA, uniform manifold approximation and projection [UMAP], or t-distributed stochastic neighbor embedding [t-SNE]). Within these neighborhoods, kBET examines the distribution of batch labels. To determine the extent of batch mixing, kBET compares the observed batch label distribution within each neighborhood against the expected uniform distribution. This comparison is performed using a statistical chi-square test. If the observed distribution significantly deviates from the expected distribution, it indicates the presence of batch effects. kBET quantifies the results by reporting the acceptance rate, which is the proportion of neighborhoods where the observed distribution does not significantly differ from the expected one. A higher acceptance rate suggests better batch mixing and minimal batch effects.

Local inverse Simpson's index

Local inverse Simpson's index (LISI) [38] is a metric used to assess cell-type preservation and batch mixing of data integration methods in single-cell studies. LISI scores offer a quantitative way to understand how well cells of similar types or from different batches are distributed within a shared embedding space. LISI scores can be computed for two distinct purposes: cell-type conservation (cLISI) and batch mixing (iLISI). The cLISI score measures how well cell-type-specific information is preserved after integration, while the iLISI score evaluates the extent to which cells from different batches are mixed. The range of LISI scores spans from 1 to the number of unique cell types or batches present in the dataset. A higher cLISI score indicates that cells from different types are grouped together, which signifies poor cell-type preservation. Ideally, cLISI scores should remain low to ensure that cell-type-specific information is retained during integration. A higher iLISI score suggests better mixing of cells from different batches, which is desirable for reducing batch effects and ensuring that the integration process does not segregate cells based on their batch origin.

LISI scores are calculated based on the local distribution of cell pairs in the shared embedding space, using the same pipeline proposed by Tran et al. [61]. The method considers the distribution of neighboring cells within a local neighborhood and computes an inverse Simpson's diversity index for each cell. These scores are then normalized to a range of 0 to 1 using the following equations, enabling direct comparisons across datasets.

$$\text{iLISI}_{\text{normalized}} = \frac{\text{iLISI} - 1}{\text{unique\_batch\_amount} - 1} \tag{10.12}$$

$$\text{cLISI}_{\text{normalized}} = \frac{\text{unique\_celltype\_amount} - \text{cLISI}}{\text{unique\_celltype\_amount} - 1}. \tag{10.13}$$

The dual utility of LISI scores makes them highly valuable in single-cell data analysis. They allow researchers to separately evaluate the quality of batch mixing and the conservation of biological information. A well-integrated dataset would exhibit high iLISI scores (indicating effective batch correction) and low cLISI scores (indicating proper cell-type preservation).

10.5.2 Metrics for Classification and Cell-Type Annotation Quality

We define True Positive (TP), True Negative (TN), False Positive (FP), and False Negative (FN) as in Table 10.1:

Table 10.1 Classification matrix between actual and predicted label.

		Predicted class	
		Class = Yes	Class = No
Actual class	Class = Yes	TP	FN
	Class = No	FP	TN

where TP is the number of true positives, TN is the number of true negatives, FP is the number of false positives, and FN is the number of false negatives. Details of the clustering metrics used in our study are shown as follows:

Rand index

Rand index (RI) represents the frequency of occurrence of agreements over the total pairs; it can also be represented as a measure of the percentage of correct decisions made by the algorithm. It can be computed using the following formula:

$$\mathrm{RI} = \frac{\mathrm{TP} + \mathrm{TN}}{\mathrm{TP} + \mathrm{FP} + \mathrm{FN} + \mathrm{TN}}. \tag{10.14}$$

Properties: the RI has a value between 0 and 1, with 0 indicating that the two clusters do not agree on any pair of points and 1 indicating that the data clusters are exactly the same [62]. The disadvantage of the RI is that it cannot handle randomly assigned clusters. As the number of clusters increases, the RI of the randomly assigned clusters will increase, which should be close to zero. Therefore, we also considered the adjusted RI (ARI).

Adjusted rand index

The ARI [63] is a modified version of the RI [62, 64, 65]. The ARI corrects for chance and will give a baseline for the randomly assigned clustering problem in the RI. It is calculated as

$$ARI = \frac{\sum_{ij}\binom{n_{ij}}{2} - \left[\sum_i \binom{a_i}{2} \sum_j \binom{b_j}{2}\right] / \binom{n}{2}}{\frac{1}{2}\left[\sum_i \binom{a_i}{2} + \sum_j \binom{b_j}{2}\right] - \left[\sum_i \binom{a_i}{2} \sum_j \binom{b_j}{2}\right] / \binom{n}{2}}. \tag{10.15}$$

Properties: the bounded range of the ARI is [−1, 1]: negative values are bad (independent labels), similar clustering results have a positive ARI, and 1.0 is the perfect match score. No assumption is made on the cluster structures [66].

Normalized mutual information

Normalized mutual information (NMI) is a normalization of the mutual information (MI) score to scale the results between 0 (no MI) and 1 (perfect correlation). MI tells us the reduction in the entropy of class labels that we get if we know the cluster labels. It is calculated as

$$\text{NMI} = \frac{H(X) + H(Y) - H(X, Y)}{(H(X) + H(Y))/2}, \tag{10.16}$$

where $H(X)/H(Y)$ is the entropy of the random variable X/Y associated with two partitions, whereas $H(X, Y)$ is the joint entropy [67].

Properties: it can be used to compare two clustering results that have different numbers of clusters.

Accuracy

Accuracy in classification problems is the number of correct predictions made by the model over the ground truth. It is calculated as

$$\text{ACC} = \frac{\text{TP} + \text{TN}}{N}, \tag{10.17}$$

N is the number of samples.

Properties: accuracy is a good measure when the target variable classes in the data are nearly balanced. Accuracy is equal to RI in the binary classification, but they are different in multiclass classification.

Precision

The precision in the binary classification is defined as the number of true positives (TP) over the number of true positives (TP) plus the number of false positives (FP) as follows

$$\text{PREC} = \frac{\text{TP}}{\text{TP} + \text{FP}}. \tag{10.18}$$

The macro-precision [67] for multiclass classification of k clusters is calculated as

$$\text{Macro-Precision} = \frac{\text{Precision}_1 + \text{Precision}_2 + ... + \text{Precision}_k}{3}. \tag{10.19}$$

Properties: the precision is a measure of result relevancy, and it is useful to measure the prediction of multiple classes.

Recall

The recall in the binary classification is defined as the number of true positives (TP) over the number of true positives (TP) plus the number of false negatives (FP) as the following

$$\mathrm{REC} = \frac{\mathrm{TP}}{\mathrm{TP} + \mathrm{FN}}. \tag{10.20}$$

The macro-recall [67] for multiclass classification of k clusters is calculated as

$$\text{Macro-Recall} = \frac{\mathrm{Recall}_1 + \mathrm{Recall}_2 + \ldots + \mathrm{Recall}_k}{3}. \tag{10.21}$$

Properties: the recall is a measure of how many truly relevant results are returned, also known as true positive rate.

Fscore

The Fscore in the binary classification is the harmonic mean of precision and recall and is calculated as

$$\mathrm{Fscore} = 2 \cdot \frac{\mathrm{Precision} \cdot \mathrm{Recall}}{\mathrm{Precision} \times \mathrm{Recall}}. \tag{10.22}$$

Macro-Fscore [67] for multiclass classification is calculated as

$$\text{Macro-Fscore} = 2 \cdot \frac{\text{Macro-Precision} \cdot \text{Macro-Recall}}{\text{Macro-Precision} \times \text{Macro-Recall}}. \tag{10.23}$$

Properties: harmonic mean is kind of an average when precision and recall are equal.

10.5.3 Benchmarking Comparison

A comprehensive benchmarking of available methods is crucial for evaluating their performance across diverse integration tasks in single-cell studies. This process enables researchers to identify the most suitable method for their specific needs, ensuring robust and biologically meaningful results. By systematically addressing key considerations, benchmarking can provide insights into method strengths and limitations while guiding researchers in method selection.

One of the primary considerations in benchmarking is the categorization of integration methods. This involves grouping methods based on their intended use cases and the modalities they support, ensuring that researchers can align their data types and research goals with appropriate approaches. For example, several methods outlined in Table 10.2 specialize in batch effect removal for scRNA-seq datasets and in integrating scRNA-seq with scATAC-seq data.

Understanding these categories simplifies the selection process and helps researchers focus on the methods best suited to their experimental design.

Another critical aspect of benchmarking is the input format and compatibility of different methods. Integration tools often require specific data structures as inputs, which can impact their performance and applicability. For instance, scJoint utilizes an ATAC gene activity score matrix for its analysis, while GLUE requires a peak matrix. These differences highlight the importance of ensuring that the chosen method aligns with the available data formats, minimizing preprocessing efforts, and avoiding potential biases introduced during data transformation.

Finally, performance evaluation metrics play a central role in benchmarking. A robust evaluation framework should comprehensively assess integration performance by addressing key aspects such as batch effect removal,

Table 10.2 Selected methods for single-cell multi-omics integration.

No.	Methods	Intra-modality	Paired	Unpaired	Input format
1	Harmony	√			Gene expression matrix
2	Conos	√			Gene expression matrix
3	Scanorama	√			Gene expression matrix
4	BBKNN	√			Gene expression matrix
5	Seurat v4	√	√	√	Gene expression + ATAC gene activity matrix
6	online iNMF	√	√	√	Gene expression count + ATAC gene activity count (same feature numbers)
7	scMVP		√		Top DEG scRNA matrix + TF-IDF normalized/binary scATAC matrix
8	scAI		√		scRNA matrix + sparse/binary scATAC matrix
9	scMVAE		√		The raw count data of scRNA-seq + scATAC data (gene activity format). Row indicates variable (genes and loci), and column indicates sample (cell).
10	scJoint		√	√	Gene expression + ATAC gene activity score matrix
11	GLUE		√	√	Gene expression profile + peak matrix with AnnData format as input
12	SCALEX		√	√	Sparse matrix of gene expression and gene activity

cell-type conservation, and modality alignment. Metrics such as kBET and iLISI can quantify the extent of batch mixing, while cLISI and Silhouette scores are useful for evaluating the preservation of cell-type-specific information post-integration. These metrics, detailed in Sections 10.5.1 and 10.5.2, provide quantitative insights into the effectiveness of integration methods, enabling researchers to make informed decisions about method selection.

10.6 Perspectives on the Design of Single-Cell Multi-Omics Integration Methods

The design of single-cell multi-omics integration methods requires a robust, forward-thinking approach that begins with meticulous data preparation and collection (Fig. 10.6). At the core of any successful integration pipeline is a well-constructed multimodal database, which serves as the foundation for meaningful analyses. Such a database should encompass a diverse array of data types, including single-cell multi-omics datasets (such as transcriptomics, epigenomics, and proteomics), spatial transcriptomics, and proteomics data, as well as high-resolution imaging data (e.g., Hematoxylin and Eosin (H&E)/Immunohistochemistry (IHC) staining). In addition, including contextual datasets, such as gene ontology annotations, biomedical knowledge derived from literature, and patient-specific data, can significantly enrich the integration process. The breadth of these data types collectively enhances the potential for deriving deeper insights into complex biological systems, addressing the multifaceted challenges inherent in biomedical research.

To effectively learn and capture the rich, nuanced representation of each data type, it is essential to employ a suite of advanced techniques designed to distill and enhance data representations. For sequential data, techniques such as bidirectional encoder representations from transformers (BERT) can be leveraged for their ability to process and contextualize sequences, while autoencoders offer powerful means for dimensionality reduction. Contrastive learning methods can be employed for robust model training, ensuring that the learned representations remain discriminative across varied datasets. In addition, manifold learning approaches are critical for uncovering the underlying geometric structures of data, aiding in the identification of key patterns across disparate data modalities.

Once robust and meaningful data representations are established, the next step involves the integration and analysis of multimodal data using advanced

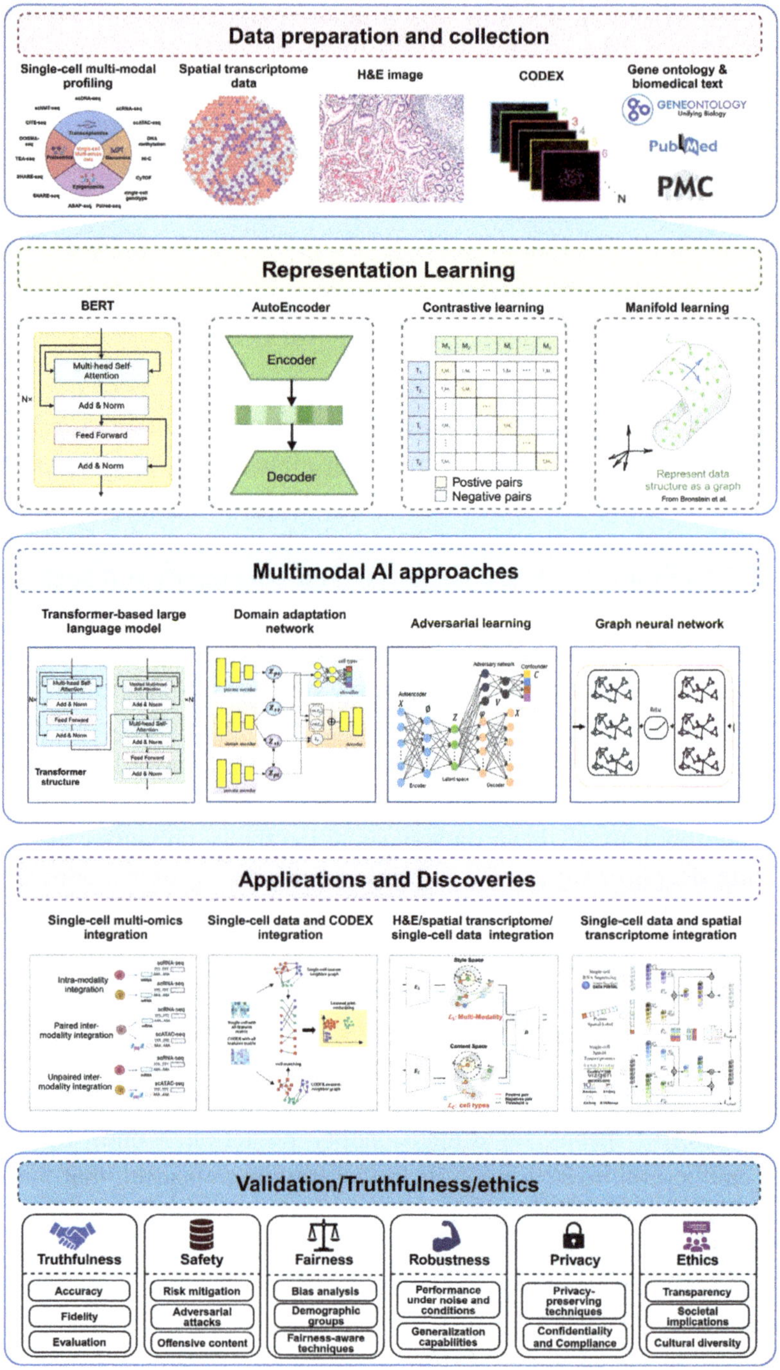

Fig. 10.6 Overall design of methods for single-cell multi-omics data integration.

multimodal AI approaches. To integrate diverse data types effectively, we recommend deploying tailored AI methodologies that account for the unique characteristics of each data modality. Among the most promising approaches are transformer-based large language models, which have demonstrated impressive capacity for handling diverse biological data; domain adaptation networks, which facilitate the transfer of knowledge between related data domains; adversarial learning, which can further refine model performance by exploiting the adversarial training paradigm; and graph neural networks, which are particularly adept at modeling complex relationships within multimodal data and can capture the intricate dependencies among different data types.

There are several key integration tasks that benefit from the application of these advanced methodologies. These tasks include (1) integrating single-cell multi-omics data: combining transcriptomic, epigenomic, and proteomic data to offer a holistic view of cellular state and function; (2) integrating single-cell and CO-Detection by indEXing (CODEX) data: incorporating spatially resolved protein expression data alongside single-cell transcriptomic data to examine cellular context within tissue architecture; and (3) integrating single-cell data with H&E images and spatial transcriptomics: merging molecular data with histopathological imagery and spatial gene expression maps to provide spatially resolved molecular insights.

In the final stage of the integration pipeline, it is crucial to emphasize validation, integrity, and ethical considerations. The reliability and robustness of the findings must be thoroughly vetted, and thus, rigorous validation processes should be implemented to minimize the risk of misinformation or model perturbations. This includes the use of cross-validation techniques, independent dataset testing, and a careful evaluation of model performance across various conditions. Furthermore, transparency in methodology, data provenance, and decision-making processes is essential to ensure scientific rigor. Ethical considerations, such as privacy concerns and the potential societal impact of the research, must also be foregrounded. We advocate for an ethically informed approach that considers stakeholder needs, including informed consent for patient data, and incorporates explicit ethical deliberations at every stage of the research process.

10.7 Conclusion

The integration of single-cell multi-omics data has become an indispensable approach for uncovering the intricate layers of cellular biology and enabling a comprehensive understanding of cellular mechanisms at unprecedented resolution. This chapter has provided a detailed overview of the various aspects of single-cell multi-omics integration, including its significance, challenges, and the innovative methodologies designed to address these challenges.

We began by highlighting the types of integration approaches, distinguishing between intra-modality, paired inter-modality, and unpaired inter-modality integration, each addressing different levels of biological complexity and experimental requirements. These methods allow researchers to combine diverse data modalities, offering holistic insights into gene expression, chromatin accessibility, DNA methylation, and other cellular processes. The development of sophisticated computational tools has been instrumental in addressing challenges such as batch effects, data sparsity, and the alignment of unmatched data sets.

This chapter also introduced CCAN, a versatile algorithm that serves as a representative approach for tackling single-cell multi-omics integration. CCAN exemplifies how advanced methodologies can simultaneously address integration challenges, remove confounding factors, and preserve biologically relevant signals, providing a robust solution for multitask integration and analysis.

To evaluate the performance of integration methods, we have discussed critical metrics and benchmarking practices. Metrics such as kBET, iLISI and cLISI, provide a quantitative foundation for assessing the quality of integration, batch effect removal, and cell-type conservation. Benchmarking comparisons emphasized the importance of categorizing methods, considering input compatibility, and leveraging appropriate evaluation metrics to ensure reliable results.

In conclusion, while the field of single-cell multi-omics integration has made remarkable progress, challenges remain in developing universal, scalable, and interpretable algorithms capable of integrating diverse modalities and handling data heterogeneity. Future research must focus on creating more generalized frameworks that address these limitations while maintaining computational efficiency and biological relevance. By advancing computational methods and evaluation strategies, single-cell multi-omics integration will continue to drive significant discoveries in cellular biology and translational research.

References

1. Kowalczyk, M.S., et al. Single-cell RNA-seq reveals changes in cell cycle and differentiation programs upon aging of hematopoietic stem cells. *Genome Res.* 2015; **25**(12): p. 1860–1872.
2. Macaulay, I. C., C.P. Ponting, and T. Voet. Single-cell multiomics: multiple measurements from single cells. *Trends Genet.* 2017; **33**(2): p. 155–168.

3. Papalexi, E. and R. Satija. Single-cell RNA sequencing to explore immune cell heterogeneity. *Nat Rev Immunol.* 2018; **18**(1): p. 35–45.
4. Tanay, A. and A. Regev. Scaling single-cell genomics from phenomenology to mechanism. *Nature.* 2017; **541**(7637): p. 331–338.
5. Gonzalez-Silva, L., L. Quevedo, and I. Varela. Tumor functional heterogeneity unraveled by scRNA-seq technologies. *Trends Cancer.* 2020; **6**(1): p. 13–19.
6. Lahnemann, D., et al. Eleven grand challenges in single-cell data science. *Genome Biol.* 2020; **21**(1): p. 31.
7. Wu, F., et al. Single-cell profiling of tumor heterogeneity and the microenvironment in advanced non-small cell lung cancer. *Nat Commun.* 2021; **12**(1): p. 2540.
8. Efremova, M. and S. A. Teichmann. Computational methods for single-cell omics across modalities. *Nat Methods.* 2020; **17**(1): p. 14–17.
9. Gao, H., et al. A universal framework for single-cell multi-omics data integration with graph convolutional networks. *Brief Bioinform.* 2023; **24**(3):bbad081.
10. Heumos, L., et al. Best practices for single-cell analysis across modalities. *Nat Rev Genet.* 2023; **24**(8): p. 550–572.
11. Athaya, T., et al. Multimodal deep learning approaches for single-cell multi-omics data integration. *Brief Bioinform.* 2023; **24**(5):bbad313.
12. Baron, M., et al. A single-cell transcriptomic map of the human and mouse pancreas reveals inter- and intra-cell population structure. *Cell Syst.* 2016; **3**(4): p. 346–360. e4.
13. Puram, S. V., et al. Single-cell transcriptomic analysis of primary and metastatic tumor ecosystems in head and neck cancer. *Cell.* 2017; **171**(7): p. 1611–1624. e24.
14. Athanasiadis, E. I., et al. Single-cell RNA-sequencing uncovers transcriptional states and fate decisions in haematopoiesis. *Nat Commun.* 2017; **8**(1): p. 2045.
15. Welch, J. D., et al. Single-cell multi-omic integration compares and contrasts features of brain cell identity. *Cell.* 2019; **177**(7): p. 1873–1887. e17.
16. Nativio, R., et al. An integrated multi-omics approach identifies epigenetic alterations associated with Alzheimer's disease. *Nat Genet.* 2020; **52**(10): p. 1024–1035.
17. Bian, S., et al. Single-cell multiomics sequencing and analyses of human colorectal cancer. *Science.* 2018; **362**(6418): p. 1060–1063.
18. Granja, J.M., et al. Single-cell multiomic analysis identifies regulatory programs in mixed-phenotype acute leukemia. *Nat Biotechnol.* 2019; **37**(12): p. 1458–1465.
19. Argelaguet, R., et al. Computational principles and challenges in single-cell data integration. *Nat Biotechnol.* 2021; **39**(10): p. 1202–1215.
20. Chen, X., S. A. Teichmann, and K. B. Meyer. From tissues to cell types and back: single-cell gene expression analysis of tissue architecture. *Annu Rev Biomed Data Sci.* 2018; **1**: p. 29–51.
21. Klemm, S. L., Z. Shipony, and W. J. Greenleaf. Chromatin accessibility and the regulatory epigenome. *Nat Rev Genet.* 2019; **20**(4): p. 207–220.
22. Berger, S. L. The complex language of chromatin regulation during transcription. *Nature.* 2007; **447**(7143): p. 407–412.
23. Ma, S., et al. Chromatin potential identified by shared single-cell profiling of RNA and chromatin. *Cell.* 2020; **183**(4): p. 1103–1116. e20.

24. Cao, J., et al. Joint profiling of chromatin accessibility and gene expression in thousands of single cells. *Science*. 2018; **361**(6409): p. 1380–1385.
25. Chen, S., B. B. Lake, and K. Zhang. High-throughput sequencing of the transcriptome and chromatin accessibility in the same cell. *Nat Biotechnol*. 2019; **37**(12): p. 1452–1457.
26. Zhu, C., et al. An ultra high-throughput method for single-cell joint analysis of open chromatin and transcriptome. *Nat Struct Mol Biol*. 2019; **26**(11): p. 1063–1070.
27. Jin, S., L. Zhang, and Q. Nie. scAI: an unsupervised approach for the integrative analysis of parallel single-cell transcriptomic and epigenomic profiles. *Genome Biol*. 2020; **21**(1): p. 25.
28. Hao, Y., et al. Integrated analysis of multimodal single-cell data. *Cell*. 2021; **184**(13): p. 3573–3587. e29.
29. Li, G., et al. A deep generative model for multi-view profiling of single-cell RNA-seq and ATAC-seq data. *Genome Biol*. 2022; **23**(1): p. 20.
30. Zuo, C. and L. Chen. Deep-joint-learning analysis model of single cell transcriptome and open chromatin accessibility data. *Brief Bioinform*. 2021; **22**(4):bbaa287.
31. Cao, Z. J. and G. Gao. Multi-omics single-cell data integration and regulatory inference with graph-linked embedding. *Nat Biotechnol*. 2022; **40**(10): p. 1458–1466.
32. Lin, Y., et al. scJoint integrates atlas-scale single-cell RNA-seq and ATAC-seq data with transfer learning. *Nat Biotechnol*. 2022; **40**(5): p. 703–710.
33. Xiong, L., et al. Online single-cell data integration through projecting heterogeneous datasets into a common cell-embedding space. *Nat Commun*. 2022; **13**(1): p. 6118.
34. Gao, C., et al. Iterative single-cell multi-omic integration using online learning. *Nat Biotechnol*. 2021; **39**(8): p. 1000–1007.
35. Mereu, E., et al. Benchmarking single-cell RNA-sequencing protocols for cell atlas projects. *Nat Biotechnol*. 2020; **38**(6): p. 747–755.
36. Barkas, N., et al. Joint analysis of heterogeneous single-cell RNA-seq dataset collections. *Nat Methods*. 2019; **16**(8): p. 695–698.
37. Hie, B., B. Bryson, and B. Berger. Efficient integration of heterogeneous single-cell transcriptomes using Scanorama. *Nat Biotechnol*. 2019; **37**(6): p. 685–691.
38. Korsunsky, I., et al. Fast, sensitive and accurate integration of single-cell data with Harmony. *Nat Methods*. 2019; **16**(12): p. 1289–1296.
39. Polanski, K., et al. BBKNN: fast batch alignment of single cell transcriptomes. *Bioinformatics*. 2020; **36**(3): p. 964–965.
40. Haghverdi, L., et al. Batch effects in single-cell RNA-sequencing data are corrected by matching mutual nearest neighbors. *Nat Biotechnol*. 2018; **36**(5): p. 421–427.
41. Shaham, U., et al. Removal of batch effects using distribution-matching residual networks. *Bioinformatics*. 2017; **33**(16): p. 2539–2546.
42. Argelaguet, R., et al. Multi-omics factor analysis—a framework for unsupervised integration of multi-omics data sets. *Mol Syst Biol*. 2018; **14**(6): p. e8124.
43. Argelaguet, R., et al. MOFA+: a statistical framework for comprehensive integration of multi-modal single-cell data. *Genome Biol*. 2020; **21**(1): p. 111.
44. Campbell, K. R., et al. clonealign: statistical integration of independent single-cell RNA and DNA sequencing data from human cancers. *Genome Biol*. 2019; **20**(1): p. 54.

45. Duan, B., et al. Model-based understanding of single-cell CRISPR screening. *Nat Commun.* 2019; **10**(1): p. 2233.
46. Cao, K., Y. Hong, and L. Wan. Manifold alignment for heterogeneous single-cell multi-omics data integration using Pamona. *Bioinformatics.* 2022; **38**: p. 211–219.
47. Song, Q., J. Su, and W. Zhang. scGCN is a graph convolutional networks algorithm for knowledge transfer in single cell omics. *Nat Commun.* 2021; **12**(1): p. 3826.
48. Buenrostro, J. D., et al. Transposition of native chromatin for fast and sensitive epigenomic profiling of open chromatin, DNA-binding proteins and nucleosome position. *Nat Methods.* 2013; **10**(12): p. 1213–1238.
49. Chen, H., et al. Assessment of computational methods for the analysis of single-cell ATAC-seq data. *Genome Biol.* 2019; **20**(1): p. 241.
50. Granja, J.M., et al. Single-cell multiomic analysis identifies regulatory programs in mixed-phenotype acute leukemia. *Nat Biotechnol.* 2019; **37**(12): p. 1458–1465.
51. Sinha, S., et al. Profiling chromatin accessibility at single-cell resolution. *Genom Proteom Bioinform.* 2021; **19**(2): p. 172–190.
52. Kawaguchi, R. K., et al. Learning single-cell chromatin accessibility profiles using meta-analytic marker genes. *Brief Bioinform.* 2023; **24**(1): p. bbac541.
53. Fang, R., et al. Comprehensive analysis of single cell ATAC-seq data with SnapATAC. *Nat Commun.* 2021; **12**(1): p. 1337.
54. Ji, Z., et al. Single-cell ATAC-seq signal extraction and enhancement with SCATE. *Genome Biol.* 2020; **21**(1): p. 161.
55. Genomics, X. Cell type annotation strategies for single cell ATAC-Seq data. Technical Note. 2020. Document Number CG000234.
56. Ma, A., G. Xin, and Q. Ma. The use of single-cell multi-omics in immuno-oncology. *Nat Commun.* 2022; **13**(1): p. 2728.
57. Bousmalis, K., et al. Domain separation networks. *Adv Neural Inf Process Syst.* 2016; **29**.
58. Liu, J., Ma, J., Wen, J., and Zhou, X. A Cell Cycle-Aware Network for Data Integration and Label Transferring of Single-Cell RNA-Seq and ATAC-Seq. *Advanced Science.* 2024; **11**(31): p. 2401815.
59. He, D., et al. A context-aware deconfounding autoencoder for robust prediction of personalized clinical drug response from cell-line compound screening. *Nat Mach Intell.* 2022; **4**(10): p. 879–892.
60. Buttner, M., et al. A test metric for assessing single-cell RNA-seq batch correction. *Nat Methods.* 2019; **16**(1): p. 43–49.
61. Tran, H. T. N., et al. A benchmark of batch-effect correction methods for single-cell RNA sequencing data. *Genome Biol.* 2020; **21**(1): p. 12.
62. Rand, W. M. Objective criteria for the evaluation of clustering methods. *J Am Stat Assoc.* 1971; **66**(336): p. 846–850.
63. Mao, Q., et al. Dimensionality reduction via graph structure learning. In *Proceedings of the 21th ACM SIGKDD International Conference on Knowledge Discovery and Data Mining*, August 10–13, 2015, Sydney, Australia.
64. Hubert, L. and P. Arabie. Comparing partitions. *J Classif.* 1985; **2**(1): p. 193–218.

65. Vinh, N. X., J. Epps, and J. Bailey. Information theoretic measures for clusterings comparison: variants, properties, normalization and correction for chance. *J Mach Learn Res*. 2010; **11**: p. 2837–2854.
66. Yeung, K. Y. and W. L. Ruzzo. An empirical study on principal component analysis for clustering gene expression data. *Bioinformatics*. 2001; **17**(9): p. 763–774.
67. Lancichinetti, A., S. Fortunato, and J. Kertész. Detecting the overlapping and hierarchical community structure in complex networks. *N J Phys*. 2009; **11**(3): p. 033015.
68. Sokolova, M. and G. Lapalme. A systematic analysis of performance measures for classification tasks. *Inf Process Manag*. 2009; **45**(4): p. 427–437.

CHAPTER 11

Systems Biology in Modeling Disease Processes from Precancer to Cancer

11.1 Introduction

Cancer is a multifaceted disease arising from complex interactions across genetic, molecular, cellular, and environmental scales. Systems biology (SB), the study of these intricate biological networks and their emergent properties, provides a holistic framework for understanding cancer development and progression. By integrating high-throughput technologies and systematically profiling data such as bulk and single-cell genomes, epigenomes, transcriptomes, epitranscriptomes, proteomes, metabolomes, lipidomes, and post-translational modification, SB enables the study of cancer as a dynamic system, where normal cellular processes are perturbed to create malignant phenotypes [1]. This chapter explores how SB approaches are applied to study tumorigenesis, offering insights into the early events of cancer initiation, progression, and the precancer-to-cancer continuum.

11.1.1 Defining Systems Biology in Cancer Progression

SB focuses on the interplay of biological components, examining how disruptions in these interactions contribute to disease. In cancer, this approach facilitates the development of predictive models that integrate data across biological scales to identify key drivers of malignancy. Technological advancements, such as single-cell and spatial multi-omics, have enhanced our ability to map cellular states and transitions, capturing the dynamic processes underpinning tumor evolution.

Understanding tumorigenesis requires unraveling the earliest events that lead tissues to deviate from normalcy. While genetic mutations are well-documented drivers of cancer, the discovery of cancer driver mutations in phenotypically normal epithelia challenges the conventional somatic mutation theory [2]. This underscores the importance of cellular plasticity, the capacity of certain mutant cells to adapt and acquire malignant potential under specific conditions. Environmental insults, such as chronic inflammation or exposure to carcinogens, also play a pivotal role in tumor initiation, often influencing genetically identical cells in heterogeneous ways [3]. Investigating how plasticity emerges during the earliest stages of tumorigenesis and contributes to neoplastic reprogramming is crucial for intercepting cancer at its origins.

Chromatin-level regulation significantly influences cellular plasticity. Stem cells, known for their open chromatin landscapes, often exhibit higher plasticity than differentiated cells. During tumor initiation, chromatin accessibility and transcriptional programs may shift, enabling cells to adapt and survive under malignant conditions. Advances in single-cell genomics and epigenomics have provided valuable insights into these processes [4, 5], yet the mechanisms underlying plasticity in precancerous tissues remain poorly understood. Addressing these knowledge gaps can illuminate strategies for early cancer detection and prevention.

11.1.2 Overview of Cancer Origin and the Precancer-to-Cancer Continuum

Theories of cancer origin provide a foundation for understanding how normal cells transition to malignant states [6]. The somatic mutation theory posits that accumulating mutations in oncogenes and tumor suppressor genes drive unregulated proliferation. In contrast, the tissue organization field theory emphasizes the role of chronic disruptions in stromal–parenchymal interactions, implicating immune cells, fibroblasts, and the extracellular matrix (ECM) in promoting malignancy. The bad luck theory highlights the stochastic nature of mutations in stem cells, while the ground state theory integrates elements of all three, underscoring the interplay of intrinsic and extrinsic factors.

Carcinogenesis, the process by which normal cells become cancerous, involves a progression through distinct stages: from normal tissue to precancerous lesions and ultimately to malignant tumors. This transformation is orchestrated by complex interactions between precancerous cells, stromal components, and immune cells within the tumor microenvironment (TME). These interactions not only influence the selection of mutant clones but also shape the cellular ecosystem that supports tumor growth. Understanding this continuum is vital for identifying opportunities to disrupt cancer initiation and progression.

11.1.3 Objectives of the Chapter

The primary goal of this chapter is to explore how SB approaches can be used to model disease processes along the precancer-to-cancer continuum. Cancer prevention remains a critical objective, necessitating a deeper understanding of precancerous biology and the mechanisms underlying tumor initiation. Currently, many breakthroughs have been made in precision prevention and therapeutics. For example, advances in precision prevention, such as Receptor Activator of Nuclear Factor Kappa-B Ligand (RANK-L) inhibitors for delaying mammary tumor onset and Human Papillomavirus (HPV) vaccines for preventing cervical lesions [7], demonstrate the potential to target early-stage diseases [8–10]. However, these interventions are limited in scope and fail to address the complexity of various cancer types.

Large-scale genomic studies have elucidated common and divergent features of cancer across tissues, yet the early stages of premalignant lesions remain underexplored [11]. Understanding neoplasia-specific communication networks offers the potential to develop therapeutics that target early tumor progression. This chapter emphasizes the importance of computational biology in unraveling tumorigenesis, focusing on specific tools and methodologies for modeling disease processes. By investigating early transitions and identifying critical molecular drivers, they aim to contribute to the development of strategies that not only prevent cancer progression but also inform treatments for advanced diseases.

11.2 Fundamentals of Systems Biology in Tumorigenesis Modeling

The study of tumorigenesis through SB relies on advanced methodologies, mathematical models, and multi-omics data to elucidate the complex mechanisms that drive cancer initiation and progression. This section introduces key tools and computational approaches, emphasizing their role in modeling tumorigenic processes and the dynamic interplay between cells and the TME.

11.2.1 Systems Biology Tools and Methodologies

The advent of cutting-edge tools, such as mutational signature analysis, single-cell next-generation sequencing (scNGS), and pseudotime trajectory analysis, has revolutionized SB by enabling researchers to analyze cancer dynamics at an unprecedented resolution.

1. **Genomics data analysis**
   Mutational signatures reveal distinct patterns of DNA alterations arising from specific mutagenic processes, such as carcinogen exposure, DNA repair defects, or aging. These signatures play a critical role in identifying the etiological factors driving cancer. For example, the ultraviolet (UV)-induced signature 7, characterized by C>T mutations at dipyrimidine sites, is strongly associated with melanoma. Analytical tools such as SigProfiler and MutationalPatterns, coupled with databases such as Catalogue of Somatic Mutations in Cancer (COSMIC), leverage techniques such as nonnegative matrix factorization (NMF) to decompose genomic data and extract actionable signatures.
2. **Single-cell data analysis**
   scNGS technologies allow for the profiling of gene expression at the resolution of individual cells, uncovering tumor heterogeneity, clonal evolution, and rare populations, such as cancer stem cells (CSCs). Platforms such as 10x Genomics and tools such as Cell Ranger [12], Seurat [13] (R), and Scanpy [14] (Python) facilitate preprocessing, clustering, and visualization of single-cell RNA sequencing (scRNA-seq) data. These technologies enable researchers to identify cell-type-specific expression profiles and discover novel biomarkers and therapeutic targets.
3. **Pseudotime analysis**
   Pseudotime analysis reconstructs the temporal progression of cellular states based on gene expression data, making it invaluable for studying dynamic processes such as differentiation, development, and tumor progression. Tools such as Monocle and RNA velocity allow researchers to order cells along trajectories, revealing transitions between states and providing insights into tumor evolution. For instance, Monocle has been used to model epithelial-to-mesenchymal transition (EMT) in lung cancer, while RNA velocity, using tools such as velocyto [15] and scVelo [16], predicts future cellular states by analyzing spliced and unspliced mRNA dynamics.

11.2.2 Mathematical Models, Computational Approaches, and Multi-Omics Integration

The complexity of tumorigenesis arises from intricate interactions between genetic, molecular, and environmental factors. SB employs mathematical models, computational tools, and multi-omics data integration to decipher these processes. Mathematical models and artificial intelligence (AI)-based algorithms are instrumental in analyzing the vast datasets generated by SB studies. Machine learning (ML) techniques, such as random forests, support vector machines (SVMs), and neural networks, enable the classification of cancer subtypes, biomarker identification, and therapy response prediction. In addition, network-based models, including gene regulatory networks (GRNs), protein-protein interaction (PPI) networks, and metabolic networks, reveal critical nodes and pathways driving tumorigenesis.

The integration of genomics, transcriptomics, proteomics, metabolomics, and epigenomics provides a systems-level view of tumorigenesis: (a) Genomics identifies driver mutations, copy number variations (CNVs), and structural alterations. (b) Transcriptomics captures dysregulated gene expression pathways. (c) Proteomics and metabolomics reveal functional changes, such as metabolic reprogramming in cancer cells. (d) Epigenomics explores chromatin accessibility, DNA methylation, and histone modifications during tumorigenesis. Multimodality integration, such as combining scRNA-seq with TME-specific changes, models stromal signaling pathways and hypoxia-induced adaptations, offering a comprehensive understanding of tumor initiation and progression.

11.2.3 Key Systems Biology Approaches for Modeling Tumorigenesis

Modeling tumorigenesis requires systems-level approaches to analyze the interactions between genetic mutations, cellular plasticity, and TME alterations.

11.2.3.1 Stem cell analysis

SB enables the study of CSCs and their role in tumor initiation and therapy resistance. Tools such as Monocle reconstruct developmental hierarchies and identify CSC transition states. For instance, systems-level analyses of glioblastoma (GBM) revealed the importance of transcription factors such as SRY (Sex Determining Region Y)-Box 2 (SOX2) and Nanog Homeobox (NANOG) in maintaining stem-like phenotypes.

11.2.3.2 Clonal evolution and fitness landscapes

Clonal expansion, genomic instability, and environmental adaptations are hallmarks of tumor progression. Computational models, such as phylogenetic trees (e.g., PyClone) [17], track the clonal architecture of evolving tumor populations. Metabolic network analyses reveal vulnerabilities in cancer metabolism, such as the Warburg effect, where tumors adapt to hypoxia by increasing glycolysis and lactate production.

11.2.3.3 Spatial and temporal dynamics

Capturing tumor progression in three dimensions over time is critical for understanding tumor evolution. Mathematical frameworks model spatial heterogeneity, while temporal analyses predict how tumors adapt to environmental stressors such as hypoxia and immune evasion.

11.3 Modeling the Malignant Transition

Understanding the transition from precancer to malignancy is central to unraveling the complexities of tumorigenesis. SB integrates advanced computational tools, mathematical modeling, and multi-omics data to examine the intricate interplay between cancer cells and their surrounding microenvironment. This section highlights key resources and methodologies that model the progression from healthy tissue to invasive carcinoma, offering novel insights into tumor initiation and malignant transformation.

11.3.1 Systems Biology Resources for Tumorigenesis

The Human Tumor Atlas Network (HTAN) provides three-dimensional (3D) atlases of tumors at crucial transition stages, utilizing single-cell and spatial methods. These resources shed light on the cooperative dynamics between transformed cells and their microenvironment during early tumorigenesis. By mapping interactions between the tumor ecosystem and microenvironmental factors, HTAN improves our understanding of precancer-to-cancer transformations, informing prevention and treatment strategies [18].

The PCTanno database is another transformative resource, compiling scRNA-seq data from 13 commonly affected tissues to analyze the continuum of changes across adjacent nontumor (ADJ), precancer, and tumor states (Fig. 11.1). PCTanno annotates tissue-specific cell types, performs differentiation trajectory inference, and maps cell-cell communication networks (Fig. 11.1). By integrating gene ontology and pathway enrichment analysis, PCTanno facilitates the identification of novel biomarkers and molecular events driving oncogenesis.

Recent modeling studies have further highlighted key transition mechanisms. For example, mathematical models of BRCA1+/mut cells have simulated the sequential mutations leading to cancer, revealing that stromal cell-induced epithelial proliferation significantly increases lifetime breast cancer risk [19]. Similarly, analyses of stem-like cells from colorectal cancer (CRC) [20] have identified transcriptional changes that drive malignancy, uncovering critical molecular targets for intervention.

11.3.2 Applications of Tumorigenesis Models

11.3.2.1 Enrichment of stem-like cells in precancer and cancer stages

Stem-like cells are increasingly implicated in the malignant transformation process. Aging, inflammation, and environmental stressors reprogram precancerous cells, endowing them with self-renewal capacity and accelerating their progression into CSCs [21]. Single-cell analyses from the PCTanno database reveal that epithelial cells from malignant stages project closer to stem cells, while benign samples are distributed more evenly across cell types (Fig. 11.2A). For instance, stem-like cells in precancerous states such as cirrhosis exhibit stemness scores comparable to or surpassing those of fully malignant cells (Fig. 11.2A). CytoTRACE predictions show high CytoTRACE scores in healthy stem cells transitioning into differentiated cells with lower scores while a skewed distribution of stem-like cells in precancers and cancers, indicating increased stem potential (Fig. 11.2B and C).

11.3.2.2 Microenvironmental changes along the malignant continuum

Stem cells, central to self-renewal and differentiation, are essential in maintaining tissue homeostasis. However, persistent stressors and aging can disrupt their regulation, fostering malignant transformations [21]. To investigate these dynamics, researchers developed a transcriptional malignancy index to characterize the progression of tissues from healthy states to invasive tumors. By analyzing differentially expressed genes (DEGs) in stem-like cells across healthy, precancerous, and cancerous tissues, they mapped a continuum of gene expression changes linked to malignancy (Fig. 11.3A). Notably, transcription factor hepatocyte nuclear factor 4 alpha (HNF4A) demonstrated distinct roles across this continuum, initially suppressing WNT signaling in precancerous stages and later promoting malignancy in CSCs. HNF4A's pivotal role as a regulator in gastrointestinal malignancies (Fig. 11.3B) underscores its potential as a therapeutic target, with studies showing that its re-expression of HNF4A reduces CSC populations and enhances chemotherapeutic responses [22]. Upregulated regulon activities of Nuclear Receptor Corepressor 1 (NCOR1), a frequently mutated gene, in stem-like cells of both cirrhosis and hepatocellular carcinoma (HCC) (Fig. 11.3C), also show increased expression along malignant progression (Fig. 11.3B).

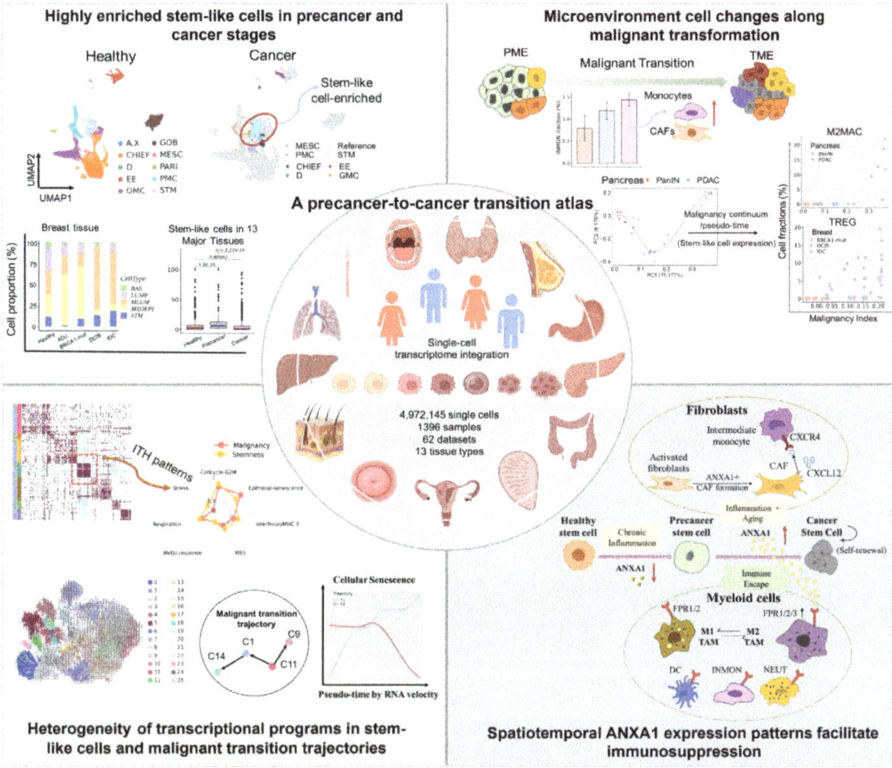

Fig. 11.1 Graphical abstract of precancer-to-cancer transition (PCT) atlas.

Beyond stem cells, the study unveiled extensive reprogramming of the TME. Immune cells displayed stage-specific distributions, with inflammatory monocytes, natural killer (NK) cells, and regulatory T cells (Treg) enriched in early lesions, while terminally exhausted CD8+ T cells and M2 macrophages dominated advanced malignancies (Fig. 11.3D). This suggests a shift from immune surveillance in early stages to immune tolerance and evasion in later stages. Transcriptional analyses identified factors such as GATA Binding Protein 3 (GATA3), Forkhead Box P3 (FOXP3), and Basic Leucine Zipper ATF-Like Transcription Factor (BATF) as critical regulators of Treg activity during disease progression. Similarly, dendritic cells and NK cells exhibited reduced cytotoxicity and stress responses as tumors advanced (Fig. 11.3E and F). Stromal cells also underwent significant changes. Myofibroblasts and pericytes emerged as key players in advanced malignancies, with fibroblast populations transitioning from inflammatory to myofibroblastic phenotypes (Fig.11.3D and E). These shifts reflect the remodeling of the TME, aligning with tumor growth and invasion.

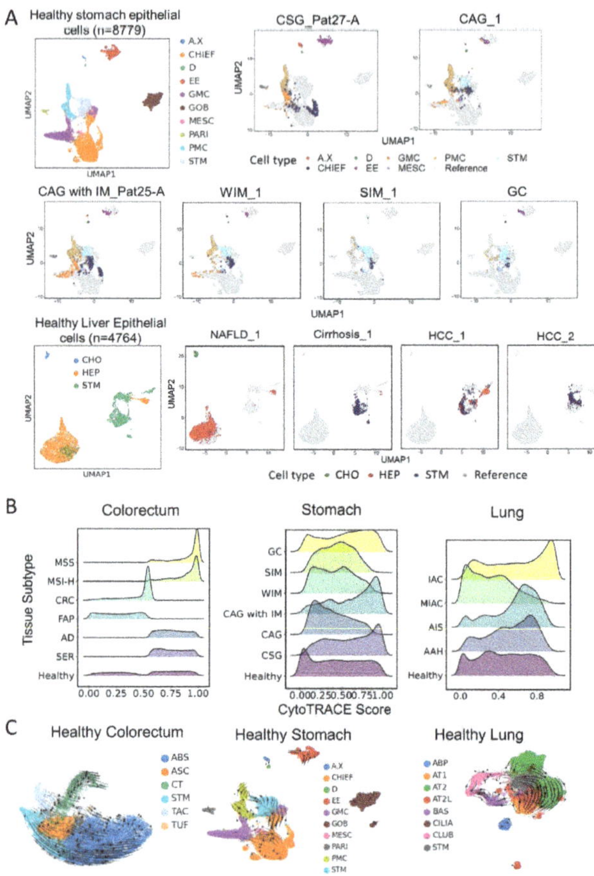

Fig. 11.2 The meticulous dissection of epithelial compartments across diverse tissues along disease progression. (A) UMAP projection of scRNA-seq epithelial cells isolated from different disease stages of stomach tissues (CSG, CAG, CAG with IM, WIM, SIM, and GC) and liver tissues (NAFLD, cirrhosis, and HCC) into the manifold of healthy epithelial cells. Projected cells are colored by the nearest normal cells in the projection, and normal epithelial cells (reference) are colored gray. (B) Ridge plot of CytoTRACE score (x-axis) distributions for different stages or subtypes (y-axis) of three representative tissues. (C) UMAP showing the inferred development dynamics of healthy epithelial cells of three representative tissues by RNA velocity. CAG: chronic atrophic gastritis; CAG with IM: chronic atrophic gastritis with intestinal metaplasia; CSG: chronic superficial gastritis; GC: gastric cancer; SIM: severe intestinal metaplasia; WIM: wild intestinal metaplasia; HCC: hepatocellular carcinoma; NAFLD: nonalcoholic fatty liver disease; A.X: A/X cell; CHIEF: chief cell; D: D cell; EE: enteroendocrine; GMC: gland mucous cell; GOB: goblet; MESC: metaplasia stem cell; PARI: parietal cell; PMC: pit mucous cell; STM: stem-like cells; CHO: cholangiocyte; HEP: hepatocyte.

In the epithelial compartment, stem-like cells increased along the malignancy continuum, while differentiated secretory cells, such as goblet cells, diminished (Fig. 11.3E). Loss of goblet cells, linked to decreased mucin production, appears to facilitate tumor progression, as observed in CRC and gastric cancer [20]. Molecularly, malignant progression was marked by heightened stress responses, including upregulated nuclear factor κB (NF-κB) signaling and heat shock protein expression (Fig. 11.3F). In T cells, this was accompanied by a metabolic switch that supported activation and effector functions in early stages, followed by exhaustion and immune evasion in advanced cancers. Interestingly, tissues such as the cervix and pancreas showed a partial reversal of these trends, highlighting the complexity of T cell dynamics across different malignancies (Fig. 11.3F).

In summary, this application illuminates the dynamic interplay between cellular subpopulations and their microenvironments during tumorigenesis. Innate immune cells dominate early responses, facilitating the removal of abnormal cells. However, chronic inflammation and persistent epithelial damage promote a pro-tumorigenic milieu. As malignancy progresses, immune suppression and stromal remodeling further support tumor growth and metastasis. This intricate balance between immunity, stromal interactions, and stemness highlights the profound plasticity of the TME and its role in cancer evolution.

11.3.2.3 Genetic diversity and mutation landscapes

Single-cell analyses of epithelial compartments reveal significant genetic diversity across tumor stages. Epithelial cells in tumors exhibited elevated CNV scores and mutation loads compared to nonmalignant counterparts (Fig. 11.4A), highlighting significant aneuploidy and mutation accumulation during malignancy. Tumor samples, such as Invasive Ductal Carcinoma (IDC), HCC, CRC, and Cutaneous Squamous Cell Carcinoma (cSCC), showed higher clonal mutations, indicating the dominance of advantageous mutated clones, often in stem-like cells. Conversely, tissues such as the cervix and lung displayed a mix of clonal and subclonal mutations, revealing dynamic mutational landscapes across stages. APOBEC mutational signatures were notably enriched in cancers, fueling subclonal expansions and tumor heterogeneity (Fig. 11.4B).

Stem-like cells frequently harbor mutations in genes such as TP53, RB1, and CDKN2A, which drive cell cycle arrest and immune evasion (Fig. 11.4C). These mutations promote malignant stem cell propagation, raising questions about the timing and origins of tumor initiation. This underscores the complexity of epithelial transformation and its impact on tumor progression and prognosis.

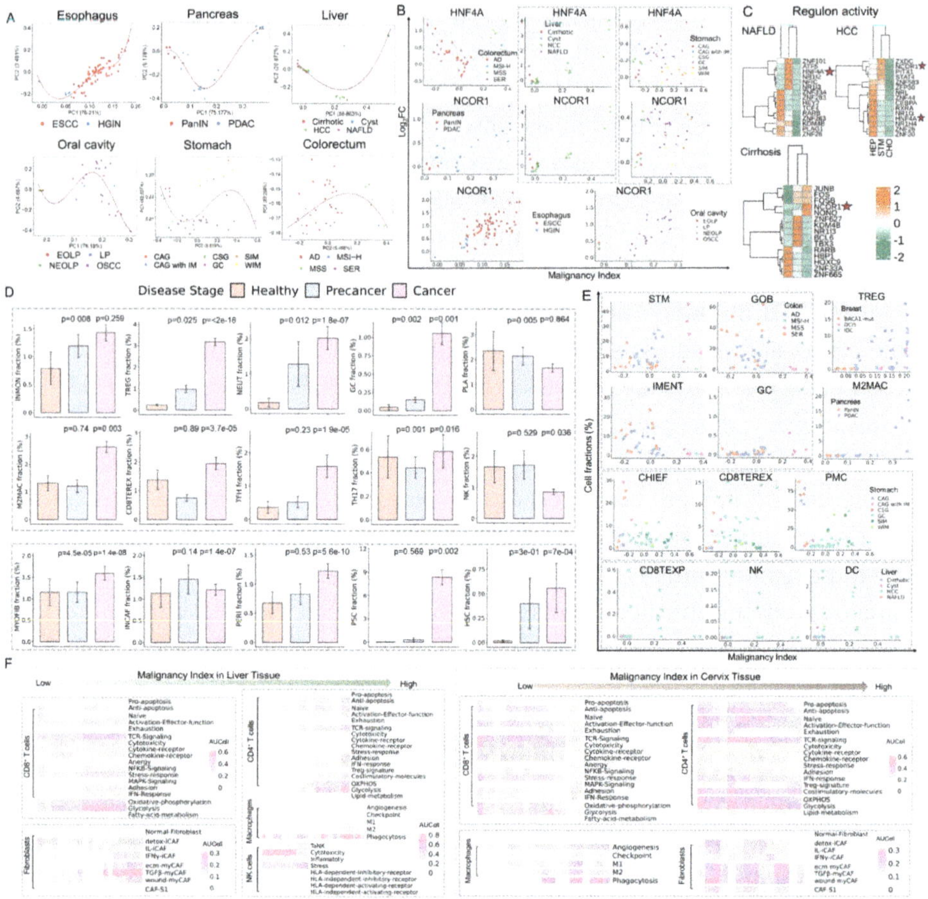

Fig. 11.3 Identification and characterization of microenvironment cell composition and molecular changes along malignancy progression. (A) Trajectories for malignant transformation in scRNA-seq datasets of four representative tissues. (B) log2FC in expression of HNF4A and GPX2 in stem-like cells from each diseased sample relative to stem-like cells in healthy samples plotted against the malignancy continuum defined in A. (C) Heatmap showing the significantly different activities of TFs among different cell types in the liver-diseased tissues. Color overlays are regulon enrichment scores. (D) Comparisons of cell proportions of immune cells (top panel) and stromal cells (bottom panel) across stage groups in 13 tissues. (E) Fraction of cell types in each sample from five representative tissues with different disease stages plotted against the position of the sample in the malignancy index defined in A. (F) Heatmap illustrating expression changes of curated gene signatures across five subclusters along malignant transitions of the liver (left) and cervix (right) tissues.

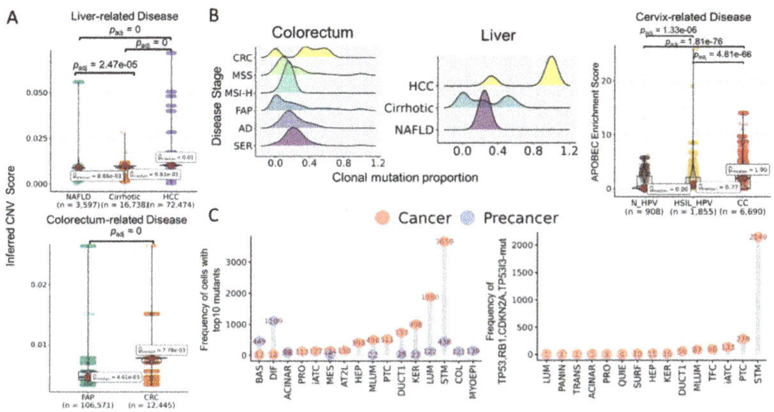

Fig. 11.4 (A) Comparisons of inferred CNV scores and mutation load among different stages and cell types of colorectal and liver tissues. (B) Left, distribution of clonal mutation proportion in colorectal and liver-related diseases; right, comparisons of APOBEC enrichment among different disease stages of the cervix. (C) Top, the frequency of the top 10 somatic mutations across different epithelial cell types detected in scRNA-seq data; bottom, the frequency of mutations from four representative cell cycle genes across cell types.

11.3.2.4 Epigenetic basis in precancer-to-cancer transitions

Emerging evidence suggests that the transition from premalignant lesions to cancer often involves profound epigenetic reprogramming, which complements or even precedes genetic alterations. For instance, widespread epigenetic field defects have been observed in normal-appearing tissues adjacent to cancers, such as in breast and inflammation-induced cancers such as *Helicobacter pylori*–associated neoplasia. In such cases, next-generation sequencing (NGS) has revealed that DNA methylation affects more cancer-related pathways than genetic mutations. This underscores the central role of epigenetic changes, such as aberrations in DNA methylation and chromatin remodeling, in early tumorigenesis.

A novel epigenetic "mitotic-like clock" provides a window into the progression of precancerous tissues. This model focuses on Polycomb group target promoter CpGs, which are typically unmethylated in fetal tissues, to establish a baseline. Deviations from this baseline correlate with stem cell divisions and cancer proliferation rates. Accelerated ticking of this epigenetic clock is observed in precancerous and carcinogen-exposed normal epithelial tissues. Smoking, for example, enhances this clock's progression. Unlike mutational signatures, which often reflect late-stage events, the epigenetic clock captures early precancerous dynamics, offering insights into clonal mosaicism and tumor heterogeneity. Furthermore, DNA methylation at "neutral" loci, not subject to selective pressure, has been used to measure mitotic divisions, revealing highly heterogeneous tumors without recent clonal expansions.

1. **TET enzymes and DNA methylation regulation**
   Epigenetic modulators, such as Ten-Eleven Translocation (family of enzymes, e.g., TET1/2/3) (TET) enzymes, play critical roles in the precancer-to-cancer transition (PCT). TET enzymes, particularly TET2, oxidize 5-methylcytosines (5mCs) to regulate locus-specific DNA methylation reversal. TET2 mutations are among the earliest epigenetic alterations observed in premalignant hematopoietic stem cells (HSCs) and are associated with disorders such as myelodysplastic syndrome (MDS) and myeloproliferative neoplasm (MPN). These mutations promote clonal HSC expansion and increase susceptibility to additional oncogenic events necessary for malignant transformation. Mouse models demonstrate that disrupting Tet2 alongside alterations in genes such as Jak2 or Ezh2 results in MDS- or MPN-like phenotypes.
   Recurrent mutations in Isocitrate Dehydrogenase 1 (IDH1) and Isocitrate Dehydrogenase 2 (IDH2), often seen in GBMs and hematologic cancers, also inhibit TET activity. These early epigenetic changes drive oncogenesis by reshaping the epigenetic landscape. Therapeutic strategies targeting TET2 and IDH enzymes, such as using IDH inhibitors or TET activators such as ascorbic acid, show promise in reactivating epigenetic regulators and enhancing immune responses in premalignant lesions.
2. **RNA editing and epigenetic dynamics**
   Another mechanism of epigenetic dysregulation involves RNA editing by Adenosine Deaminase Acting on RNA (ADAR) enzymes. These enzymes catalyze adenosine-to-inosine (A-to-I) conversions, mimicking adenosine-to-guanine mutations in RNA. RNA editing can induce protein-coding mutations or alter microRNA susceptibility, influencing cancer progression. For instance, aberrant editing of AZIN1 by ADAR1 promotes HCC development, while APOBEC-1-mediated editing leads to dysplasia in liver tissue. Furthermore, ADAR1 activity is linked to immune evasion, particularly during leukemia progression, through its regulation of cytokine networks and JAK2/STAT signaling. These findings highlight the intricate interplay between RNA editing, epigenetic reprogramming, and immune regulation in tumor initiation.

3. **Epigenetic plasticity and tumor microenvironment interactions**
   Epigenetic plasticity, the ability of cells to adopt diverse transcriptional states, further drives malignant transitions. For instance, in pancreatic tumorigenesis, Kras-mutant cells exhibit distinct chromatin accessibility patterns early in tumorigenesis, enabling inflammation-driven cell fate transitions. These plastic cell states allow cells to interact dynamically with their microenvironment, particularly through receptor-ligand communication networks. Using advanced algorithms such as calligraphy, researchers identified an interleukin (IL)-33-mediated feedback loop between epithelial and immune cells that drives neoplastic progression. Such findings underscore the importance of epigenetic plasticity and microenvironmental crosstalk in shaping the malignant transition.
4. **Case studies: Insights into epigenetic regulation in cancer stem cells**
   a. Pancreatic tumorigenesis: Early in tumorigenesis, Kras-mutant cells acquire plastic cell states defined by open chromatin near cell-cell communication genes. This enables these cells to establish modular receptor-ligand networks that facilitate immune crosstalk. Targeting key mediators, such as IL-33, disrupts inflammation-induced plasticity, impeding neoplastic progression.
   b. GBM stem cells: GBM brain tumor stem cells (BTSCs) rely on epigenetic regulators, such as Histone Deacetylase 2 (HDAC2), to maintain self-renewal and tumorigenicity. HDAC2 collaborates with SMAD Family Member 3 - SKI Proto-Oncogene (SMAD3-SKI) to organize chromatin landscapes, supporting CSC stemness and therapeutic resistance. Inhibition of HDAC2 disrupts these epigenetic programs, offering a targeted approach to GBM treatment.

   These studies highlight how epigenetic mechanisms underpin the malignant transition by regulating cellular plasticity, microenvironmental interactions, and CSC maintenance. Summarizing, epigenetic reprogramming is a pivotal driver of the PCT, linking early mutations to malignant phenotypes. From DNA methylation and chromatin remodeling to RNA editing and TME crosstalk, epigenetic alterations create a fertile ground for cancer initiation and progression. While significant strides have been made in understanding these mechanisms, further exploration is required to translate these insights into precision prevention and therapeutic strategies.

11.4 Spatial and Temporal Analysis of Precancer-to-Cancer Transition

In cancer progression, cellular heterogeneity plays a pivotal role in the acquisition of malignancy. This heterogeneity, particularly within stem-like subpopulations, manifests as a complex array of transcriptional programs that influence tumorigenesis and disease progression. A comprehensive understanding of this dynamic process requires an in-depth analysis of both spatial and temporal changes occurring during the transition from precancerous to cancerous states. By leveraging advanced methodologies, such as transcriptomic profiling, RNA velocity analysis, and computational modeling, recent studies have identified crucial mechanisms underlying this transition, providing valuable insights into how tumors evolve from benign lesions to highly malignant forms.

11.4.1 Transitional Stem-like Cell Subpopulations and Heterogeneity of Transcriptional Programs in Tumorigenesis

Tumor cells exhibit varying degrees of differentiation, ranging from undifferentiated, stem-like cells to fully differentiated cancer cells. This differentiation spectrum contributes to the functional and phenotypic heterogeneity of tumors. Gene expression programs, used to characterize cellular identities and activities, have revealed that the process of malignant transformation is accompanied by significant intratumor heterogeneity (ITH). This heterogeneity is particularly pronounced in stem-like cells, which are thought to drive tumor initiation, progression, and recurrence.

To investigate the role of stem-like cells in tumorigenesis, researchers utilized transcriptomic profiling to identify the expression programs within stem-like populations from a variety of tumors and their precursor lesions. Using NMF [23, 24], they uncovered multiple gene expression programs, with the number of programs varying across different cancer types. Through clustering analysis (K-means) of co-expressed genes, they identified 30 distinct ITH programs, referred to as meta-programs (MPs), across 13 different tissue types (Fig. 11.5A). The activity of these MPs was quantified using AUCell scores, which were then correlated with stemness, as measured by CytoTRACE. Notably, several MPs, including MYC Proto-Oncogene, BHLH Transcription Factor (MYC), hypoxia, cell cycle (G2M and G0 arrest), translation–initiation, and epithelial senescence (EpiSen), exhibited strong correlations with stemness and tumor malignancy (Fig. 11.5A). These findings underscore the role of these MPs in facilitating the acquisition of stemness and tumorigenic properties during malignant transformation.

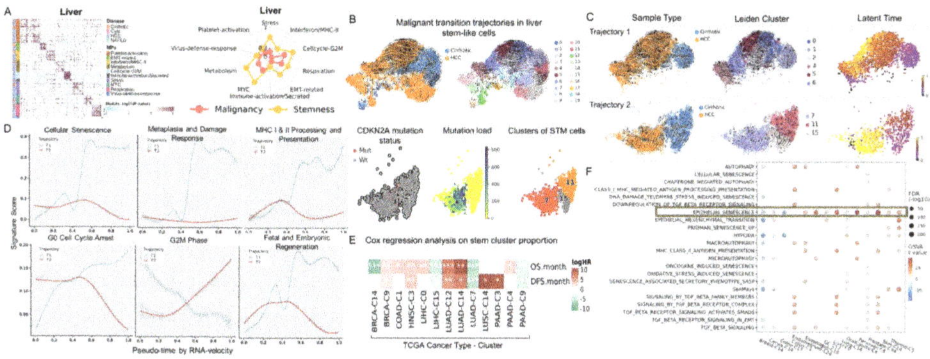

Fig. 11.5 Transcriptional programs and precancer-to-cancer transformation routes. (A) Left, heatmaps of the significance of the overlap between individual tissue modules (hypergeometric test *p*-value) in liver tissues. The bar indicates the annotation for disease states and modules. Programs are ordered by clustering and grouped into MPs. Right, correlation of MPs with stemness and malignancy index. (B) UMAP plots showing the inferred development dynamics of stem-like cells colored by sample type (top left) and Leiden subclusters (top right); bottom, stem-like cells from trajectory 2 of liver tissues labeled with genomic alterations in CDKN2A (left), mutation load (middle), and Leiden clusters (right). (C) RNA velocity indicating two trajectories of stem cells involved in cirrhotic-to-HCC transition. From left to right, colored by sample or cell type, Leiden clusters, and latent time. (H) Simplified trajectories across breast and liver tissues inferred by RNA velocity. (D) Line plots of changes for signatures over pseudotime. (E) Cox regression analysis showing the OS and DFS of TCGA patients significantly stratified by the median proportion of stem subclusters. (F) Representative pathways enriched in malignant or benign stem clusters at the end point in transition paths.

Interestingly, the expression of these MPs was found to vary across tissue types and stages of tumorigenesis. The RNA velocity analysis of these subpopulations provided additional insights into the directionality of cell transitions within tissues. For instance, in liver disease, subclusters of stem-like cells were identified using the Leiden algorithm [25], revealing significant increases in ITH as cells transitioned from benign to malignant states (Fig. 11.5B). Two pronounced directional flows were observed from cirrhotic stem-like cells (Cluster1 and Cluster11) toward HCC sub-clusters (Cluster0 and Cluster7) (Fig. 11.5C). Along trajectory 1 (Fig. 11.5C), there is a significant increase in cellular senescence, damage response signatures, Major Histocompatibility Complex I (MHC I) I and II processing and presentation, and cell cycle G0 arrest scores (Fig. 11.5D). Interestingly, a notable uptick in mutational load exists within Cluster0 compared with Cluster1. In trajectory 2, genomic events regarding HCC cells in Cluster7 harboring mutations in CDKN2A and showing the highest mutation load further supported that stem-like cells in a quiescent (or G0)-like state restarted proliferation (Fig. 11.5B). Based on cellular components of corresponding The Cancer Genome Atlas (TCGA) samples by utilizing CIBERSORTx, the inferred cell composition through deconvolution was found to relate to clinical features. Notably, the Overall Survival (OS) and Disease-Free Survival (DFS) of TCGA patients were predicted by the percentage of several transitional populations, such as Cluster0 and Cluster15 in the liver (Fig. 11.5E). When compared with healthy stem cells, several transitioning clusters from precancerous tissues were enriched with signaling pathways regulating inflammation initiation, such as NF-kB signaling and Janus Kinase – Signal Transducer and Activator of Transcription (JAK-STAT) pathways. This implicated the crosstalk of pre-CSCs between tumorigenesis and inflammatory processes. Most intriguingly, EpiSen was inevitably enriched in critical transitional stem cell populations involved in malignant transformation across 13 tissues (Fig. 11.5F), reminiscent of senescence-associated stemness. This suggests that precancerous stem-like cells may evolve into fully malignant cells, acquiring stem-like properties along the way. Further analysis of stem-like cell subpopulations across various tissues revealed distinct transitional trajectories that highlighted the dynamic nature of tumorigenesis.

11.4.2 Temporal Alterations in Cell–cell Interactions and Core Pro-tumorigenic Mediators

As the TME evolves during malignant transformation, the temporal changes in cell-cell interactions and the activation of pro-tumorigenic mediators play a critical role in sustaining tumor growth and progression. By analyzing different stem cell subpopulations at various stages of transition, researchers were able to identify key genes and signaling pathways involved in the progression from precancerous lesions to fully malignant tumors.

Researchers analyzed dynamics in the stem cell crosstalk with the microenvironment along an identified pseudotime axis to seek gene(s) in 30 core sets of PCT signatures that may drive stem cell inflammation in the loss of tissue homeostasis and precancer development (Fig. 11.6A). In particular, the activation of pro-inflammatory cytokines such as Tumor Necrosis Factor - TNF Receptor Superfamily Member 1A (TNF-TNFRSF1A), C-C Motif Chemokine Ligand 2 - Atypical Chemokine Receptor 1 (CCL2-ACKR1), and C-X-C Motif Chemokine Ligand 2 - Atypical Chemokine Receptor 1 (CXCL2-ACKR1) was observed in the transition from precancerous to malignant states. This highlights the crosstalk between transitioning stem-like cells and immune cells, which likely contributes to the establishment of a pro-inflammatory environment that facilitates tumor progression (Fig. 11.6B).

Moreover, Fibroblasts (FIB), Inflammatory Monocytes (INMON), and Dendritic Cells (DC) expressed TGFB1 to communicate with ECM receptors, indicating that the Transforming Growth Factor Beta (TGF-β) signaling of C4 tumors may regulate fibroblast activity and modulate ECM production and components (Fig. 11.6B). This indicated that extensive stromal remodeling occurs during tumorigenesis of injured tissues with a gain of EMT-like features (e.g., FN1) in transitioning stem cells, which is linked to aggressive phenotypes and poor survival (Figs. 11.5E and 11.6B). This finding suggests that under physiochemical stress, adjacent tumor-promoting fibroblasts can provide fertile "soil" for cancer "seeds," further affecting platelet activation. In turn, activated platelets produce pro-angiogenic and growth factors to facilitate tumor growth and survival, as well as promote the metastatic potential of tumor cells [26].

Notably, several ligand–receptor pairs were significantly enriched in communications between transitional stem cells and myeloid subsets. MIF-CD74+CXCR4 interaction may play an immunosuppressive role by affecting the downstream Mitogen-Activated Protein Kinase (MAPK) signaling pathway effectors. Most interestingly, Annexin A1 - Formyl Peptide Receptor 1/2 (ANXA1-FPR1/2) was activated in the malignant transition of 11 tissues (Fig. 11.6C). ANXA1, one of the representative genes in four stem cell MPs (i.e., EpiSen, stress, and unfolded protein response), is highly expressed in most healthy human tissues. The expression levels of ANXA1 in stem cells decreased from the healthy to the precancerous stage but gradually increased along malignant transformation paths (Fig. 11.6D). A pronounced elevation was observed in the inflammatory response scores of transitioning stem cells at the starting point of trajectories when ANXA1 RNA expression was depressed during tissue damage (Fig. 11.6E). Conversely, as malignancy progressed, positive correlations were noted between higher ANXA1 expression levels and the Gene Set Variation Analysis (GSVA) score of TGF-β signaling, implying the involvement of ANXA1 in promoting TGF-β activation from epithelial stem cells (Fig. 11.6F).

11.4.3 Spatiotemporal Expression Patterns in Heterogeneous Tumor Ecosystems Facilitate Immunosuppression of the Microenvironment

In addition to temporal changes, spatial heterogeneity within tumors also plays a critical role in determining the behavior of transitioning stem-like subpopulations. Spatial transcriptomics (ST) and advanced computational tools, such as SpaCET [27], were used to map the expression of key genes within tumor tissues, providing insights into the spatial arrangement of stem-like cells and their interactions with the TME.

The analysis of ANXA1, a key gene involved in the regulation of stemness and immune response, revealed distinct spatiotemporal patterns in tumor tissues. In breast, cervix, and liver cancers, the expression of ANXA1 was found to be elevated in both epithelial and stromal regions, with its levels increasing as the disease progressed from precancerous to malignant states (Fig. 11.7A). This finding was further supported by spatial analysis, which demonstrated that ANXA1 was highly expressed in immune and mesenchymal compartments, particularly at the tumor boundary regions. These spatial patterns suggest that ANXA1 plays a critical role in the immune modulation of the TME, potentially influencing the polarization of macrophages toward the M2 phenotype and promoting the transition of fibroblasts into Gene Set Variation Analysis (CAFs) (Fig. 11.7B and C). The activation of the ANXA1-FPR1/2 pathway in transitioning stem cells likely facilitates the creation of an immunosuppressive microenvironment, which aids in the survival and progression of tumor cells.

Furthermore, spatially resolved analysis revealed that the distribution of stem-like cells and their interactions with myeloid and mesenchymal cells vary across tumor stages. In the case of CRC, ANXA1-expressing cells were observed to be dispersed throughout the tumor, with distinct clusters of cells exhibiting different signaling profiles. This suggests that transitioning stem-like cells not only contribute to tumor growth but also influence the surrounding TME, further promoting tumorigenesis through complex cellular interactions. The spots of FIB–INMON interactions with co-expression of ANXA1-FPR3 and TGFB1-TGFBR1/2 in a breast tumor sample are close to the C9 population but distant from the C4 population (Fig. 11.7C and D). Two different cancer cell states in ST data displayed distinct enrichment pathways that were consistent with GSVA analysis on scRNA-seq data. Close C9 cells clustered separately from other tumor cells and showed senescent phenotypes, while distant C4 spots demonstrated hypermetabolic features and higher activity of oncogenic pathways, such as Kirsten Rat Sarcoma Viral Oncogene (KRAS) and Epidermal Growth Factor Receptor (EGFR) signaling as well as DNAX-Activating Protein of 12 kDa (DAP12) interactions (with myeloid cells) (Fig. 11.7E).

The integration of ST with scRNA-seq data provides a powerful approach to unraveling the intricate spatiotemporal dynamics of tumor progression. By mapping the expression of key genes and signaling pathways within specific regions of the tumor, researchers can gain deeper insights into how malignant transformation occurs and how the TME evolves in response to the presence of stem-like cells.

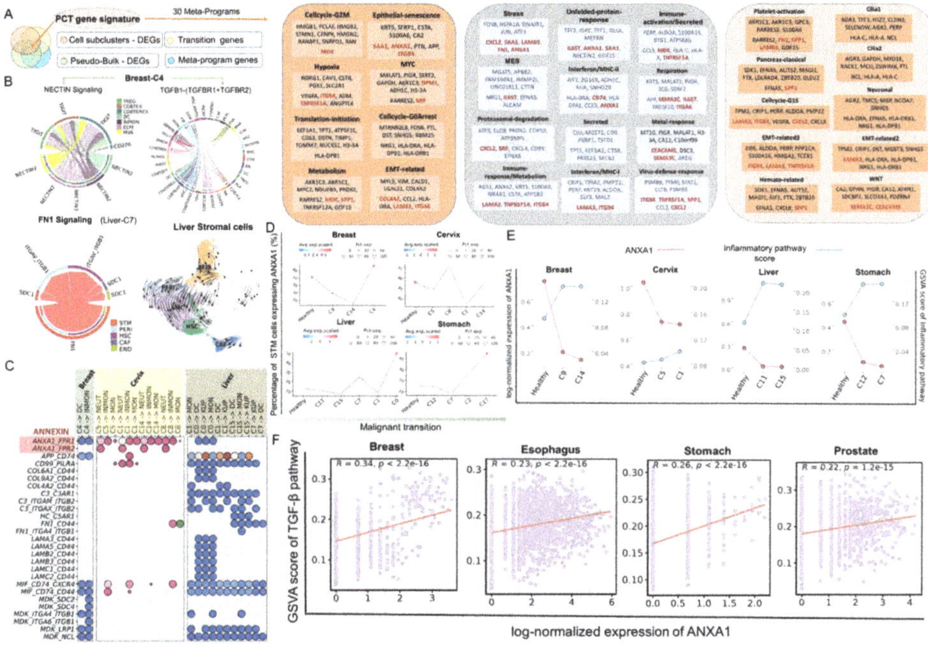

Fig. 11.6 Alterations in cell–cell communication along pseudotime axis and identification of master pro-tumorigenic mediators. (A) The left diagram shows the generation process of PCT gene signature. Right, 10 representative genes in 30 defined MPs. (B) Cell–cell interactions of NECTIN and TGF-β (left) pathways identified in TME cells from breast C4 stage (top) and of FN1 signaling identified in TME cells from malignant stages of liver tissues (bottom left); RNA velocities overlaid on the UMAP of stromal cells from liver tissues (bottom right). Arrows show the RNA velocity field. Dots are colored by stromal cell subsets. (C) The dot plot showing representative ligand–receptor pairs significantly enriched in communications between transitioning stem-like cells and myeloid subsets. (D) The dot-line plot shows the dynamics of log-normalized ANXA1 RNA level in transitional stem cells along malignant transformation. (E) The dot-line plot shows the inflammatory pathway scores and the ANXA1 RNA levels in stem clusters from the starting point of transition trajectories. (F) The scatterplots for correlations of ANXA1 expression levels with the GSVA scores of TGF-β signaling.

11.5 Challenges and Future Directions

SB has revolutionized our understanding of disease processes, providing tools to model complex dynamics in tumorigenesis. However, challenges remain in fully leveraging these tools for comprehensive and predictive modeling of disease processes. This section outlines the limitations of current SB approaches, highlights emerging technologies poised to address these gaps, and explores the potential of integrating AI with SB to advance disease modeling.

11.5.1 Limitations of Current Systems Biology Approaches in Modeling Disease Processes

Despite the transformative power of SB, several limitations hinder its full application in disease modeling. (a) Lack of longitudinal data for PCT. Single-cell lineage tracing of genetic subclones requires matched data from precancerous and cancer samples within the same patient. The absence of such datasets across multiple tissues limits our ability to map clonal dynamics and infer the evolutionary trajectories of subclones, especially using inferred CNVs or mutations. (b) Impact of pretreatment and sample variability. While sample sizes are often adequate for analyzing tumor initiation mechanisms, prior treatments (e.g., chemotherapy and radiation) in cancer samples may alter the tumorigenesis landscape. This variability can obscure the natural progression of disease and complicate the interpretation of results. (c) Intra-patient heterogeneity. High variability within a single patient, particularly a modest population of epithelial stem cells in certain tissues, poses challenges for understanding malignancy transformation. This limitation complicates the modeling of stem cell–driven tumorigenesis, as rare cell populations may not be sufficiently captured in single-cell analyses. (d) Narrow focus on specific tumorigenic phases. Current models often emphasize the PCT, leaving critical phases such as local tumor expansion, metastasis, and therapeutic resistance underexplored. Expanding the scope of SB models to encompass these aspects is necessary for a more holistic understanding of cancer progression. (e) Lack of experimental validation. Many SB studies rely heavily on computational or inferred observations. The absence of experimental validation limits the confidence in these findings and their translation into actionable insights. Integrating experimental models, such as organoids, patient-derived xenografts (PDXs), and in vivo animal studies, is essential for corroborating computational predictions. (f) Data integration and standardization challenges. The integration of multi-omics data remains technically challenging due to differences in data formats, batch effects, and incomplete datasets. Effective standardization and cross-platform compatibility are needed to create reliable, reproducible models.

11.5.2 Emerging Technologies Addressing Current Challenges

Emerging technologies are transforming the landscape of SB and hold great promise for addressing these limitations. (a) Single-cell sequencing technologies. Advances in scRNA-seq, Single-Cell Assay for Transposase-Accessible Chromatin using Sequencing (scATAC-seq), and multimodal single-cell platforms (e.g., Cellular Indexing of Transcriptomes and Epitopes by Sequencing (CITE-seq) and Multiome) now enable simultaneous measurement of gene expression, chromatin accessibility, and surface protein markers. These technologies can improve our ability to trace lineage dynamics and uncover rare cell populations critical to tumorigenesis. (b) ST and spatial proteomics. The spatial context of gene and protein expression is crucial for understanding TME interactions. Technologies such as 10x Visium, MERFISH, GeoMx Digital Spatial Profiler, and CODEX allow for high-resolution spatial mapping of molecular features within tissues. These tools facilitate the study of cellular niches and their remodeling during PCTs. (c) Organoid and 3D culture models. Organoids and advanced 3D culture systems provide experimental platforms for studying tumor dynamics in a more physiologically relevant context. Combined with single-cell sequencing and spatial profiling, these models allow for functional validation of SB predictions.

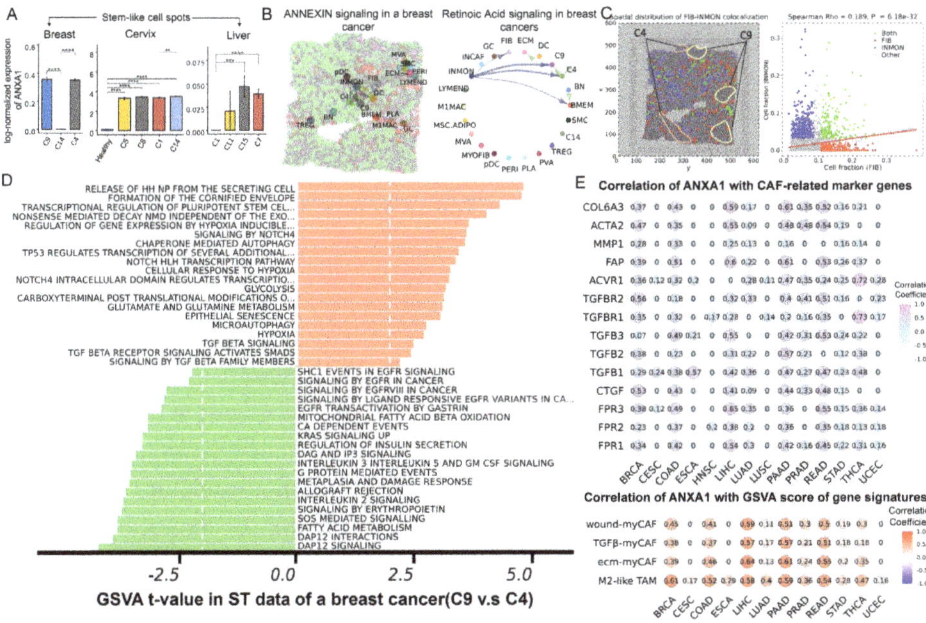

Fig. 11.7 Spatially resolved transcriptomics data revealing the transitional cell heterogeneity and potential mechanisms responsible for the mediation of stem cell senescence and immune suppression induction. (A) Barplots showing log-normalized ANXA1 RNA level in spatial stem-like spots of different stage samples from three tissue types. (B) Spatial plots presenting significant ANNEXIN and retinoic acid signaling networks identified in breast tissues. (C) Spatial distribution of FIB-INMON (right) colocalization in a breast cancer sample. Each dot represents an ST spot. (D) GSVA showing the significant differential pathways between C9 (close) and C4 (distant) spots of breast cancer samples. (E) Correlation analysis on expression levels of ANXA1 with CAF-related markers (top) and the GSVA scores of four gene signatures (bottom) based on TCGA bulk data.

(d) CRISPR-based lineage tracing. CRISPR/Cas9 tools enable precise lineage tracing by introducing heritable markers into genomes. This approach allows researchers to experimentally track clonal dynamics and validate computational lineage inference models. (e) Multimodal data integration. Emerging platforms, such as Seurat [13], Scanpy [14], and MOFA+ [28], are designed to integrate multi-omics data, enabling a comprehensive view of tumorigenic processes. Combined with advanced network analysis, these tools can reveal hidden molecular relationships.

11.5.3 Integrating Artificial Intelligence and Systems Biology for Disease Modeling

AI and ML are emerging as transformative forces in SB, offering new opportunities for disease modeling. (a) AI-driven data integration. AI tools excel at synthesizing and interpreting multidimensional datasets. For example, deep learning approaches are being used to integrate single-cell transcriptomics, spatial data, and proteomics to identify novel cell types and interactions in tumorigenesis. Graph neural networks (GNNs) can analyze complex interaction networks, such as GRNs or PPI networks. (b) Predictive modeling of disease progression. AI models can predict future cellular states or disease trajectories based on input data. For instance, reinforcement learning is used to simulate cancer progression and optimize therapeutic strategies. Generative adversarial networks (GANs) can predict tumor evolution by generating synthetic data that models clonal dynamics. (c) Drug discovery and personalized medicine. By combining SB models with AI, researchers can identify potential drug targets and predict therapeutic responses. Virtual screening algorithms powered by AI can rapidly test thousands of compounds against specific tumor models. AI-integrated SB approaches can identify patient-specific vulnerabilities by analyzing tumor heterogeneity and predicting therapy resistance mechanisms. (d) Real-time SB analysis. AI tools enabe real-time analysis of high-throughput data from single-cell and spatial platforms, allowing researchers to dynamically adapt experimental designs. (e) Challenges in AI integration. While promising, integrating AI with SB faces several challenges. For data scarcity and quality, AI models require large, high-quality datasets for training, which are not always available for rare cancer types or precancerous stages. For interpretability, many AI models, especially deep learning methods, operate as "black boxes," making their predictions challenging to interpret biologically. For bias and generalizability,

AI models trained on specific datasets may not generalize well across diverse populations or experimental conditions.

11.5.4 Future Directions in Tumorigenesis Modeling

To advance tumorigenesis modeling and address current challenges, future research should focus on the following: (a) Creating longitudinal and multi-tissue datasets. Developing standardized, multimodal datasets that capture the PCT across diverse tissues and patient cohorts. (b) Enhancing experimental validation. Integrating patient-derived organoids, animal models, and functional genomics approaches to validate predictions from computational models. (c) Expanding beyond early tumorigenesis. Modeling later stages of cancer progression, including metastasis and therapeutic resistance, by leveraging advanced spatial and temporal SB tools. (d) Ethical and inclusive data practices. Ensuring equitable representation in multi-omics datasets to improve the applicability of SB models across diverse populations. (e) Bridging AI, SB, and clinical applications. Developing clinically translatable AI-SB frameworks to guide personalized therapy and early cancer detection.

11.6 Conclusion

While current SB approaches have made significant strides in modeling tumorigenesis, addressing their limitations requires the adoption of emerging technologies, the integration of experimental validation, and the fusion of AI with SB. By overcoming these challenges, they can unlock deeper insights into tumorigenesis and pave the way for transformative advances in cancer prevention, diagnosis, and treatment.

References

1. Du, W. and O. Elemento. Cancer systems biology: embracing complexity to develop better anticancer therapeutic strategies. *Oncogene*. 2015; **34**: p. 3215–3225. doi:10.1038/onc.2014.291.
2. Wijewardhane, N., L. Dressler, and F. D. Ciccarelli. Normal somatic mutations in cancer transformation. *Cancer Cell*. 2021; **39**: p. 125–129. doi:10.1016/j.ccell.2020.11.002.
3. Nam, A. S., R. Chaligne, and D. A. Landau. Integrating genetic and non-genetic determinants of cancer evolution by single-cell multi-omics. *Nat Rev Genet*. 2021; **22**: p. 3–18. doi:10.1038/s41576-020-0265-5.
4. Marjanovic, N. D., et al. Emergence of a high-plasticity cell state during lung cancer evolution. *Cancer Cell*. 2020; **38**: p. 229–246. e213. doi:10.1016/j.ccell.2020.06.012.

5. LaFave, L. M., et al. Epigenomic state transitions characterize tumor progression in mouse lung adenocarcinoma. *Cancer Cell*. 2020; **38**: p. 212–228. e213. doi:10.1016/j.ccell.2020.06.006.
6. Jassim, A., et al. Cancers make their own luck: theories of cancer origins. *Nat Rev Cancer*. 2023; **23**: p. 710–724. doi:10.1038/s41568-023-00602-5.
7. Widschwendter, M., et al. Osteoprotegerin (OPG), the endogenous inhibitor of receptor activator of NF-kappaB ligand (RANKL), is dysregulated in BRCA mutation carriers. *EBioMed*. 2015; **2**: p. 1331–1339. doi:10.1016/j.ebiom.2015.08.037.
8. Mariz, F. C., et al. Peak neutralizing and cross-neutralizing antibody levels to human papillomavirus types 6/16/18/31/33/45/52/58 induced by bivalent and quadrivalent HPV vaccines. *NPJ Vaccines*. 2020; **5**: p. 14. doi:10.1038/s41541-020-0165-x.
9. Iheagwara, U. K., et al. Influenza virus infection elicits protective antibodies and T cells specific for host cell antigens also expressed as tumor-associated antigens: a new view of cancer immunosurveillance. *Cancer Immunol Res*. 2014; **2**: p. 263–273. doi:10.1158/2326-6066.CIR-13-0125.
10. Zhai, K., et al. Mitochondrial C150T polymorphism increases the risk of cervical cancer and HPV infection. *Mitochondrion*. 2011; **11**: p. 559–563. doi:10.1016/j.mito.2011.02.005.
11. The ICGC/TCGA Pan-Cancer Analysis of Whole Genomes Consortium. Pan-cancer analysis of whole genomes. *Nature*. 2020; **578**: p. 82–93. doi:10.1038/s41586-020-1969-6.
12. Zheng, G. X., et al. Massively parallel digital transcriptional profiling of single cells. *Nat Commun*. 2017; **8**: p. 14049. doi:10.1038/ncomms14049.
13. Butler, A., et al. Integrating single-cell transcriptomic data across different conditions, technologies, and species. *Nat Biotechnol*. 2018; **36**: p. 411–420. doi:10.1038/nbt.4096.
14. Wolf, F. A., P. Angerer, and F. J. Theis. SCANPY: large-scale single-cell gene expression data analysis. *Genome Biol*. 2018; **19**: p. 15. doi:10.1186/s13059-017-1382-0.
15. La Manno, G., et al. RNA velocity of single cells. *Nature*. 2018; **560**: p. 494–498. doi:10.1038/s41586-018-0414-6.
16. Bergen, V., et al. Generalizing RNA velocity to transient cell states through dynamical modeling. *Nat Biotechnol*. 2020; **38**: p. 1408–1414. doi:10.1038/s41587-020-0591-3.
17. Roth, A., et al. PyClone: statistical inference of clonal population structure in cancer. *Nat Methods*. 2014; **11**: p. 396–398. doi:10.1038/nmeth.2883.
18. Zhang, S., et al. Tumor initiation and early tumorigenesis: molecular mechanisms and interventional targets. *Signal Transduct Target Ther*. 2024; **9**: p. 149. doi:10.1038/s41392-024-01848-7.
19. Nee, K., et al. Preneoplastic stromal cells promote BRCA1-mediated breast tumorigenesis. *Nat Genet*. 2023; **55**: p. 595–606. doi:10.1038/s41588-023-01298-x.
20. Becker, W. R., et al. Single-cell analyses define a continuum of cell state and composition changes in the malignant transformation of polyps to colorectal cancer. *Nat Genet*. 2022; **54**: p. 985–995. doi:10.1038/s41588-022-01088-x.
21. Jamieson, C. H. M. and I. L. Weissman. Stem-cell aging and pathways to precancer evolution. *N Engl J Med*. 2023; **389**: p. 1310–1319. doi:10.1056/NEJMra2304431.
22. Zolfaghari, R., et al. Hepatocyte nuclear factor 4alpha (HNF4alpha) plays a controlling role in expression of the retinoic acid receptor beta (RARbeta) gene in hepatocytes. *Int J Mol Sci*. 2023; **24**: p. 8608. doi:10.3390/ijms24108608.

23. Jin, S., et al. Single-cell transcriptomic analysis defines the interplay between tumor cells, viral infection, and the microenvironment in nasopharyngeal carcinoma. *Cell Res*. 2020; **30**: p. 950–965. doi:10.1038/s41422-020-00402-8.
24. Gavish, A., et al. Hallmarks of transcriptional intratumour heterogeneity across a thousand tumours. *Nature*. 2023; **618**: p. 598–606. doi:10.1038/s41586-023-06130-4.
25. Traag, V. A., L. Waltman, and N. J. van Eck. From Louvain to Leiden: guaranteeing well-connected communities. *Sci Rep*. 2019; **9**: p. 5233. doi:10.1038/s41598-019-41695-z.
26. Ramadori, P., et al. Platelets in chronic liver disease, from bench to bedside. *JHEP Rep*. 2019; **1**: p. 448–459. doi:10.1016/j.jhepr.2019.10.001.
27. Ru, B., et al. Estimation of cell lineages in tumors from spatial transcriptomics data. *Nat Commun*. 2023; **14**: p. 568. doi:10.1038/s41467-023-36062-6.
28. Argelaguet, R., et al. MOFA+: a statistical framework for comprehensive integration of multi-modal single-cell data. *Genome Biol*. 2020; **21**: p. 111. doi:10.1186/s13059-020-020 15-1.

CHAPTER 12

Systems Biology in Cancer Immunology

12.1 Introduction

In recent years, the cancer treatment landscape has undergone a paradigm shift with the advent of immunotherapy. Harnessing the immune system to combat cancer, particularly through immune checkpoint blockade (ICB) therapies, has emerged as a transformative approach [1]. However, the inherent complexity of the immune responses against tumors poses significant challenges to patient care [2]. While immunotherapy represents a revolutionary approach, it remains a challenge to accurately predict individual responses to immunotherapy. Traditional biomarkers, such as programmed death-ligand 1 (PD-L1) expression and tumor mutational burden (TMB), have limitations in predictive accuracy. Due to the intricate interactions between tumors and their microenvironment and the dynamic nature of immune responses, innovative approaches are needed to enhance the ability to predict treatment outcomes accurately.

This chapter focuses on the burgeoning field of computational immunology, which integrates advanced computational methodologies with sophisticated cancer immunotherapy. The central goal is to explore how computational approaches can unravel the complexities of the tumors, tumor microenvironment (TME), and microbiomes and provide insights into embedding these factors in predicting responses to immunotherapy. The exploration of computational immunology's intricacies underscores its potential to transform our comprehension of immunotherapy outcomes, thereby forging a path toward cancer treatments that are both more personalized and efficacious [3]. This chapter will delve into specific computational approaches, from investigating tumor intrinsic factors to analyzing the TME and integrating multi-omics data, such as genomics, transcriptomics, and microbiomes (Fig. 12.1). By understanding the challenges and future directions in this field, we aim to elucidate the transformative potential of computational immunology in predicting immunotherapy responses and shaping the future of clinical oncology.

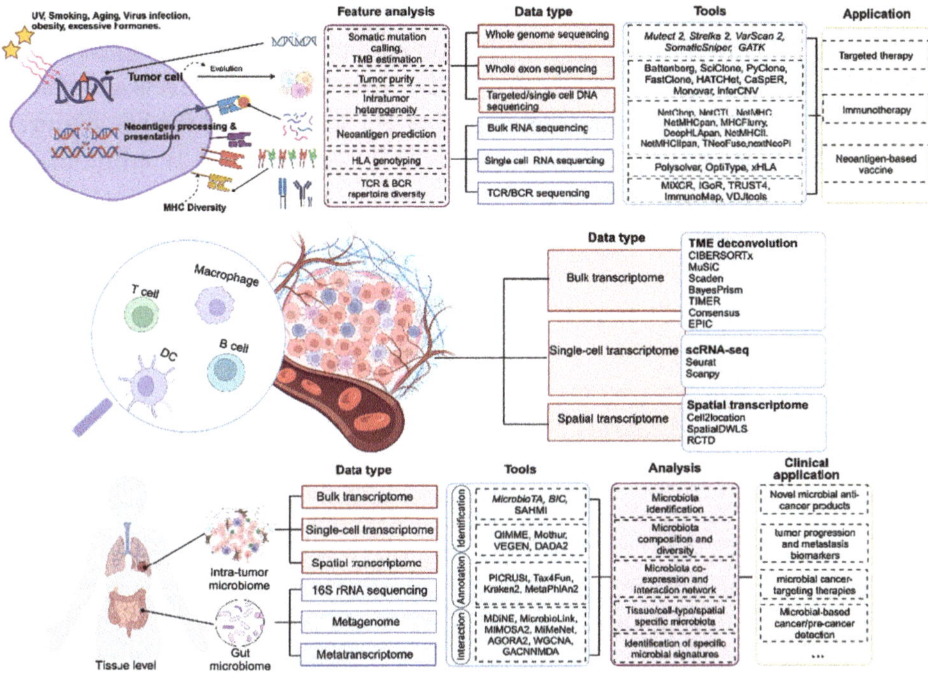

Fig. 12.1 Computational approaches that developed to predict the immunotherapy response from various perspectives. Advanced computational methods extensively leverage tumor-related markers, tumor microenvironment, and microbiome data to predict immunotherapy responses. This figure summarizes the state-of-the-art computational tools to investigate the omics data and predict immunotherapy response.

12.2 Computational Approaches in Cancer Immunology

The advent of next-generation sequencing (NGS) technologies and advanced algorithms has transformed our understanding of omics data, offering extensive datasets and powerful computational tools to predict immunotherapy responses. Raw sequencing data usually undergo rigorous preprocessing to ensure their utility in downstream analyses. For genomic data, preprocessing starts with quality controls using tools such as FastQC [4], followed by trimming and filtering steps with software such as Trimmomatic [5] or Cutadapt [6]. Cleaned reads are then aligned to a reference genome using aligners such as BWA [7] or Bowtie2 [8], and post-alignment, variant calling tools such as GATK [9] or Samtools [10] identify genetic variations such as single-nucleotide polymorphisms (SNPs) and insertions/deletions (indels). Transcriptomic data preprocessing involves similar quality control and trimming steps, with subsequent alignment often performed using STAR [11] or HISAT2 [12] (Fig. 12.2). Gene expression levels are quantified with tools such as FeatureCounts [13] or HTSeq [14] and further analyzed using methods such as DESeq2 [15] or edgeR [16] (Fig. 12.2). Microbiome sequencing can be processed using tools such as Kraken2 and QIIME2 [17]. These tools also perform quality control, cluster sequences into operational taxonomic units (OTUs), and assign taxonomic labels to the OTUs based on reference databases such as Greengenes [18] (Fig. 12.2).

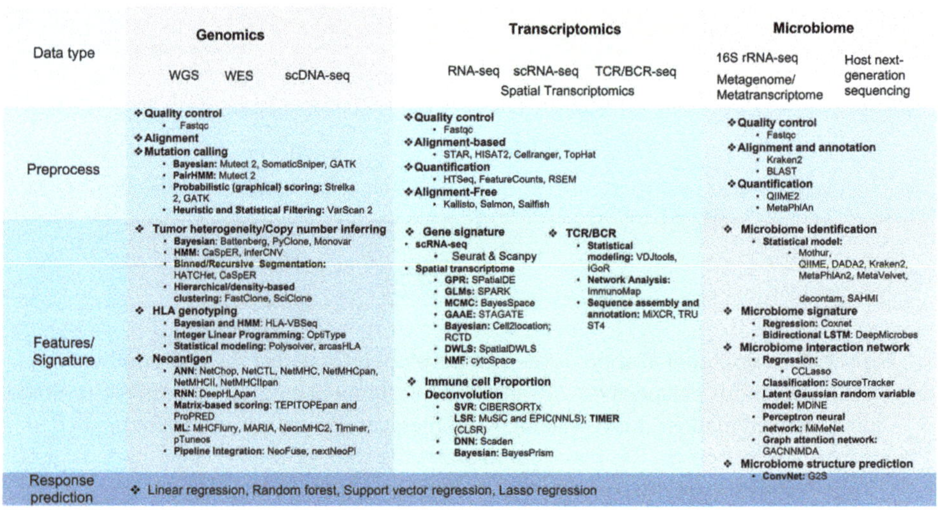

Fig. 12.2 Computational tools and their algorithms are involved in the process of the omics data for predicting immunotherapy response. PairHMM: pair hidden Markov model; ANN: artificial neural network; RNN: recurrent neural network; ML: machine learning; GPR: Gaussian process regression; GLMs: generalized linear models; MCMC: Markov chain Monte Carlo; GAAE: graph attention autoencoder; DWLS: dampened weighted least squares; NMF: nonnegative matrix factorization; SVR: support vector regression; LSR: least squares regression; DNN: deep neural network; LSTM: long short-term memory.

Extracting meaningful predictive features from processed data is essential for immunotherapy modeling. In genomics, TMB, neoantigens, and genetic alterations are significant markers. Transcriptomic features include immune gene signatures and infiltration profiles, assessed using tools such as CIBERSORTx [19] and xCell [20] (Fig. 12.2). Microbiome diversity and spatial omics data provide additional insights, while imaging techniques such as computed tomography (CT), magnetic resonance imaging (MRI), and histology offer additional predictive features.

Artificial intelligence (AI) significantly enhances immunotherapy prediction by efficiently analyzing complex data and revealing actionable insights. Supervised learning models (e.g., logistic regression and neural networks) classify patients as responders or nonresponders, while unsupervised methods (e.g., clustering) uncover novel patient subgroups. Ensemble methods improve predictive accuracy, and deep learning (DL) advances areas such as protein structure modeling and drug discovery. Graph-based approaches analyze biological networks, while sequence-based models such as transformers identify key biomarkers. These computational methods bridge complex data with clinical applications, driving precision medicine and improving patient outcomes in cancer immunology.

12.3 DNA-Sequencing Data Analysis in Onco-Immunology

Large-scale genomics studies have uncovered various genetic alterations related to immunotherapy response [21, 22]. Advances in NGS and bioinformatics have enhanced mutation detection, enabling precise neoantigen targeting [21]. High TMB and elevated PD-L1 expression are associated with better outcomes with ICB therapies [23, 24], although challenges such as drug resistance and nonstandardized thresholds for these biomarkers persist [25, 26]. Major histocompatibility complex (MHC) molecule genotyping and neoantigen burdens also correlate with immunotherapy efficacy, yet systematic identification of patient-specific neoantigens remains difficult. Hence, we will introduce advances in neoantigen identification tools accordingly, focusing on different neoantigen origins and feature integration methods. Moreover, tumor purity and heterogeneity influence neoantigen identification and therapeutic outcomes, with heterogeneity often driving resistance. This section explores computational approaches utilizing bulk DNA sequencing data to profile tumor-intrinsic biomarkers, offering strategies to enhance immunotherapy efficacy.

12.3.1 Genomic Alterations and Signatures in Immunotherapy Predictions

Variant calling tools such as Genome Analysis Toolkit (GATK) [9] enable the identification of genomic alterations and gene signatures that predict response to immunotherapy, as summarized in Supplementary Table S1. Notable findings include the impact of microsatellite instability (MSI) on immunotherapy outcomes and the correlation between PD-L1 overexpression and intratumoral inflammatory environment [27, 28, 29]. Mutant genes such as TP53 [30], KRAS [30], ARID1A [31], CDK12 [32], PBRM1 [33], and SMARCA4 [34] have been associated with enhanced responses to ICB. Moreover, genomic features associated with exogenous mutagens, such as ultraviolet (UV) light exposure, aflatoxin B1, tobacco smoke, and oncogenic bacteria or viruses, can modulate the effectiveness of ICB therapies [35]. Conversely, alterations in genes such as B2M, JAK, STK11, KEAP1, and EGFR can predict poor outcomes or resistance to ICB therapy [36–39]. Furthermore, amplification of the Genome Analysis Toolkit (MDM2) family and Epidermal Growth Factor Receptor (EGFR) mutations has been correlated with hyper-progression following ICB treatment [40].

12.3.2 Tumor Mutational Burden

TMB was initially characterized using whole-exome sequencing (WES). Based on the association and Cox regression analysis, clinical studies have shown that cancer patients with high TMB often have improved response rates and survival [41]. The assessment of TMB involves counting nonsynonymous single-nucleotide variants (nsSNVs) and short indels within coding regions, excluding germline alterations identified in matched normal samples [42]. In addition to WES, FoundationOne11 and MSK-Impact [3] are among the U.S. Food and Drug Administration (FDA)-approved tests that utilize targeted NGS panels for the genomic profiling of tumors. These panels detect various genomic alterations, including point mutations, insertions, deletions, copy number variations (CNVs), and gene rearrangements across cancer-associated genes. However, TMB alone may not fully capture the complexity of factors influencing immunotherapy response [43], [44]. Therefore, integrating TMB with various other biomarkers and genomic features is crucial for more accurate predictions of ICB efficacy [26].

12.3.3 Computational Approaches for MHC/HLA Genotyping

Human leukocyte antigen class I (HLA-I), the human counterpart of the MHC, plays a pivotal role in the immune system by presenting peptide fragments derived from intracellular proteins, including neoantigens, to cytotoxic T cells [45]. The polymorphism of HLA-I significantly impacts immune responses [45]. For instance, HLA-I homozygosity has been associated with both overall survival and progression-free survival in patients undergoing immunotherapy for non-small cell lung cancer [46]. In addition, studies suggest that the loss of HLA expression or alterations in MHC-I components contribute to resistance against ICB therapies [47]. Notably, the combination of HLA heterozygosity and high TMB has been linked to improved clinical outcomes and enhanced responses to cancer immunotherapy [45]. MHC class II (MHC-II) molecules, which present antigens to helper T cells, complement the immune activity of MHC-I and contribute to the overall anticancer immune response [48].

Table 12.1 Computational tools to predict HLA genotyping.

Tools	HLA type	Resolution (digits)	Methods	Assay data	Input data format	Publications
PolySolver	Class I	4	Bayesian classification scoring and alignment to the IMGT sequence library	WGS, WES	BAM	PMID: 26372948
OptiType	Class I	4	Construct a hit binary matrix with matched IMGT reference sequences	WGS, WES, RNA-seq	FASTQ	PMID: 25143287
xHLA	Class I, II	4	Alignment to IMGT sequence library + HLA amino acids	WGS, WES	FASTQ	PMID: 28674023
seq2HLA	Class I	4	Variability defined by Shannon's entropy at exon 2 and 3 positions of aligned reads to IMGT reference sequences	RNA-seq	FASTQ	PMID: 23259685
HLA-VBSeq	Class I, II	8	Variational Bayesian approach for aligned reads to IMGT reference + human decoy sequences (hs37d5)	WGS, WES	FASTQ	PMID: 25708870
arcasHLA	Class I, II	4	HLA de Bruijn graph–based alignment to IMGT reference sequences	RNA-seq	BAM	PMID: 31173059
HLA reporter	Class I, II	4	Reads mapped to a comprehensive reference panel with HLA genes and score the assembled contigs	WGS, WES	FASTQ	PMID: 25908942

(*Continued*)

Table 12.1 Computational tools to predict HLA genotyping. (*Continued*)

Tools	HLA type	Resolution (digits)	Methods	Assay data	Input data format	Publications
ATHLATES	Class I, II	4	Quantify aligned reads to exons using Hamming distance	WES	FASTQ	PMID: 23748956
HLAMiner	Class I, II	2	Scoring system on assembled contigs with IMGT reference sequences	WES, RNA-seq	FASTQ	PMID: 23228053

Computational tools used for HLA genotyping from whole-genome sequencing (WGS), WES, and RNA-seq data generally follow two main approaches: alignment-based and de novo assembly (Table 12.1). Among the popular tools, OptiType [49] is widely used for class I loci, achieving over 95% accuracy at 50X coverage. Its robustness extends to short read lengths and damaged DNA, provided coverage depth is high [50]. Benchmarking studies have shown that OptiType and arcasHLA perform comparably well, while HLA-VBseq struggles due to the advantages of integer linear programming over Bayesian methods (Fig. 12.2). Despite their high accuracy for class I typing, OptiType and similar tools underperform in class II genotyping compared to traditional methods. Moreover, these tools exhibit higher accuracy in European samples than African samples [51], underscoring the need for further refinement to improve performance across diverse populations [52].

12.3.4 Neoantigen-Based Approaches

12.3.4.1 Identification of neoantigens

Neoantigen-targeting therapeutics are an area of great promise in cancer immunology, yet accurately predicting neoantigens for treatment remains a complex challenge [53]. The process typically involves several computational steps: identifying somatic mutations, detecting proteasomal cleavage sites, and predicting peptide-MHC (pMHC) binding affinity (BA) and stability. While earlier algorithms struggled due to limited training data quality, modern tools have improved, although interpretation remains cautious, particularly in diverse datasets (Fig. 12.3a). NetCTL [54] integrates predictions for proteasomal cleavage, transporter associated with antigen processing (TAP) efficiency, and MHC-neoantigen binding. NetChop predicts regions within proteins susceptible to proteasomal cleavage, particularly at the C-terminal end [55]. NetMHC v.4.0 [56], NetMHCpan-4.1 and NetMHCIIpan-4.0 [57], and MHCflurry [58] predict pMHC BA using artificial neural networks (ANNs) trained on epitope databases. However, biases persist in these models due to HLA allele prevalence in training data [59]. Neoantigens arise not only from single nucleotide variants (SNVs) in coding regions [60] but also from splicing events, post-translational modifications [40, 60], long noncoding RNAs (lncRNAs) [61], transposable elements (TEs), and so on [60–68]. Computational tools such as NeoSplice [69], IRIS [70], NeoSV [71], and ASNEO [72] analyze these alternative sources. For instance, TEs can generate peptides presented as neoantigens, while tools such as LENS [73] focus on structural variants (Fig. 12.3a).

In benchmark studies, MHCflurry 2.0 has demonstrated superior performance for MHC I binding prediction, outperforming tools such as NetMHCpan [58]. Similarly, for HLA class II predictions, MARIA has shown high accuracy, achieving an area under the curve (AUC) of 0.89-0.92 [74] in validation datasets.

12.3.4.2 Integrated neoantigen models

Neoantigen burden, which quantifies the total neoantigens in a tumor, is a critical determinant of ICB efficacy. Neoantigen presentation varies between tumor regions, and clonal neoantigens, even in smaller fractions, are associated with stronger and more durable immunotherapy responses [75]. Simultaneous prediction of pMHC BA and immunogenicity has advanced therapy prediction. For instance (in Fig. 12.3b), the CSiN Scoring System accounts for neoantigen clonality and MHC BA, improving ICB response prediction [76]. NEOPS combines HLA loss-of-heterozygosity detection and MHC class I prediction using advanced models such as XGBoost [77]. NEO2IS and NEO2ITHS are DL-based tools that enhance neoantigen prediction accuracy [78]. NetBio is a network-based method that integrates protein–protein interaction networks and transcriptome features for personalized ICB predictions [79]. For evolutionary dynamics models, Monte Carlo simulations can analyze the immune escape and neoantigen clonality [80]. The hierarchical Bayesian model, NETIE, is designed to track neoantigen–T cell interactions and predict immune selection pressure [81]. In general, these integrated neoantigen-based methods inferred the antineoantigen immune pressure using Bayesian or described the evolutionary dynamics of neoantigens in growing tumors using machine learning/DL models, providing insights into the immunotherapy response prediction.

12.3.5 Tumor Heterogeneity and Treatment Resistance

Understanding tumor heterogeneity is crucial to addressing treatment resistance in cancer. Tumor heterogeneity encompasses both the cellular composition of tumors and the genomic diversity within cancer cell populations. This section delves into computational advancements in assessing tumor purity, estimating genomic intratumor heterogeneity (ITH), and exploring their implications for cancer therapy.

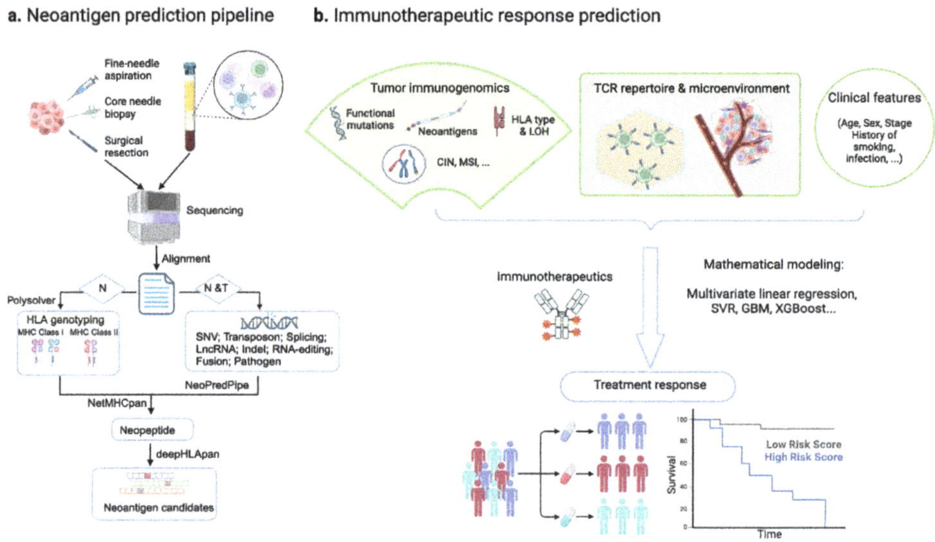

Fig. 12.3 Computational approaches and pipelines to predict neoantigens and evaluate ICB efficacy from NGS data. (a) Neoantigen prediction pipeline and (b) prediction of ICB treatment response using mathematical models.

12.3.5.1 Tumor purity analysis

Tumor purity, the proportion of cancer cells within a tumor sample [82], significantly influences the accuracy of computational analyses of both tumor-intrinsic and tumor-extrinsic immunogenomic features [83]. Numerous computational methods leverage bulk DNA-seq [84], RNA-seq [85], DNA methylation data [82], and integrated consensus approaches [82] to estimate tumor purity (as illustrated in Fig. 12.1). These methods demonstrate strong concordance across different molecular datasets [82]. Several strategies have been used to enhance the accuracy of profiling both tumor-intrinsic and tumor-extrinsic immunogenomic signatures, such as integration of multi-omics, including single-cell analysis, spatial transcriptomics, and imaging mass cytometry, using AI tools [82].

12.3.5.2 Tumor heterogeneity estimation

Genomic ITH, characterized by variations in clonal populations within tumors [86], poses significant challenges for therapeutic efficacy [87, 88]. Bulk sequencing methods such as WGS and WES often fail to capture rare subclonal populations, particularly when their distribution is uneven. Emerging techniques, such as circulating tumor DNA (ctDNA) and cell-free DNA (cfDNA) derived from liquid biopsies, enable the detection of minor subclones, including those associated with treatment resistance.

Advanced computational tools address these challenges. For instance, copy number callers such as Sclust [89], FACETS [90], Sequenza [91], and TITAN [92] analyze copy number alterations, while algorithms such as Battenberg [93], SciClone [94], PyClone [95], and FastClone [96] deconvolve subclonal structures using Bayesian models. Single-cell sequencing methods provide finer resolution, revealing genetic and transcriptomic variations at the cellular level. Benchmark studies [97, 98] suggest that pipelines combining Mutect2, FACETS, and PyClone achieve optimal accuracy, although scalability issues persist for larger datasets [97–100]. Tools such as Monovar [101] and LiRA [102], InferCNV [103], and CONICS [104] estimate somatic mutations, CNVs, and allele-specific CNVs, while methods such as B-SCITE [105] and SCARLET [106] reconstruct clonal lineages to study tumor evolution. Together, these methods advance our understanding of ITH, offering opportunities to refine diagnostics and develop tailored therapeutic strategies.

12.3.6 Biological and Clinical Significance of Genomics-Based Computational Tools

Computational tools have transformed cancer research by enabling sophisticated analyses of genomic data, advancing immunotherapy strategies, and guiding precision medicine. Mutation-calling algorithms such as Mutect 2 identify driver mutations linked to cancer progression and therapeutic resistance, while tools for inferring ITH provide insights into treatment resistance and tumor evolution.

HLA genotyping algorithms, essential for immunotherapy, facilitate personalized vaccine design and enhance our understanding of immune responses. Neoantigen-based approaches illuminate immune escape mechanisms and support the development of therapeutic vaccines, such as the HER-2/neu peptide vaccine currently in phase III trials [107]. In addition, tumor-infiltrating lymphocyte (TIL)-based strategies, involving the ex vivo expansion of TILs, are being optimized for clinical use to target clonal neoantigens [107]. By integrating computational and experimental approaches, these methods advance the identification and validation of therapeutic targets, driving innovations in personalized oncology.

12.3.7 Real-World Perspectives

The integration of computational models into clinical workflows has revolutionized cancer diagnosis and treatment. Tools such as Mutect 2 are incorporated into NGS assays for tumor profiling, while HLA typing tools such as OptiType have gained clinical validation. Cloud-based platforms, such as pVACtools and CloudNeo, democratize access to genomics analyses, enabling smaller laboratories to adopt these technologies. The use of multi-omics approaches, combining genomic, transcriptomic, and proteomic data, provides a comprehensive view of patient health. For example, neural network ensemble models trained on clinical trial data predict survival outcomes based on immune-related adverse events (irAEs) [108]. These innovations enhance immunotherapy outcome predictions and highlight the potential of nonlinear modeling in clinical applications. Despite challenges in scalability and data interpretation, integrative approaches, such as combining DNA and RNA long-read sequencing, promise to improve our ability to analyze tumor dynamics. These advancements set the stage for more robust studies in immuno-oncology, ultimately transforming cancer care.

This section introduces cutting-edge computational methods for assessing tumor heterogeneity and treatment resistance, with a focus on their clinical implications. Building on these insights, the following section will explore RNA-sequencing data analysis and its role in uncovering immune-related transcriptomic dynamics in cancer.

12.4 RNA-Sequencing Data Analysis for Immune Profiling

RNA sequencing (RNA-seq) technologies have significantly enhanced our ability to investigate tumor immune microenvironments. This section highlights computational approaches for analyzing single-cell RNA sequencing (scRNA-seq), bulk RNA-seq, spatial transcriptomics data, T cell receptor (TCR) sequencing, and B cell receptor (BCR) sequencing. These techniques are pivotal in understanding tumor–immune interactions, predicting responses to immunotherapy, and addressing irAEs, which are critical for optimizing clinical outcomes.

12.4.1 Single-Cell RNA-seq in Profiling Immune Microenvironment

scRNA-seq has transformed the study of cellular heterogeneity by revealing the transcriptomic profiles of individual cells. It provides a detailed view of the TME, uncovering mechanisms of response and resistance to immunotherapy [109]. Computational workflows for scRNA-seq involve several stages, including quality control, alignment, and gene quantification (Fig. 12.2). Tools such as Cell Ranger [110], STARsolo [111], and Alevin [112] facilitate these processes efficiently.

Downstream analyses include normalization, feature selection, clustering, differential expression analysis, and cell-type annotation. Seurat [113], a widely used R package, provides an integrated pipeline for scRNA-seq analysis and visualization, catering to users with limited computational expertise. Scanpy [114], a Python-based framework, is optimized for handling large datasets, offering flexibility in analysis and integration with other scientific libraries. scRNA-seq has been instrumental in identifying key immune cell populations within the TME, such as B cells, cytotoxic T cells, regulatory T cells (Tregs), and macrophages, and their roles in immune suppression [115]. For example, studies have characterized TILs, which play critical roles in antitumor immunity. scRNA-seq has also been applied to identify biomarkers for predicting irAEs, helping clinicians manage these adverse events proactively [116, 117].

In scenarios lacking scRNA-seq data, deconvolution tools for bulk RNA-seq have proven invaluable. These tools estimate cell-type proportions within bulk samples, providing insights into immune infiltration and enabling retrospective analysis of archived datasets. Tools such as CIBERSORTx [19], MuSiC [118], and Bisque [119] utilize scRNA-seq-derived reference datasets to refine cell proportion estimations. Bayesian models, as used in MuSiC and Bisque, further enhance accuracy by accounting for sample variability. Other methods, such as BayesPrism's [120], xCELL [20], TIMER [121], MCP-counter [122], and EPIC [123], employ distinct statistical approaches to model immune cell landscapes. Benchmark studies [124] indicate that CIBERSORT and EPIC achieve high accuracy in estimating immune cell fractions, while TIMER and xCELL are robust for larger datasets. Deconvolution analysis has also been used to identify immune cell signatures associated with irAEs, improving our understanding of immune dysregulation in these contexts [125].

12.4.2 Spatial Transcriptomics in Dissecting the Distribution of the Immune Microenvironment

Spatial transcriptomics is an innovative technology that integrates gene expression data with spatial context, revealing the spatial organization of the TME. This spatial resolution is crucial for understanding cellular interactions and immunotherapy responses. The computational workflow for spatial transcriptomics involves aligning tissue sections, filtering low-quality data, normalizing sequencing depth, and mapping spatial gene expression [126].

Tools such as Space Ranger provide streamlined workflows for processing spatial data, while Seurat integrates spatial transcriptomics with scRNA-seq analysis for a comprehensive understanding of the TME. Spatial transcriptomics has illuminated the spatial arrangement of immune cells, such as macrophages, T cells, and fibroblasts, offering strategies to overcome immunosuppressive niches and enhance immunotherapy efficacy.

Numerous computational tools have been developed to analyze spatial transcriptomics data. For instance, SpatialDE [127] identifies spatially variable genes using robust statistical methods, while Giotto [128] integrates spatial and transcriptomics data for advanced visualization. Bayesian-based tools such as BayesSpace [129] and Cell2location [130] enhance the resolution of spatial transcriptomics through probabilistic modeling. Squidpy [131] facilitates spatial transcriptomics data analysis by integrating with Scanpy and providing modules for spatial data handling, visualization, and analysis. Tangram [132], a tool for integrating spatial transcriptomics and scRNA-seq data, allows for spatial mapping of single-cell data onto tissue sections using DL algorithms, offering flexibility but requiring comprehensive knowledge of both data types.

Other tools, such as SPOTlight [133], Stereoscope [134], and Robust Cell Type Decomposition (RCTD) [135], focus on deconvolution of spatial spots to infer cell-type proportions, integrating scRNA-seq data to improve accuracy. Benchmark studies have identified top-performing tools such as Tangram [132], gimVI, and SpaGE for predicting spatial RNA distributions, as well as Cell2location, SpatialDWLS, and RCTD for cell-type deconvolution. These computational advancements provide high-resolution maps of cellular organization within tissues, revealing the dynamics of immune interactions in the TME.

Due to the limited resolution of spatial transcriptomics, deconvolution tools are often employed to infer cell types and their proportions at each spatial spot. Methods such as SPOTlight, Cell2location, and DestVI [136] use non-negative matrix factorization, probabilistic modeling, or variational inference to achieve accurate deconvolution. For dynamic spatial mapping, CellTrek integrates scRNA-seq with spatial data, offering insights into developmental trajectories [137].

Comparative analyses of these methods have demonstrated their efficacy in various tasks, such as identifying immune cell niches, mapping differentiation states, and predicting immunotherapy outcomes. For example, CytoTRACE [138] and CytoSPACE [139] integrate spatial and single-cell data to provide detailed maps of cell types and differentiation states, advancing our understanding of tumor evolution and therapy resistance.

Benchmark studies have been extensively conducted to evaluate the performance of these tools. For instance, a study compared 16 integration methods using 45 paired datasets (combining spatial transcriptomics and scRNA-seq data) alongside 32 simulated datasets [140]. The results indicate that Tangram, gimVI, and SpaGE provided superior performance in predicting RNA transcript spatial distributions, while Cell2location, SpatialDWLS, and RCTD emerged as the leading methods for cell-type deconvolution in spatial spots. Another study benchmarked 18 existing methods for cellular deconvolution using 50 real-world and simulated datasets, assessing their accuracy, robustness, and usability. The comparison of these methods was thorough, utilizing various metrics, resolutions, spatial transcriptomics platforms, spot counts, and gene numbers. Based on performance, CARD, Cell2location, and Tangram proved to be the most effective for the cellular deconvolution task [141].

12.4.3 Integrating Bulk and Single-Cell RNA-seq

12.4.3.1 Immune gene signatures

Immune gene signatures are critical tools for predicting immunotherapy responses, as they capture the immune cell dynamics within the TME. These signatures, derived from both bulk sequencing and scRNA-seq, reflect the interplay between cancer and immune cells, enabling precise patient stratification and personalized treatments. Advanced machine learning models, such as SVMs and neural networks, are often employed to identify these signatures.

T cell-specific gene signatures have been pivotal in understanding T cell activation, exhaustion, and functional diversity in cancer. For instance, the TIDE [142] tool evaluates tumor immune evasion by analyzing T cell dysfunction and pretreatment RNA profiles, predicting outcomes for melanoma patients treated with Anti-Programmed Cell Death Protein 1 (anti-PD1) or Anti-Cytotoxic T-Lymphocyte-Associated Protein 4 (anti-CTLA4) therapies [143]. Similarly, scRNA-seq analyses in hepatocellular carcinoma (HCC) tissues identified 181 T cell-related genes for stratifying patients based on immunotherapy responses [144]. These insights have advanced the development of pan-cancer stemness signatures that predict treatment outcomes and present potential targets for CRISPR-Cas9 gene-editing technologies to enhance immune responses [145].

Macrophage-related signatures (MRSs) have also demonstrated significant clinical utility [146]. M2-like macrophage signatures correlate with immunotherapy efficacy in HCC [147], while other macrophage-related gene clusters predict drug resistance in gliomas. Similarly, dendritic cell (DC)-related gene signatures emphasize the importance of DC maturation and antigen-presentation capacity in promoting favorable immunotherapy outcomes, as evidenced by studies in colon cancer and bladder urothelial carcinoma [148].

12.4.3.2 Nonimmune gene signatures

Nonimmune gene signatures offer an additional dimension for predicting immunotherapy responses. These signatures integrate genetic, metabolic, and functional data to provide a holistic view of tumor behavior and therapeutic outcomes [149]. Aging-associated gene (ARG) signatures, for example, have been linked to prognostic predictions in lung squamous carcinoma and acute myeloid leukemia, highlighting aging biomarkers' significance in cancer treatment [150].

Metabolism-related gene signatures also play a key role in predicting immunotherapy efficacy. For instance, glutamine metabolomics and other metabolic markers have been associated with improved responses in lung adenocarcinoma. Moreover, reactive oxygen species (ROS), cuproptosis, and cellular senescence influence immune cell functions and tumor responses to therapy [151]. Computational models integrating machine learning algorithms have identified ROS gene signatures, cuproptosis-related lncRNAs, and cellular senescence genes, demonstrating their utility in predicting immunotherapy outcomes in diverse cancers such as clear cell renal cell carcinoma and melanoma [152]. By combining immune and nonimmune gene signatures with advanced computational tools, researchers can better understand the intricate molecular landscape of the TME, improving therapeutic strategies and patient outcomes [153].

12.4.4 Computational Measurements of TCR and BCR Repertoire Diversity

The diversity of an individual's TCR and BCR repertoires is critical in recognizing antigens and mounting an effective immune response, particularly in cancer [154, 155]. Profiling TCR and BCR sequences allows for the assessment of clonal composition, expansion, and heterogeneity within the TME. Computational methods are extensively used to infer TCR and BCR structures from both bulk and single-cell sequencing data, providing insights into repertoire diversity and immune responses during ICB therapy.

Several tools have been developed to analyze TCR and BCR repertoires from bulk RNA-seq, Single-Cell T Cell Receptor Sequencing (scTCR-seq), and Single-Cell B Cell Receptor Sequencing (scBCR-seq) data, including sequence recovery, clustering, and annotation [156–158] (Fig. 12.1). Tools such as MiXCR and TRUST4 are commonly used tools for TCR sequence reconstruction. In addition, advanced methods such as elastic net models with Monte Carlo cross-validation have been developed to integrate TCR-modeled entropy into biomarker development, improving the discrimination of responders in cancer immunotherapy datasets [159]. Recently, predicTCR [160] has been introduced as the first machine learning classifier for tumor TCR reactivity, demonstrating highly accurate identification of tumor-reactive TCR clonotypes in TILs originating from various cancer types. As single-cell technologies continue to advance, the methods and tools discussed are expected to evolve for deeper analyses. For example, studies developed tools that can identify tissue-resident microbial sequences from host sequences, which show the contribution of the microbiome to tumor progression and treatment response [161]. Therefore, we further describe the computational methods in microbiome data related to cancer immunotherapy.

12.5 Imaging Data Analysis for Immunotherapy Response Predictions

Medical imaging technologies, including ultrasound, multidetector CT, diffusion-weighted MRI, positron emission tomography (PET), and near-infrared imaging, facilitate tumor assessment and therapeutic monitoring by leveraging qualitative features such as tumor volume and density, intratumoral cellular and acellular components (e.g., blood, necrosis, and mineralization), tumor margin regularity, metabolic activity, anatomical relationships with surrounding tissues, and their impact on adjacent structures. Based on these features, Response Evaluation Criteria in Solid Tumors (RECIST) have been foundational but are limited in capturing the complex biological responses induced by immunotherapies [162]. The development of computational methods finds great utility in performing three clinical tasks: detection, characterization, and monitoring of tumors in the treatment process. For example, radiomics analysis, a computational approach, converts these imaging datasets into quantitative features, enabling the extraction of patterns associated with tumor heterogeneity, immune infiltration, and response to immunotherapies [163]. Advanced AI methods, such as Convolutional Neural Networks (CNNs) and Support Vector Machine (SVM), are employed to classify imaging phenotypes, predict immune biomarkers (e.g., PD-L1 expression), and evaluate treatment responses. For instance, CNNs trained on radiomic features can discern complex spatial relationships and intensities that correlate with immunogenicity, while feature selection algorithms prioritize reproducible markers such as entropy or dissimilarity indices [164]. Integrating radiomics with multi-omics data, such as genomic and transcriptomic profiles, through machine learning pipelines enables a deeper understanding of tumor–immune dynamics and offers predictive models that can guide personalized immunotherapy. Several reviews have previously summarized the contributions and applications of computational tools and models in tumor medical imaging [165, 166]. In this section, we focus on the application of computational models in molecular/pathology imaging and spatial omics for immunotherapy.

Recent advances in DL have shown great promise in predicting mutation status and phenotypic biomarkers in cancer using imaging data, offering potential noninvasive prognostic tools. In prostate cancer, the TiDo model [167] accurately predicted TP53 mutation–linked phenotypes using whole slide images (WSIs), indicating that WSI-based DL models may capture histological patterns associated with aggressive disease, such as stromal composition changes, even in the presence of tumor heterogeneity. Similarly, in colorectal cancer, a DL model [168] using contrast-enhanced CT scans successfully predicted KRAS mutations, with the axial 2D ResNet configuration reaching the highest predictive accuracy. Another study developed a radiomics imaging biomarker neutrophil-to-lymphocyte ratio (NLR) [169] to predict the immunotherapy response of gastric cancer using immunohistochemistry and CT-based features via maximum relevance minimum redundancy (mRMR) and LASSO regression.

In addition, imaging-based genomics and spatial proteomics technologies offer valuable insights into protein expression with subcellular resolution, enhancing the characterization of cancers and identifying treatment potentials. Deep visual proteomics (DVP) [170] integrates AI-driven cellular phenotyping with image-guided microdissection and ultrasensitive mass spectrometry to measure single-cell protein abundance while preserving spatial context. DVP has been applied to classify cell states in culture and identify spatial proteomics changes in archived melanoma tissue, revealing pathways linked to cancer progression, such as mRNA splicing dysregulation. Similarly, another study [171] leveraged MSI combined with optical microscopy, producing single-cell molecular profiles that reveal inter- and intracellular heterogeneity, crucial for understanding complex tissue structures. Based on super-resolution microscopy images of histone H3, RNA polymerase II, or DNA, AI of the nucleus (AINU) [172] identifies nuclear signatures at nanoscale resolution to distinguish cell states, accurately classifying somatic, Induced Pluripotent Stem Cell (iPSC), virally infected, and cancer cells. AI interpretability further highlights RNA polymerase II localizations as markers of iPSC states, positioning AINU as a promising tool for studying cellular heterogeneity in diagnostics and therapeutic applications.

12.6 Conclusions and Future Directions

The integration of computational approaches into cancer immunotherapy represents a beacon of hope for enhancing our ability to predict treatment responses accurately. From predicting neoantigens to unraveling the complexities of the TME and microbiome, computational approaches offer a comprehensive toolkit for personalized and effective cancer immunotherapy. Looking ahead, leveraging multi-omics data, including genomics, transcriptomics, characteristics of the TME, tumor imaging features, and host–microbiome interactions, through computational tools that integrate knowledge-based graphs and neural networks will enhance our understanding of the various factors that influence treatment responses. This approach promises to deepen insights into the complex interplay of biological variables impacting therapeutic outcomes (Fig. 12.4). In the rapidly evolving field of computational immuno-oncology, generative AI and large language models (LLMs) offer a transformative approach to predicting immunotherapy responses. By leveraging their ability to process and integrate multimodal data, including text-based clinical records, biological sequences, and imaging data, LLMs can generate nuanced predictions that consider the intricate interplay between tumor biology and immune response. Furthermore, generative AI models can simulate various therapeutic scenarios, providing insights into potential outcomes and identifying novel biomarkers or therapeutic targets. As these models continue to improve, they may significantly enhance the precision of immunotherapy by offering personalized predictions that guide treatment decisions, ultimately leading to better patient outcomes.

Despite the immense promise of computational immunology, challenges persist in translating these approaches into routine clinical practice. Neoantigen-related studies highlight that one of the central challenges is the accurate prioritization of neoantigen candidates capable of eliciting a robust

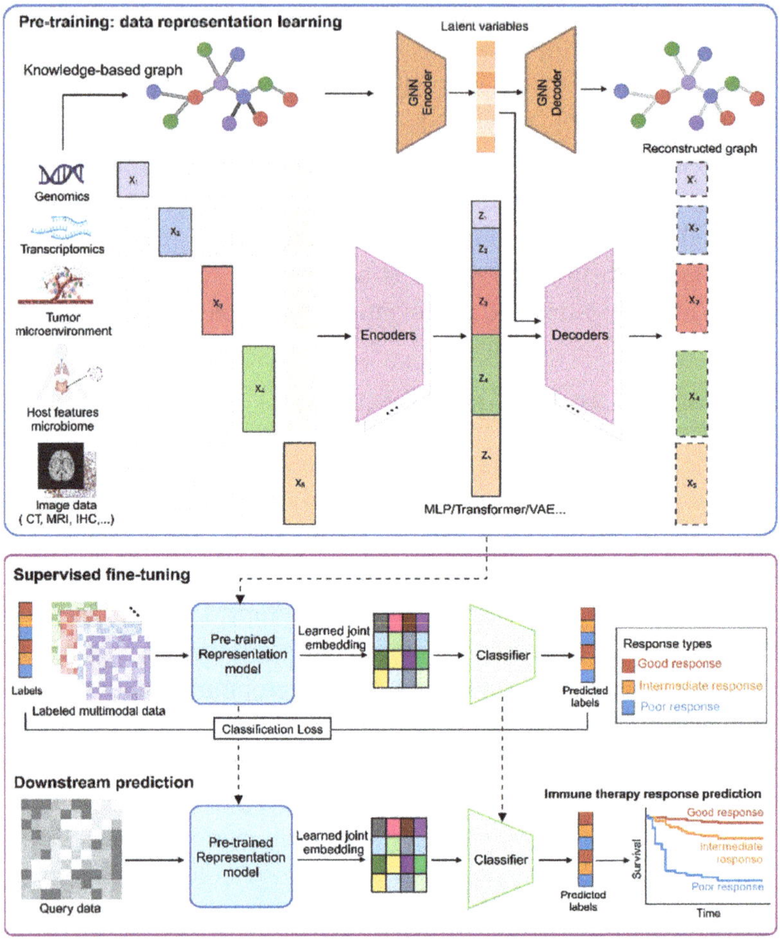

Fig. 12.4 Predict immunotherapy response by integrating multimodal data. Computational models, such as multi-layer perception (MLP), transformer, and variational autoencoder (VAE), have been utilized to integrate the data from different levels. This computational approach can be achieved through a "pre-training+fine-tuning" methodology. In the pre-training step, multi-omics data with various features are utilized to construct an initial knowledge-based graph using the K-nearest neighbors (KNN) algorithm, which is then inputted into graph neural networks (GNN) for training. Different modalities undergo separate autoencoders and are aligned and integrated in the latent space. Autoencoders such as MLP-based, transformer-based, and VAE are trained iteratively with GNN to learn representations of both the knowledge-based graph and the integrated embeddings. In the fine-tuning step, a classifier based on MLP can be incorporated into the pretrained model to predict immunotherapy response. A loss function is defined to measure the discrepancy between the predicted labels and the actual labels of multi-omics data. After supervised training of the fine-tuned models on a large set of labeled multi-omics data, the fine-tuned model can be directly applied to predict immune therapy response on query data, followed by further survival analysis on the prediction results.

immune response. While early computational tools focused on sequence-based approaches, they often faced limitations in accurately predicting binding outcomes. Furthermore, the identification of truly immunogenic neoantigens is complicated by tumor heterogeneity, as subclonal neoantigens may vary across tumor regions, potentially evading immune surveillance. In silico predictions also face difficulty in distinguishing between neoantigens that generate immune tolerance versus those that trigger a cytotoxic T-cell response.

Future directions involve integrating pMHC structural data with machine learning models to refine predictions and enhance understanding of neoantigen–MHC–TCR interactions. Incorporating features such as post-translational modifications, peptide processing kinetics, and TME factors could further improve prediction accuracy. Data standardization across different platforms and institutions remains a significant hurdle, and it requires concerted efforts from researchers and clinicians. Model interpretability is another critical aspect, as the black-box nature of some machine learning algorithms hinders the understanding of the biological rationale behind predictions.

As we navigate the challenges and envision future directions, the synergy between computational approaches and clinical insights will undoubtedly shape the next frontier of precision oncology.

Table S1 Genes/proteins/signatures associated with clinical benefit of immunotherapy.

Biomarkers	**Assay Data; Computational methods**	**Publications**
dMMR/MSI-H/MMR gene mutation	WES, MSI-PCR & IHC;	PMID: 9850653
Indel mutation load	WES;	PMID: 31048490
DDR gene mutation	MSK-IMPACT (panel);	PMID: 29489427
PD-L1 expression	IHC, Quantitative Immunofluorescence (QIF);	PMID: 29426340, PMID: 31615933
PD-L1 amplification	FoundationOne (panel);	PMID: 29902298
TP53/KRAS co-mutation	WGS & WES;	PMID: 28039262
KRAS mutation	Mouse model;	PMID: 31644929
ARID1A mutation	Foundation Medicine (DNA-seq);	PMID: 32111729
CDK12 mutation	WGS;	PMID: 29906450
PBRM1 loss	WES;	PMID: 29301960
SMARCA4	IHC;	PMID: 32156746
APOBEC-related mutagenesis signature	FoundationOne (panel); APOBEC mutational signature enrichment (AMSE) method	PMID: 30723579

(Continued)

Table S1 Genes/proteins/signatures associated with clinical benefit of immunotherapy. (*Continued*)

Biomarkers	Assay Data; Computational methods	Publications
UV-related signature	FoundationOne (panel); UV mutational signature enrichment (UVMSE)	PMID: 32530570
B2M mutation	WES;	PMID: 27903500
JAK1/2 mutations	WES;	PMID: 7903500
STK11/LKB1 mutations	WES;	PMID: 9773717
KEAP1 mutation	MSK-IMPACT;	PMID: 2175433
MDM2/MDM4 amplification and EGFR mutation	DNA-seq (panel);	PMID: 28351930
Clonal/subclonal neoantigen	WES;	PMID: 6940869
Functional immune/ anti-immunogenic mutations & MHC-binding peptides/Neoantigens	WES, IEDB peptide-MHC binding data; Convolutional neural network (CNN) model	PMID: 32075964
Cauchy-Schwarz index of Neoantigens (CSiN) score	WES; Cauchy chwarz index of Neoantigens	PMID: 32086382
NEOPS	WES, RNA-seq; XGBoost model	PMID: 34341053
NEO2IS and NEO2ITHS	WES, RNA-seq, scRNA-seq; RNN & multivariate linear regression	PMID: 37100920
Neoantigen fitness model	Exome sequencing & T-cell antigens data of IEDB; MHC binding affinity and a non-linear dependence on its sequence similarity to known antigens	PMID: 29132144
Network-based approach (NetBio)	Expression data; Amachine learning (ML) framework that leverages network-based analyses to identify ICI treatment biomarkers	PMID: 35764641
lose HLA expression	WES, RNA-seq, IHC;	PMID: 30021886
HLA-I homozygosity	Exome sequencing;	PMID: 29217585
TCR repertoire clonality	TCR_ Sequencing; Clonality was defined as 1- Pielou evenness	PMID: 30635271

References

1. Abbosh, C., et al. Phylogenetic ctDNA analysis depicts early-stage lung cancer evolution. *Nature*. 2017; **545**(7655): p. 446–451.
2. Hugo, W., et al. Genomic and transcriptomic features of response to anti-PD-1 therapy in metastatic melanoma. *Cell*. 2016; **165**(1): p. 35–44.
3. Samstein, R.M., et al. Tumor mutational load predicts survival after immunotherapy across multiple cancer types. *Nat Genet*. 2019; **51**(2): p. 202–206.
4. Andrews, S. FastQC: a quality control tool for high throughput sequence data. Online, 2010.
5. Bolger, A.M., M. Lohse, and B. Usadel. Trimmomatic: a flexible trimmer for Illumina sequence data. *Bioinformatics*. 2014; **30**(15): p. 2114–2120.
6. Martin, M. Cutadapt removes adapter sequences from high-throughput sequencing reads. 2011; **17**(1): p. 3.
7. Li, H. and R. Durbin. Fast and accurate short read alignment with Burrows-Wheeler transform. *Bioinformatics*. 2009; **25**(14): p. 1754–1760.
8. Langmead, B. and S. L. Salzberg. Fast gapped-read alignment with Bowtie 2. *Nat Methods*. 2012; **9**(4): p. 357–359.
9. McKenna, A., et al. The genome analysis toolkit: a MapReduce framework for analyzing next-generation DNA sequencing data. *Genome Res*. 2010; **20**(9): p. 1297–1303.
10. Li, H., et al. The sequence alignment/map format and SAMtools. *Bioinformatics*. 2009; **25**(16): p. 2078–2079.
11. Dobin, A., et al. STAR: ultrafast universal RNA-seq aligner. *Bioinformatics*. 2013; **29**(1): p. 15–21.
12. Kim, D., et al. Graph-based genome alignment and genotyping with HISAT2 and HISAT-genotype. *Nat Biotechnol*. 2019; **37**(8): p. 907–915.
13. Liao, Y., G. K. Smyth, and W. Shi. FeatureCounts: an efficient general purpose program for assigning sequence reads to genomic features. *Bioinformatics*. 2014; **30**(7): p. 923–930.
14. Anders, S., P. T. Pyl, and W. Huber. HTSeq—a Python framework to work with high-throughput sequencing data. *Bioinformatics*. 2015; **31**(2): p. 166–169.
15. Love, M. I., W. Huber, and S. Anders. Moderated estimation of fold change and dispersion for RNA-seq data with DESeq2. *Genome Biol*. 2014; **15**(12): p. 550.
16. Robinson, M. D., D. J. McCarthy, and G. K. Smyth. edgeR: a Bioconductor package for differential expression analysis of digital gene expression data. *Bioinformatics*. 2010; **26**(1): p. 139–140.
17. Bolyen, E., et al. Reproducible, interactive, scalable and extensible microbiome data science using QIIME 2. *Nat Biotechnol*. 2019; **37**(8): p. 852–857.
18. Callahan, B. J., et al. DADA2: high-resolution sample inference from Illumina amplicon data. *Nat Methods*. 2016; **13**(7): p. 581–583.
19. Newman, A. M., et al. Determining cell type abundance and expression from bulk tissues with digital cytometry. *Nat Biotechnol*. 2019; **37**(7): p. 773–782.
20. Aran, D., Z. Hu, and A. J. Butte. xCell: digitally portraying the tissue cellular heterogeneity landscape. *Genome Biol*. 2017; **18**(1): p. 220.
21. Blass, E. and P. A. Ott. Advances in the development of personalized neoantigen-based therapeutic cancer vaccines. *Nat Rev Clin Oncol*. 2021; **18**(4): p. 215–229.

22. Leko, V. and S. A. Rosenberg. Identifying and targeting human tumor antigens for T cell-based immunotherapy of solid tumors. *Cancer Cell.* 2020; **38**(4): p. 454–472.
23. Chan, T. A., et al. Development of tumor mutation burden as an immunotherapy biomarker: utility for the oncology clinic. *Ann Oncol.* 2019; **30**(1): p. 44–56.
24. Davis, A. A. and V. G. Patel. The role of PD-L1 expression as a predictive biomarker: an analysis of all US Food and Drug Administration (FDA) approvals of immune checkpoint inhibitors. *J Immunother Cancer.* 2019; **7**(1): p. 278.
25. Lesterhuis, W. J., et al. Dynamic versus static biomarkers in cancer immune checkpoint blockade: unravelling complexity. *Nat Rev Drug Discov.* 2017; **16**(4): p. 264–272.
26. Jardim, D. L., et al. The challenges of tumor mutational burden as an immunotherapy biomarker. *Cancer Cell.* 2021; **39**(2): p. 154–173.
27. Mandal, R., et al. Genetic diversity of tumors with mismatch repair deficiency influences anti-PD-1 immunotherapy response. *Science.* 2019; **364**(6439): p. 485–491.
28. Binnewies, M., et al. Understanding the tumor immune microenvironment (TIME) for effective therapy. *Nat Med.* 2018; **24**(5): p. 541–550.
29. Liu, Y., et al. Immune cell PD-L1 colocalizes with macrophages and is associated with outcome in PD-1 pathway blockade therapy. *Clin Cancer Res.* 2020; **26**(4): p. 970–977.
30. Dong, Z. Y., et al. Potential predictive value of TP53 and KRAS mutation status for response to PD-1 blockade immunotherapy in lung adenocarcinoma. *Clin Cancer Res.* 2017; **23**(12): p. 3012–3024.
31. Okamura, R., et al. ARID1A alterations function as a biomarker for longer progression-free survival after anti-PD-1/PD-L1 immunotherapy. *J Immunother Cancer.* 2020; **8**(1): p. e000438.
32. Wu, Y. M., et al. Inactivation of CDK12 delineates a distinct immunogenic class of advanced prostate cancer. *Cell.* 2018; **173**(7): p. 1770–1782. e14.
33. Miao, D., et al. Genomic correlates of response to immune checkpoint therapies in clear cell renal cell carcinoma. *Science.* 2018; **359**(6377): p. 801–806.
34. Tischkowitz, M., et al. Small-cell carcinoma of the ovary, hypercalcemic type-genetics, new treatment targets, and current management guidelines. *Clin Cancer Res.* 2020; **26**(15): p. 3908–3917.
35. Pham, T. V., et al. Role of ultraviolet mutational signature versus tumor mutation burden in predicting response to immunotherapy. *Mol Oncol.* 2020; **14**(8): p. 1680–1694.
36. Shin, D. S., et al. Primary resistance to PD-1 blockade mediated by JAK1/2 mutations. *Cancer Discov.* 2017; **7**(2): p. 188–201.
37. Sun, J. Y., et al. Resistance to PD-1/PD-L1 blockade cancer immunotherapy: mechanisms, predictive factors, and future perspectives. *Biomark Res.* 2020; **8**: p. 35.
38. Skoulidis, F., et al. STK11/LKB1 mutations and PD-1 inhibitor resistance in KRAS-mutant lung adenocarcinoma. *Cancer Discov.* 2018; **8**(7): p. 822–835.
39. Chen, X., et al. Pan-cancer analysis of KEAP1 mutations as biomarkers for immunotherapy outcomes. *Ann Transl Med.* 2020; **8**(4): p. 141.
40. Kato, S., et al. Hyperprogressors after immunotherapy: analysis of genomic alterations associated with accelerated growth rate. *Clin Cancer Res.* 2017; **23**(15): p. 4242–4250.
41. Kruger, S., et al. Advances in cancer immunotherapy 2019—latest trends. *J Exp Clin Cancer Res.* 2019; **38**(1): p. 268.

42. Chalmers, Z. R., et al. Analysis of 100,000 human cancer genomes reveals the landscape of tumor mutational burden. *Genome Med.* 2017; **9**(1): p. 34.
43. Diaz, L. A., Jr., et al. Pembrolizumab versus chemotherapy for microsatellite instability-high or mismatch repair-deficient metastatic colorectal cancer (KEYNOTE-177): final analysis of a randomised, open-label, phase 3 study. *Lancet Oncol.* 2022; **23**(5): p. 659–670.
44. Le, D. T., et al. Mismatch repair deficiency predicts response of solid tumors to PD-1 blockade. *Science.* 2017; **357**(6349): p. 409–413.
45. Chowell, D., et al. Patient HLA class I genotype influences cancer response to checkpoint blockade immunotherapy. *Science.* 2018; **359**(6375): p. 582–587.
46. Abed, A., et al. Prognostic value of HLA-I homozygosity in patients with non-small cell lung cancer treated with single agent immunotherapy. *J Immunother Cancer.* 2020; **8**(2): p. e001620.
47. Rodig, S. J., et al. MHC proteins confer differential sensitivity to CTLA-4 and PD-1 blockade in untreated metastatic melanoma. *Sci Transl Med.* 2018; **10**(450): p. eaar3342.
48. Marty Pyke, R., et al. Evolutionary pressure against MHC class II binding cancer mutations. *Cell.* 2018; **175**(2): p. 416–428. e13.
49. Yi, J., et al. Investigations of sequencing data and sample type on HLA class Ia typing with different computational tools. *Brief Bioinform.* 2021; **22**(3): p. bbaa143.
50. Thuesen, N. H., et al. Benchmarking freely available HLA typing algorithms across varying genes, coverages and typing resolutions. *Front Immunol.* 2022; **13**: p. 987655.
51. Yu, D., et al. A rigorous benchmarking of alignment-based HLA typing algorithms for RNA-seq data. *bioRxiv*, 2024.
52. Lang, K., et al. Full-length HLA class I genotyping with the MinION Nanopore sequencer. *Methods Mol Biol.* 2018; **1802**: p. 155–162.
53. Garcia-Garijo, A., C. A. Fajardo, and A. Gros. Determinants for neoantigen identification. *Front Immunol.* 2019; **10**: p. 1392.
54. Larsen, M. V., et al. Large-scale validation of methods for cytotoxic T-lymphocyte epitope prediction. *BMC Bioinformatics.* 2007; **8**: p. 424.
55. Nielsen, M., et al. The role of the proteasome in generating cytotoxic T-cell epitopes: insights obtained from improved predictions of proteasomal cleavage. *Immunogenetics.* 2005; **57**(1-2): p. 33–41.
56. Andreatta, M., and M. Nielsen. Gapped sequence alignment using artificial neural networks: application to the MHC class I system. *Bioinformatics.* 2016; **32**(4): p. 511–517.
57. Reynisson, B., et al. NetMHCpan-4.1 and NetMHCIIpan-4.0: improved predictions of MHC antigen presentation by concurrent motif deconvolution and integration of MS MHC eluted ligand data. *Nucleic Acids Res.* 2020; **48**(W1): p. W449–W454.
58. O’Donnell, T. J., A. Rubinsteyn, and U. Laserson. MHCflurry 2.0: improved pan-allele prediction of MHC class I-presented peptides by incorporating antigen processing. *Cell Syst.* 2020; **11**(1): p. 42–48. e7.
59. Frahm, N., et al. Extensive HLA class I allele promiscuity among viral CTL epitopes. *Eur J Immunol.* 2007; **37**(9): p. 2419–2433.
60. Xie, N., et al. Neoantigens: promising targets for cancer therapy. *Signal Transduct Target Ther.* 2023; **8**(1): p. 9.

61. Barczak, W., et al. Long non-coding RNA-derived peptides are immunogenic and drive a potent anti-tumour response. *Nat Commun*. 2023; **14**(1): p. 1078.
62. Turajlic, S., et al. Insertion-and-deletion-derived tumour-specific neoantigens and the immunogenic phenotype: a pan-cancer analysis. *Lancet Oncol*. 2017; **18**(8): p. 1009–1021.
63. Yang, W., et al. Immunogenic neoantigens derived from gene fusions stimulate T cell responses. *Nat Med*. 2019; **25**(5): p. 767–775.
64. Kahles, A., et al. Comprehensive analysis of alternative splicing across tumors from 8,705 patients. *Cancer Cell*. 2018; **34**(2): p. 211–224. e6.
65. Park, J., and Y. J. Chung. Identification of neoantigens derived from alternative splicing and RNA modification. *Genomics Inform*. 2019; **17**(3): p. e23.
66. Shen, L., et al. RNA transcription and splicing errors as a source of cancer frameshift neoantigens for vaccines. *Sci Rep*. 2019; **9**(1): p. 14184.
67. Zhang, M., et al. RNA editing derived epitopes function as cancer antigens to elicit immune responses. *Nat Commun*. 2018; **9**(1): p. 3919.
68. Smith, C. C., et al. Alternative tumour-specific antigens. *Nat Rev Cancer*. 2019; **19**(8): p. 465–478.
69. Chai, S., et al. NeoSplice: a bioinformatics method for prediction of splice variant neoantigens. *Bioinform Adv*. 2022; **2**(1): p. vbac032.
70. Pan, Y., et al. IRIS: discovery of cancer immunotherapy targets arising from pre-mRNA alternative splicing. *Proc Natl Acad Sci U S A*. 2023; **120**(21): p. e2221116120.
71. Shi, Y., B. Jing, and R. Xi. Comprehensive analysis of neoantigens derived from structural variation across whole genomes from 2528 tumors. *Genome Biol*. 2023; **24**(1): p. 169.
72. Zhang, Z., et al. ASNEO: identification of personalized alternative splicing based neoantigens with RNA-seq. *Aging (Albany NY)*. 2020; **12**(14): p. 14633–14648.
73. Vensko, S. P., et al. LENS: landscape of effective neoantigens software. *Bioinformatics*. 2023; **39**(6): p. btad322.
74. Jensen, K. K., et al. Improved methods for predicting peptide binding affinity to MHC class II molecules. *Immunology*. 2018; **154**(3): p. 394–406.
75. Vitale, I., et al. Intratumoral heterogeneity in cancer progression and response to immunotherapy. *Nat Med*. 2021; **27**(2): p. 212–224.
76. Lu, T., et al. Tumor neoantigenicity assessment with CSiN score incorporates clonality and immunogenicity to predict immunotherapy outcomes. *Sci Immunol*. 2020; **5**(44): p. eaaz3199.
77. Abbott, C. W., et al. Prediction of immunotherapy response in melanoma through combined modeling of neoantigen burden and immune-related resistance mechanisms. *Clin Cancer Res*. 2021; **27**(15): p. 4265–4276.
78. Luo, R., et al. A novel integrated approach to predicting cancer immunotherapy efficacy. *Oncogene*. 2023; **42**(23): p. 1913–1925.
79. Kong, J., et al. Network-based machine learning approach to predict immunotherapy response in cancer patients. *Nat Commun*. 2022; **13**(1): p. 3703.
80. Lakatos, E., et al. Evolutionary dynamics of neoantigens in growing tumors. *Nat Genet*. 2020; **52**(10): p. 1057–1066.

81. Lu, T., et al. Netie: inferring the evolution of neoantigen-T cell interactions in tumors. *Nat Methods*. 2022; **19**(11): p. 1480–1489.
82. Aran, D., M. Sirota, and A. J. Butte. Systematic pan-cancer analysis of tumour purity. *Nat Commun*. 2015; **6**: p. 8971.
83. Kalbasi, A., and A. Ribas. Tumour-intrinsic resistance to immune checkpoint blockade. *Nat Rev Immunol*. 2020; **20**(1): p. 25–39.
84. Carter, S. L., et al. Absolute quantification of somatic DNA alterations in human cancer. *Nat Biotechnol*. 2012; **30**(5): p. 413–421.
85. Yoshihara, K., et al. Inferring tumour purity and stromal and immune cell admixture from expression data. *Nat Commun*. 2013; **4**: p. 2612.
86. Jamal-Hanjani, M., et al. Tracking the evolution of non-small-cell lung cancer. *N Engl J Med*. 2017; **376**(22): p. 2109–2121.
87. Abecassis, J., et al. Assessing reliability of intra-tumor heterogeneity estimates from single sample whole exome sequencing data. *PLoS One*. 2019; **14**(11): p. e0224143.
88. Liu, L. Y., et al. Quantifying the influence of mutation detection on tumour subclonal reconstruction. *Nat Commun*. 2020; **11**(1): p. 6247.
89. Cun, Y., et al. Copy-number analysis and inference of subclonal populations in cancer genomes using Sclust. *Nat Protoc*. 2018; **13**(6): p. 1488–1501.
90. Shen, R., and V. E. Seshan. FACETS: allele-specific copy number and clonal heterogeneity analysis tool for high-throughput DNA sequencing. *Nucleic Acids Res*. 2016; **44**(16): p. e131.
91. Favero, F., et al. Sequenza: allele-specific copy number and mutation profiles from tumor sequencing data. *Ann Oncol*. 2015; **26**(1): p. 64–70.
92. Ha, G., et al. TITAN: inference of copy number architectures in clonal cell populations from tumor whole-genome sequence data. *Genome Res*. 2014; **24**(11): p. 1881–1893.
93. Nik-Zainal, S., et al. The life history of 21 breast cancers. *Cell*. 2012; **149**(5): p. 994–1007.
94. Miller, C. A., et al. SciClone: inferring clonal architecture and tracking the spatial and temporal patterns of tumor evolution. *PLoS Comput Biol*. 2014; **10**(8): p. e1003665.
95. Roth, A., et al. PyClone: statistical inference of clonal population structure in cancer. *Nat Methods*. 2014; **11**(4): p. 396–398.
96. Xiao, Y., et al. FastClone is a probabilistic tool for deconvoluting tumor heterogeneity in bulk-sequencing samples. *Nat Commun*. 2020; **11**(1): p. 4469.
97. Salcedo, A., et al. A community effort to create standards for evaluating tumor subclonal reconstruction. *Nat Biotechnol*. 2020; **38**(1): p. 97–107.
98. Tanner, G., et al. Benchmarking pipelines for subclonal deconvolution of bulk tumour sequencing data. *Nat Commun*. 2021; **12**(1): p. 6396.
99. Dentro, S. C., et al. Characterizing genetic intra-tumor heterogeneity across 2,658 human cancer genomes. *Cell*. 2021; **184**(8): p. 2239–2254. e39.
100. Martinez-Ruiz, C., et al. Genomic-transcriptomic evolution in lung cancer and metastasis. *Nature*. 2023; **616**(7957): p. 543–552.
101. Zafar, H., et al. Monovar: single-nucleotide variant detection in single cells. *Nat Methods*. 2016; **13**(6): p. 505–507.

102. Bohrson, C. L., et al. Linked-read analysis identifies mutations in single-cell DNA-sequencing data. *Nat Genet*. 2019; **51**(4): p. 749–754.
103. Tickle, T., et al. *inferCNV of the Trinity CTAT Project*. Klarman Cell Observatory, Broad Institute of MIT and Harvard; 2019. https://github.com/broadinstitute/inferCNV
104. Muller, S., et al. CONICS integrates scRNA-seq with DNA sequencing to map gene expression to tumor sub-clones. *Bioinformatics*. 2018; **34**(18): p. 3217–3219.
105. Malikic, S., et al. Integrative inference of subclonal tumour evolution from single-cell and bulk sequencing data. *Nat Commun*. 2019; **10**(1): p. 2750.
106. Satas, G., et al. SCARLET: single-cell tumor phylogeny inference with copy-number constrained mutation losses. *Cell Syst*. 2020; **10**(4): p. 323–332. e8.
107. Clifton, G. T., et al. Clinical development of the E75 vaccine in breast cancer. *Breast Care (Basel*). 2016; **11**(2): p. 116–121.
108. Sun, G. J., et al. Machine learning modeling of patient health signals informs long-term survival on immune checkpoint inhibitor therapy. *iScience*. 2024; **27**(9): p. 110634.
109. Liu, C. C., C. B. Steen, and A. M. Newman. Computational approaches for characterizing the tumor immune microenvironment. *Immunology*. 2019; **158**(2): p. 70–84.
110. Zheng, G. X., et al. Massively parallel digital transcriptional profiling of single cells. *Nat Commun*. 2017; **8**: p. 14049.
111. Kaminow, B., D. Yunusov, and A. Dobin. STARsolo: accurate, fast and versatile mapping/quantification of single-cell and single-nucleus RNA-seq data. *bioRxiv*. 2021: p. 2021.05.05.442755.
112. Srivastava, A., et al. Alevin efficiently estimates accurate gene abundances from dscRNA-seq data. *Genome Biol*. 2019; **20**(1): p. 65.
113. Butler, A., et al. Integrating single-cell transcriptomic data across different conditions, technologies, and species. *Nat Biotechnol*. 2018; **36**(5): p. 411–420.
114. Wolf, F. A., P. Angerer, and F. J. Theis. SCANPY: large-scale single-cell gene expression data analysis. *Genome Biol*. 2018; **19**(1): p. 15.
115. Zhang, S., et al. Single cell transcriptomic analyses implicate an immunosuppressive tumor microenvironment in pancreatic cancer liver metastasis. *Nat Commun*. 2023; **14**(1): p. 5123.
116. Liu, S., et al. Single-cell RNA sequencing analysis of peripheral blood mononuclear cells in PD-1-induced renal toxicity in patients with lung cancer. *BMC Nephrol*. 2024; **25**(1): p. 307.
117. Lozano, A. X., et al. T cell characteristics associated with toxicity to immune checkpoint blockade in patients with melanoma. *Nat Med*. 2022; **28**(2): p. 353–362.
118. Wang, X., et al. Bulk tissue cell type deconvolution with multi-subject single-cell expression reference. *Nat Commun*. 2019; **10**(1): p. 380.
119. Jew, B., et al. Accurate estimation of cell composition in bulk expression through robust integration of single-cell information. *Nat Commun*. 2020; **11**(1): p. 1971.
120. Chu, T., et al. Cell type and gene expression deconvolution with BayesPrism enables Bayesian integrative analysis across bulk and single-cell RNA sequencing in oncology. *Nat Cancer*. 2022; **3**(4): p. 505–517.
121. Li, T., et al. TIMER: a web server for comprehensive analysis of tumor-infiltrating immune cells. *Cancer Res*. 2017; **77**(21): p. e108–e110.

122. Becht, E., et al. Estimating the population abundance of tissue-infiltrating immune and stromal cell populations using gene expression. *Genome Biol.* 2016; **17**(1): p. 218.
123. Racle, J., and D. Gfeller. EPIC: a tool to estimate the proportions of different cell types from bulk gene expression data. *Methods Mol Biol.* 2020; **2120**: p. 233–248.
124. Sturm, G., et al. Comprehensive evaluation of transcriptome-based cell-type quantification methods for immuno-oncology. *Bioinformatics.* 2019; **35**(14): p. i436–i445.
125. Bolshakov, E. S., et al. 144 Identifying a composite signature for predicting immune-related adverse events in advanced melanoma patients treated with immune checkpoint blockade. *J ImmunoTherapy Cancer.* 2024; **12**(Suppl 2): p. A161–A162.
126. Jiang, Y., et al. Biology-guided deep learning predicts prognosis and cancer immunotherapy response. *Nat Commun.* 2023; **14**(1): p. 5135.
127. Svensson, V., S. A. Teichmann, and O. Stegle. SpatialDE: identification of spatially variable genes. *Nat Methods.* 2018; **15**(5): p. 343–346.
128. Dries, R., et al. Giotto: a toolbox for integrative analysis and visualization of spatial expression data. *Genome Biol.* 2021; **22**(1): p. 78.
129. Zhao, E., et al. Spatial transcriptomics at subspot resolution with BayesSpace. *Nat Biotechnol.* 2021; **39**(11): p. 1375–1384.
130. Kleshchevnikov, V., et al. Cell2location maps fine-grained cell types in spatial transcriptomics. *Nat Biotechnol.* 2022; **40**(5): p. 661–671.
131. Palla, G., et al. Squidpy: a scalable framework for spatial omics analysis. *Nat Methods.* 2022; **19**(2): p. 171–178.
132. Biancalani, T., et al. Deep learning and alignment of spatially resolved single-cell transcriptomes with Tangram. *Nat Methods.* 2021; **18**(11): p. 1352–1362.
133. Elosua-Bayes, M., et al. SPOTlight: seeded NMF regression to deconvolute spatial transcriptomics spots with single-cell transcriptomes. *Nucleic Acids Res.* 2021; **49**(9): p. e50.
134. Andersson, A., et al. Single-cell and spatial transcriptomics enables probabilistic inference of cell type topography. *Commun Biol.* 2020; **3**(1): p. 565.
135. Cable, D. M., et al. Robust decomposition of cell type mixtures in spatial transcriptomics. *Nat Biotechnol.* 2022; **40**(4): p. 517–526.
136. Lopez, R., et al. DestVI identifies continuums of cell types in spatial transcriptomics data. *Nat Biotechnol.* 2022; **40**(9): p. 1360–1369.
137. Wei, R., et al. Spatial charting of single-cell transcriptomes in tissues. *Nat Biotechnol.* 2022; **40**(8): p. 1190–1199.
138. Gulati, G. S., et al. Single-cell transcriptional diversity is a hallmark of developmental potential. *Science.* 2020; **367**(6476): p. 405–411.
139. Vahid, M. R., et al. High-resolution alignment of single-cell and spatial transcriptomes with CytoSPACE. *Nat Biotechnol.* 2023; **41**(11): p. 1543–1548.
140. Li, B., et al. Benchmarking spatial and single-cell transcriptomics integration methods for transcript distribution prediction and cell type deconvolution. *Nat Methods.* 2022; **19**(6): p. 662–670.
141. Li, H., et al. A comprehensive benchmarking with practical guidelines for cellular deconvolution of spatial transcriptomics. *Nat Commun.* 2023; **14**(1): p. 1548.
142. Brinkman, E. K., et al. Easy quantification of template-directed CRISPR/Cas9 editing. *Nucleic Acids Res.* 2018; **46**(10): p. e58.

143. Jiang, P., et al. Signatures of T cell dysfunction and exclusion predict cancer immunotherapy response. *Nat Med*. 2018; **24**(10): p. 1550–1558.
144. Chen, X., et al. A T-cell-related signature for prognostic stratification and immunotherapy response in hepatocellular carcinoma based on transcriptomics and single-cell sequencing. *BMC Bioinformatics*. 2023; **24**(1): p. 216.
145. Zhang, Z., et al. Integrated analysis of single-cell and bulk RNA sequencing data reveals a pan-cancer stemness signature predicting immunotherapy response. *Genome Med*. 2022; **14**(1): p. 45.
146. Feng, Q., H. Lu, and L. Wu. Identification of M2-like macrophage-related signature for predicting the prognosis, ecosystem and immunotherapy response in hepatocellular carcinoma. *PLoS One*. 2023; **18**(9): p. e0291645.
147. Wang, T., et al. Comprehensive molecular analyses of a macrophage-related gene signature with regard to prognosis, immune features, and biomarkers for immunotherapy in hepatocellular carcinoma based on WGCNA and the LASSO algorithm. *Front Immunol*. 2022; **13**: p. 843408.
148. An, B., et al. Derivation and external validation of dendritic cell-related gene signatures for predicting prognosis and immunotherapy efficacy in bladder urothelial carcinoma. *Front Immunol*. 2022; **13**: p. 1080947.
149. Shi, H., et al. Identification and validation of a siglec-based and aging-related 9-gene signature for predicting prognosis in acute myeloid leukemia patients. *BMC Bioinformatics*. 2022; **23**(1): p. 284.
150. Xu, Q., and Y. Chen. An aging-related gene signature-based model for risk stratification and prognosis prediction in lung adenocarcinoma. *Front Cell Dev Biol*. 2021; **9**: p. 685379.
151. Li, J., et al. Integrating machine learning algorithms to systematically assess reactive oxygen species levels to aid prognosis and novel treatments for triple-negative breast cancer patients. *Front Immunol*. 2023; **14**: p. 1196054.
152. Xiao, Z., et al. Prediction of the prognosis of clear cell renal cell carcinoma by cuproptosis-related lncRNA signals based on machine learning and construction of ceRNA network. *J Oncol*. 2023; **2023**: p. 4643792.
153. Liang, X., et al. Genes associated with cellular senescence favor melanoma prognosis by stimulating immune responses in tumor microenvironment. *Comput Biol Med*. 2023; **158**: p. 106850.
154. Linnemann, C., R. Mezzadra, and T.N. Schumacher. TCR repertoires of intratumoral T-cell subsets. *Immunol Rev*. 2014; **257**(1): p. 72–82.
155. Yusko, E., et al. Association of tumor microenvironment T-cell repertoire and mutational load with clinical outcome after sequential checkpoint blockade in melanoma. *Cancer Immunol Res*. 2019; **7**(3): p. 458–465.
156. Bolotin, D. A., et al. MiXCR: software for comprehensive adaptive immunity profiling. *Nat Methods*. 2015; **12**(5): p. 380–381.
157. Marcou, Q., T. Mora, and A. M. Walczak. High-throughput immune repertoire analysis with IGoR. *Nat Commun*. 2018; **9**(1): p. 561.
158. Song, L., et al. TRUST4: immune repertoire reconstruction from bulk and single-cell RNA-seq data. *Nat Methods*. 2021; **18**(6): p. 627–630.

159. Bortone, D. S., et al. Improved T-cell receptor diversity estimates associate with survival and response to anti-PD-1 therapy. *Cancer Immunol Res.* 2021; **9**(1): p. 103–112.
160. Tan, C. L., et al. Prediction of tumor-reactive T cell receptors from scRNA-seq data for personalized T cell therapy. *Nat Biotechnol.* 2025; **43**(1): p. 134–142.
161. Dohlman, A. B., et al. The cancer microbiome atlas: a pan-cancer comparative analysis to distinguish tissue-resident microbiota from contaminants. *Cell Host & Microbe.* 2021; **29**(2): p. 281–298. e5.
162. Eisenhauer, E. A., et al. New response evaluation criteria in solid tumours: revised RECIST guideline (version 1.1). *Eur J Cancer.* 2009; **45**(2): p. 228–247.
163. Kothari, G. Role of radiomics in predicting immunotherapy response. *J Med Imaging Radiat Oncol.* 2022; **66**(4): p. 575–591.
164. McGale, J., et al. Artificial intelligence and radiomics: clinical applications for patients with advanced melanoma treated with immunotherapy. *Diagnostics.* 2023; **13**(19): p. 3065.
165. Wu, M., et al. Imaging-based biomarkers for predicting and evaluating cancer immunotherapy response. *Radiol: Imaging Cancer.* 2019; **1**(2): p. e190031.
166. Butner, J. D., et al. Mathematical modeling of cancer immunotherapy for personalized clinical translation. *Nat Comput Sci.* 2022; **2**(12): p. 785–796.
167. Pizurica, M., et al. Whole slide imaging-based prediction of TP53 mutations identifies an aggressive disease phenotype in prostate cancer. *Cancer Res.* 2023; **83**(17): p. 2970–2984.
168. He, K., et al. Noninvasive KRAS mutation estimation in colorectal cancer using a deep learning method based on CT imaging. *BMC Med Imaging.* 2020; **20**(1): p. 59.
169. Huang, W., et al. Noninvasive imaging of the tumor immune microenvironment correlates with response to immunotherapy in gastric cancer. *Nat Commun.* 2022; **13**(1): p. 5095.
170. Mund, A., et al. Deep visual proteomics defines single-cell identity and heterogeneity. *Nat Biotechnol.* 2022; **40**(8): p. 1231–1240.
171. Bien, T., et al. Mass spectrometry imaging to explore molecular heterogeneity in cell culture. *Proc Natl Acad Sci USA.* 2022; **119**(29): p. e2114365119.
172. Carnevali, D., et al. A deep learning method that identifies cellular heterogeneity using nanoscale nuclear features. *Nat Mach Intell.* 2024; **6**(9): p. 1021–1033.

CHAPTER 13

Systems Biology for Cancer Drug Resistance Research

13.1 Introduction to Cancer Drug Resistance and Systems Biology

13.1.1 Overview of Cancer Drug Resistance

Cancer drug resistance is a significant challenge in modern oncology, contributing to treatment failure and poor patient outcomes. Despite advancements in targeted therapies and immunotherapies, many cancers develop resistance initially to effective treatments. This phenomenon often leads to disease recurrence and progression, posing a critical barrier to achieving durable cancer control. Furthermore, cancer cells with resistance characteristics frequently display cross-resistance to a range of structurally unrelated anticancer drugs, a phenomenon known as multidrug resistance (MDR) [1].

Cancer drug resistance can be classified as intrinsic or acquired based on the timing of its development and as genetic or nongenetic based on the underlying resistance mechanisms (Fig. 13.1). Intrinsic resistance, also referred to as primary resistance, arises from inherent factors within tumor cells or tissues that exist before treatment. These factors provide a survival advantage, allowing cancer cells to adapt to the stresses of the early stages of treatment. Acquired resistance

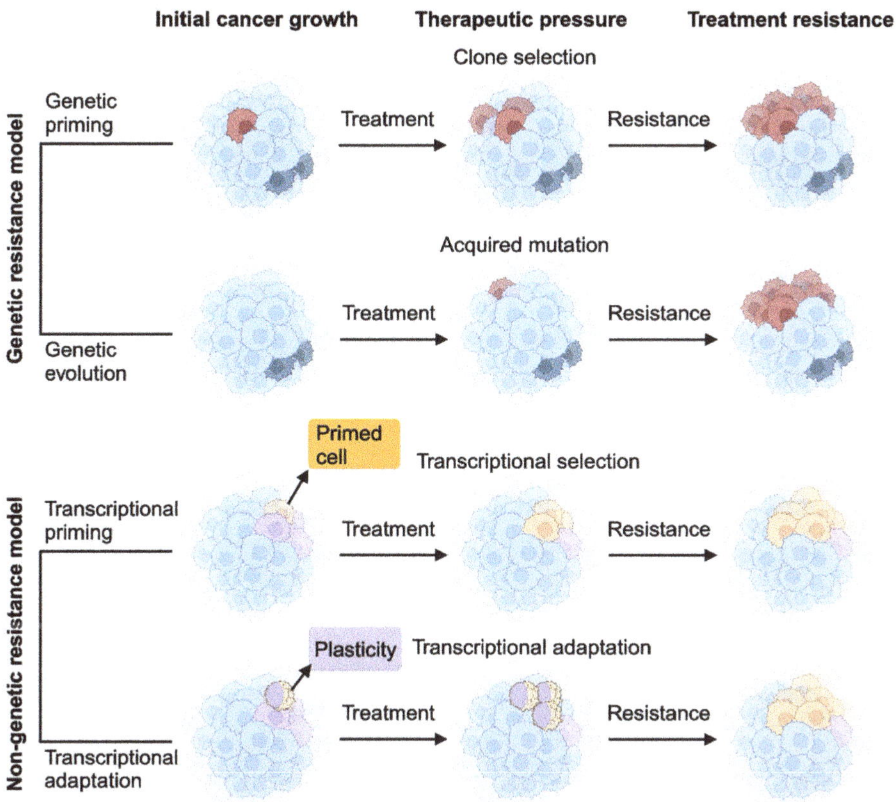

Fig. 13.1 Genetic mechanisms of drug resistance include mechanisms by which a preexisting genetic mutation primes cancer, conferring intrinsic resistance to a particular therapy. Alternatively, cancer cells may acquire a mutation during therapy that affords a proliferative or survival advantage by negating the effects of the treatment. Nongenetic mechanisms may result in the expression of a particular transcriptional program that provides intrinsic resistance to a particular drug. Cancer cells can rapidly adapt to therapeutic pressure by rewiring their gene expression to acquire a program that offers a selective advantage in the context of therapy, enabling the cells to escape the therapeutic pressure. It is likely that all cancers, to a greater or lesser degree, display all of these mechanisms of resistance [5].

arises after cancer treatment and is typically driven by adaptive changes in initially sensitive tumors, reducing the effectiveness of the therapy [2–4].

Cancer cells may become less responsive to treatment due to different reasons, usually involving both genetic and non-genetic factors (Fig. 13.1). Genetic elements within tumor cells are key drivers of therapeutic resistance, including genetic mutations in drug targets, oncogene amplification in compensatory or alternative pathways, and epigenetic changes. These factors contribute to intratumor heterogeneity (ITH), tumor cell adaptability, enhanced DNA repair, and altered vulnerability to cell death pathways, ultimately leading to resistance driven by multiple factors [6, 7]. However, resistant cases in the absence of genetic alteration have been extensively recognized in different types of cancers [5]. Phenotype changes may be independent of genotype alteration in resistance mediated by metabolic inactivation of drugs [8, 9], reduced intracellular drug concentration by transporters [10–12], drug compartmentation [13],

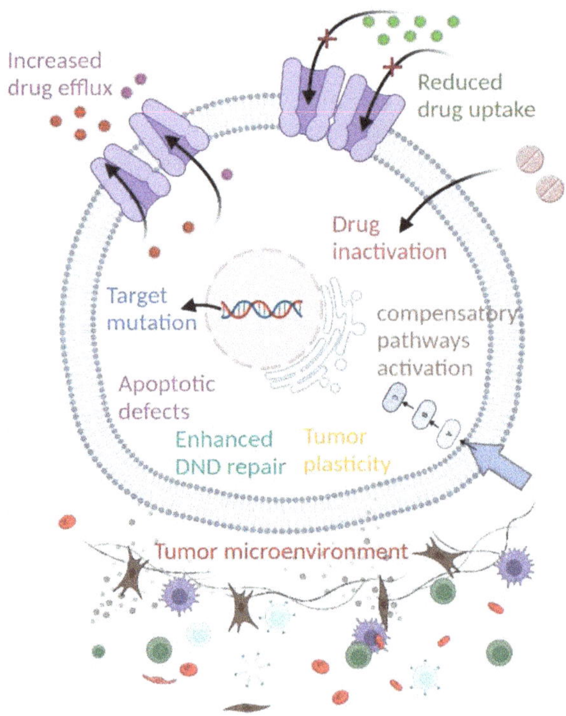

Fig. 13.2 Cancer resistance mechanisms, including drug target mutation, reduced drug uptake, increased drug efflux, drug inactivation, apoptotic defects, enhanced DNA repair, compensatory pathways activation, tumor plasticity, and tumor microenvironment. Source: This figure was created with Biorender.com.

and drug-induced reversible transcriptional or post-translational regulations on adaptive pathways. In addition to the factors within tumor cells, the tumor microenvironment (TME) is also considered to be involved in the development of resistance in some cancers [14, 15] (Fig. 13.2).

Importantly, resistance mechanisms in cancer are not independent of one another and can work together to drive therapeutic resistance. Consequently, advancing effective strategies to overcome cancer resistance requires a comprehensive understanding of these mechanisms in their entirety.

13.1.2 Importance of Integrative Systems Biology Approaches to Tackle Drug Resistance

Understanding and combating cancer drug resistance requires systems biology approaches. Systems biology integrates data from multiple biological scales, including genomics, transcriptomics, proteomics, and metabolomics, to construct a comprehensive view of cancer as a dynamic and complex system.

Systems biology offers transformative potential in addressing drug resistance. (1) Identifying biomarkers. Multi-omics approaches can identify predictive biomarkers for drug resistance, enabling personalized treatment strategies. (2) Developing combination therapies. Network analysis helps design combination treatments that simultaneously target multiple resistance mechanisms. (3) Revealing evolutionary dynamics. Systems biology elucidates how resistance evolves, guiding adaptive treatment protocols to delay or prevent resistance. (4) Optimizing drug development. Integrative models streamline the drug discovery process by identifying and validating novel therapeutic targets.

Cancer drug resistance remains a formidable challenge in oncology, but the application of systems biology offers a promising pathway to unraveling its complexities. By adopting a holistic and integrative perspective, researchers and clinicians can develop more effective strategies to predict, prevent, and overcome resistance, ultimately improving patient outcomes. As technological advancements continue to expand the capabilities of systems biology, the prospects for transforming cancer treatment grow ever brighter.

13.2 Mechanisms of Cancer Drug Resistance

Cancer drug resistance can be caused by genetic, epigenetic, and microenvironmental factors. A better understanding of the molecular mechanisms of cancer drug resistance is necessary to develop effective strategies to overcome drug resistance. In the following section, we will discuss the identified mechanisms of resistance and review recent findings before proposing appropriate responses, such as novel drug design, target-specific combination therapies, addressing alternative signaling pathways, and modulation of TME.

13.2.1 Reduced Drug Uptake

Drugs can cross the cell membrane through diffusion, endocytosis, or transporters, which can be affected by the permeability and lipid composition of the plasma membrane and the functions or expression levels of membrane transporters. Alteration of cell membrane structure can impair drug diffusion across the plasma membrane and the endocytosis process. The plasma membranes from drug-resistant cancer cells have a different lipid composition compared to those from the parental drug-sensitive cells: the cholesterol and phospholipid levels are elevated, and the protein-to-lipid ratio is increased by up to 60% in MDR cells compared to sensitive cells [16].

The intracellular concentration of drugs is largely influenced by the activity of uptake transporters, primarily the solute carrier (SLC) family. This superfamily comprises over 400 members grouped into 52 families [17]. Key SLCs associated with chemoresistance include organic anion transporting polypeptides (OATPs), organic cation transporters, concentrative nucleoside transporters, equilibrative nucleoside transporters, and copper transporters [18, 19]. Reduced uptake of chemotherapeutic agents can result from genetic variants that produce truncated uptake transporter proteins with diminished or lost function or from changes in transporter expression caused by pharmacological selection pressures. Among these, the solute carrier organic anion transporter (SLCO) family of OATPs is the most extensively studied. Notably, increased expression of OATP1A2, OATP1B1, and OATP1B3 has been observed in various cancers [20]. This amplified expression is thought to enhance chemosensitivity, as these transporters facilitate the uptake of multiple anticancer drugs, including taxanes, platinum-based agents, camptothecin derivatives, methotrexate, and certain tyrosine kinase inhibitors (TKIs) [21, 22].

In a patient-derived docetaxel-resistant prostate tumor xenograft model, OATP1B3 downregulation was observed, resulting in lower intratumoral concentrations of docetaxel and cabazitaxel compared to chemotherapy-naive tumors. Conversely, prostate tumors expressing OATP1B3 demonstrated increased uptake of both taxanes, leading to greater sensitivity to chemotherapy [23]. While OATP1B3 overexpression in prostate cancer may enhance the efficacy of chemotherapeutic treatments, it can negatively impact hormone therapy. This is because OATP1B3 facilitates testosterone uptake in cancer cells, contributing to resistance against androgen deprivation therapy (ADT) [24].

Interestingly, ADT itself can induce the upregulation of OATP1B3 expression, offering a potential strategy for sequential therapy: increased SLCO expression following ADT could enhance taxane uptake and improve therapeutic outcomes [25]. Furthermore, genetic variants in OATP transporters may promote resistance to TKIs through altered uptake activity, rather than by increasing sensitivity.

Recent findings by Haberkorn et al. revealed that the cancer-type OATP1B3 protein, a splice variant of the liver-type OATP1B3, is located in the lysosomal membrane of colorectal cancer (CRC) cells. This variant facilitates the transport of encorafenib and vemurafenib into lysosomes, leading to reduced cytoplasmic drug concentrations and diminished therapeutic efficacy [26]. These observations underscore the importance of identifying the genotypes of intratumorally expressed SLCs to better understand their roles in drug resistance.

While enhancing drug uptake by targeting SLC transporters could potentially increase chemosensitivity, this approach may not be practical for cancer treatment. Given the widespread expression of SLCs across various tissues, altering their function might disrupt the transport of nutrients and xenobiotics in normal tissues, potentially causing adverse side effects and drug–drug interactions. Furthermore, modulating SLCs could inadvertently affect nutrient uptake in cancer cells, potentially fueling tumor growth and metastasis. A more effective strategy to address resistance mediated by uptake transporters in cancer may involve developing drugs or delivery systems capable of bypassing the dysfunctional transporter, either by enhancing passive diffusion or by utilizing alternative transport mechanisms.

13.2.2 Increased Drug Efflux

Apart from decreased drug uptake, the overexpression of adenosine triphosphate-binding cassette (ABC) transporters plays a significant role in mediating drug efflux, which is a primary mechanism leading to insufficient intracellular drug concentrations. The ABC transporter superfamily in humans comprises 49 members divided into seven subfamilies (ABCA to ABCG). These transporters are responsible for moving a diverse range of substrates, such as lipids, ions, peptides, and xenobiotics [27]. The involvement of ABC transporters in MDR was first discovered in 1976 when ABCB1, which is also known as *P*-glycoprotein (*P*-gp) or multidrug resistance 1 (MDR1) protein, was discovered in mouse MDR cell lines [28]. So far, at least 13 ABC transporters have been identified that can directly mediate chemoresistance through the efflux of anticancer drugs [29].

These ABC transporters are primarily located on the plasma membrane and become activated upon binding to substrate drugs. This activation triggers an ATP hydrolysis-driven conformational change in the transporter, leading to the extrusion of the substrate drugs [30]. As a result, the overexpression of these transporters has been associated with poor responses to chemotherapy and worse patient prognoses across various cancer types. Research has largely concentrated on ABC transporters with broad substrate specificity, such as ABCB1, ABCG2 (also known as breast cancer resistance protein), and members of the ABCC subfamily, commonly referred to as MDR-associated proteins (MRPs), with ABCC1 and ABCC10 being key representatives. ABCB1 and ABCCs are capable of effluxing a wide range of cytotoxic chemotherapeutic agents, including paclitaxel, periwinkle alkaloids, and anthracyclines, from cancer cells. The range of anticancer drugs impacted by ABCG2 overlaps significantly with that of ABCB1 and ABCCs, although it also has specific distinctions. For instance, ABCG2 shows a strong efflux activity against anthracenediones (e.g., mitoxantrone) and camptothecins (e.g., topotecan and SN-38), while its effect on taxanes and vinca alkaloids is relatively limited [31]. Moreover, some kinase inhibitors used in clinical settings, such as imatinib, nilotinib, dasatinib, and palbociclib, along with newly developed targeted therapies such as OTS964 and ARS-1620, have also been identified as substrates of ABC transporters [32–35]. Moreover, ABCCs generally have high affinities to glutathione (GSH)-conjugated or glucuronate-conjugated forms of drugs; therefore, the efflux activity of ABCC transporters may be regulated by phase II metabolic enzymes [36]. In addition, for brain tumors and metastatic cancer in the brain, chemosensitivity is also affected by the expression of ABCB1 and ABCG2 transporters on the luminal membrane of endothelial cells of brain microvessels, which are known to impede anticancer drug delivery across the blood-brain barrier (BBB) [37, 38].

Ongoing efforts have been taken to develop modulators of ABC transporters in order to overcome this resistance mechanism. Numerous small-molecule compounds have been developed and tested as inhibitors against ABC transporters, particularly ABCB1. There have been three generations of ABCB1 inhibitors; however, these inhibitors have failed to enter clinical use as combination therapy with substrate anticancer drugs due to different drawbacks. The first-generation ABCB1 inhibitors, such as verapamil, erythromycin, and cyclosporine A, mostly lack sufficient therapeutic efficacy, which results in considerable side effects from the required high dosages. Although the second-generation inhibitors, such as dexverapamil and valspodar, have exhibited improved potency, they failed to proceed to clinical application because of their unwanted interactions with the CYP450 enzymes, leading to unfavorable pharmacokinetic profiles. The third-generation inhibitors, for example, tariquidar and zosuquidar, are able to overcome the selectivity problem; however, the performance in clinical settings turns out to be unsuccessful due to interpatient variability [39]. In recent decades, some repurposing targeted anticancer drugs, such as selonsertib, tepotinib, and poziotinib, have been discovered as dual inhibitors for ABCB1 and ABCG2 [40–42]. Inhibitors that can antagonize both ABCB1- and ABCC1-mediated resistance have also been reported, for example, cediranib and CBT-1 (tetrandrine) [43, 44]. Targeting multiple ABC transporters may be more prospective, given that co-expression of different ABC transporters is common in tumor tissues and multitargeting drugs may be beneficial in avoiding compensatory upregulation of another transporter induced by selective inhibition of a single transporter [45].

Silencing the efflux transporters using gene-editing technologies, such as small interfering RNA (siRNA) and CRISPR/Cas9 system, has been increasingly investigated as a novel approach to reducing drug resistance, which can potentially benefit genetic variants with intrinsic ABC transporter-related resistance [46]. However, many studies conducted gene editing in cell line-based settings before the step of tumor xenograft model establishment, which is unlikely to represent gene therapy for patients. Cancer-targeted delivery remains a major challenge in developing gene therapies against ABC transporters. Another trending strategy against ABC transporter-dependent MDR is to target the regulatory factors of the ABC transporter expression, including transcriptional regulation and epigenetic modifications on the gene and post-translational modifications on the protein. Inhibiting the cancer-specific regulators of ABC transporters may render an advantage to targeting tumor tissues selectively and increasing safety to normal tissues [39]. Dysregulated noncoding RNAs (ncRNAs), including microRNAs (miRNAs) and long noncoding RNAs (lncRNAs), have been found to be associated with the overexpression of ABCB1, ABCG2, and ABCCs in chemoresistant cancers [47]. The effect of suppressing ABC transporters and reversing cancer MDR by nanoparticle-delivered ncRNAs or inhibitors has been verified from in vitro and in vivo studies [48, 49]. Another epigenetic factor-linked overexpression of ABC transporters in cancer is the hypomethylation of the transporter gene promoter, and evidence has shown that agonists of DNA methyltransferases can induce the hypermethylation of the *ABCG2* promoter, resulting in lower ABCG2 expression and increased intracellular concentration of substrate drugs [50, 51]. Alternatively, the expression of ABC transporters can be modulated by targeting the upstream regulatory pathways. For instance, in breast cancer cells, inhibiting the overexpressed receptor tyrosine kinase (RTK)-like orphan receptor 1 (ROR1), which is an upstream regulator of ABCB1 transcription via mitogen-activated protein kinase (MAPK)/extracellular signal-regulated kinase (ERK) and p53 pathways, can reduce ABCB1-mediated drug efflux and resensitize breast cancer cells to doxorubicin [52]. Another recent study showed that suppressing mitochondrial respiration using methylation-controlled J protein mimetics can decrease ATP production, thereby decreasing the energy for ABCB1 and ABCG2 efflux activity and overcoming chemoresistance in vitro and in vivo [11].

Indeed, the direct approach to circumventing this resistance mechanism is to avoid using substrate drugs or develop novel anticancer drugs that can bypass the efflux transporters. This may rely on advanced computer-aided drug designs to modify drug structures of the current known and actively reported substrates of ABC transporters.

13.2.3 Metabolism-Associated Drug Inactivation

The activation and deactivation processes of many chemotherapeutic agents are regulated by drug-metabolizing enzymes (DMEs) [53]. Dysregulation of DMEs and metabolic signaling pathways, which can lead to the detoxification of drugs or failure in the conversion of drugs into active metabolites, is one of the major mechanisms for chemoresistance in cancer [54].

Some anticancer drugs require activation by metabolic enzymes. For example, the biotransformation of irinotecan into the active metabolite SN-38 is catalyzed by carboxylesterase [55], and thymidine phosphorylase is responsible for the activating metabolism of 5-fluorouracil (5-FU) into fluorodeoxyuridine monophosphate [56]. Cytarabine (AraC), a nucleoside drug used in patients with acute myeloid leukemia (AML) and non-Hodgkin's lymphoma, relies on the phosphorylation catalyzed by deoxycytidine kinase (DCK) to become the cytotoxic form cytarabine triphosphate [57], and the deficiency of DCK has been considered to be associated with AraC resistance in AML [58]. Wu et al. reported that DCK mutations were found in 4 of 10 patients with AML relapse after complete remission and high-dose AraC post-remission treatment [59]. In vitro studies using drug-selected AraC-resistant AML and lymphoma cell models indicated that mutation or deficient expression and function of DCK could be acquired after receiving AraC therapy, resulting in a reduced cellular response to AraC and possible cross-resistance to other nucleoside drugs such as gemcitabine [59–61]. A potential strategy to overcome DCK deficiency-mediated nucleoside drug resistance is using nucleoside analog phosphate prodrugs with anticancer efficacy without the demand for DCK phosphorylation [62], among which NUC-1031, a gemcitabine phosphoramidate prodrug, has entered clinical trials for the treatment of gemcitabine-resistant cancers [63]. Other reported attempts in preclinical evaluations include enhancing DCK activity by etoposide in combination with AraC for the treatment of AML [64] and bypassing the drug resistance mechanism associated with downregulated DCK expression using second-generation deoxyadenosine analog clofarabine in acute lymphoblastic leukemia cells [65].

The other aspect of metabolism-associated anticancer drug inactivation is the detoxification of drugs by metabolizing enzymes. For instance, certain isoforms of aldehyde dehydrogenases (ALDHs), including ALDH1A1 and ALDH3A1, have been reported to specifically cause resistance against chemotherapeutic drugs of the nitrogen mustard type, such as cyclophosphamide, mafosfamide, and ifosfamide [53]. In contrast, cytochrome P450 (CYP450) in phase I metabolism and glutathione-*S*-transferase (GST) as well as uridine diphosphoglucuronosyltransferase (UGT) in phase II conjugating biotransformation are involved in the inactivation of a broader spectrum of anticancer drugs [66, 67].

The CYP450 enzyme superfamily consists of 57 members that are responsible for the phase I oxidation of most clinically used drugs [68]. Of the CYP450 enzymes, CYP1B1, 2C8, 3A4, and 3A5 have been reported to be correlated to cancer resistance [69, 70], with CYP1B1 being mostly studied, which is found to be exclusively overexpressed in various types of cancers but has relatively low expression in normal tissues [71]. Intratumoral CYP1B1 overexpression may contribute to the diminished effectiveness of a diversity of chemotherapeutic drugs, such as paclitaxel and docetaxel, mitoxantrone, flutamide, and gemcitabine [72–75]. However, whether the resistance against these drugs is contributed by CYP1B1-catalyzed drug inactivation remains controversial. McFadyen et al. suggested that docetaxel was metabolized by CYP1B1 and the docetaxel resistance in CYP1B1-overexpressing Chinese hamster ovary cancer cells could be reversed by a CYP1 inhibitor [76, 77]. However, Martinez et al. later demonstrated that CYP1B1 did not directly inactivate docetaxel because the cytotoxicity of docetaxel in MCF-7 breast cancer cells was not affected by silencing CYP1B1 or adding recombinant CYP1B1 [78]. Although it is possible that CYP1B1 may play different roles in regulating drug resistance in different cancer types, the inhibition of CYP1B1 activity has been considered a therapeutic target for improving chemotherapy [79]. Although lacking high selectivity, phytochemicals are the most common source of CYP1B1 inhibitors, including stilbene, flavonoids, coumarins, anthraquinones, and alkaloids [80, 81]. By modifying the structures of phytochemicals, highly potent and selective inhibitors of CYP1B1 have been developed and are under preclinical evaluation, such as TMS ((*E*)-2,3′,4,5′-tetramethoxystilbene) [82] and α-naphthoflavone derivatives [83].

GSTs are involved in drug detoxification in phase II metabolism and catalyze the GSH conjugation to drugs [84]. Overexpression of GSTs in cancer cells may enhance the detoxification of anticancer drugs [85]. Moreover, after GSH conjugation, the conjugated drugs may become substrates of ABC transporters, particularly the MDR-associated proteins (MRPs, which belong to the ABCC subfamily), and are actively pumped out of the cancer cells [86–88]. Among the GST superfamily, GST alpha 1 (GSTA1), GST Mu 2 (GSTM2), and GST Pi 1 (GSTP1) have been found to be associated with cisplatin resistance in ovarian, lung, and gastric cancers [89]. Notably, the overexpression of GSTP1 has been considered a factor impairing the efficacy of platinum drugs such as cisplatin, carboplatin, and oxaliplatin by promoting the formation of platinum–GSH conjugates [90, 91]. The I150V polymorphism of GSTP1 with a phenotype of reduced enzymatic capacity has been correlated with better therapeutic outcomes in gastric cancer patients receiving oxaliplatin-based chemotherapy [92, 93]. Besides platinum drugs, the sensitivity to doxorubicin and various alkylating agents in cancer may also be affected by the detoxification process mediated by GSTA subclass, GSTP1, and GSTM1 enzymes [94, 95]. However, additional mechanisms other than GSH conjugation may be involved in GST-related chemoresistance to doxorubicin and alkylators, such as free radical scavenging and apoptosis suppression via the inhibition of the MAPK pathway [96–98]. Thus, there has been an increasing focus on targeting the GSTs to overcome chemoresistance. Inhibitors of GSTs have been shown to be effective in treating resistant cancer cells. Ethacrynic acid and analogs are among the earliest investigated GST inhibitors that exhibited resensitizing effects on tumor cells to alkylating agents via covalent binding to GSTs and reducing the enzyme activity [99, 100]. NBDHEX (6-(7-nitro-2,1,3-benzoxadiazol-4-ylthio)hexanol) is another potent inhibitor against GSTP1, GSTM2, and other GST isoenzymes [101], which has been found beneficial in combating cisplatin-resistant osteosarcoma when used in combination with cisplatin [102]. Apart from inhibiting GST activity, NBDHEX can also induce the dissociation of GSTP1 from its complex with c-Jun N-terminal kinase or tumor necrosis factor receptor-associated factor 2, thereby promoting the activation of apoptosis pathways in cancer cells [103].

However, one of the significant roadblocks that GST inhibitors encounter in clinical trials is their insufficient specificity, prompting the development of novel NBDHEX analogs with improved selectivity, among which MC3181 was screened out as a specific GSTP1-1 inhibitor and tested highly efficient in inhibiting vemurafenib-resistant melanoma in vitro and in vivo [104, 105]. Although the GST inhibitors showed promising effects on inducing cell apoptosis via inhibiting GST activities, investigations on combinations with chemotherapeutic drugs are lacking; thus, whether the inhibitors can reduce drug inactivation remains to be determined. On the other hand, utilizing prodrugs such as canfosfamide (TLK286) and brostallicin, which are activated by GSTs to become cytotoxic, has entered clinical evaluations as a practical strategy to circumvent the GST-mediated resistance mechanism [106, 107].

Similar to GSTs, the UGT enzymes are involved in phase II metabolism, and they catalyze the glucuronidation process. Among the three subfamilies (UGT1A, 2A, and 2B) of the UGTs, UGT1A enzymes, particularly UGT1A1, have been shown to overexpress in tumor tissues and play a role in resistance to anticancer drugs [8]. UGT1A1 is physiologically responsible for facilitating bilirubin elimination by catalyzing the glucuronic acid conjugation of bilirubin. Elevated tumoral UGT1A1 levels can increase the glucuronidation of SN-38, the active metabolite of the anticancer drug irinotecan, resulting in inactive SN-38 glucuronide and reduced therapeutic efficacy [108]. UGT1A enzymes have also been reported to be associated with primary resistance to some targeted drugs, such as heat shock protein HSP90 inhibitors, ganetespib and luminespib [109], and epidermal growth factor receptor (EGFR) inhibitor (EGFRi), erlotinib [110]. Drug-induced increase of UGT1A expression has been reported in AML patients treated with ribavirin, which resulted in acquired resistance [111]. Apart from substrate drugs, the efficacy of monoclonal antibody anticancer drugs may be affected by UGT1As. For instance, UGT1A6 has been found to be correlated with resistance to programmed cell death 1 (PD-1) antibody nivolumab in patients with advanced renal clear-cell cancer [112]. As UGTs are not known to conjugate proteins such as antibodies, UGT-mediated resistance to nivolumab may not be related to directed drug inactivation but to other mechanisms, such as indirect metabolism regulations by UGTs and other resistance-correlated signaling regulations involving UGTs [113]. Because of their association with hampered drug response, the expression of UGT1As might be a useful indicator for patient stratification so as to avoid the application of substrate drugs to patients who are likely to have low response due to high UGT1A levels. Meanwhile, UGTs can be potential pharmacological targets to overcome drug resistance in cancer. However, the use of UGT inhibitors has been limited by toxic side effects, and selectivity remains the major challenge in developing novel inhibitors of UGTs [114, 115].

Notably, direct detoxification is not the only mechanism of DME-associated cancer resistance; other involved factors such as drug-drug interactions and signaling molecules inherent in cancer cells should also be taken into consideration in developing strategies targeting metabolic enzymes in cancer. Regulating the expression and activity of DMEs through regulatory signaling transductions can be beneficial to overcoming cancer resistance not only by means of retaining drug bioavailability but also by means of other aspects such as inhibiting cancer cell proliferation or invasion.

13.2.4 Target Mutation

The leading limitation of targeted cancer therapies is that cancer cells can be intrinsically irresponsive or acquire resistance after a period of treatment because of either the mutation of target molecules ("on-target" mechanism) or tumor cells gaining survival advantages from a new mechanism independent of the target ("off-target" mechanism), such as the activation of downstream signaling pathways or compensatory pathways [116, 117].

The primary approach to overcome the "on-target" mechanism is to develop new generations of targeted drugs against drug-resistant tumors. For instance, the first- and second-generation EGFR-TKIs, such as gefitinib, erlotinib, and dacomitinib, are less effective in treating patients with additional EGFR mutation T790M [118]. This has led to structural modification and the development of the third-generation EGFR-TKI therapy. Osimertinib (AZD9291), a third-generation EGFR-TKI, presented a superior clinical response and outcome in EGFR-mutated non-small cell lung cancer (NSCLC). However, the rapidly acquired resistance to osimertinib conferred by the EGFR C797S mutation has been observed [119]. Besides, some patients who exhibit primary resistance remain unresponsive to the third-generation and other newly developed EGFR-TKIs. Complete remission is rare, and all patients eventually develop resistance, suggesting that both primary and acquired resistance mechanisms reduce the efficacy of the drug [119, 120]. To reverse these types of resistance, more recently, the fourth-generation EGFRis, which can inhibit both T790M and C797S signaling, have been introduced into clinical evaluation [118]. So far, EAI045 is the first allosteric TKI developed for this purpose. The C797S mutation is unlikely to impair the efficacy of EAI045 because its allosteric binding pocket is not affected by this cysteine residue. EGFR dimerization invalidates drug-mediated inhibition alone. The activity against T790M and C797S can be restored by a combination regimen with cetuximab, an antibody against EGFR dimerization [121]. Other fourth-generation EGFR-TKIs, such as JND3229 and JBJ-04-125-02, were recently found to be active in EGFR C797S-T790M-L858R signal transduction in vitro and in vivo [122, 123].

For the first-generation tropomyosin receptor kinase (TRK) kinase inhibitors, such as entrectinib and larotrectinib, the secondary mutations occurring at the ATP binding pocket of the TRK kinase domain, including G667C and G595R mutations in *NTRK1* gene and G696A and G623R mutations in *NTRK3* gene, are the commonly acquired resistance mechanisms [124, 125]. In addition, second-generation TRK inhibitors such as selitrectinib have been developed to overcome this acquired resistance.

The multitargeted TKI crizotinib was approved by the Food and Drug Administration (FDA) in 2011 to treat patients with advanced NSCLC harboring anaplastic lymphoma kinase (ALK) rearrangements [126]. However, despite a high response rate of 60% in ALK-rearranged NSCLC, most patients develop resistance to crizotinib, typically within 1-2 years. Studies of ALK-rearranged lung cancers with acquired resistance to crizotinib have identified ALK fusion gene amplification and secondary ALK tyrosine kinase domain mutations in about one-third of cases [127, 128]. To date, seven different acquired resistance mutations have been identified among crizotinib-resistant patients. The most frequently identified secondary mutations are L1196M and G1269A. In addition to these mutations, the 1151T-ins, L1152R, C1156Y, G1202R, and S1206Y mutations have also been detected in crizotinib-resistant cancers [129–131]. In approximately one-third of crizotinib-resistant tumors, there is evidence of activation of bypass signaling tracts such as EGFR [131]. In the remaining one-third of crizotinib-resistant tumors, the resistance mechanisms remain to be identified. Next-generation ALK inhibitors with improved potency and selectivity compared with crizotinib have been developed to overcome crizotinib resistance in the clinic. The ability of several ALK TKIs (TAE684, AP26113, ASP3026, and CH5424802) to inhibit ALK activity was evaluated in models harboring different ALK secondary mutations [127, 132]. These studies revealed variable sensitivity to these ALK inhibitors depending on the specific resistance mutation present. For example, the gatekeeper L1196M mutation was sensitive to TAE684, AP26113, and ASP3026, whereas 1151T-ins conferred resistance to all next-generation ALK TKIs. Ceritinib, an ATP-competitive, potent, and selective next-generation ALK inhibitor, has shown favorable selectivity to ALK [133]. In clinical studies involving ALK-positive NSCLC patients, ceritinib has exhibited marked antitumor activity and has been observed in both crizotinib-relapsed and crizotinib-naive patients [134, 135]. On the basis of this impressive clinical activity, ceritinib received FDA approval on April 29, 2014.

13.2.5 Signaling Pathway Alterations

Despite increasing successes in targeting oncogenic driver amplifications or mutations, several of the most formidable oncogenes and tumor suppressor genes remain undruggable. These include the rat sarcoma virus oncogene (RAS) family, comprising Kirsten rat sarcoma virus oncogene (KRAS), Neuroblastoma RAS viral oncogene (NRAS), and Harvey rat sarcoma virus oncogene (HRAS), and the rapidly accelerated fibrosarcoma (RAF) family, including A-Raf proto-oncogene, serine/threonine kinase (ARAF), B-Raf proto-oncogene, serine/threonine kinase (BRAF), and RAF1 proto-oncogene, serine/threonine kinase (CRAF). Effective targeting of KRAS signaling has been tough to realize in patients [136, 137]. Blocking the localization of KRAS on the plasma membrane, a vital element for its activation, has been ineffective because multiple compensatory pathways modulate this process. Similarly, targeting effector signaling downstream of KRAS has not achieved remarkable clinical benefits on account of paradoxical signaling activation generated by the inhibitor or because of on-target toxicity limiting the maximum tolerated dose in patients [138, 139]. Tumor cells with KRAS or $BRAF^{V600E}$ mutations are addicted to a downstream ERK signaling cascade for their growth, viability, and other malignant properties [140]. The adaption of selective inhibitor target $KRAS^{G12C}$ only briefly inhibits KRAS within 1-3 days of treatment, and initial signaling suppression is accompanied by the re-storage of active KRAS and reactivation of the ERK1/2 signaling pathway [141, 142]. Accumulation of active KRAS illustrates that compensatory activation of upstream RTKs, including EGFR, is to a large extent responsible for the adaptive alterations noted during $KRAS^{G12C}$ inhibitor treatment. In RAS^{Mut} tumors, RAF inhibitors (RAFis) are ineffective because they drive paradoxical ERK1/2 pathway activation and adventitious tumor progression [143]. However, MAPK kinase (MEK) inhibitors (MEKis) do not present the same limitations as RAFis, but the relief of feedback inhibition and pathway reactivation restricts MEKi monotherapy in RAS^{Mut}-driven tumors [140, 144]. By the amplification of $KRAS^{G13D}$ or $BRAF^{V600E}$, CRC cells acquire resistance to the MEKis [145]. Furthermore, MEKi resistance driven by $BRAF^{V600E}$ amplification is thoroughly reversible upon prolonged drug withdrawal, as $BRAF^{V600E}$ amplification brings about a selective disadvantage in the absence of MEKis, and MEKi withdrawal drives ERK1/2 activation beyond a key point that is optimal for cell proliferation and viability [146]. Remarkably,

MEKi resistance driven by $KRAS^{G13D}$ amplification is not reversible. Upon MEKi withdrawal, CRC cells experience an ERK1/2-dependent epithelial-to-mesenchymal transition (EMT) and emerge resistant to frequently used chemotherapeutics instead of exhibiting growth defects [146]. Consequently, the appearance of MEKi resistance, drug addiction, and the possible options for intermittent dosing schedules may depend on the nature of the amplified oncogenes (e.g., $KRAS^{G13D}$ $BRAF^{V600E}$), further highlighting the hardships of targeting RAS- and RAF-mutant tumors. Mutations of tumor suppressors can also result in resistance to targeted therapies; for example, mutations in phosphatase and tensin homolog deleted on chromosome 10 (PTEN) can activate phosphatidylinositol-3-kinase (PI3K) β signaling in response to PI3Kα inhibitors [147], as well as reverse mutations in breast cancer 1 (BRCA1) or BRCA2 in response to poly(ADP-ribose) polymerase (PARP) inhibitors [148].

In NSCLC, the majority of all EGFR mutations are $EGFR^{L858R}$ in exon 21 and activating EGFR exon 19 deletions, and tumor cells harboring the specific activating mutations exhibit high sensitivity to EGFR-TKIs [149, 150]. These constitutively activate mutant EGFR oncoproteins signaling by regulating MAPK and PI3K/AKT/mechanistic target of rapamycin (mTOR) signaling cascades to promote tumorigenesis [151]. In a series of EGFR-TKI-resistant tumor samples, 1% of patients exhibited resistance to early-generation EGFR-TKIs via the acquisition of $BRAF^{V600E}$ or $BRAF^{G469A}$ [152]. Increased mTOR levels were related to EGFR-TKI resistance in clinical samples [153], and in mouse models, the application of rapamycin (mTOR inhibitor) presented positive progression of $EGFR^{mut}$ lung tumors [154]. EGFR-TKIs in combination with PI3K/AKT/mTOR signaling pathway inhibitors have proved improved tolerability and efficacy. Another pattern of acquired resistance to kinase inhibition is the amplification of upstream genes, which can be exemplified by mesenchymal–epithelial transition (MET) amplification leading to resistance to EGFR-TKIs [155].

The clinical value of proto-oncogene 1 (ROS1)-directed TKIs was first prospectively explored in NSCLC patients, and in 2016, the FDA and European Medicines Agency (EMA) approved crizotinib for the treatment of advanced-stage ROS1-rearranged NSCLC [156]. Afterward, new ROS1-TKIs had gotten into clinical trials, leading to the FDA and Ministry of Health, Labor and Welfare of Japan approvals of entrectinib for the treatment of ROS1-aberrant NSCLC. So far, all developed ROS1-TKIs are multikinase inhibitors, which can also target MET (e.g., crizotinib and cabozantinib), ALK (e.g., ceritinib and crizotinib), TRK (e.g., entrectinib and repotrectinib), and several other kinases (e.g., SRC, EGFR, JAK2, and FLT3) with equivalent or lower potency [157–159]. Various compensatory pathways are involved in TKI therapy of ROS1-rearranged tumor progression, such as KRAS, NRAS, BRAF, HER2, EGFR, MEK, and MET. For example, mutations of $KRAS^{G12D}$ and $BRAF^{V600E}$ have emerged with crizotinib treatment in the clinical testing [160], and $NRAS^{Q61K}$ has occurred with entrectinib treatment [161]. Upregulated human epidermal growth factor receptor 2 (HER2) phosphorylation has been identified in patient-derived, crizotinib-resistant CD74-ROS1-positive CUTO23 (NSCLC cell line), and afatinib combined application recovered crizotinib sensitivity in CUTO23 [162]. Usually, ROS1 fusions de novo barely co-occur with EGFR mutations in clinics [163], but ROS1 fusions come into sight as resistance mechanisms to EGFR-TKI therapy in EGFR-mutant NSCLC [164], suggesting feedback or potential signaling reciprocity among ROS1 and EGFR. In patients treated with ROS1-TKIs, activating PIK3CA mutations have also been reported [162, 165]. In ROS1 fusion-positive NSCLC patients treated with ROS1-TKIs, 11% were found to have concurrent MEK/MAPK alterations [166]. Subsequently, it was proved that the deletion of MEK1 or MEKK1 in cells conferred resistance to ROS1-TKIs, and combined ROS1 and MEKis inhibited cell growth [166]. Until 2020, MET amplification was found to act as an important participator of ROS1-TKI (e.g., lorlatinib) resistance [116]. ROS1 kinase domain mutations that trigger the resistance of ROS1 fusion-positive tumors to ROS1-TKIs have been identified by various preclinical and clinical studies [162]. Some of these mutations are paralogous to ALK resistance mutations that emerge with targeted therapy for ALK fusions [167], among which $ROS1^{G2032R}$ is the most common resistance substitution observed in patients treated with crizotinib [168]. Considering its clinical benefits but its lack of activity against $ROS1^{G2032R}$, iorlatinib can rebuild enduring disease control in ROS1-rearranged tumors with the limited spectrum of resistance mutations [169]. However, the FDA has conferred repotrectinib as a Fast Track Designation agent in patients who have previously experienced one ROS1-TKI and platinum-doublet chemotherapy. Up to now, next-generation ROS1-TKIs have not been approved for the treatment of ROS1 rearrangement-positive tumors.

13.2.6 Enhanced DNA Repair

The DNA damage response (DDR) is a mechanism for repairing drug-induced DNA damage. Mutation or dysregulation of certain genes and DDR mechanisms is common in many types of cancer and can influence the sensitivity to DNA-damaging chemotherapy drugs, such as cisplatin and 5-FU [170, 171]. The nucleotide excision repair (NER) responsible for repairing DNA damage and mismatch repair (MMR) responsible for maintaining genomic integrity are known to be related to 5-FU resistance. It is demonstrated by Liu et al. that the acquired upregulation of the excision repair cross-complementing 1 (ERCC1), a component of the NER system, was induced by 5-FU treatment in gastric cancer cells, which can subsequently weaken the anticancer activity of 5-FU. Further mechanism studies revealed that 5-FU-induced ERCC1 overexpression may be regulated by ERK 1/2 and p38 signaling-mediated activation of the transcription factor c-jun/activator protein-1, which provides a clue to overcome this resistance mechanism using ERKis or p38 kinase inhibitors [172]. In contrast, the role of MMR in 5-FU resistance remains uncertain. It was recently reported by Oliver et al. that both MMR-proficient and MMR-deficient CRC cells exhibited 5-FU resistance after a period of treatment, suggesting that 5-FU chemoresistance in CRC cells may be independent of MMR status. Nevertheless, MMR gene human homolog 1 (hMLH1), a component of MMR, was upregulated in an MMR-deficient cell line, which indicated a potential involvement in 5-FU sensitivity [173].

In addition to gene mutations that activate DNA repair genes, epigenetic regulations, including DNA methylation, histone modifications, and miRNAs, can all affect the expression of DNA repair genes, contributing to drug resistance.

DNA methylation can suppress gene expression. In the context of DNA repair–associated drug resistance, DNA methylation has been implicated in the regulation of key DNA repair genes, such as O^6-methylguanine-DNA methyltransferase (*MGMT*), *BRCA1/2*, and *MLH1*. *MGMT* promoter methylation is associated with a loss of MGMT protein expression and activity in the tumor and has been shown to correlate with a better outcome of therapy in several studies [174, 175]. Consequently, the methylation status of the *MGMT* promoter has emerged as a useful biomarker for predicting the treatment response of glioblastoma (GBM) patients to methylating chemotherapeutic agents such as temozolomide [176]. In breast cancer, the hypermethylation of the *BRCA1* promoter has been associated with decreased BRCA1 expression and sensitivity to DNA-damaging drugs and PARP inhibitors [177]. In contrast, the hypermethylation of the *MLH1* promoter has been associated with decreased MLH1 expression and DNA repair capacity in cancer cells, which can increase the accumulation of somatic mutations in cancer cells and induce drug resistance [178]. These findings highlight the complex interplay among DNA methylation, DNA repair, and cancer resistance and suggest that targeting DNA methylation may be a promising strategy for overcoming drug resistance in cancer.

Histone modifications also play an important role in DNA repair and cancer resistance. A recent study demonstrated that, in high-grade serous ovarian carcinoma cell lines and patient-derived xenograft models, the upregulation of histone methyltransferases euchromatic histone lysine methyltransferase 1 (EHMT1) and euchromatic histone lysine methyltransferase 2 (EHMT2) is responsible for the trimethylation of histone H3 lysine 9 (H3K9me3) and contributes to PARP inhibitor resistance through inhibiting the expression of the tumor suppressor gene *CDKN1A*, which is involved in the regulation of the cell cycle and DNA repair [179]. However, another study showed that the loss of the histone methyltransferase, enhancer of zeste homolog 2 (EZH2), has led to decrease in global H3K27 methylation, resulting in the upregulation of genes associated with cisplatin resistance and the downregulation of genes involved in DDR. This study also showed that treatment with EZH2 inhibitors could reverse chemotherapy resistance in testicular germ cell tumors (TGCTs), indicating the potential for epigenetic therapies in treating drug-resistant tumors [180].

miRNAs and lncRNAs have been implicated in the regulation of multiple pathways, including DNA repair. For example, miR-15a and miR-16 are found to downregulate the expression of B-lymphoma Moloney murine leukemia virus insertion region-1 protein (BMI1), a protein involved in the ubiquitin-mediated DNA repair pathway, thereby decreasing DNA repair capacity and increasing the sensitivity of breast cancer cells to doxorubicin treatment [181]. Upregulation of lncRNA HOTAIR has been observed in breast cancer cells following radiation therapy, which promotes DNA repair and radioresistance by interacting with the EZH2 protein [182]. These findings have implicated that the specific miRNAs and lncRNAs could serve as biomarkers for predicting therapeutic response in cancer patients and could be potential targets for overcoming drug resistance.

13.2.7 Apoptotic Defects

Cancerous cells are known to highly regulate the apoptotic pathways, and apoptosis plays different roles in tumor: eliminates infected cells from the human body, assists the functioning of the immune system, and contributes to the maintenance of homeostasis [183]. Generally, those signals induce the activation of effector caspases that mediate intracellular signaling, resulting in DNA fragmentation. Caspase mutations occurring in tumor therapy chemoresistance have been widely reported. For example, caspase-8 mutations are mainly detected in gastric cancers, and the procaspase-8^{Q482H} mutation abrogates apoptosis by leading to the dimerization attenuation of the procaspase-8 protein monomers, resulting in resistance to chemotherapy [184]. The intrinsic apoptosis pathway by mitochondrial damage is activated through numerous exogenous or endogenous stimuli that induce DNA damage [185]. The activation of this pathway involves the activation or inhibition of proteins from the B-cell lymphoma 2 (Bcl-2) family, and multiple drugs are known to induce apoptosis of cancerous cells through this mechanism. For example, the small-molecule RAF kinase inhibitor sorafenib induces apoptosis of AML through the downregulation of Mcl-1 (an antiapoptotic protein) and the upregulation of BIM (an activator of the intrinsic pathway) [186]. Previous studies (in vitro and in vivo) have suggested that upregulating Bcl-2 in breast cancer cells promotes metastasis to the lung in mice through the EMT process and triggers resistance to chemotherapy (PD168393, a specific inhibitor of EGFR; AG490, an inhibitor of JAK2) [187, 188]. In the treatment of hematological malignancies, the aim to integrate the application of Bcl-2 inhibitors has led to further research and development of Bcl-2-selective and dual Bcl-2 and Bcl-xL inhibitors. Similar to ABT-199 (the first-ever Bcl-2 inhibitor known as Venetoclax), Navitoclax (ABT-263) is one of the most studied dual inhibitors; it acts as a BH3-mimetic, permitting the release of BH3-only proteins and further MOMP by Bax/Bak activation [189]. One clinical trial detected the effect of ABT-263 in 118 chronic lymphocytic leukemia (CLL) patients and reported partial responses in 34.6% of patients in phase I trials [190]. Rituximab, a monoclonal antibody targeting the protein CD-20, was combined with ABT-263 and showed a statistically significant overall response rate of 70% in untreated patients with CLL compared to rituximab alone [191]. However, because of the characteristics of Bcl-xL in platelet survival, a main concern with ABT-263 is the negative effect: inducing thrombocytopenia in a dose-dependent manner. Furthermore, an additional concern is regarding the emergence of tumor cell resistance to this agent through the upregulation of Mcl-1 phosphorylation and further sequestration of the proapoptotic protein Bim [192]. Overall, the clinical use of ABT-263 is under restriction, and further studies are necessary before its approval.

Autophagy is a conserved and tightly regulated catabolic self-degrading mechanism whereby cells maintain homeostasis and respond to stress through recycling damaged cellular organelles, proteins, and other cellular components [193]. In tumors, autophagy is a double-edged sword, and its anti- or pro-tumorigenic characteristic relies on the oncogenic context and stage of tumorigenesis [194]. In the early stages of tumorigenesis, autophagy exhibits tumor suppression function by removing aggregated and misfolded proteins and damaged organelles, limiting cell growth, necrosis, and chronic inflammation [195]. Upon advanced periods of tumor progression, autophagy helps maintain tumor cell survival and increases cellular resistance by conferring stress tolerance by providing the recycling substrate to promote cancer cell survival and contributes to cancer progression and drug resistance [196]. For example, in ovarian cancer [197], leukemia [198], hepatocellular carcinoma (HCC) [199], and CRC [200], the increased levels of basal autophagy contribute to tumor cell growth and increase aggressiveness. Another study showed that autophagy was activated in mice with Myc-induced lymphomas when treated with chemotherapy; autophagy inhibition sensitizes tumor cells to cell death induced by chemotherapeutic agents, suggesting that autophagy induction leads to a self-defense effect in this case [201]. Several barriers result from TKI resistances that emerge in a later period of tumor attributed to the mutation in the kinase domain or function of basal autophagy as housekeeping as well as the regulator of RTK activity, which challenges effective treatment against cancer-targeted therapy. Therefore, the combination of autophagy modulators with TKIs has been considered in cancer therapy. For example, in NSCLC, erlotinib and gefitinib can induce a high level of autophagy as a drug resistance and cytoprotective mechanism that is accompanied by the suppression of the PI3K/AKT/mTOR signaling cascade. In addition, autophagy inhibition by the pharmacological inhibitor chloroquine, and siRNAs targeting autophagy-related gene 5 (ATG5) and ATG7, enhances the erlotinib and gefitinib cytotoxicity [202]. Gastrointestinal stromal tumors (GISTs), in which the activation of CD117, stem cell growth factor receptor, or platelet-derived growth factor receptor-α mutations comes up, rarely respond to only-imatinib mesylate (IM) treatment. IM triggers autophagy as a survival mechanism in quiescent GIST cells, resulting in acquired resistance and ineffective treatment. Nevertheless, the combination therapy by which IM was used with siRNAs targeting ATGs potentiates GIST cytotoxicity [203]. Another study revealed that IM induces autophagy in chronic myeloid leukemia by induction of endoplasmic reticulum (ER) stress in a different way from apoptosis induction, and the combination therapy with autophagy inhibition using either siRNA targeting ATGs or pharmacological inhibitors promotes cell death. Moreover, other TKIs such as dasatinib or nilotinib, combined with autophagy suppression treatment, exhibited similar effects [204].

Ferroptosis is an intracellular iron-dependent mechanism of cell death that is distinct from apoptosis, necrosis, and autophagy [205]. Extensive preclinical and clinical studies have concentrated on conquering drug resistance, and inducing ferroptosis has been proven to reverse drug resistance. According to the theoretical framework for initiating the process of ferroptosis, there are three diverse pathways to reverse chemotherapy resistance: the canonical glutathione peroxidase 4 (GPX4)-regulated pathway, iron metabolism pathway, and lipid metabolism pathway [206]. Genetic inhibition of GPX4 can induce tumor cell ferroptosis and suppress tumor growth in vivo [207]. In addition, the inactivation of GPX4 results in the accumulation of phospholipid hydroperoxides to induce cell membrane damage and ferroptotic death [208]. In GBM, androgen receptor ubiquitination induced by the curcumin analog was found to inhibit GPX4, generating ferroptosis and reversing temozolomide resistance [209]. In CRC, elevated kinesin family member 20A (KIF20A) expression was associated with oxaliplatin resistance; cellular ferroptosis may be induced by disturbing the KIF20A/NUAK1/GPX4 signaling pathway to reverse the resistance of CRC to oxaliplatin [210]. Cystine/glutamate antiporter (xCT) system, as an essential component of the GPX4-regulated pathway, has also been revealed to be capable of triggering ferroptosis. A study reported that erastin and sulfasalazine (xCT system inhibitors) could induce head-and-neck cancer cell ferroptosis and overcome cisplatin resistance [211]. Another study showed that in gastric cancer, inducing ferroptosis by restraining Nrf2/Keap1/xCT signaling can sensitize cisplatin-resistant cells to cisplatin treatment [212]. Ferroptosis is defined as an iron-catalyzed form of regulated necrosis [213]. The upregulation of cellular labile iron pool (LIP) leads to increased vulnerability to ferroptosis. Du et al. reported that tumor cell ferroptosis contributed to catastrophic LIP accumulation after dihydroartemisinin treatment in pancreatic ductal adenocarcinoma, and dihydroartemisinin treatment could overcome cisplatin resistance by triggering ferroptosis [213]. Divalent metal transporter 1 (DMT1) and secreted glycoprotein lipocalin-2 are both key proteins in regulating iron homeostasis. Turcu et al. found that elevated cellular LIP caused by DMT1 inhibition induced ferroptosis, thereby wiping out breast cancer stem cells (CSCs) and reversing MDR [214]. A recent report by Chaudhary et al. indicated that targeting lipocalin-2 overcame 5-FU resistance in CRC by elevating intracellular iron levels, which in turn resulted in tumor cell ferroptosis [215]. The lipid metabolism pathway is also tightly associated with cell vulnerability to ferroptosis. Acyl-CoA synthetase long-chain family member 4 (ACSL4), an important participant in ferroptosis execution, is involved in producing phospholipid hydroperoxides in an enzymatic manner [216]. In pancreatic cancer, the activation of ACSL4 conquers gemcitabine resistance by inducing ferroptosis [217].

Recently, Tsvetkov et al. revealed a previously uncharacterized cell death mechanism termed cuproptosis [218]. Cuproptosis is a mechanism distinct from all other known manners of regulated cell death, such as apoptosis, pyroptosis, necroptosis, and ferroptosis. Cuproptosis is regulated by protein lipoylation in the citric acid cycle (or tricarboxylic acid cycle (TCA) cycle), where lipoylation is necessary for enzymatic function [219, 220]. Their research illuminates the connection between the sensitivity to copper-mediated cell death and mitochondrial metabolism: respiration. TCA-cycle active cells have upregulated levels of lipoylated TCA enzymes (particularly, the pyruvate dehydrogenase complex), and the lipoyl moiety acts as a direct copper binder, inducing the aggregation of lipoylated proteins and destabilization of Fe-S cluster proteins, ultimately leading to proteotoxic stress and cell death [218]. Genetic variation in copper homeostasis leads to serious disease, and copper chelators and copper ionophores have been proposed as anticancer agents [221–224]. For genetic disorders of copper homeostasis such as Menke's disease and Wilson's disease, copper chelation is an effective treatment [225]. However, in tumors, copper ionophores such as elesclomol have been tested in clinical trials, but neither the benefit of a biomarker of the appropriate patient population nor an understanding of the drug's mechanism of action is considered in such testing. For example, a phase III combination clinical trial of elesclomol in melanoma patients indicated the absence of efficacy in this unselected population, yet a post hoc analysis of samples with low plasma lactate dehydrogenase (LDH) levels displayed evidence of antitumor activity [226]. Low LDH represents a higher cellular dependency on mitochondrial metabolism, in accordance with Tsvetkov's finding that cells more reliant on mitochondrial respiration are nearly 1,000-fold more sensitive to copper ionophores than cells going through glycolysis. Considering the distinct mechanism of cuproptosis from other forms of cell death, deeply understanding how cuproptosis is initiated, progressed, and ultimately executed may exhibit great significance for specific therapeutic interventions and workable combination therapies [227].

13.2.8 Tumor Plasticity

CSCs are a small population of cells within a tumor that possess stem cell–like properties and are thought to be responsible for tumor initiation, growth, and metastasis [228, 229]. These cells have been identified in various cancers, including breast cancer, colon cancer, and GBM. CSCs are less sensitive to chemotherapy and radiation therapy due to their ability to activate DNA damage repair and antiapoptotic signaling pathways, which may lead to incomplete tumor eradication and recurrence after treatment. Based on these findings, the researchers put forward the "CSC theory" and found evidence of the existence of CSCs in some cancer tissues [230, 231]. CSCs can resist therapy mainly because of the overexpression of MDR transporters that mediate drug efflux, the more active DNA repair capacity, and the tendency to form new microvasculature for the tumor [232]. These characteristics enable CSCs to tolerate treatment, maintain the tumor with nutrients and oxygen, and rapidly repopulate the tumor. This mechanism is similar to that of normal tissue stem cells in response to trauma, which may explain why the bladder CSCs actively contribute to the cause of chemotherapy resistance after several cycles of chemotherapy [233, 234]. Targeting this trauma response of CSCs can become a new therapeutic intervention. CSCs between different treatment cycles actively regenerate and respond to chemotherapy-induced injury or apoptosis, much as normal tissue stem cells respond to trauma-induced damage. Dying cells release a metabolite called prostaglandin E2 (PGE2) that stimulates proliferation and causes CSC to repopulate cancers shrunk by chemotherapy [235, 236]. In normal cells, this is an integral part of the wound repair process, where PGE2 induces the regeneration of tissue stem cells. Ironically, PGE2 induces more CSC regeneration between chemotherapy cycles in cancers. An essential characteristic of CSC is maintaining self-renewal ability, and the mechanism may be one key factor in promoting cancer development and metastasis [237, 238]. CSC theory helps people understand the occurrence and development of cancers. In addition, CSCs can activate signaling pathways that promote cell survival and proliferation, such as the Wnt/β-catenin and Notch pathways [239–241]. Therefore, understanding the mechanisms underlying CSC-mediated resistance to therapy is essential for developing effective cancer treatments. It provides a new perspective on cancer development, metastasis, drug resistance, and recurrence. Leukemia therapy targeting leukemia stem cells (LSCs) has achieved good clinical efficacy, and beneficial research has also been carried out in solid cancers [242].

EMT is one of the essential mechanisms of cancer metastasis. The latest research suggests that cancer cells with dual characteristics of EMT and stem cells are the key to cancer metastasis and drug resistance [243]. There is a direct link between EMT and CSCs, but the molecular mechanism is still unclear. The EMT phenomenon refers to a reversible process in which relatively stable epithelial cells lose cell polarity and intercellular adhesion and transform into spindle-shaped mesenchymal cells with migration ability. It is ubiquitous in epithelial cancer cells. Its essential feature is the loss of the expression of epithelial markers, such as E-cadherin, β-catenin, tight junction protein, and epithelial cell adhesion molecules on the cell membrane of cancer cells [244–247], and meanwhile obtaining a mesenchymal phenotype with the increased expression of vimentin, N-cadherin, fibronectin, and β1 and β3 integrins [248–250]. The mechanism of EMT is mainly due to changes in the epithelial cells themselves or the surrounding microenvironment, leading to the activation of a series of signal transduction pathways, and related transcription factors in the nucleus play a regulatory role [251]. Precise intracellular signaling mechanisms regulate different degrees of epithelial cell transformation. Various extracellular signals activate different nuclear transcription factors by binding to specific receptors on the cell surface. The common feature of these transcription factors is that they can recognize the DNA binding sequence of the E-box motif on the target gene promoter, thereby regulating the expression of the target gene and initiating EMT [252]. Loss of E-cadherin expression is currently considered the most prominent feature of EMT. Decreased E-cadherin levels can lead to decreased cell adhesion and make cells easily invade and metastasize [245]. Cancer cells lose some of the characteristics of epithelial cells and acquire the characteristics of mesenchymal cells through EMT, so cancer cells can obtain stronger invasion and migration capabilities [253, 254].

Snail is involved in triggering EMT during the progression of epithelial cancers and is a crucial point in the occurrence of EMT. Snail expression correlates with the reduced loss of E-cadherin and acts as a direct repressor of E-cadherin transcription [255]. Twist is another essential transcription factor that regulates EMT. Its mechanism of inducing EMT is to directly or indirectly bind the E-cadherin promoter through the E-box motif, thereby inhibiting the expression of E-cadherin [256]. Snail/Twist1-knockout breast cancer model demonstrates that chemotherapy resistance is associated with EMT [257]. Studies have shown that EMT may be involved in drug resistance of cancer cells, such as lung, pancreatic, and breast cancer [258–260]. Numerous studies have shown that cancer cells develop resistance to carboplatin or paclitaxel by acquiring a mesenchymal cell phenotype, indicating that EMT may be the instigator of chemotherapy resistance [261, 262]. The latest research shows that cancers may have a type of partial EMT, which means the cancer cells have both epithelial and mesenchymal cell phenotypes, and these cancer cells are more aggressive than the others [262]. EMT has been described as a significant cause of EGFRi failure in EGFR-mutated NSCLC [253]. Cellular transdifferentiation, a phenotypic shift from adenocarcinoma to squamous cell carcinoma or neuroendocrine carcinoma, occurs in 3%-14% of EGFR-mutant NSCLC patients, and approximately 17% of prostate cancer patients failed with abiraterone/enzalutamide therapy [263].

From the discussion earlier, it is evident that targeting EMT processes or cellular plasticity has great potential to circumvent drug resistance. However, only a few compounds designed to inhibit the EMT process are currently in clinical trials.

Inhibitors targeting Notch, transforming growth factor-beta (TGF-β), and Wnt signaling pathways are also promising candidates for inhibiting EMT and cellular plasticity [264]. TGF-β heterogeneity in the cancer microenvironment creates rapidly dividing CSCs that accelerate cancer growth and others that invade surrounding healthy tissue and evade treatment [265, 266]. For example, PF-03446962 and galunisertib are antagonists designed to inhibit TGF-β receptors, currently in phase I clinical studies in solid cancers (NCT00557856 and NCT02423343) [267]. Both PF-03446962 and galunisertib inhibit the EMT program, thereby preventing cancer development. In addition, Wnt inhibitors such as ETC-1922159 and OMP-54F28 have been reported to inhibit the EMT program and are currently in phase I clinical trials in cancer (NCT02521844 and NCT01608867) [268].

13.2.9 Regulation from Tumor Microenvironment

How tumor cells react to therapy not merely relies on the genomic aberrations they harbor but also is mediated by the TME [269]. TME is a complicated element of tumors and is quite heterogeneous [270]. The significant character of the TME in changing tumor activity has been stated by plentiful studies uncovering how the TME can influence the malignant behavior and therapeutic resistance of tumor cells [271]. The TME contains a multitudinous cellular and acellular milieu; diverse stromal and immune cells are recruited to develop and maintain such self-sustained circumstances [272]. The extracellular matrix (ECM), including proteins such as laminins, fibronectin, proteoglycans, vitronectin, tenascin-C, and collagen, is devoted to the majority component of the TME and is vital for the maintenance of TME and the induction of cellular adhesion, migration, invasion, and metastasis [273]. Moreover, the composition and organization of ECM can also influence the sensitivity to drug therapy. For example, it has been reported that ECM proteins (laminin, fibronectin, and vitronectin) mediated cell adhesion–mediated drug resistance (cilengitide, an integrin inhibitor, and/or carmustine, an alkylating chemotherapy) in GBM. Enrichment of the above proteins in the TME facilitates GBM cell proliferation via the integrin α_v-mediated focal adhesion kinase (FAK)/paxillin/AKT signaling cascade and suppresses p53-involved tumor apoptosis [274]. In breast cancer, cancer-associated fibroblasts (CAFs) promote drug resistance by increasing hyaluronan production [275]. Another study showed that the inhibition of the β1 integrin activity by monoclonal antibody AIIB2 (anti-integrin β1 antibody) markedly promotes radiotherapy efficacy and elevates sensitivity to HER2-targeting agents of breast cancer cells [276, 277]. In melanoma, CAFs promote cell metastasis and drug resistance by upregulating the level of matrix metalloproteinase 1 (MMP1) and MMP2 [278]. As a significant aspect of the TME, the cell-to-ECM interaction and cellular crosstalk induce the release of soluble factors (such as angiogenic factors and chemokines) in charge of ECM remodeling and immune evasion, which further bring about therapy resistance. For example, the interaction between α4β1 integrin on cancer cells and fibronectin induces drug resistance in AML and CLL through the PI3K/AKT/BCL2 signaling pathway [279, 280]. Stromal cell-derived factor 1 (SDF1) can interplay with CXC motif chemokine receptor type 4 (CXCR4) and activate AKT and ERK1/2 signaling pathways, resulting in antiapoptotic effects and contributing to tumor cell survival in CLL cells [281]. In AML cells, CXCR4 activation induced by SDF1-mediated leads to resistance to cytarabine through reducing miRNA let-7a and facilitating transcriptional activation of MYC and Bcl-xl [282]. In CRC, CXCR4 is upregulated in chemoresistant tumor cells, and lymph node-derived stromal cells enhance resistance to oxaliplatin and 5-FU in an SDF1/CXCR4-dependent manner [283].

Exosomes, intraluminal vesicles of multivesicular bodies with a diameter of 30-100 nm, are ubiquitously present in most body fluids such as blood, cerebrospinal fluid, urine, saliva, and lymphatic fluid [284, 285]. Exosome contents not only reflect the composition of the donor cell but also mirror a regulated sorting mechanism [286]. A complication of multifarious proteins, including ECM proteins, enzymes, transcription factors, receptors, lipids, and nucleic acids (DNA, mRNA, and miRNA) inside and on the surface of the exosomes, constitutes their content [287, 288]. Extensive evidence has shown that exosome-mediated factors can promote tumorigenesis, metastasis, and therapeutic resistance of cancer cells via intercellular communication within TME [284, 289, 290]. Tumor-derived exosomes (TDEs) are involved in the initiation, development, and progression of diverse tumor courses, including angiogenesis, TME remodeling, metastasis, and therapy resistance [284, 290, 291]. TDEs induce the differentiation of various kinds of TME cells to CAFs that are the main cell population of TME in the majority of tumors; thus, exosomes play a vital role in ECM remodeling and TME reprogramming [292, 293]. Exosomes derived from CAFs contain different molecules such as miRNAs and growth factors that have distinct influences on the target cells of TME. For example, the gemcitabine therapy of CAFs in pancreatic cancer stimulates the expression of miR-146a and snail and prolonged exosome secretion, thereby promoting epithelial cell proliferation [294]. Drug-resistant tumor cells can pack the chemotherapeutic drugs in exosomes and shuttle therapeutic agents out of cancer cells [294]. Besides, the delivery of exosomal cargo containing proteins, mRNA, and miRNA to cancerous cells is related to therapeutic resistance [295, 296]. For instance, GBM cell-derived exosomes, which comprise MET-protein and tyrosine phosphatase receptor type Z (PTPRZ1) fusion proteins, obtain temozolomide resistance via EMT [297]. TDEs from gemcitabine-resistant pancreatic cancer cells induce chemoresistance by trapping multidrug resistance-associated protein 5 (MRP5) and *P*-gp or letting gemcitabine flow back to TME [298]. Exosomes derived from $MCF7^{WT}$ breast cancer cells contain *P*-gp and ubiquitin C-terminal hydrolase-L1 proteins that are able to induce doxorubicin resistance by elevating the level of *P*-gp [299]. TDEs promote platinum resistance by upregulating EMT markers and changing TGF-β/SMAD signaling pathway in ovarian cancer cells, and exosomes from epithelial ovarian cancer A2780 platinum-resistant cells attain resistance via promoting EMT [300]. In MDA-MB-231 and MCF-7 breast cancer cells, exosome-mediated miR-155 induces chemoresistance via upregulating EMT markers and targeting CCAAT/enhancer-binding protein β (C/EBP-β), TGF-β, and forkhead box O 3α (FOXO-3α) mRNA [301]. In NSCLC, exosome-mediated miR-21 delivery induces the upregulation of p-AKT levels, thereby resulting in increased gefitinib resistance [302]. HCC-derived exosomes containing miR-221 induce sorafenib resistance through modulating apoptosis inhibition and caspase-3 activity [303]. Considering that exosomes are involved in various pathophysiological conditions, understanding the molecular mechanisms underlying exosome biogenesis and chemoresistance will benefit in developing novel therapeutics targeting exosome-mediated tumorigenesis, progression, and chemoresistance.

One of the important functions of TDEs is to induce tumor vascular development. Angiogenesis, a multistep course by which cancers form new vasculature, is essential for tumor progression [304]. Since Folkman revealed the significant character of vascular networks for the proliferation and progression of solid tumors, establishing angiogenesis as a therapeutic target has become a focal aim [305]. Up to now, antiangiogenic therapy has emerged as a significant targeted therapeutic, and diverse antiangiogenic agents are currently used in combination with other chemotherapeutic drugs for the treatment of various malignant tumors. Nowadays, over 11 antiangiogenic drugs have been approved by the U.S. FDA, including bevacizumab, aflibercept, sorafenib, ramucirumab, sunitinib, pazopanib, axitinib, vandetanib, lenvatinib, and regorafenib [306]. However, the treatment is in doubt on the grounds of sustainable efficacy, side effects, off-target toxicities, and therapy resistance [307, 308]. Clinical and experimental studies have illustrated that cancers employ compensatory/

alternative/bypass angiogenic pathways and other adaptive mechanisms for their sustained growth, proliferation, and metastasis after undergoing a therapy episode(s) with antiangiogenic drugs. That is, various compensatory signaling pathways driving tumor growth and metastasis invariably become the underlying cause of tumor refractoriness [309]. Several preclinical and clinical studies have linked poor performance and resistance to antiangiogenic agents to the activation of a range of compensatory angiogenic signaling pathways and various angiogenic factors that support the angiogenic bypass mechanism [310]. The vascular endothelial growth factor (VEGF)/VEGF receptor signaling pathway is the most promising angiogenic target due to its pivotal role in angiogenesis and tumor growth [311, 312]. Previous studies have suggested that revascularization appears even after blocking VEGF signaling pathways because of the activation of redundant angiogenic pathways.

Serial evidence accumulated in recent years linked the function of various compensatory/alternative/bypass angiogenic canonical mechanisms to sustaining the progression of cancers while exposed to antiangiogenic drugs [308, 309, 313]. The current experimental, clinical, and epidemiological data have definitely outlined at least four potential different mechanisms, which can be considered for an explanation of evasive resistance to antiangiogenic therapies. The first mechanism refers to the activation and/or upregulation of compensatory pro-angiogenic signaling pathways within cancer [314]. The second mechanism is mainly driven by myeloid/stromal cells, which compensate for the requirement of the VEGF-mediated pathway, thereby promoting tumor angiogenesis [313]. The third manner is attributed to the dual role of pericytes; first, in establishing the increased pericyte coverage of the tumor vasculature, and second, in their potential angiogenic attributes, both serving as escape mechanisms from VEGF-mediated angiogenesis [315]. The fourth way is associated with remodeling and accessing normal vasculature for the invasion and metastasis of tumors in lieu of obligate neovascularization [314].

As an important component of TME, the immune system plays a critical role in regulating the response of cancer cells to therapies, particularly immunotherapies. Immune-targeted therapies approved for cancer include monoclonal antibodies against immune checkpoints such as cytotoxic T-lymphocyte-associated protein 4 (CTLA4), PD-1, and programmed cell death ligand 1 (PD-L1), which aim to modulate the antitumor T cell immune response. The resistance to immune-checkpoint inhibitors (ICIs) has been attributed to genomic and non-genomic mechanisms that are influenced by the tumor-host-microenvironment relationship [316]. One mechanism is the upregulation of alternative immune checkpoint pathways. Koyama et al. revealed that, in mouse xenograft models with lung adenocarcinoma, other immune checkpoint proteins, including T cell immunoglobulin mucin 3, lymphocyte activation gene 3, and CTLA-4, were upregulated after PD-1 blockade therapy. The increased expression of alternative immune checkpoints in PD-1 antibody-bound T cells can inhibit T cell activity and confer resistance to PD-1 blockade [317]. Other mechanisms include the presence of immunosuppressive cells, such as regulatory T cells (Tregs) or myeloid-derived suppressor cells (MDSCs), which can inhibit T cell function [318]. Besides, the presence of stromal cells in TME, such as CAFs, has also been found to be associated with suppressive $CD8^+$ T cell infiltration and insensitivity to αPD-L1 antibody plus αCTLA-4 antibody dual immune checkpoint blockade [319].

Combination treatments appear to be the promising strategy to overcome resistance to ICIs. Combinations of ICIs with radiation, chemotherapy, targeted drugs, or tumor vaccines have been proposed and tested. Radiation can trigger an adaptive mechanism to increase the expression of PD-L1 on tumor cells [320]. In addition, radiation may increase the depth and duration of immune responses by promoting a more diverse adaptive antitumor immune response. As reported by Victor et al., the combination of radiation with dual checkpoint blockade (anti-CTLA-4 and anti-PD-1) can enhance tumor control and survival in a preclinical model of melanoma, where radiation primes immune response by increasing T cell receptor repertoire of intratumoral T cells and expanding peripheral clones [321]. Combination with chemotherapy aims to assist ICI treatment by releasing neoantigens or modulating the TME via depleting Tregs and MDSCs [322]. However, it has been reported that the efficacy of this combination therapy is model (cancer-type)-dependent and regimen-type-dependent. Combination with chemotherapeutic drugs is not always able to tune the TME or enhance antitumor efficacy. Even with the effect of depleting MDSCs, the combination may not be beneficial to enhance treatment responsiveness [322]. Additional studies combining single agents with ICIs are still required to better understand potential interactions between drugs and the heterogeneity of TME. Compared to the less predictable ICI-chemotherapy combination, the combination of TME-targeted drugs with immune checkpoint blockade is considered a relatively logical approach to enhance the stimulation of antitumor immune response. TME-targeted therapies, such as angiogenesis inhibitors, stromal cell depletion agents, and immunomodulatory agents, can modulate the TME and enhance the immune response against the tumor. Combining these agents with ICIs can lead to synergistic effects of blocking inhibitory signals in T cells and activating the immune response, thereby improving treatment outcomes [323]. For example, the VEGF inhibitor bevacizumab can augment intratumoral $CD8^+$ T cell infiltration through vascular normalization and endothelial cell activation, thereby potentiating PD-L1 checkpoint inhibition with the PD-L1 inhibitor atezolizumab [324]. This combination has shown promising anticancer efficacy and has been granted FDA approval for the treatment of patients with advanced unresectable or metastatic HCC [325].

Combining oncolytic vaccines with ICIs is an emerging approach to overcome ICI resistance. Oncolytic vaccines are a type of oncolytic virus that is engineered to express tumor-specific antigens, which can stimulate an immune response against cancer cells [326]. As ICI-resistant tumors are with suppressed immune microenvironments, the application of an oncolytic virus can reshape TME and assist ICIs to boost the immune system, leading to a stronger antitumor response [327]. This combination therapy has shown promising results in preclinical and clinical studies in a variety of cancers, including melanoma, breast cancer, and bladder cancer [328–330].

Recent studies have shown that bacteria can reside within tumors and play a role in controlling cancer response to therapy by modulating the TME [331]. For example, *Pseudomonas aeruginosa* and *Escherichia coli* are two types of bacteria found to colonize tumors in various types of cancer, including lung, breast, and pancreatic cancer [332–334]. These bacteria can metabolize chemotherapy drugs such as gemcitabine, reducing their concentration within the tumor and rendering them less effective [335, 336]. Moreover, certain commensal bacteria or pathogens within the TME may interact with immune cells and alter the production of cytokines and chemokines, which are involved in the immune response to cancer [337, 338]. The influence of gut microbiota on cancer resistance has also gained extensive attention. A recent study found that in mice and patients with castration-resistant prostate cancer, an adaptive change in the gut commensal microbiota was observed after ADT: there was an expansion of commensal microbiota species that can convert androgen precursors to produce androgens, which were absorbed into the systemic circulation, thus inducing castration or endocrine therapy resistance [339]. This type of resistance could be delayed by applying antibiotics to reduce gut microbiota or could be reversed by fecal microbiota transplantation in hormone-sensitive prostate cancer patients [339], implicating two possible ways to overcome bacterial-mediated resistance. Although using antibiotics can directly eradicate the bacteria within the TME, this approach may have limitations due to the risk of antibiotic resistance [336]. Another approach is to develop drug delivery systems specifically targeting the tumor cells while avoiding bacterial-mediated metabolism [340, 341]. For example, nanotechnology-based drug delivery systems can be engineered to release drugs only in the tumor cells, minimizing the exposure of the drug to the bacteria [331, 340]. The presence of bacteria within the TME and the possible influences from gut microbiota are thought to be significant factors contributing to treatment resistance in cancer. Understanding the complex interactions between bacteria and tumors is essential for developing effective cancer treatments to overcome this resistance.

13.3 Public Databases Related to Drug Resistance and Application

13.3.1 Cancer Cell Line Encyclopedia

The Cancer Cell Line Encyclopedia (CCLE, https://sites.broadinstitute.org/ccle/) is a comprehensive database containing omics data from a wide range of cancer cell lines [342]. Currently, this database is integrated into the DepMap portal [343](https://depmap.org/portal/). The latest version of CCLE performed RNA sequencing (RNA-seq) for 1,019 cell lines, whole-exome sequencing (WES) for 326 cell lines, whole-genome sequencing (WGS) for 329 cell lines, reverse-phase protein array (RPPA) for 899 cell lines, reduced representation bisulfite sequencing (RRBS) for 843 cell lines, miRNA expression profiling for 954 cell lines, and global histone modification profiling for 897 cell lines. In addition, CCLE provides drug–response data for 15 compounds. A parallel study also reports the abundance of 225 metabolites across 928 cell lines [344]. This database frequently collaborates with other drug response resources to explore molecular features influencing drug response and sensitivity.

13.3.2 Genomics of Drug Sensitivity in Cancer

The Genomics of Drug Sensitivity in Cancer (GDSC) database (https://www.cancerrxgene.org/) is a comprehensive public resource providing information on drug sensitivity in cancer cells and molecular markers associated with drug response [345]. This database is now also integrated into the DepMap portal. The GDSC database conducts drug sensitivity experiments, with a primary focus on identifying genomic molecular markers of drug response. To achieve this, cell line drug sensitivity data are combined with extensive genomic datasets obtained from the Catalogue of Somatic Mutations in Cancer (COSMIC) [346]. These datasets include detailed information on somatic mutations in cancer genes, gene amplifications and deletions, tissue types, and transcriptional profiles.

The GDSC database comprises two datasets, each based on different drug sensitivity measurement assays. GDSC1 utilized resazurin or Syto60, while GDSC2 employed CellTiter-Glo. The GDSC1 dataset was generated collaboratively by the Wellcome Sanger Institute and Massachusetts General Hospital between 2009 and 2015, using a matched set of cancer cell lines known as the GDSC1000. This dataset includes drug sensitivity data from approximately 333,292 experiments (IC50s), encompassing the responses of around 970 cancer cell lines to 403 anticancer compounds.

Based on the drug sensitivity data from the GDSC1 dataset, researchers mapped cancer-associated alterations, identified in 11,289 tumors across 29 tissue types, onto 1,001 molecularly annotated human cancer cell lines. These alterations were integrated with data on somatic mutations, copy number alterations, DNA methylation, and gene expression and correlated with sensitivity to 265 drugs. The results of this analysis are stored in the GDSC1000 database [347].

GDSC2 has been developed at the Wellcome Sanger Institute since 2015, incorporating improvements to both the screening design and assay methodology. This dataset includes drug sensitivity data from approximately 243,466 experiments (IC50s), detailing the responses of nearly 969 cancer cell lines to 297 anticancer compounds. Several experiments from GDSC1 have been repeated in GDSC2, and it is recommended to prioritize the results from GDSC2 when duplicate IC50 values are available, as they reflect the updated and optimized methodology.

13.3.3 Profiling Relative Inhibition Simultaneously in Mixtures Drug Repurposing Resource

The Profiling Relative Inhibition Simultaneously in Mixtures (PRISM) method, a molecular barcoding approach, was utilized to screen drugs against cell lines in pooled experiments. This involved mixing multiple cell lines to conduct drug response assays. The resulting data have been made available through https://depmap.org/repurposing and https://figshare.com/articles/dataset/Repurposing_Public_24Q2/25917643. This data release contains the two most recent PRISM

Repurposing screens. The primary PRISM Repurposing dataset contains the results of pooled-cell line chemical-perturbation viability screens for 4,518 compounds (3,350 of the compounds [74%] are either approved for clinical use in the USA or Europe or are in clinical development. The remaining 1,168 [26%] are tool compounds with known activities. Most compounds, 3,466 [77%], were non-oncology-related, with the remaining compounds being either chemotherapeutics [2%] or targeted oncology agents [21%]) screened against 578 or 562 cell lines at a single dose (2.5 μM). The secondary PRISM Repurposing dataset contains the results of pooled-cell line chemical-perturbation viability screens for 1,448 compounds that were positives in primary screening screened against 499 cell lines in an eight-point dose–response ranging from 10 μM to 610pM [348].

13.3.4 Cancer Therapeutics Response Portal

The Cancer Therapeutics Response Portal (CTRP, https://portals.broadinstitute.org/ctrp.v2.1/) connects the genetic, lineage, and other cellular characteristics of cancer cell lines to small-molecule sensitivity, aiming to accelerate the discovery of patient-matched cancer therapeutics. This database is now integrated into the DepMap portal. CTRP generated an "Informer Set" consisting of 481 small-molecule probes and drugs, including 70 FDA-approved agents, 100 clinical candidates, and 311 small-molecule probes. These compounds selectively target specific nodes in cell circuitry and collectively modulate a wide array of cellular processes. The database quantitatively assessed the sensitivity of 860 deeply characterized cancer cell lines to these compounds. It also includes analyses correlating drug sensitivity with cancer features from the CCLE database, such as mutations, gene expression, copy number variations, and lineage [349, 350].

13.3.5 DepMap Portal and Comparison of Cell Line Drug Sensitivity Database

The DepMap project [343] (https://depmap.org/portal/), building off of the original CCLE project, generates data and tools that can be used and shared by researchers. New DepMap data is released twice a year, in May and November.

Table 13.1 shows the differences and similarities between the four cell line-related drug response screening databases mentioned earlier.

Table 13.1 Details about four cell line-related drug response screening databases.

Feature	PRISM	CCLE	GDSC	CTRP
Focus	Drug repurposing, drug sensitivity	Broad molecular profiling, drug sensitivity	Drug sensitivity, genomics	Drug sensitivity, genomics
Cell line coverage	1,000+ cell lines	1,000+ cell lines	1,000+ cell lines (GDSC1000)	~1,000 cell lines
Genomic data	From the CCLE database	Gene expression, mutations, CNVs, epigenetics, proteomics, metabolomics	From CCLE and COSMIC databases	From CCLE database
Drug data	Drug response data	No	Drug response data	Drug response data
Data sources	Derived from large-scale screenings	Broad omics molecular profiling	Derived from large-scale screenings	Derived from large-scale screenings
Primary use	Drug discovery, repurposing	Cancer cell line omics profiling	Drug sensitivity, cancer genomics	Drug sensitivity, cancer genomics

13.3.6 Literature Mining and Patient's Tissue-Based Database

The DRMref is a reference atlas of drug resistance mechanisms derived from 42 single-cell datasets spanning 30 studies, including 14 with both pre- and post-treatment samples and representing 26 cancer subtypes [351]. These datasets, sourced from the Gene Expression Omnibus (GEO) database and PubMed, encompass drug response information for chemotherapy, targeted therapy, and immunotherapy. The study identified resistance-related differentially expressed genes (R-DEGs) by comparing resistant and sensitive cells across cell types and performed pathway enrichment, transcription factor (TF), and miRNA regulatory analyses for detailed R-DEG annotation. ITH and EMT scores were compared to assess malignant cell activity, while alterations in the TME were analyzed to evaluate cell-type composition changes during drug resistance. In addition, in-depth cell-cell interaction studies highlighted how the surrounding microenvironment influences cellular drug responses. By integrating these findings, DRMref offers a comprehensive understanding of drug resistance mechanisms, providing valuable insights into cancer biology and therapeutic strategies.

The DRESIS database collects validated drug resistance molecules belonging to different drug resistance mechanisms by retrieving published literature and explicitly describes clinically/experimentally verified resistance data for a large number of drugs [352]. The CTR-DB database utilizes pretreatment bulk RNA-seq data to characterize cancer drug response [353]. CeDR Atlas is primarily based on scRNA-seq data to utilize the connectivity map resource for drug response prediction [354]. CREAMMIST provides an integrative summary of the dose-response curve across datasets based on cancer cell line data [355]. CancerDR is a resource encompassing 148 anticancer drugs and their pharmacological profiles across 952 cancer cell lines [356]. GEAR presents genomic elements, including genes, single-nucleotide polymorphisms, and miRNAs that are associated with drug resistance via literature mining [357]. KinaseMD mines the existing literature and provides annotations of mutations and their corresponding kinase inhibitor responses in four types of protein substructures that have been associated with kinase inhibitor resistance [358].

13.3.7 Conclusion

These drug response-related databases fulfill diverse roles, ranging from predicting personalized drug responses to uncovering the genetic and molecular mechanisms underlying drug efficacy and toxicity. They are instrumental in advancing pharmacogenomics, optimizing cancer therapies, and driving personalized medicine by providing insights that enable the development of safer and more effective treatments tailored to individual genetic profiles.

13.4 Systems Biology Approaches for Drug Resistance Research

While various bioinformatic analyses are available for analyzing omics data, such as enrichment analyses, gene co-expression networks (GCNs) [359, 360], and their variants (e.g., a network of lncRNAs and mRNAs) [361, 362], as well as dimensionality reduction like t-distributed Stochastic Neighbor Embedding (t-SNE) and Uniform Manifold Approximation and Projection (UMAP), omics data have also been subjected to computational modeling. These approaches make predictions for discovering novel mechanisms and devising treatment strategies for drug-resistant cancers. The use of survival analysis in combination with GCNs and the development of a gene regulatory network (GRN) model using a set of ordinary differential equations (ODEs), machine learning models, and genome-scale metabolic models (GEMs) are representative computational modeling approaches that have recently been considered for studying drug-resistant cancer cells.

13.4.1 Network-Based Modeling

GCN has been a popular analysis for understanding gene expression patterns from transcriptome data. GCN is an undirected graph that can be constructed from transcriptome data (e.g., RNA-seq) and connects pairs of genes (nodes in a GCN) with an edge if each pair of genes shows significant co-expression patterns across the transcriptome data. GCN analysis, such as identifying hub genes and/or modules, allows prioritizing candidate genes that may be highly associated with the drug resistance of cancer cells. Weighted GCN additionally considers the level of significance in the co-expression relationship between genes in a pair. Often, outcomes from (weighted) GCN analysis are further subjected to other computational analyses, for example, survival analysis, to validate the biological and/or clinical significance of the candidate genes. As a recent example, Li et al. focused on PPP2R2B, encoding serine/threonine–protein phosphatase 2A 55 kDa regulatory subunit B beta isoform, as a potential prognostic biomarker for triple-negative breast cancer (TNBC) on the basis of a series of bioinformatic analyses involving a GCN [363]. Kaplan-Meier survival analysis for this gene revealed that patients with a low expression level of PPP2R2B showed shorter survival times than those with a high expression level of PPP2R2B. Interestingly, PPP2R2B upregulation could attenuate the resistance of TNBC cells to doxorubicin. Likewise, the Cox proportional hazards regression model (Cox regression model) was used for genes selected from GCNs to predict prognostic biomarkers for breast cancer and to suggest genes (e.g., CCNE2 and KIF14) that may help overcome drug resistance [364].

While GCNs can provide clinically important information when combined with additional predictive models, such as survival analysis above, they have limitations in generating clues on a molecular mechanism associated with the development of drug resistance, in particular, dynamic interactions between genes. To address this problem, Zhang et al. developed a time-course RNA-seq data-driven computational framework (DryNetMC) to construct GRNs that help elucidate dynamic interactions between genes and identify key genes associated with mechanisms of drug resistance [365]. DryNetMC involves a set of ODEs, a regularized regression method, and a series of network analyses. Using DryNetMC, GRNs were constructed for bucladesine-sensitive and bucladesine-resistant glioma cells based on their time-course RNA-seq data. These differential GRNs were subsequently subjected to a systematic characterization to identify their unique network properties (e.g., node importance) that helped identify key genes (e.g., KIF2C, CCNA2, NDC80, KIF11, and KIF23) that are predictive of a cancer cell's drug response. Because network-based models, either using a GCN or other methods (e.g., ODEs), can visualize a biological context (e.g., association between genes), they will continue to be actively used in the analysis of omics data and likely along with additional predictive models.

13.4.2 Machine Learning

Increasing availability of omics data for drug-resistant cancer cells has also provided unprecedented opportunities for building machine learning models. In general, machine learning models perform classification or regression, depending on a given problem. Recently, the prediction of anticancer drug response was attempted using various types of machine learning methods, such as logistic regression [366, 367], random forest [368], and deep neural network (DNN; e.g., multilayer perceptron) [369] on the basis of a range of omics and drug response data. When developing these machine learning models, transcriptome (RNA-seq or mRNA microarray) was the most frequently adopted dataset, but other types of datasets were also considered, including genome (e.g., gene mutations) [367], proteome [370], epigenome [368], mass spectrometry data [371], and molecular features of a target drug [372].

In a recent study by Kong et al., a machine learning model was developed that can predict a patient's drug response on the basis of the analysis of protein–protein interaction (PPI) network and pharmacogenomic data from three-dimensional (3D) organoid culture models [373]. Specifically, potential biomarkers were first inferred from the PPI network analysis, and their corresponding expression profiles along with drug response data (IC_{50}) were used to train a machine learning model (e.g., ridge regression). The resulting drug responses were validated using survival analysis by focusing on CRC and bladder cancer patients treated with 5-FU and cisplatin, respectively. The predicted drug responses also appeared to be consistent with transcriptome profiles from drug-sensitive and drug-resistant isogenic cancer cell lines as well as data on somatic mutations associated with already-known biomarkers. In this study, consideration of the network analysis not only helped improve the performance of the developed machine learning model but also facilitated the interpretation of model prediction outcomes. Likewise, in another study, elastic net and random forest regression were used to predict drug responses from abundance data of proteins and their phosphorylation sites in cancer cell lines [370].

Among machine learning methods, DNNs are increasingly used for various predictions, and they have also been used to predict drug responses. Sakellaropoulos et al. developed a DNN model by using GDSC datasets (i.e., transcriptomic data for 1,001 cancer cell lines and IC_{50} values of 251 drugs) to predict drug responses [374]. Across several datasets tested, the DNN model showed consistently better performance than elastic net and random forest models. The DNN model was validated by conducting survival analyses for the model-predicted IC_{50} values, which split patients based on their drug responsiveness. Importantly, pathway enrichment analysis using information from the DNN model (i.e., weights that connect the input layer and the first hidden layer) appeared to associate specific biological pathways with mechanisms of action for drugs. In a more recent study, predicting drug response was also attempted by using a DNN model combined with multiple elastic nets [375], referred to as a reference drug-based neural network (RefDNN). RefDNN was developed more in the context of drug resistance, which predicts whether a given cell line is resistant to a target drug by processing gene expression profiles and molecular structure of a drug. RefDNN was also shown to help identify biomarker genes associated with drug resistance and explore a novel anticancer drug via drug repositioning.

Despite its demonstrated performance, machine learning is often challenged with the limited availability of training datasets for many technical fields. This challenge can be addressed to a certain extent by employing transfer learning, as recently demonstrated [372]. Zhu et al. demonstrated that ensemble transfer learning can improve the prediction of drug responses in the context of drug repositioning (i.e., use of a drug for another cancer that is already known), precision oncology (i.e., use of a drug for a new cancer that has never been treated before), and new drug development (i.e., use of a new drug for already known cancer). In this particular study, light gradient boosting machine (LightGBM) and two different DNN models were considered for ensemble transfer learning; larger datasets from the CTRP and GDSC were used as source data for initial training of models, and smaller datasets from CCLE and the Genentech cell line screening initiative (gCSI) served as target data for further refinement and testing of the models. It was shown that ensemble transfer learning–based models almost always outperformed models that were not developed using transfer learning. This study suggests the use of transfer learning for other drug-resistant cancer cells where a training dataset is not sufficiently available.

13.4.3 Genome-Scale Metabolic Modeling

GEM is a computational model that describes gene-protein-reaction (GPR) associations and can be simulated to predict genome-scale metabolic flux distributions [376]. GEMs are now available for an increasing number of organisms that are important in biotechnology and biomedicine. Several versions of human

GEMs [377–379] are currently available, which have been used to examine a target cell's metabolism and to predict biomarkers and drug targets for various diseases [376, 380]. For a medical application, a generic human GEM, covering all the known GPR associations in human metabolism, is initially integrated with omics data, often transcriptome (e.g., RNA-seq), to build a context-specific GEM, a GEM that is specific to a target cell or tissue [381, 382]. The resulting context-specific GEM is then simulated for various metabolic studies.

Human GEMs have recently been used to study radiation-resistant tumors [383, 384] but not drug-resistant cancer cells, to the best of our knowledge. Lewis et al. newly constructed GEMs for radiation-sensitive and radiation-resistant tumors through multi-omics integration (i.e., transcriptome data, mutational data, kinetic data, and thermodynamic data) [383]. These context-specific GEMs were used to identify changes in redox cofactor production that give resistance to radiation therapy. In the other study, ensemble machine learning classifiers were developed to predict whether an individual is responsive or resistant to radiation therapy by considering data of metabolite production rates predicted from context-specific GEMs as well as mutation data, transcriptome data, and clinical data from The Cancer Genome Atlas (TCGA) [384]. These two studies obviously suggest that GEM-based approaches can also be considered to identify metabolic signatures of drug-resistant cancer cells and to predict effective drug targets for these cancer cells.

13.4.4 Outlook

Understanding genotype-phenotype associations in drug-resistant cancer cells is a highly complex problem, and therefore, the use of multi-omics data has been considered to capture various aspects of these troubling cancer cells. In particular, multi-omics analyses along with additional tools, such as genome engineering (e.g., CRISPR-Cas9), will continue to play an important role in the thorough characterization of drug-resistant cancer cells. Furthermore, an increasing volume of omics data will facilitate the development of various types of computational models. As a consequence, prediction outcomes from computational models will allow more systematic design of experiments for drug-resistant cancer cells.

Despite the promises of omics data and computational models, technical challenges exist. First, current coverage of multi-omics data is not sufficient for thoroughly studying a range of drug-resistant cancer cells. In particular, generation of a consistent set of multi-omics data from each single cell is necessary for in-depth study of a target cancer cell and comparison of different types of cancer cells. In addition, it will be interesting to examine the effects of using datasets obtained from patients having a specific disease instead of publicly available datasets (e.g., GDSC and CTRP). While currently available machine learning models have been rigorously validated using public datasets, they might reveal previously unnoticed limitations in a clinical setting because the public datasets are often generated in a highly controlled condition. In particular, additional consideration of nongenetic factors (e.g., age, gender, and lifestyle) may help reveal new insights on drug-resistant cancer cells. The use of patient-specific datasets will allow more widespread use of state-of-the-art computational models in a clinical setting.

For network-based modeling, including both GCN and GRN, a breakthrough is needed that allows efficiently developing a cell-specific large-scale GRN that can be simulated under various conditions (e.g., gene perturbation). For machine learning, despite its high predictive performance, there is always a challenge of avoiding overfitting and achieving explainability. Explainability in terms of biological processes is particularly important in the field of biomedicine in order to explain prediction outcomes and make medical decisions. In the case of human GEMs, because patient-specific omics data (e.g., RNA-seq) are available to a certain extent, human GEMs should be more actively considered to systematically examine the metabolism of drug-resistant cancer cells. The availability of multi-omics data will be particularly useful for interpreting human GEMs and their prediction outcomes; because human GEMs only cover a metabolic network, the use of multi-omics data can help explain a complex interplay between metabolic and regulatory networks. Prediction outcomes from the simulation of human GEMs will in turn help explain the insights reaped from omics analyses.

Taken together, advances in omics technologies and computational modeling will bring about positive impacts in understanding and treating drug-resistant cancer cells. Feedback from clinicians and biomedical researchers will be additionally useful for the efficient development and clinical application of computational models.

13.5 Strategies to Overcome Drug Resistance Using Systems Biology

A range of strategies have emerged to address the various mechanisms of resistance to tumor drugs. The choice of which approach to pursue depends heavily on knowledge of the specific resistance mechanisms at each target, and therefore, a detailed understanding of the resistance mechanisms is essential. The purpose of this section is to exemplify the most typical chemical approaches to customer service drug resistance and to draw lessons from them.

13.5.1 Template or Binding Mode Optimization

When the impact of the amino acid mutation is largely steric, the resistance mechanism is likely to be driven by a reduction in the target affinity of the drug for the mutant form. A number of approaches have been highlighted to target such resistance mechanisms and improve the potency against the mutated form. One approach has been to drive target potency (or selectivity) to higher levels as a way of overcoming the impact of the mutation. Such an approach was reported in ALK, whereby more recent compounds such as alectinib, certinib, and brigatinib have increased target potency and/or kinase selectivity compared to crizotinib [385]. A more rational approach has been to optimize only the part of the inhibitor that is near, or in contact with, the mutation in an attempt to negate or reduce the impact of the mutation. An example of this approach was toward the breakpoint cluster region/Abelson (BCR/ABL) fusion oncogene, where imatinib was further optimized into nilotinib, which made differential interactions with the mutations causing resistance [386].

13.5.2 Target Alternative Protein Conformations

Where amino acid mutations may impact the preferred conformation of a target (such as the equilibrium of an inactive versus active conformation), there may be a rationale to target an alternative conformation for the mutant form. For example, dasatinib is a second-generation BCR/ABL inhibitor and may target the active Aspartate-Phenylalanine-Glycine-in (DFG-in) conformation, whereas imatinib and ponatinib both bind to DFG-out forms [387]. These compounds are able to cover different mutations due to these differences in binding modes. Another example can be observed in the kinase mesenchymal-epithelial transition factor (cMET), where the reported D1228V and Y1230C render the majority of the "type I" inhibitors, such as savolitinib, significantly less active in the treatment of NSCLC. It has been observed, however, that "type II" inhibitors such as cabozantinib are able to overcome this mutation and maintain target activity [388].

13.5.3 Change in Drug Mechanism of Action

One of the most notable examples of a change in the mechanism of a compound to overcome a resistance mechanism is for the kinase EGFR in the treatment of NSCLC. Although the first-generation inhibitors such as gefitinib and erlotinib were effective and showed high response rates, resistance often emerged within

9-12 months due to the T790M gatekeeper mutation [389]. Initially assumed to be a steric effect similar to BCR/ABL, it was subsequently shown that the primary effect of this mutation was the increase of the affinity of ATP, which rendered all reversible ATP pocket inhibitors significantly less active [390]. Subsequently, irreversible covalent inhibitors were discovered based on different chemical templates to target the mutations specifically and have a selectivity margin to the wild-type form of the protein. The first approved inhibitor with such a profile was osimertinib [391].

PROTACs are being developed to target a number of clinically relevant targets where resistance has emerged [392]. One rationale here is that degrading the target may delay or avoid the emergence of the resistance mutations arising from more traditional small-molecule inhibitors [393]. Compounds have been disclosed targeting EGFR (based on ATP competitive [394] and allosteric molecules), Bruton's tyrosine kinase (BTK) [395], and BRC/ABL [396] with this mechanism of action aimed at targeting various resistance forms. Interestingly, such an approach may be vulnerable to specific resistance mechanisms, either from compound binding to the primary target or via the ability of the compound to recruit the E3 ligase. In a recent publication evaluating BET family-targeted PROTACs, drug-induced resistance was observed not from mutations that impact the ability for the compounds to bind to BRD4 but via genomic alterations that impact core components of the E3 ligase complex [397].

13.5.4 Target Alternative Binding Site

An additional approach to avoid the impact of specific mechanisms of resistance to one binding site could be to target an alternative pocket of the drug target. Binding distal from such mutations would be expected to maintain target activity, assuming that there is no long-range impact or communication between the different sites. A specific example of this approach comes from BCR/ABL, where researchers identified a novel scaffold that did not bind to the ATP pocket where imatinib binds. This approach resulted in the discovery of asciminib, targeting the myristate pocket [398]. The identification of allosteric inhibitors provides options to sequence not only drug treatments when one fails but also a "double hit" strategy where two compounds are tested in combination to avoid additional resistance mutations [399].

13.5.5 Improvement in Blood–Brain Barrier Permeability

One of the more generic resistance mechanisms to targeted cancer drugs can be the appearance of brain metastasis as an escape mechanism in response to drugs that are unable to penetrate the BBB [400]. This has been observed in not only broader agents such as chemotherapies but also more targeted agents [401]. As with the emergence of resistance from amino acid mutations, understanding these cases has enabled drug discovery scientists to develop additional agents that are able to target the central nervous system (CNS) and achieve brain penetration. The key to achieving such a profile appears to be maintaining high intrinsic permeability while minimizing compound efflux [402]. A range of preclinical models are available to guide the evolution of chemical series toward improved brain penetration. Examples of later-generation drugs with improved BBB permeability, which may enable the improved treatment of patients with brain metastasis or avoid the appearance of brain metastasis, include the ALK inhibitor lorlatinib [403] and the EGFR-mutant inhibitors AZD3759 [404] and osimertinib [405].

13.5.6 Combination Therapy

In response to the emerging challenges around cancer drug resistance, the use of drug combinations has become a highly important area in which progress is currently being made [406]. This approach is not without challenges, as there are a vast number of possible combination studies that could be evaluated in a clinical setting; therefore, understanding the optimal combinations informed by preclinical studies is essential. The aim of such drug combinations could be synergistic or additive combinations, synthetic lethality, or the reversal of resistance through additional targeting of alternative pathways/mutations. To help enable this, there are a number of preclinical methods, both experimental [407] and computational [408], that can be utilized to identify rational drug combination treatments for evaluating in a clinical trial.

13.6 Challenges and Future Directions in Systems Biology for Cancer Drug Resistance

The future of overcoming drug resistance in cancer treatment is promising, with ongoing research and innovation poised to significantly enhance therapeutic efficacy. One of the most compelling future perspectives is the potential for drug modification to play a pivotal role in overcoming resistance. Advances in molecular biology, nanotechnology, and computational sciences are expected to converge, leading to the development of more sophisticated and targeted drug modifications. The development of next-generation inhibitors that can specifically target resistance-causing mutations is a key area of focus. These inhibitors are designed to overcome mutations that render the first-line treatments ineffective, thereby restoring the sensitivity of cancer cells to therapy. For example, the third-generation EGFRis have shown efficacy against resistant mutations in NSCLC.

Drug modification strategies: Nanotechnology will likely play a significant role in drug modification by enabling the creation of nanoparticle-based delivery systems. These systems can enhance the delivery and bioavailability of anticancer drugs, allowing them to bypass mechanisms such as drug efflux pumps. Nanoparticles can be engineered to release drugs in a controlled manner, directly at the tumor site, thereby reducing systemic toxicity and improving therapeutic outcomes. The use of prodrugs, which are inactive precursors that convert into active drugs within cancer cells, offers another innovative approach [409, 410]. Prodrugs can be designed to become activated only in the presence of specific cancer cell enzymes, ensuring that the active drug exerts its effect predominantly within the tumor, thereby minimizing resistance and side effects. Modifying existing drugs to include epigenetic components could also be a promising strategy. By incorporating DNA methyltransferase inhibitors or histone deacetylase inhibitors, it is possible to reverse epigenetic changes that contribute to drug resistance, enhancing the efficacy of conventional therapies. Future therapies are expected to leverage combination drug formulations, where multiple agents are combined into a single treatment regimen. This approach can simultaneously target multiple pathways involved in resistance, reducing the likelihood of cancer cells developing resistance to any single agent. Personalized medicine will continue to evolve, utilizing comprehensive genomic and proteomic profiling to tailor drug modifications for individual patients. This approach ensures that treatments are specifically designed to target the unique genetic and molecular landscape of each patient's cancer, maximizing efficacy and minimizing resistance [411]. The future will also see the refinement of adaptive therapy regimens that dynamically adjust based on real-time monitoring of tumor response. This adaptive approach can involve modifying drug dosages, schedules, or combinations based on the evolving characteristics of the tumor, thus preemptively addressing the emergence of resistance.

13.7 Conclusion

Exploring new strategies to overcome drug resistance in cancer cells involves a multifaceted approach that includes targeted therapies, advancements in immunotherapy, combination treatments, nanotechnology-based drug delivery systems, and precision medicine guided by biomarker identification [412]. Targeted therapies such as small-molecule inhibitors and monoclonal antibodies target specific genetic or molecular changes in cancer cells to make treatments more effective. Immunotherapy uses the immune system to identify and destroy drug-resistant cancer cells, including treatments such as ICIs, CAR T cell therapy, and therapeutic vaccines [413]. Combining different treatments targets multiple pathways to overcome resistance and improve treatment response. Nanotechnology helps deliver drugs precisely to tumor sites, avoiding drug pumps and improving treatment results. Precision medicine uses biomarkers to create personalized treatment plans based on individual molecular profiles. These different approaches show a variety of ways to overcome drug resistance and improve outcomes for cancer patients [414].

Continued research and collaboration are crucial for advancing cancer treatment outcomes, especially for drug-resistant cancers. Collaboration among researchers, doctors, pharmaceutical companies, regulators, and patient advocates helps drive innovation, speed up drug development, and turn scientific discoveries into practical treatments. Research focuses on understanding drug resistance mechanisms, finding biomarkers, improving treatment combinations, developing targeted therapies, refining drug delivery, using immunotherapy, employing artificial intelligence, running clinical trials, and gathering real-world evidence [415]. This collaborative effort leads to a better understanding of cancer, more precise treatments, improved effectiveness, fewer side effects, longer survival, and a better quality of life for cancer patients. Stressing the importance of ongoing research and collaboration is vital for navigating the complexities of cancer care, addressing treatment challenges, overcoming drug resistance, and ultimately achieving better results in the fight against cancer.

References

1. Catalano, A., et al. Multidrug resistance (MDR): a widespread phenomenon in pharmacological therapies. *Molecules*. 2022; **27**(3): p. 616.
2. Sharma, P., et al. Primary, adaptive, and acquired resistance to cancer immunotherapy. *Cell*. 2017; **168**(4): p. 707–723.
3. Lippert, T. H., H. J. Ruoff, and M. Volm. Intrinsic and acquired drug resistance in malignant tumors. The main reason for therapeutic failure. *Arzneimittelforschung*. 2008; **58**(6): p. 261–264.
4. Wang, X., H. Zhang, and X. Chen. Drug resistance and combating drug resistance in cancer. *Cancer Drug Resist*. 2019; **2**(2): p. 141–160.
5. Marine, J. C., S. J. Dawson, and M. A. Dawson. Non-genetic mechanisms of therapeutic resistance in cancer. *Nat Rev Cancer*. 2020; **20**(12): p. 743–756.

6. Hu, X., and Z. Zhang. Understanding the genetic mechanisms of cancer drug resistance using genomic approaches. *Trends Genet.* 2016; **32**(2): p. 127–137.
7. Salgia, R., and P. Kulkarni. The genetic/non-genetic duality of drug 'resistance' in cancer. *Trends Cancer.* 2018; **4**(2): p. 110–118.
8. Pathania, S., et al. Drug metabolizing enzymes and their inhibitors' role in cancer resistance. *Biomed Pharmacother.* 2018; **105**: p. 53–65.
9. Allain, E. P., et al. Emerging roles for UDP-glucuronosyltransferases in drug resistance and cancer progression. *Br J Cancer.* 2020; **122**(9): p. 1277–1287.
10. Buß, I., et al. Relevance of copper transporter 1 and organic cation transporters 1–3 for oxaliplatin uptake and drug resistance in colorectal cancer cells. *Metallomics.* 2018; **10**(3): p. 414–425.
11. Giddings, E. L., et al. Mitochondrial ATP fuels ABC transporter-mediated drug efflux in cancer chemoresistance. *Nat Commun.* 2021; **12**(1): p. 2804.
12. Robey, R. W., et al. Revisiting the role of ABC transporters in multidrug-resistant cancer. *Nat Rev Cancer.* 2018; **18**(7): p. 452–464.
13. Zhitomirsky, B., and Y. G. Assaraf. Lysosomes as mediators of drug resistance in cancer. *Drug Resist Updat.* 2016; **24**: p. 23–33.
14. Salemme, V., et al. The role of tumor microenvironment in drug resistance: emerging technologies to unravel breast cancer heterogeneity. *Front Oncol.* 2023; **13**: p. 1170264.
15. Wu, P., et al. Adaptive mechanisms of tumor therapy resistance driven by tumor microenvironment. *Front Cell Dev Biol.* 2021; **9**: p. 641469.
16. Peetla, C., S. Vijayaraghavalu, and V. Labhasetwar. Biophysics of cell membrane lipids in cancer drug resistance: implications for drug transport and drug delivery with nanoparticles. *Adv Drug Deliv Rev.* 2013; **65**(13–14): p. 1686–1698.
17. Perland, E., and R. Fredriksson. Classification systems of secondary active transporters. *Trends Pharmacol Sci.* 2017; **38**(3): p. 305–315.
18. Al-Abdulla, R., et al. Unraveling 'The Cancer Genome Atlas' information on the role of SLC transporters in anticancer drug uptake. *Expert Rev Clin Pharmacol.* 2019; **12**(4): p. 329–341.
19. Wu, Z., et al. Emerging roles of the solute carrier family in pancreatic cancer. *Clin Transl Med.* 2021; **11**(3): p. e356.
20. Buxhofer-Ausch, V., et al. Tumor-specific expression of organic anion-transporting polypeptides: transporters as novel targets for cancer therapy. *J Drug Deliv.* 2013; **2013**: p. 863539.
21. Sun, R., et al. The emerging role of the SLCO1B3 protein in cancer resistance. *Protein Pept Lett.* 2020; **27**(1): p. 17–29.
22. Zimmerman, E. I., et al. Contribution of OATP1B1 and OATP1B3 to the disposition of sorafenib and sorafenib-glucuronide. *Clin Cancer Res.* 2013; **19**(6): p. 1458–1466.
23. de Morrée, E. S., et al. Loss of SLCO1B3 drives taxane resistance in prostate cancer. *Br J Cancer.* 2016; **115**(6): p. 674–681.
24. Sissung, T. M., et al. Differential expression of OATP1B3 mediates unconjugated testosterone influx. *Mol Cancer Res.* 2017; **15**(8): p. 1096–1105.

25. Alsinnawi, M., et al. Association of prostate cancer SLCO gene expression with Gleason grade and alterations following androgen deprivation therapy. *Prostate Cancer Prostatic Dis*. 2019; **22**(4): p. 560–568.
26. Haberkorn, B., et al. Cancer-type organic anion transporting polypeptide 1B3 (Ct-OATP1B3) is localized in lysosomes and mediates resistance against kinase inhibitors. *Mol Pharmacol*. 2022; **102**: p. MOLPHARM-AR-2022-000539.
27. Falguières, T. ABC transporters in human diseases: future directions and therapeutic perspectives. *Int J Mol Sci*. 2022; **23**(8): p. 4250.
28. Juliano, R. L., and V. Ling. A surface glycoprotein modulating drug permeability in Chinese hamster ovary cell mutants. *Biochim Biophys Acta*. 1976; **455**(1): p. 152–162.
29. Wang, J. Q., et al. ATP-binding cassette (ABC) transporters in cancer: a review of recent updates. *J Evid Based Med*. 2021; **14**(3): p. 232–256.
30. Thomas, C., and R. Tampé. Structural and mechanistic principles of ABC transporters. *Annu Rev Biochem*. 2020; **89**: p. 605–636.
31. Chen, X. Y., et al. Natural product as substrates of ABC transporters: a review. *Recent Pat Anticancer Drug Discov*. 2021; **16**(2): p. 222–238.
32. Anreddy, N., et al. Tyrosine kinase inhibitors as reversal agents for ABC transporter mediated drug resistance. *Molecules*. 2014; **19**(9): p. 13848–13877.
33. Dong, X. D., et al. Overexpression of ABCB1 associated with the resistance to the KRAS-G12C specific inhibitor ARS-1620 in cancer cells. *Front Pharmacol*. 2022; **13**: p. 843829.
34. Yang, Y., et al. PBK/TOPK inhibitor OTS964 resistance is mediated by ABCB1-dependent transport function in cancer: in vitro and in vivo study. *Mol Cancer*. 2022; **21**(1): p. 40.
35. Fu, H., et al. The resistance of cancer cells to palbociclib, a cyclin-dependent kinase 4/6 inhibitor, is mediated by the ABCB1 transporter. *Front Pharmacol*. 2022; **13**: p. 861642.
36. Wang, J. Q., et al. Multidrug resistance proteins (MRPs): structure, function and the overcoming of cancer multidrug resistance. *Drug Resist Updat*. 2021; **54**: p. 100743.
37. Parrish, K. E., et al. Efflux transporters at the blood–brain barrier limit delivery and efficacy of cyclin-dependent kinase 4/6 inhibitor palbociclib (PD-0332991) in an orthotopic brain tumor model. *J Pharmacol Exp Ther*. 2015; **355**(2): p. 264–271.
38. Mittapalli, R. K., et al. ABCG2 and ABCB1 limit the efficacy of dasatinib in a PDGF-B-driven brainstem glioma model. *Mol Cancer Ther*. 2016; **15**(5): p. 819–829.
39. Engle, K., and G. Kumar. Cancer multidrug-resistance reversal by ABCB1 inhibition: a recent update. *Eur J Med Chem*. 2022; **239**: p. 114542.
40. Ji, N., et al. Selonsertib (GS-4997), an ASK1 inhibitor, antagonizes multidrug resistance in ABCB1- and ABCG2-overexpressing cancer cells. *Cancer Lett*. 2019; **440-441**: p. 82–93.
41. Wu, Z. X., et al. MET inhibitor tepotinib antagonizes multidrug resistance mediated by ABCG2 transporter: in vitro and in vivo study. *Acta Pharm Sin B*. 2022; **12**(5): p. 2609–2618.
42. Zhang, Y., et al. Poziotinib inhibits the efflux activity of the ABCB1 and ABCG2 transporters and the expression of the ABCG2 transporter protein in multidrug resistant colon cancer cells. *Cancers (Basel)*. 2020; **12**(11): p. 3249.
43. Fanelli, M., et al. Targeting ABCB1 and ABCC1 with their specific inhibitor CBT-1® can overcome drug resistance in osteosarcoma. *Curr Cancer Drug Targets*. 2016; **16**(3): p. 261–274.

44. Tao, L. Y., et al. Cediranib (recentin, AZD2171) reverses ABCB1- and ABCC1-mediated multidrug resistance by inhibition of their transport function. *Cancer Chemother Pharmacol*. 2009; **64**(5): p. 961–969.
45. Stefan, S. M. Multi-target ABC transporter modulators: what next and where to go? *Future Med Chem*. 2019; **11**(18): p. 2353–2358.
46. Tiash, S., and E. H. Chowdhury. siRNAs targeting multidrug transporter genes sensitise breast tumour to doxorubicin in a syngeneic mouse model. *J Drug Target*. 2019; **27**(3): p. 325–337.
47. Wang, Y., et al. The role of non-coding RNAs in ABC transporters regulation and their clinical implications of multidrug resistance in cancer. *Expert Opin Drug Metab Toxicol*. 2021; **17**(3): p. 291–306.
48. Yi, H., et al. Synergistic therapy of doxorubicin and miR-129-5p with self-cross-linked bioreducible polypeptide nanoparticles reverses multidrug resistance in cancer cells. *Biomacromolecules*. 2016; **17**(5): p. 1737–1747.
49. Yan, L., et al. Inhibition of SMYD2 suppresses tumor progression by down-regulating microRNA-125b and attenuates multi-drug resistance in renal cell carcinoma. *Theranostics*. 2019; **9**(26): p. 8377–8391.
50. Wang, X. K., et al. Afatinib enhances the efficacy of conventional chemotherapeutic agents by eradicating cancer stem-like cells. *Cancer Res*. 2014; **74**(16): p. 4431–4445.
51. Martín, V., et al. Melatonin-induced methylation of the ABCG2/BCRP promoter as a novel mechanism to overcome multidrug resistance in brain tumour stem cells. *Br J Cancer*. 2013; **108**(10): p. 2005–2012.
52. Fultang, N., et al. ROR1 regulates chemoresistance in breast cancer via modulation of drug efflux pump ABCB1. *Sci Rep*. 2020; **10**(1): p. 1821.
53. Verma, H., et al. Drug metabolizing enzymes-associated chemo resistance and strategies to overcome it. *Drug Metab Rev*. 2019; **51**(2): p. 196–223.
54. Rahman, M., and M. R. Hasan. Cancer metabolism and drug resistance. *Metabolites*. 2015; **5**(4): p. 571–600.
55. Xu, Y., and M. A. Villalona-Calero. Irinotecan: mechanisms of tumor resistance and novel strategies for modulating its activity. *Ann Oncol*. 2002; **13**(12): p. 1841–1851.
56. Miura, K., et al. 5-FU metabolism in cancer and orally-administrable 5-FU drugs. *Cancers (Basel)*. 2010; **2**(3): p. 1717–1730.
57. Galmarini, C. M., J. R. Mackey, and C. Dumontet. Nucleoside analogues and nucleobases in cancer treatment. *Lancet Oncol*. 2002; **3**(7): p. 415–424.
58. Medina-Sanson, A., et al. Role of genetic polymorphisms of deoxycytidine Kinase and cytidine deaminase to predict risk of death in children with acute myeloid leukemia. *Biomed Res Int*. 2015; **2015**: p. 309491.
59. Wu, B., et al. Deoxycytidine kinase (DCK) mutations in human acute myeloid leukemia resistant to cytarabine. *Acta Haematol*. 2021; **144**(5): p. 534–541.
60. Song, J. H., et al. Defective expression of deoxycytidine kinase in cytarabine-resistant acute myeloid leukemia cells. *Int J Oncol*. 2009; **34**(4): p. 1165–1171.
61. Klanova, M., et al. Downregulation of deoxycytidine kinase in cytarabine-resistant mantle cell lymphoma cells confers cross-resistance to nucleoside analogs gemcitabine, fludarabine

and cladribine, but not to other classes of anti-lymphoma agents. *Mol Cancer*. 2014; **13**: p. 159.
62. Cahard, D., C. McGuigan, and J. Balzarini. Aryloxy phosphoramidate triesters as pro-tides. *Mini Rev Med Chem*. 2004; **4**(4): p. 371–381.
63. Slusarczyk, M., et al. Application of ProTide technology to gemcitabine: a successful approach to overcome the key cancer resistance mechanisms leads to a new agent (NUC-1031) in clinical development. *J Med Chem*. 2014; **57**(4): p. 1531–1542.
64. Spasokoukotskaja, T., et al. Treatment of normal and malignant cells with nucleoside analogues and etoposide enhances deoxycytidine kinase activity. *Eur J Cancer*. 1999; **35**(13): p. 1862–1867.
65. Huang, M., et al. Clofarabine exerts antileukemic activity against cytarabine-resistant B-cell precursor acute lymphoblastic leukemia with low deoxycytidine kinase expression. *Cancer Med*. 2018; **7**(4): p. 1297–1316.
66. Riddick, D. S., et al. Cancer chemotherapy and drug metabolism. *Drug Metab Dispos*. 2005; **33**(8): p. 1083–1096.
67. Kaur, G., et al. Drug-metabolizing enzymes: role in drug resistance in cancer. *Clin Transl Oncol*. 2020; **22**(10): p. 1667–1680.
68. Esteves, F., J. Rueff, and M. Kranendonk. The central role of cytochrome P450 in Xenobiotic metabolism-a brief review on a fascinating enzyme family. *J Xenobiot*. 2021; **11**(3): p. 94–114.
69. Hofman, J., et al. Roles of CYP3A4, CYP3A5 and CYP2C8 drug-metabolizing enzymes in cellular cytostatic resistance. *Chem Biol Interact*. 2021; **340**: p. 109448.
70. García-Martín, E., et al. Acquired resistance to the anticancer drug paclitaxel is associated with induction of cytochrome P450 2C8. *Pharmacogenomics*. 2006; **7**(4): p. 575–585.
71. Morvan, V. L., et al. Cytochrome P450 1B1 polymorphism drives cancer cell stemness and patient outcome in head-and-neck carcinoma. *Br J Cancer*. 2020; **123**(5): p. 772–784.
72. Yada, E., et al. Possible role of cytochrome P450 1B1 in the mechanism of gemcitabine resistance in pancreatic cancer. *Biomedicines*. 2021; **9**(10): p. 1396.
73. McFadyen, M. C., and G. I. Murray. Cytochrome P450 1B1: a novel anticancer therapeutic target. *Future Oncol*. 2005; **1**(2): p. 259–263.
74. Pastina, I., et al. Cytochrome 450 1B1 (CYP1B1) polymorphisms associated with response to docetaxel in castration-resistant prostate cancer (CRPC) patients. *BMC Cancer*. 2010; **10**: p. 511.
75. Norouzi-Barough, L., et al. Molecular mechanisms of drug resistance in ovarian cancer. *J Cell Physiol*. 2018; **233**(6): p. 4546–4562.
76. McFadyen, M. C., et al. Cytochrome P450 CYP1B1 protein expression: a novel mechanism of anticancer drug resistance. *Biochem Pharmacol*. 2001; **62**(2): p. 207–212.
77. Rochat, B., et al. Human CYP1B1 and anticancer agent metabolism: mechanism for tumor-specific drug inactivation? *J Pharmacol Exp Ther*. 2001; **296**(2): p. 537–541.
78. Martinez, V. G., et al. CYP1B1 expression is induced by docetaxel: effect on cell viability and drug resistance. *Br J Cancer*. 2008; **98**(3): p. 564–570.
79. Carrera, A. N., M. K. O. Grant, and B. N. Zordoky. CYP1B1 as a therapeutic target in cardio-oncology. *Clin Sci (Lond)*. 2020; **134**(21): p. 2897–2927.

80. Dutour, R., and D. Poirier. Inhibitors of cytochrome P450 (CYP) 1B1. *Eur J Med Chem*. 2017; **135**: p. 296–306.
81. Zhu, Z., et al. CYP1B1 enhances the resistance of epithelial ovarian cancer cells to paclitaxel in vivo and in vitro. *Int J Mol Med*. 2015; **35**(2): p. 340–348.
82. Lin, H., et al. Overcoming taxol-resistance in A549 cells: a comprehensive strategy of targeting P-gp transporter, AKT/ERK pathways, and cytochrome P450 enzyme CYP1B1 by 4-hydroxyemodin. *Biochem Pharmacol*. 2020; **171**: p. 113733.
83. Cui, J., et al. Design and synthesis of new α-naphthoflavones as cytochrome P450 (CYP) 1B1 inhibitors to overcome docetaxel-resistance associated with CYP1B1 overexpression. *J Med Chem*. 2015; **58**(8): p. 3534–3547.
84. Chatterjee, A., and S. Gupta. The multifaceted role of glutathione S-transferases in cancer. *Cancer Lett*. 2018; **433**: p. 33–42.
85. Townsend, D. M., and K. D. Tew. The role of glutathione-S-transferase in anti-cancer drug resistance. *Oncogene*. 2003; **22**(47): p. 7369–7375.
86. Smitherman, P. K., et al. Role of multidrug resistance protein 2 (MRP2, ABCC2) in alkylating agent detoxification: MRP2 potentiates glutathione S-transferase A1-1-mediated resistance to chlorambucil cytotoxicity. *J Pharmacol Exp Ther*. 2004; **308**(1): p. 260–267.
87. Cole, S. P., and R. G. Deeley. Transport of glutathione and glutathione conjugates by MRP1. *Trends Pharmacol Sci*. 2006; **27**(8): p. 438–446.
88. Brechbuhl, H. M., et al. Glutathione transport is a unique function of the ATP-binding cassette protein ABCG2. *J Biol Chem*. 2010; **285**(22): p. 16582–16587.
89. Zou, M., et al. Glutathione S-transferase isozyme alpha 1 is predominantly involved in the cisplatin resistance of common types of solid cancer. *Oncol Rep*. 2019; **41**(2): p. 989–998.
90. Sawers, L., et al. Glutathione S-transferase P1 (GSTP1) directly influences platinum drug chemosensitivity in ovarian tumour cell lines. *Br J Cancer*. 2014; **111**(6): p. 1150–1158.
91. Mathieu, A., et al. Development of a chemoresistant orthotopic human nonsmall cell lung carcinoma model in nude mice: analyses of tumor heterogenity in relation to the immunohistochemical levels of expression of cyclooxygenase-2, ornithine decarboxylase, lung-related resistance protein, prostaglandin E synthetase, and glutathione-S-transferase-alpha (GST)-alpha, GST-mu, and GST-pi. *Cancer*. 2004; **101**(8): p. 1908–1918.
92. Li, Q. F., et al. Genetic polymorphism of GSTP1: prediction of clinical outcome to oxaliplatin/5-FU-based chemotherapy in advanced gastric cancer. *J Korean Med Sci*. 2010; **25**(6): p. 846–852.
93. Huang, Z. H., D. Hua, and X. Du. Polymorphisms in p53, GSTP1 and XRCC1 predict relapse and survival of gastric cancer patients treated with oxaliplatin-based adjuvant chemotherapy. *Cancer Chemother Pharmacol*. 2009; **64**(5): p. 1001–1007.
94. Wang, Z., et al. Identification of proteins responsible for adriamycin resistance in breast cancer cells using proteomics analysis. *Sci Rep*. 2015; **5**: p. 9301.
95. Pljesa-Ercegovac, M., et al. Glutathione transferases: potential targets to overcome chemoresistance in solid tumors. *Int J Mol Sci*. 2018; **19**(12): p. 3785.
96. Saleh, E. M., et al. Altered expression of proliferation-inducing and proliferation-inhibiting genes might contribute to acquired doxorubicin resistance in breast cancer cells. *Cell Biochem Biophys*. 2009; **55**(2): p. 95–105.

97. Tew, K. D. and D. M. Townsend. Regulatory functions of glutathione S-transferase P1-1 unrelated to detoxification. *Drug Metab Rev.* 2011; **43**(2): p. 179–193.
98. Attaoua, C., et al. Differential involvement of glutathione S-transferase mu 1 and multidrug resistance protein 1 in melanoma acquired resistance to vinca alkaloids. *Fundam Clin Pharmacol.* 2015; **29**(1): p. 62–71.
99. Hansson, J., et al. Sensitization of human melanoma cells to the cytotoxic effect of melphalan by the glutathione transferase inhibitor ethacrynic acid. *Cancer Res.* 1991; **51**(1): p. 94–98.
100. Tew, K. D., A. M. Bomber, and S. J. Hoffman. Ethacrynic acid and piriprost as enhancers of cytotoxicity in drug resistant and sensitive cell lines. *Cancer Res.* 1988; **48**(13): p. 3622–3625.
101. Federici, L., et al. Structural basis for the binding of the anticancer compound 6-(7-nitro-2,1,3-benzoxadiazol-4-ylthio)hexanol to human glutathione s-transferases. *Cancer Res.* 2009; **69**(20): p. 8025–8034.
102. Pasello, M., et al. Overcoming glutathione S-transferase P1-related cisplatin resistance in osteosarcoma. *Cancer Res.* 2008; **68**(16): p. 6661–6668.
103. Sau, A., et al. Targeting GSTP1-1 induces JNK activation and leads to apoptosis in cisplatin-sensitive and -resistant human osteosarcoma cell lines. *Mol Biosyst.* 2012; **8**(4): p. 994–1006.
104. De Luca, A., et al. The nitrobenzoxadiazole derivative MC3181 blocks melanoma invasion and metastasis. *Oncotarget.* 2017; **8**(9): p. 15520–15538.
105. Di Paolo, V., et al. Characterization of water-soluble esters of nitrobenzoxadiazole-based GSTP1-1 inhibitors for cancer treatment. *Biochem Pharmacol.* 2020; **178**: p. 114060.
106. Dourado, D. F., et al. Mechanism of glutathione transferase P1-1-catalyzed activation of the prodrug canfosfamide (TLK286, TELCYTA). *Biochemistry.* 2013; **52**(45): p. 8069–8078.
107. Gelderblom, H., et al. Brostallicin versus doxorubicin as first-line chemotherapy in patients with advanced or metastatic soft tissue sarcoma: an European Organisation for Research and Treatment of Cancer Soft Tissue and Bone Sarcoma Group randomised phase II and pharmacogenetic study. *Eur J Cancer.* 2014; **50**(2): p. 388–396.
108. Raynal, C., et al. Pregnane X receptor (PXR) expression in colorectal cancer cells restricts irinotecan chemosensitivity through enhanced SN-38 glucuronidation. *Mol Cancer.* 2010; **9**: p. 46.
109. Landmann, H., et al. UDP glucuronosyltransferase 1A expression levels determine the response of colorectal cancer cells to the heat shock protein 90 inhibitor ganetespib. *Cell Death Dis.* 2014; **5**(9): p. e1411.
110. López-Ayllón, B. D., et al. Biomarkers of erlotinib response in non-small cell lung cancer tumors that do not harbor the more common epidermal growth factor receptor mutations. *Int J Clin Exp Pathol.* 2015; **8**(3): p. 2888–2898.
111. Zahreddine, H. A., et al. The sonic hedgehog factor GLI1 imparts drug resistance through inducible glucuronidation. *Nature.* 2014; **511**(7507): p. 90–93.
112. Ascierto, M. L., et al. The intratumoral balance between metabolic and immunologic gene expression is associated with anti-PD-1 response in patients with renal cell carcinoma. *Cancer Immunol Res.* 2016; **4**(9): p. 726–733.

113. Bueschbell, B., et al. Network biology and artificial intelligence drive the understanding of the multidrug resistance phenotype in cancer. *Drug Resist Updat*. 2022; **60**: p. 100811.
114. Miners, J. O., et al. Inhibition of human UDP-glucuronosyltransferase enzymes by lapatinib, pazopanib, regorafenib and sorafenib: implications for hyperbilirubinemia. *Biochem Pharmacol*. 2017; **129**: p. 85–95.
115. Osborne, M. J., et al. Overcoming drug resistance through the development of selective inhibitors of UDP-glucuronosyltransferase enzymes. *J Mol Biol*. 2019; **431**(2): p. 258–272.
116. Liu, W. J., et al. Drug resistance to targeted therapeutic strategies in non-small cell lung cancer. *Pharmacol Ther*. 2020; **206**: p. 107438.
117. Sarmento-Ribeiro, A. B., et al. The emergence of drug resistance to targeted cancer therapies: clinical evidence. *Drug Resist Updat*. 2019; **47**: p. 100646.
118. Di Noia, V., et al. Treating disease progression with osimertinib in EGFR-mutated non-small-cell lung cancer: novel targeted agents and combination strategies. *ESMO Open*. 2021; **6**(6): p. 100280.
119. Piotrowska, Z., et al. Heterogeneity underlies the emergence of EGFRT790 wild-type clones following treatment of T790M-positive cancers with a third-generation EGFR inhibitor. *Cancer Discov*. 2015; **5**(7): p. 713–722.
120. Thress, K. S., et al. Acquired EGFR C797S mutation mediates resistance to AZD9291 in non-small cell lung cancer harboring EGFR T790M. *Nat Med*. 2015; **21**(6): p. 560–562.
121. Jia, Y., et al. Overcoming EGFR(T790M) and EGFR(C797S) resistance with mutant-selective allosteric inhibitors. *Nature*. 2016; **534**(7605): p. 129–132.
122. Lu, X., et al. Discovery of JND3229 as a new EGFR(C797S) mutant inhibitor with in vivo monodrug efficacy. *ACS Med Chem Lett*. 2018; **9**(11): p. 1123–1127.
123. To, C., et al. Single and dual targeting of mutant EGFR with an allosteric inhibitor. *Cancer Discov*. 2019; **9**(7): p. 926–943.
124. Drilon, A., et al. Efficacy of larotrectinib in TRK fusion-positive cancers in adults and children. *N Engl J Med*. 2018; **378**(8): p. 731–739.
125. Drilon, A., et al. What hides behind the MASC: clinical response and acquired resistance to entrectinib after ETV6-NTRK3 identification in a mammary analogue secretory carcinoma (MASC). *Ann Oncol*. 2016; **27**(5): p. 920–926.
126. Kwak, E. L., et al. Anaplastic lymphoma kinase inhibition in non-small-cell lung cancer. *N Engl J Med*. 2010; **363**(18): p. 1693–1703.
127. Katayama, R., et al. Mechanisms of acquired crizotinib resistance in ALK-rearranged lung cancers. *Sci Transl Med*. 2012; **4**(120): p. 120ra17.
128. Gainor, J. F., et al. ALK rearrangements are mutually exclusive with mutations in EGFR or KRAS: an analysis of 1,683 patients with non-small cell lung cancer. *Clin Cancer Res*. 2013; **19**(15): p. 4273–4281.
129. Tibaldi, C. Mechanisms of resistance to crizotinib in patients with ALK gene rearranged non-small-cell lung cancer. *Pharmacogenomics*. 2014; **15**(2): p. 133–135.
130. Sasaki, T., et al. The neuroblastoma-associated F1174L ALK mutation causes resistance to an ALK kinase inhibitor in ALK-translocated cancers. *Cancer Res*. 2010; **70**(24): p. 10038–10043.

131. Sasaki, T., et al. A novel ALK secondary mutation and EGFR signaling cause resistance to ALK kinase inhibitors. *Cancer Res*. 2011; **71**(18): p. 6051–6060.
132. Katayama, R., et al. Therapeutic strategies to overcome crizotinib resistance in non-small cell lung cancers harboring the fusion oncogene EML4-ALK. *Proc Natl Acad Sci U S A*. 2011; **108**(18): p. 7535–7540.
133. Marsilje, T. H., et al. Synthesis, structure-activity relationships, and in vivo efficacy of the novel potent and selective anaplastic lymphoma kinase (ALK) inhibitor 5-chloro-N2-(2-isopropoxy-5-methyl-4-(piperidin-4-yl)phenyl)-N4-(2-(isopropylsulfonyl)phenyl)pyrimidine-2,4-diamine (LDK378) currently in phase 1 and phase 2 clinical trials. *J Med Chem*. 2013; **56**(14): p. 5675–5690.
134. Friboulet, L., et al. The ALK inhibitor ceritinib overcomes crizotinib resistance in non-small cell lung cancer. *Cancer Discov*. 2014; **4**(6): p. 662–673.
135. Shaw, A. T., et al. Ceritinib in ALK-rearranged non-small-cell lung cancer. *N Engl J Med*. 2014; **370**(13): p. 1189–1197.
136. Ostrem, J. M. and K. M. Shokat. Direct small-molecule inhibitors of KRAS: from structural insights to mechanism-based design. *Nat Rev Drug Discov*. 2016; **15**(11): p. 771–785.
137. Papke, B., and C. J. Der. Drugging RAS: know the enemy. *Science*. 2017; **355**(6330): p. 1158–1163.
138. Karoulia, Z., E. Gavathiotis, and P. I. Poulikakos. New perspectives for targeting RAF kinase in human cancer. *Nat Rev Cancer*. 2017; **17**(11): p. 676–691.
139. Lito, P., N. Rosen, and D. B. Solit. Tumor adaptation and resistance to RAF inhibitors. *Nat Med*. 2013; **19**(11): p. 1401–1409.
140. Caunt, C. J., et al. MEK1 and MEK2 inhibitors and cancer therapy: the long and winding road. *Nat Rev Cancer*. 2015; **15**(10): p. 577–592.
141. Ryan, M. B., et al. Vertical pathway inhibition overcomes adaptive feedback resistance to KRAS(G12C) inhibition. *Clin Cancer Res*. 2020; **26**(7): p. 1633–1643.
142. Xue, J. Y., et al. Rapid non-uniform adaptation to conformation-specific KRAS(G12C) inhibition. *Nature*. 2020; **577**(7790): p. 421–425.
143. Poulikakos, P. I., et al. RAF inhibitors transactivate RAF dimers and ERK signalling in cells with wild-type BRAF. *Nature*. 2010; **464**(7287): p. 427–430.
144. Lito, P., et al. Disruption of CRAF-mediated MEK activation is required for effective MEK inhibition in KRAS mutant tumors. *Cancer Cell*. 2014; **25**(5): p. 697–710.
145. Little, A. S., et al. Amplification of the driving oncogene, KRAS or BRAF, underpins acquired resistance to MEK1/2 inhibitors in colorectal cancer cells. *Sci Signal*. 2011; **4**(166): p. ra17.
146. Sale, M. J., et al. MEK1/2 inhibitor withdrawal reverses acquired resistance driven by BRAF(V600E) amplification whereas KRAS(G13D) amplification promotes EMT-chemoresistance. *Nat Commun*. 2019; **10**(1): p. 2030.
147. Juric, D., et al. Convergent loss of PTEN leads to clinical resistance to a PI(3)Kα inhibitor. *Nature*. 2015; **518**(7538): p. 240–244.
148. Patch, A. M., et al. Corrigendum: whole-genome characterization of chemoresistant ovarian cancer. *Nature*. 2015; **527**(7578): p. 398.
149. Lee, C. K., et al. Impact of specific epidermal growth factor receptor (EGFR) mutations and clinical characteristics on outcomes after treatment with EGFR tyrosine kinase inhibitors

versus chemotherapy in EGFR-mutant lung cancer: a meta-analysis. *J Clin Oncol*. 2015; **33**(17): p. 1958–1965.

150. Kancha, R. K., et al. Functional analysis of epidermal growth factor receptor (EGFR) mutations and potential implications for EGFR targeted therapy. *Clin Cancer Res*. 2009; **15**(2): p. 460–467.
151. Yarden, Y. and M. X. Sliwkowski. Untangling the ErbB signalling network. *Nat Rev Mol Cell Biol*. 2001; **2**(2): p. 127–137.
152. Ohashi, K., et al. Lung cancers with acquired resistance to EGFR inhibitors occasionally harbor BRAF gene mutations but lack mutations in KRAS, NRAS, or MEK1. *Proc Natl Acad Sci U S A*. 2012; **109**(31): p. E2127–2133.
153. Karachaliou, N., et al. BIM and mTOR expression levels predict outcome to erlotinib in EGFR-mutant non-small-cell lung cancer. *Sci Rep*. 2015; **5**: p. 17499.
154. Kawabata, S., et al. Rapamycin prevents the development and progression of mutant epidermal growth factor receptor lung tumors with the acquired resistance mutation T790M. *Cell Rep*. 2014; **7**(6): p. 1824–1832.
155. Engelman, J. A., et al. MET amplification leads to gefitinib resistance in lung cancer by activating ERBB3 signaling. *Science*. 2007; **316**(5827): p. 1039–1043.
156. Shaw, A. T., et al. Crizotinib in ROS1-rearranged non-small-cell lung cancer. *N Engl J Med*. 2014; **371**(21): p. 1963–1971.
157. Drilon, A., et al. Repotrectinib (TPX-0005) is a next-generation ROS1/TRK/ALK inhibitor that potently inhibits ROS1/TRK/ALK solvent- front mutations. *Cancer Discov*. 2018; **8**(10): p. 1227–1236.
158. Katayama, R., et al. The new-generation selective ROS1/NTRK inhibitor DS-6051b overcomes crizotinib resistant ROS1-G2032R mutation in preclinical models. *Nat Commun*. 2019; **10**(1): p. 3604.
159. Drilon, A., et al. Safety and antitumor activity of the multitargeted pan-TRK, ROS1, and ALK inhibitor entrectinib: combined results from two phase I trials (ALKA-372-001 and STARTRK-1). *Cancer Discov*. 2017; **7**(4): p. 400–409.
160. Watanabe, J., N. Furuya, and Y. Fujiwara. Appearance of a BRAF mutation conferring resistance to crizotinib in non-small cell lung cancer harboring oncogenic ROS1 fusion. *J Thorac Oncol*. 2018; **13**(4): p. e66–e69.
161. Zhu, Y. C., et al. Concurrent ROS1 gene rearrangement and KRAS mutation in lung adenocarcinoma: a case report and literature review. *Thorac Cancer*. 2018; **9**(1): p. 159–163.
162. McCoach, C. E., et al. Resistance mechanisms to targeted therapies in ROS1(+) and ALK(+) non-small cell lung cancer. *Clin Cancer Res*. 2018; **24**(14): p. 3334–3347.
163. He, Y., et al. Different types of ROS1 fusion partners yield comparable efficacy to crizotinib. *Oncol Res*. 2019; **27**(8): p. 901–910.
164. Zhu, V. W., S. J. Klempner, and S. I. Ou. Receptor tyrosine kinase fusions as an actionable resistance mechanism to EGFR TKIs in EGFR-mutant non-small-cell lung cancer. *Trends Cancer*. 2019; **5**(11): p. 677–692.
165. Dagogo-Jack, I., et al. Molecular analysis of plasma from patients with ROS1-positive NSCLC. *J Thorac Oncol*. 2019; **14**(5): p. 816–824.

166. Sato, H., et al. MAPK pathway alterations correlate with poor survival and drive resistance to therapy in patients with lung cancers driven by ROS1 fusions. *Clin Cancer Res*. 2020; **26**(12): p. 2932–2945.
167. Cocco, E., M. Scaltriti, and A. Drilon. NTRK fusion-positive cancers and TRK inhibitor therapy. *Nat Rev Clin Oncol*. 2018; **15**(12): p. 731–747.
168. Gainor, J. F., et al. Patterns of metastatic spread and mechanisms of resistance to crizotinib in ROS1-positive non-small-cell lung cancer. *JCO Precis Oncol*. 2017; 2017: p. PO.17.00063.
169. Sehgal, K., et al. Cases of ROS1-rearranged lung cancer: when to use crizotinib, entrectinib, lorlatinib, and beyond? *Precis Cancer Med*. 2020; **3**.
170. Li, L. Y., et al. DNA repair pathways in cancer therapy and resistance. *Front Pharmacol*. 2020; **11**: p. 629266.
171. Sethy, C. and C. N. Kundu. 5-Fluorouracil (5-FU) resistance and the new strategy to enhance the sensitivity against cancer: Implication of DNA repair inhibition. *Biomed Pharmacother*. 2021; **137**: p. 111285.
172. Liu, J. L., et al. Effect of 5-fluorouracil on excision repair cross-complementing 1 expression and consequent cytotoxicity regulation in human gastric cancer cells. *J Cell Biochem*. 2018; **119**(10): p. 8472–8480.
173. Oliver, J. A., et al. MMR-proficient and MMR-deficient colorectal cancer cells: 5-fluorouracil treatment response and correlation to CD133 and MGMT expression. *J Biosci*. 2020; **45**: p. 121.
174. Weller, M., et al. MGMT promoter methylation is a strong prognostic biomarker for benefit from dose-intensified temozolomide rechallenge in progressive glioblastoma: the DIRECTOR trial. *Clin Cancer Res*. 2015; **21**(9): p. 2057–2064.
175. Abe, H., et al. MGMT expression contributes to temozolomide resistance in H3K27M-mutant diffuse midline gliomas. *Front Oncol*. 2019; **9**: p. 1568.
176. Brandner, S., et al. MGMT promoter methylation testing to predict overall survival in people with glioblastoma treated with temozolomide: a comprehensive meta-analysis based on a cochrane systematic review. *Neuro Oncol*. 2021; **23**(9): p. 1457–1469.
177. Jacot, W., et al. BRCA1 promoter hypermethylation is associated with good prognosis and chemosensitivity in triple-negative breast cancer. *Cancers* (*Basel*). 2020; **12**(4): p. 828.
178. Mazloumi, Z., et al. Effect of aberrant DNA methylation on cancer stem cell properties. *Exp Mol Pathol*. 2022; **125**: p. 104757.
179. Watson, Z. L., et al. Histone methyltransferases EHMT1 and EHMT2 (GLP/G9A) maintain PARP inhibitor resistance in high-grade serous ovarian carcinoma. *Clin Epigenetics*. 2019; **11**(1): p. 165.
180. Singh, R., et al. Epigenetic remodeling through downregulation of polycomb repressive complex 2 mediates chemotherapy resistance in testicular germ cell tumors. *Cancers* (*Basel*). 2019; **11**(6): p. 796.
181. Patel, N., et al. Erratum: miR-15a/miR-16 down-regulates BMI1, impacting Ub-H2A mediated DNA repair and breast cancer cell sensitivity to doxorubicin. *Sci Rep*. 2017; **7**(1): p. 12932.
182. Qian, L., et al. lncRNA HOTAIR promotes DNA repair and radioresistance of breast cancer via EZH2. *DNA Cell Biol*. 2020. Online ahead of print.

183. Edinger, A. L. and C. B. Thompson. Death by design: apoptosis, necrosis and autophagy. *Curr Opin Cell Biol*. 2004; **16**(6): p. 663–669.
184. Li, M., et al. The M476W/Q482H mutation of procaspase-8 restored caspase-8-mediated apoptosis. *Biochem Biophys Res Commun*. 2019; **514**(3): p. 653–658.
185. Roos, W. P., A. D. Thomas, and B. Kaina. DNA damage and the balance between survival and death in cancer biology. *Nat Rev Cancer*. 2016; **16**(1): p. 20–33.
186. Zhang, W., et al. Sorafenib induces apoptosis of AML cells via Bim-mediated activation of the intrinsic apoptotic pathway. *Leukemia*. 2008; **22**(4): p. 808–818.
187. Zuo, J., et al. Bcl-2 overexpression induces a partial epithelial to mesenchymal transition and promotes squamous carcinoma cell invasion and metastasis. *Mol Cancer Res*. 2010; **8**(2): p. 170–182.
188. Real, P. J., et al. Resistance to chemotherapy via Stat3-dependent overexpression of Bcl-2 in metastatic breast cancer cells. *Oncogene*. 2002; **21**(50): p. 7611–7618.
189. Ritter, V., et al. Bcl-2/Bcl-xL inhibitor ABT-263 overcomes hypoxia-driven radioresistence and improves radiotherapy. *Cell Death Dis*. 2021; **12**(7): p. 694.
190. Roberts, A. W., et al. Substantial susceptibility of chronic lymphocytic leukemia to BCL2 inhibition: results of a phase I study of navitoclax in patients with relapsed or refractory disease. *J Clin Oncol*. 2012; **30**(5): p. 488–496.
191. Kipps, T. J., et al. A phase 2 study of the BH3 mimetic BCL2 inhibitor navitoclax (ABT-263) with or without rituximab, in previously untreated B-cell chronic lymphocytic leukemia. *Leuk Lymphoma*. 2015; **56**(10): p. 2826–2833.
192. Bose, P., V. Gandhi, and M. Konopleva. Pathways and mechanisms of venetoclax resistance. *Leuk Lymphoma*. 2017; **58**(9): p. 1–17.
193. Salimi-Jeda, A., et al. Autophagy modulation and cancer combination therapy: a smart approach in cancer therapy. *Cancer Treat Res Commun*. 2022; **30**: p. 100512.
194. Liu, X. L., J. Ding, and L. H. Meng. Oncogene-induced senescence: a double edged sword in cancer. *Acta Pharmacol Sin*. 2018; **39**(10): p. 1553–1558.
195. Lorente, J., et al. The interplay between autophagy and tumorigenesis: exploiting autophagy as a means of anticancer therapy. *Biol Rev Camb Philos Soc*. 2018; **93**(1): p. 152–165.
196. Yang, Z. J., et al. The role of autophagy in cancer: therapeutic implications. *Mol Cancer Ther*. 2011; **10**(9): p. 1533–1541.
197. Wang, J., and G. S. Wu. Role of autophagy in cisplatin resistance in ovarian cancer cells. *J Biol Chem*. 2014; **289**(24): p. 17163–17173.
198. Castro, I., B. Sampaio-Marques, and P. Ludovico. Targeting metabolic reprogramming in acute myeloid leukemia. *Cells*. 2019; **8**(9): p. 967.
199. Wong, M. M., et al. Interplay of autophagy and cancer stem cells in hepatocellular carcinoma. *Mol Biol Rep*. 2021; **48**(4): p. 3695–3717.
200. Wen, Y. A., et al. Adipocytes activate mitochondrial fatty acid oxidation and autophagy to promote tumor growth in colon cancer. *Cell Death Dis*. 2017; **8**(2): p. e2593.
201. Amaravadi, R. K., et al. Autophagy inhibition enhances therapy-induced apoptosis in a Myc-induced model of lymphoma. *J Clin Invest*. 2007; **117**(2): p. 326–336.
202. Han, W., et al. EGFR tyrosine kinase inhibitors activate autophagy as a cytoprotective response in human lung cancer cells. *PLoS One*. 2011; **6**(6): p. e18691.

203. Gupta, A., et al. Autophagy inhibition and antimalarials promote cell death in gastrointestinal stromal tumor (GIST). *Proc Natl Acad Sci USA*. 2010; **107**(32): p. 14333–14338.
204. Bellodi, C., et al. Targeting autophagy potentiates tyrosine kinase inhibitor-induced cell death in Philadelphia chromosome-positive cells, including primary CML stem cells. *J Clin Invest*. 2009; **119**(5): p. 1109–1123.
205. Dixon, S. J., et al. Ferroptosis: an iron-dependent form of nonapoptotic cell death. *Cell*. 2012; **149**(5): p. 1060–1072.
206. Zhang, C., et al. Ferroptosis in cancer therapy: a novel approach to reversing drug resistance. *Mol Cancer*. 2022; **21**(1): p. 47.
207. Yang, W. S., et al. Regulation of ferroptotic cancer cell death by GPX4. *Cell*. 2014; **156**(1-2): p. 317–331.
208. Forcina, G. C., and S. J. Dixon. GPX4 at the crossroads of lipid homeostasis and ferroptosis. *Proteomics*. 2019; **19**(18): p. e1800311.
209. Chen, T. C., et al. AR ubiquitination induced by the curcumin analog suppresses growth of temozolomide-resistant glioblastoma through disrupting GPX4-Mediated redox homeostasis. *Redox Biol*. 2020; **30**: p. 101413.
210. Yang, C., et al. Suppressing the KIF20A/NUAK1/Nrf2/GPX4 signaling pathway induces ferroptosis and enhances the sensitivity of colorectal cancer to oxaliplatin. *Aging (Albany NY)*. 2021; **13**(10): p. 13515–13534.
211. Roh, J. L., et al. Induction of ferroptotic cell death for overcoming cisplatin resistance of head and neck cancer. *Cancer Lett*. 2016; **381**(1): p. 96–103.
212. Fu, D., et al. Induction of ferroptosis by ATF3 elevation alleviates cisplatin resistance in gastric cancer by restraining Nrf2/Keap1/xCT signaling. *Cell Mol Biol Lett*. 2021; **26**(1): p. 26.
213. Du, J., et al. DHA exhibits synergistic therapeutic efficacy with cisplatin to induce ferroptosis in pancreatic ductal adenocarcinoma via modulation of iron metabolism. *Cell Death Dis*. 2021; **12**(7): p. 705.
214. Turcu, A. L., et al. DMT1 inhibitors kill cancer stem cells by blocking lysosomal iron translocation. *Chemistry*. 2020; **26**(33): p. 7369–7373.
215. Chaudhary, N., et al. Lipocalin 2 expression promotes tumor progression and therapy resistance by inhibiting ferroptosis in colorectal cancer. *Int J Cancer*. 2021; **149**(7): p. 1495–1511.
216. Doll, S., et al. ACSL4 dictates ferroptosis sensitivity by shaping cellular lipid composition. *Nat Chem Biol*. 2017; **13**(1): p. 91–98.
217. Ye, Z., et al. Abrogation of ARF6 promotes RSL3-induced ferroptosis and mitigates gemcitabine resistance in pancreatic cancer cells. *Am J Cancer Res*. 2020; **10**(4): p. 1182–1193.
218. Tsvetkov, P., et al. Copper induces cell death by targeting lipoylated TCA cycle proteins. *Science*. 2022; **375**(6586): p. 1254–1261.
219. Mayr, J. A., et al. Lipoic acid biosynthesis defects. *J Inherit Metab Dis*. 2014; **37**(4): p. 553–563.
220. Solmonson, A., and R. J. DeBerardinis. Lipoic acid metabolism and mitochondrial redox regulation. *J Biol Chem*. 2018; **293**(20): p. 7522–7530.

221. Gaggelli, E., et al. Copper homeostasis and neurodegenerative disorders (Alzheimer's, prion, and Parkinson's diseases and amyotrophic lateral sclerosis). *Chem Rev*. 2006; **106**(6): p. 1995–2044.
222. Bandmann, O., K. H. Weiss, and S. G. Kaler. Wilson's disease and other neurological copper disorders. *Lancet Neurol*. 2015; **14**(1): p. 103–113.
223. Tsang, T., et al. Copper is an essential regulator of the autophagic kinases ULK1/2 to drive lung adenocarcinoma. *Nat Cell Biol*. 2020; **22**(4): p. 412–424.
224. Cui, L., et al. Mitochondrial copper depletion suppresses triple-negative breast cancer in mice. *Nat Biotechnol*. 2021; **39**(3): p. 357–367.
225. Aggarwal, A., and M. Bhatt. Advances in treatment of Wilson disease. *Tremor Other Hyperkinet Mov (N Y)*. 2018; **8**: p. 525.
226. O'Day, S. J., et al. Final results of phase III SYMMETRY study: randomized, double-blind trial of elesclomol plus paclitaxel versus paclitaxel alone as treatment for chemotherapy-naive patients with advanced melanoma. *J Clin Oncol*. 2013; **31**(9): p. 1211–1218.
227. Tang, D., X. Chen, and G. Kroemer. Cuproptosis: a copper-triggered modality of mitochondrial cell death. *Cell Res*. 2022; **32**(5): p. 417–418.
228. Lawson, D. A., et al. Single-cell analysis reveals a stem-cell program in human metastatic breast cancer cells. *Nature*. 2015; **526**(7571): p. 131–135.
229. Mani, S. A., et al. The epithelial-mesenchymal transition generates cells with properties of stem cells. *Cell*. 2008; **133**(4): p. 704–715.
230. Shimokawa, M., et al. Visualization and targeting of LGR5(+) human colon cancer stem cells. *Nature*. 2017; **545**(7653): p. 187–192.
231. Soliman, A. A., A. A. Elzarkaa, and E. Malik. Epithelial ovarian cancer and cancer stem cells. *Adv Exp Med Biol*. 2021; **1330**: p. 21–32.
232. Zhou, H. M., et al. Targeting cancer stem cells for reversing therapy resistance: mechanism, signaling, and prospective agents. *Signal Transduct Target Ther*. 2021; **6**(1): p. 62.
233. Wang, H., et al. Single-cell analyses reveal mechanisms of cancer stem cell maintenance and epithelial-mesenchymal transition in recurrent bladder cancer. *Clin Cancer Res*. 2021; **27**(22): p. 6265–6278.
234. Shi, Z. D., et al. Targeting HNRNPU to overcome cisplatin resistance in bladder cancer. *Mol Cancer*. 2022; **21**(1): p. 37.
235. He, X., et al. Tumor-initiating stem cell shapes its microenvironment into an immunosuppressive barrier and pro-tumorigenic niche. *Cell Rep*. 2021; **36**(10): p. 109674.
236. Wang, D., et al. Prostaglandin E2 promotes colorectal cancer stem cell expansion and metastasis in mice. *Gastroenterology*. 2015; **149**(7): p. 1884–1895.e4.
237. Sun, L., et al. Gastric cancer mesenchymal stem cells regulate PD-L1-CTCF enhancing cancer stem cell-like properties and tumorigenesis. *Theranostics*. 2020; **10**(26): p. 11950–11962.
238. Wei, F., et al. PD-L1 promotes colorectal cancer stem cell expansion by activating HMGA1-dependent signaling pathways. *Cancer Lett*. 2019; **450**: p. 1–13.
239. Ouadah, Y., et al. Rare pulmonary neuroendocrine cells are stem cells regulated by Rb, p53, and Notch. *Cell*. 2019; **179**(2): p. 403–416. e23.
240. Novellasdemunt, L., et al. NEDD4 and NEDD4L regulate Wnt signalling and intestinal stem cell priming by degrading LGR5 receptor. *Embo J*. 2020; **39**(3): p. e102771.

241. Zhang, L., et al. Single-cell transcriptomics in medulloblastoma reveals tumor-initiating progenitors and oncogenic cascades during tumorigenesis and relapse. *Cancer Cell.* 2019; **36**(3): p. 302–318. e7.
242. Deslauriers, A. G., A. G. Kotini, and E. P. Papapetrou. Modeling leukemia stem cells with patient-derived induced pluripotent stem cells. *Methods Mol Biol.* 2021; **2185**: p. 411–422.
243. Chaffer, C. L., et al. EMT, cell plasticity and metastasis. *Cancer Metastasis Rev.* 2016; **35**(4): p. 645–654.
244. Na, T. Y., et al. The functional activity of E-cadherin controls tumor cell metastasis at multiple steps. *Proc Natl Acad Sci USA.* 2020; **117**(11): p. 5931–5937.
245. Han, L. L., et al. Sirtuin6 (SIRT6) promotes the EMT of hepatocellular carcinoma by stimulating autophagic degradation of E-cadherin. *Mol Cancer Res.* 2019; **17**(11): p. 2267–2280.
246. Zhang, J., et al. LGR5, a novel functional glioma stem cell marker, promotes EMT by activating the Wnt/β-catenin pathway and predicts poor survival of glioma patients. *J Exp Clin Cancer Res.* 2018; **37**(1): p. 225.
247. Tang, Q., et al. TM4SF1 promotes EMT and cancer stemness via the Wnt/β-catenin/SOX2 pathway in colorectal cancer. *J Exp Clin Cancer Res.* 2020; **39**(1): p. 232.
248. Zhang, N., et al. Isorhapontigenin (ISO) inhibits EMT through FOXO3A/METTL14/VIMENTIN pathway in bladder cancer cells. *Cancer Lett.* 2021; **520**: p. 400–408.
249. Tian, H., et al. AKT-induced lncRNA VAL promotes EMT-independent metastasis through diminishing Trim16-dependent Vimentin degradation. *Nat Commun.* 2020; **11**(1): p. 5127.
250. Luo, J., et al. 14, 15-EET induces breast cancer cell EMT and cisplatin resistance by up-regulating integrin αvβ3 and activating FAK/PI3K/AKT signaling. *J Exp Clin Cancer Res.* 2018; **37**(1): p. 23.
251. Kim, T. W., et al. MicroRNA-17-5p regulates EMT by targeting vimentin in colorectal cancer. *Br J Cancer.* 2020; **123**(7): p. 1123–1130.
252. Sinha, S., S. Pattnaik, and G. K. Aradhyam. Molecular evolution guided functional analyses reveals nucleobindin-1 as a canonical E-box binding protein promoting epithelial-to-mesenchymal transition (EMT). *Biochim Biophys Acta Proteins Proteom.* 2019; **1867**(9): p. 765–775.
253. Raoof, S., et al. Targeting FGFR overcomes EMT-mediated resistance in EGFR mutant non-small cell lung cancer. *Oncogene.* 2019; **38**(37): p. 6399–6413.
254. Torzilli, P. A., et al. A new paradigm for mechanobiological mechanisms in tumor metastasis. *Semin Cancer Biol.* 2012; **22**(5-6): p. 385–395.
255. Zhu, Y., et al. HIF-1α regulates EMT via the Snail and β-catenin pathways in paraquat poisoning-induced early pulmonary fibrosis. *J Cell Mol Med.* 2016; **20**(4): p. 688–697.
256. Zhang, P., Y. Sun, and L. Ma. ZEB1: at the crossroads of epithelial-mesenchymal transition, metastasis and therapy resistance. *Cell Cycle.* 2015; **14**(4): p. 481–487.
257. Xu, Y., et al. Breast tumor cell-specific knockout of Twist1 inhibits cancer cell plasticity, dissemination, and lung metastasis in mice. *Proc Natl Acad Sci USA.* 2017; **114**(43): p. 11494–11499.

258. Zhang, W., et al. Chemoresistance to 5-fluorouracil induces epithelial-mesenchymal transition via up-regulation of Snail in MCF7 human breast cancer cells. *Biochem Biophys Res Commun*. 2012; **417**(2): p. 679–685.
259. Wang, H., et al. Acquisition of epithelial-mesenchymal transition phenotype and cancer stem cell-like properties in cisplatin-resistant lung cancer cells through AKT/β-catenin/Snail signaling pathway. *Eur J Pharmacol*. 2014; **723**: p. 156–166.
260. Namba, T., et al. Zidovudine, an anti-viral drug, resensitizes gemcitabine-resistant pancreatic cancer cells to gemcitabine by inhibition of the Akt-GSK3β-Snail pathway. *Cell Death Dis*. 2015; **6**(6): p. e1795.
261. Fischer, K. R., et al. Epithelial-to-mesenchymal transition is not required for lung metastasis but contributes to chemoresistance. *Nature*. 2015; **527**(7579): p. 472–476.
262. Jin, H., et al. Targeting lipid metabolism to overcome EMT-associated drug resistance via integrin β3/FAK pathway and tumor-associated macrophage repolarization using legumain-activatable delivery. *Theranostics*. 2019; **9**(1): p. 265–278.
263. Sheng, W., et al. Calreticulin promotes EGF-induced EMT in pancreatic cancer cells via Integrin/EGFR-ERK/MAPK signaling pathway. *Cell Death Dis*. 2017; **8**(10): p. e3147.
264. Wu, Q., et al. lncRNA SNHG11 promotes gastric cancer progression by activating the Wnt/β-catenin pathway and oncogenic autophagy. *Mol Ther*. 2021; **29**(3): p. 1258–1278.
265. Cheng, F., et al. Vimentin coordinates fibroblast proliferation and keratinocyte differentiation in wound healing via TGF-β-Slug signaling. *Proc Natl Acad Sci USA*. 2016; **113**(30): p. E4320–4327.
266. Kariya, Y., et al. αvβ3 Integrin induces partial EMT independent of TGF-β signaling. *Commun Biol*. 2021; **4**(1): p. 490.
267. Gachpazan, M., et al. Therapeutic Potential of targeting transforming growth factor-beta in colorectal cancer: Rational and progress. *Curr Pharm Des*. 2019; **25**(38): p. 4085–4089.
268. Katoh, M., and M. Katoh. Molecular genetics and targeted therapy of WNT-related human diseases (Review). *Int J Mol Med*. 2017; **40**(3): p. 587–606.
269. Vasan, N., J. Baselga, and D. M. Hyman. A view on drug resistance in cancer. *Nature*. 2019; **575**(7782): p. 299–309.
270. Lambrechts, D., et al. Phenotype molding of stromal cells in the lung tumor microenvironment. *Nat Med*. 2018; **24**(8): p. 1277–1289.
271. Burgos-Panadero, R., et al. The tumour microenvironment as an integrated framework to understand cancer biology. *Cancer Lett*. 2019; **461**: p. 112–122.
272. Leroy, J., et al. Laparoscopic total mesorectal excision (TME) for rectal cancer surgery: long-term outcomes. *Surg Endosc*. 2004; **18**(2): p. 281–289.
273. Henke, E., R. Nandigama, and S. Ergün. Extracellular matrix in the tumor microenvironment and its impact on cancer therapy. *Front Mol Biosci*. 2019; **6**: p. 160.
274. Yu, Q., et al. Extracellular matrix proteins confer cell adhesion-mediated drug resistance through integrin α (v) in glioblastoma cells. *Front Cell Dev Biol*. 2021; **9**: p. 616580.
275. Misra, S., et al. Regulation of multidrug resistance in cancer cells by hyaluronan. *J Biol Chem*. 2003; **278**(28): p. 25285–25288.
276. Park, C. C., et al. Beta1 integrin inhibition dramatically enhances radiotherapy efficacy in human breast cancer xenografts. *Cancer Res*. 2008; **68**(11): p. 4398–4405.

277. Weigelt, B., et al. HER2 signaling pathway activation and response of breast cancer cells to HER2-targeting agents is dependent strongly on the 3D microenvironment. *Breast Cancer Res Treat*. 2010; **122**(1): p. 35–43.
278. Zigrino, P., R. Nischt, and C. Mauch. The disintegrin-like and cysteine-rich domains of ADAM-9 mediate interactions between melanoma cells and fibroblasts. *J Biol Chem*. 2011; **286**(8): p. 6801–6807.
279. de la Fuente, M. T., et al. Engagement of alpha4beta1 integrin by fibronectin induces in vitro resistance of B chronic lymphocytic leukemia cells to fludarabine. *J Leukoc Biol*. 2002; **71**(3): p. 495–502.
280. Matsunaga, T., et al. Interaction between leukemic-cell VLA-4 and stromal fibronectin is a decisive factor for minimal residual disease of acute myelogenous leukemia. *Nat Med*. 2003; **9**(9): p. 1158–1165.
281. Nishio, M., et al. Nurselike cells express BAFF and APRIL, which can promote survival of chronic lymphocytic leukemia cells via a paracrine pathway distinct from that of SDF-1alpha. *Blood*. 2005; **106**(3): p. 1012–1020.
282. Chen, Y., et al. CXCR4 downregulation of let-7a drives chemoresistance in acute myeloid leukemia. *J Clin Invest*. 2013; **123**(6): p. 2395–2407.
283. Margolin, D. A., et al. Lymph node stromal cells enhance drug-resistant colon cancer cell tumor formation through SDF-1α/CXCR4 paracrine signaling. *Neoplasia*. 2011; **13**(9): p. 874–886.
284. Azmi, A. S., B. Bao, and F. H. Sarkar. Exosomes in cancer development, metastasis, and drug resistance: a comprehensive review. *Cancer Metastasis Rev*. 2013; **32**(3-4): p. 623–642.
285. Colombo, M., G. Raposo, and C. Théry. Biogenesis, secretion, and intercellular interactions of exosomes and other extracellular vesicles. *Annu Rev Cell Dev Biol*. 2014; **30**: p. 255–289.
286. Valadi, H., et al. Exosome-mediated transfer of mRNAs and microRNAs is a novel mechanism of genetic exchange between cells. *Nat Cell Biol*. 2007; **9**(6): p. 654–659.
287. Mathivanan, S., H. Ji, and R. J. Simpson. Exosomes: extracellular organelles important in intercellular communication. *J Proteomics*. 2010; **73**(10): p. 1907–1920.
288. D'Asti, E., et al. Oncogenic extracellular vesicles in brain tumor progression. *Front Physiol*. 2012; **3**: p. 294.
289. Zhang, H. G., and W. E. Grizzle. Exosomes: a novel pathway of local and distant intercellular communication that facilitates the growth and metastasis of neoplastic lesions. *Am J Pathol*. 2014; **184**(1): p. 28–41.
290. Quail, D. F., and J. A. Joyce. Microenvironmental regulation of tumor progression and metastasis. *Nat Med*. 2013; **19**(11): p. 1423–1437.
291. Deep, G., and G. K. Panigrahi. Hypoxia-induced signaling promotes prostate cancer progression: exosomes role as messenger of hypoxic response in tumor microenvironment. *Crit Rev Oncog*. 2015; **20**(5-6): p. 419–434.
292. Webber, J., et al. Cancer exosomes trigger fibroblast to myofibroblast differentiation. *Cancer Res*. 2010; **70**(23): p. 9621–9630.
293. Zhang, Q., and C. Peng. Cancer-associated fibroblasts regulate the biological behavior of cancer cells and stroma in gastric cancer. *Oncol Lett*. 2018; **15**(1): p. 691–698.

294. Richards, K. E., et al. Cancer-associated fibroblast exosomes regulate survival and proliferation of pancreatic cancer cells. *Oncogene*. 2017; **36**(13): p. 1770–1778.
295. Shedden, K., et al. Expulsion of small molecules in vesicles shed by cancer cells: association with gene expression and chemosensitivity profiles. *Cancer Res*. 2003; **63**(15): p. 4331–4337.
296. Corrado, C., et al. Exosomes as intercellular signaling organelles involved in health and disease: basic science and clinical applications. *Int J Mol Sci*. 2013; **14**(3): p. 5338–5366.
297. Maleki, S., J. Jabalee, and C. Garnis. The role of extracellular vesicles in mediating resistance to anticancer therapies. *Int J Mol Sci*. 2021; **22**(8): p. 4166.
298. Muralidharan-Chari, V., et al. Microvesicle removal of anticancer drugs contributes to drug resistance in human pancreatic cancer cells. *Oncotarget*. 2016; **7**(31): p. 50365–50379.
299. Ning, K., et al. UCH-L1-containing exosomes mediate chemotherapeutic resistance transfer in breast cancer. *J Surg Oncol*. 2017; **115**(8): p. 932–940.
300. Crow, J., et al. Exosomes as mediators of platinum resistance in ovarian cancer. *Oncotarget*. 2017; **8**(7): p. 11917–11936.
301. Santos, J. C., et al. Exosome-mediated breast cancer chemoresistance via miR-155 transfer. *Sci Rep*. 2018; **8**(1): p. 829.
302. Jing, C., et al. Exosome-mediated gefitinib resistance in lung cancer HCC827 cells via delivery of miR-21. *Oncol Lett*. 2018; **15**(6): p. 9811–9817.
303. Fornari, F., et al. In hepatocellular carcinoma miR-221 modulates sorafenib resistance through inhibition of caspase-3-mediated apoptosis. *Clin Cancer Res*. 2017; **23**(14): p. 3953–3965.
304. Hicklin, D. J., and L. M. Ellis. Role of the vascular endothelial growth factor pathway in tumor growth and angiogenesis. *J Clin Oncol*. 2005; **23**(5): p. 1011–1027.
305. Folkman, J. Tumor angiogenesis: therapeutic implications. *N Engl J Med*. 1971; **285**(21): p. 1182–1186.
306. Liang, H., and M. Wang. Prospect of immunotherapy combined with anti-angiogenic agents in patients with advanced non-small cell lung cancer. *Cancer Manag Res*. 2019; **11**: p. 7707–7719.
307. Gacche, R. N. Compensatory angiogenesis and tumor refractoriness. *Oncogenesis*. 2015; **4**(6): p. e153.
308. Jayson, G. C., et al. Antiangiogenic therapy in oncology: current status and future directions. *Lancet*. 2016; **388**(10043): p. 518–529.
309. Gacche, R. N., and Y. G. Assaraf. Redundant angiogenic signaling and tumor drug resistance. *Drug Resist Updat*. 2018; **36**: p. 47–76.
310. Ribatti, D. The chick embryo chorioallantoic membrane (CAM). A multifaceted experimental model. *Mech Dev*. 2016; **141**: p. 70–77.
311. Lohela, M., et al. VEGFs and receptors involved in angiogenesis versus lymphangiogenesis. *Curr Opin Cell Biol*. 2009; **21**(2): p. 154–165.
312. Ellis, L. M., and D. J. Hicklin. Pathways mediating resistance to vascular endothelial growth factor-targeted therapy. *Clin Cancer Res*. 2008; **14**(20): p. 6371–6375.
313. Crawford, Y., et al. PDGF-C mediates the angiogenic and tumorigenic properties of fibroblasts associated with tumors refractory to anti-VEGF treatment. *Cancer Cell*. 2009; **15**(1): p. 21–34.

314. Vasudev, N. S., and A. R. Reynolds. Anti-angiogenic therapy for cancer: current progress, unresolved questions and future directions. *Angiogenesis*. 2014; **17**(3): p. 471–494.
315. Bergers, G., and S. Song. The role of pericytes in blood-vessel formation and maintenance. *Neuro Oncol*. 2005; **7**(4): p. 452–464.
316. Aldea, M., et al. Overcoming resistance to tumor-targeted and immune-targeted therapies. *Cancer Discov*. 2021; **11**(4): p. 874–899.
317. Koyama, S., et al. Adaptive resistance to therapeutic PD-1 blockade is associated with upregulation of alternative immune checkpoints. *Nat Commun*. 2016; **7**: p. 10501.
318. Saleh, R., and E. Elkord. Treg-mediated acquired resistance to immune checkpoint inhibitors. *Cancer Lett*. 2019; **457**: p. 168–179.
319. Jenkins, L., et al. Cancer-associated fibroblasts suppress CD8+ T-cell infiltration and confer resistance to immune-checkpoint blockade. *Cancer Res*. 2022; **82**(16): p. 2904–2917.
320. Deng, L., et al. Irradiation and anti-PD-L1 treatment synergistically promote antitumor immunity in mice. *J Clin Invest*. 2014; **124**(2): p. 687–695.
321. Twyman-Saint Victor, C., et al. Radiation and dual checkpoint blockade activate non-redundant immune mechanisms in cancer. *Nature*. 2015; **520**(7547): p. 373–377.
322. Galluzzi, L., et al. Immunostimulation with chemotherapy in the era of immune checkpoint inhibitors. *Nat Rev Clin Oncol*. 2020; **17**(12): p. 725–741.
323. Tang, T., et al. Advantages of targeting the tumor immune microenvironment over blocking immune checkpoint in cancer immunotherapy. *Signal Transduct Target Ther*. 2021; **6**(1): p. 72.
324. Wallin, J. J., et al. Atezolizumab in combination with bevacizumab enhances antigen-specific T-cell migration in metastatic renal cell carcinoma. *Nat Commun*. 2016; **7**: p. 12624.
325. Casak, S. J., et al. FDA approval summary: atezolizumab plus bevacizumab for the treatment of patients with advanced unresectable or metastatic hepatocellular carcinoma. *Clin Cancer Res*. 2021; **27**(7): p. 1836–1841.
326. Harrington, K., et al. Optimizing oncolytic virotherapy in cancer treatment. *Nat Rev Drug Discov*. 2019; **18**(9): p. 689–706.
327. Martin, N. T., and J. C. Bell. Oncolytic virus combination therapy: killing one bird with two stones. *Mol Ther*. 2018; **26**(6): p. 1414–1422.
328. Bourgeois-Daigneault, M. C., et al. Neoadjuvant oncolytic virotherapy before surgery sensitizes triple-negative breast cancer to immune checkpoint therapy. *Sci Transl Med*. 2018; **10**(422): p. eaao1641.
329. Li, R., et al. Using oncolytic viruses to ignite the tumour immune microenvironment in bladder cancer. *Nat Rev Urol*. 2021; **18**(9): p. 543–555.
330. Ribas, A., et al. Oncolytic virotherapy promotes intratumoral T cell infiltration and improves anti-PD-1 immunotherapy. *Cell*. 2017; **170**(6): p. 1109–1119. e10.
331. Zhang, X., et al. Dual gate-controlled therapeutics for overcoming bacterium-induced drug resistance and potentiating cancer immunotherapy. *Angew Chem Int Ed Engl*. 2021; **60**(25): p. 14013–14021.
332. Arthur, J. C., et al. Intestinal inflammation targets cancer-inducing activity of the microbiota. *Science*. 2012; **338**(6103): p. 120–123.
333. Dejea, C. M., et al. Patients with familial adenomatous polyposis harbor colonic biofilms containing tumorigenic bacteria. *Science*. 2018; **359**(6375): p. 592–597.

334. Zheng, D. W., et al. Phage-guided modulation of the gut microbiota of mouse models of colorectal cancer augments their responses to chemotherapy. *Nat Biomed Eng*. 2019; **3**(9): p. 717–728.
335. Han, H., et al. Metformin-induced stromal depletion to enhance the penetration of gemcitabine-loaded magnetic nanoparticles for pancreatic cancer targeted therapy. *J Am Chem Soc*. 2020; **142**(10): p. 4944–4954.
336. Geller, L. T., et al. Potential role of intratumor bacteria in mediating tumor resistance to the chemotherapeutic drug gemcitabine. *Science*. 2017; **357**(6356): p. 1156–1160.
337. Dong, X., et al. Bioinorganic hybrid bacteriophage for modulation of intestinal microbiota to remodel tumor-immune microenvironment against colorectal cancer. *Sci Adv*. 2020; **6**(20): p. eaba1590.
338. Iida, N., et al. Commensal bacteria control cancer response to therapy by modulating the tumor microenvironment. *Science*. 2013; **342**(6161): p. 967–970.
339. Pernigoni, N., et al. Commensal bacteria promote endocrine resistance in prostate cancer through androgen biosynthesis. *Science*. 2021; **374**(6564): p. 216–224.
340. Zhu, Y. X., et al. Nanomedicines for combating multidrug resistance of cancer. *Wiley Interdiscip Rev Nanomed Nanobiotechnol*. 2021; **13**(5): p. e1715.
341. Xiong, M. H., et al. Differential anticancer drug delivery with a nanogel sensitive to bacteria-accumulated tumor artificial environment. *ACS Nano*. 2013; **7**(12): p. 10636–10645.
342. Ghandi, M., et al. Next-generation characterization of the cancer cell line encyclopedia. *Nature*. 2019; **569**(7757): p. 503–508.
343. Tsherniak, A., et al. Defining a cancer dependency map. *Cell*. 2017; **170**(3): p. 564–576. e16.
344. Li, H., et al. The landscape of cancer cell line metabolism. *Nat Med*. 2019; **25**(5): p. 850–860.
345. Yang, W., et al. Genomics of drug sensitivity in cancer (GDSC): a resource for therapeutic biomarker discovery in cancer cells. *Nucleic Acids Res*. 2013; **41**(Database issue): p. D955–961.
346. Forbes, S. A., et al. COSMIC: mining complete cancer genomes in the Catalogue of Somatic Mutations in Cancer. *Nucleic Acids Res*. 2011; **39**(Database issue): p. D945–950.
347. Iorio, F., et al. A landscape of pharmacogenomic interactions in cancer. *Cell*. 2016; **166**(3): p. 740–754.
348. Corsello, S. M., et al. Discovering the anti-cancer potential of non-oncology drugs by systematic viability profiling. *Nat Cancer*. 2020; **1**(2): p. 235–248.
349. Basu, A., et al. An interactive resource to identify cancer genetic and lineage dependencies targeted by small molecules. *Cell*. 2013; **154**(5): p. 1151–1161.
350. Seashore-Ludlow, B., et al. Harnessing connectivity in a large-scale small-molecule sensitivity dataset. *Cancer Discov*. 2015; **5**(11): p. 1210–1223.
351. Liu, X., et al. DRMref: comprehensive reference map of drug resistance mechanisms in human cancer. *Nucleic Acids Res*. 2024; **52**(D1): p. D1253–d1264.
352. Sun, X., et al. DRESIS: the first comprehensive landscape of drug resistance information. *Nucleic Acids Res*. 2023; **51**(D1): p. D1263–d1275.
353. Liu, Z., et al. CTR-DB, an omnibus for patient-derived gene expression signatures correlated with cancer drug response. *Nucleic Acids Res*. 2022; **50**(D1): p. D1184–d1199.

354. Wang, Y. Y., et al. CeDR Atlas: a knowledgebase of cellular drug response. *Nucleic Acids Res*. 2022; **50**(D1): p. D1164–d1171.
355. Yingtaweesittikul, H., et al. CREAMMIST: an integrative probabilistic database for cancer drug response prediction. *Nucleic Acids Res*. 2023; **51**(D1): p. D1242–d1248.
356. Kumar, R., et al. CancerDR: cancer drug resistance database. *Sci Rep*. 2013; **3**: p. 1445.
357. Wang, Y. Y., et al. GEAR: a database of genomic elements associated with drug resistance. *Sci Rep*. 2017; **7**: p. 44085.
358. Hu, R., et al. KinaseMD: kinase mutations and drug response database. *Nucleic Acids Res*. 2021; **49**(D1): p. D552–d561.
359. Cui, H., et al. Inferences of individual drug response-related long non-coding RNAs based on integrating multi-omics data in breast cancer. *Mol Ther Nucleic Acids*. 2020; **20**: p. 128–139.
360. Qi, W., and Q. Zhang. Gene's co-expression network and experimental validation of molecular markers associated with the drug resistance of gastric cancer. *Biomark Med*. 2020; **14**(9): p. 761–773.
361. Huang, P., et al. lncRNA profile study reveals the mRNAs and lncRNAs associated with docetaxel resistance in breast cancer cells. *Sci Rep*. 2018; **8**(1): p. 17970.
362. Liu, H., et al. Drug resistance-related competing interactions of lncRNA and mRNA across 19 cancer types. *Mol Ther Nucleic Acids*. 2019; **16**: p. 442–451.
363. Li, Z., et al. PPP2R2B downregulation is associated with immune evasion and predicts poor clinical outcomes in triple-negative breast cancer. *Cancer Cell Int*. 2021; **21**(1): p. 13.
364. Li, Y. K., et al. Genetic co-expression networks contribute to creating predictive model and exploring novel biomarkers for the prognosis of breast cancer. *Sci Rep*. 2021; **11**(1): p. 7268.
365. Zhang, J., et al. Differential regulatory network-based quantification and prioritization of key genes underlying cancer drug resistance based on time-course RNA-seq data. *PLoS Comput Biol*. 2019; **15**(11): p. e1007435.
366. Frejno, M., et al. Pharmacoproteomic characterisation of human colon and rectal cancer. *Mol Syst Biol*. 2017; **13**(11): p. 951.
367. Yu, L., et al. Prediction of drug response in multilayer networks based on fusion of multiomics data. *Methods*. 2021; **192**: p. 85–92.
368. Xu, Y., et al. Identifying subpathway signatures for individualized anticancer drug response by integrating multi-omics data. *J Transl Med*. 2019; **17**(1): p. 255.
369. Malik, V., Y. Kalakoti, and D. Sundar. Deep learning assisted multi-omics integration for survival and drug-response prediction in breast cancer. *BMC Genomics*. 2021; **22**(1): p. 214.
370. Frejno, M., et al. Proteome activity landscapes of tumor cell lines determine drug responses. *Nat Commun*. 2020; **11**(1): p. 3639.
371. Liu, R., G. Zhang, and Z. Yang. Towards rapid prediction of drug-resistant cancer cell phenotypes: single cell mass spectrometry combined with machine learning. *Chem Commun (Camb)*. 2019; **55**(5): p. 616–619.
372. Zhu, Y., et al. Ensemble transfer learning for the prediction of anti-cancer drug response. *Sci Rep*. 2020; **10**(1): p. 18040.
373. Kong, J., et al. Network-based machine learning in colorectal and bladder organoid models predicts anti-cancer drug efficacy in patients. *Nat Commun*. 2020; **11**(1): p. 5485.

374. Sakellaropoulos, T., et al. A deep learning framework for predicting response to therapy in cancer. *Cell Rep*. 2019; **29**(11): p. 3367–3373. e4.
375. Choi, J., S. Park, and J. Ahn. RefDNN: a reference drug based neural network for more accurate prediction of anticancer drug resistance. *Sci Rep*. 2020; **10**(1): p. 1861.
376. Gu, C., et al. Current status and applications of genome-scale metabolic models. *Genome Biol*. 2019; **20**(1): p. 121.
377. Ryu, J. Y., H. U. Kim, and S. Y. Lee. Framework and resource for more than 11,000 gene-transcript-protein-reaction associations in human metabolism. *Proc Natl Acad Sci U S A*. 2017; **114**(45): p. E9740–e9749.
378. Brunk, E., et al. Recon3D enables a three-dimensional view of gene variation in human metabolism. *Nat Biotechnol*. 2018; **36**(3): p. 272–281.
379. Robinson, J. L., et al. An atlas of human metabolism. *Sci Signal*. 2020; **13**(624): p. eaaz1482.
380. Cook, D. J., and J. Nielsen. Genome-scale metabolic models applied to human health and disease. *Wiley Interdiscip Rev Syst Biol Med*. 2017; **9**(6).
381. Ryu, J. Y., H. U. Kim, and S. Y. Lee. Reconstruction of genome-scale human metabolic models using omics data. *Integr Biol* (*Camb*). 2015; **7**(8): p. 859–868.
382. Opdam, S., et al. A systematic evaluation of methods for tailoring genome-scale metabolic models. *Cell Syst*. 2017; **4**(3): p. 318–329. e6.
383. Lewis, J. E., et al. Personalized genome-scale metabolic models identify targets of redox metabolism in radiation-resistant tumors. *Cell Syst*. 2021; **12**(1): p. 68–81. e11.
384. Lewis, J. E., and M. L. Kemp. Integration of machine learning and genome-scale metabolic modeling identifies multi-omics biomarkers for radiation resistance. *Nat Commun*. 2021; **12**(1): p. 2700.
385. Rodon Ahnert, J., et al. What it takes to improve a first-generation inhibitor to a second- or third-generation small molecule. *Am Soc Clin Oncol Educ Book*. 2019; **39**: p. 196–205.
386. Blay, J. Y., and M. von Mehren. Nilotinib: a novel, selective tyrosine kinase inhibitor. *Semin Oncol*. 2011; **38 Suppl 1**(0 1): p. S3–9.
387. Laurini, E., et al. Through the open door: preferential binding of dasatinib to the active form of BCR-ABL unveiled by in silico experiments. *Mol Oncol*. 2013; **7**(5): p. 968–975.
388. Bahcall, M., et al. Acquired METD1228V mutation and resistance to MET inhibition in lung cancer. *Cancer Discov*. 2016; **6**(12): p. 1334–1341.
389. Lynch, T. J., et al. Activating mutations in the epidermal growth factor receptor underlying responsiveness of non-small-cell lung cancer to gefitinib. *N Engl J Med*. 2004; **350**(21): p. 2129–2139.
390. Yun, C. H., et al. The T790M mutation in EGFR kinase causes drug resistance by increasing the affinity for ATP. *Proc Natl Acad Sci U S A*. 2008; **105**(6): p. 2070–2075.
391. Cross, D. A., et al. AZD9291, an irreversible EGFR TKI, overcomes T790M-mediated resistance to EGFR inhibitors in lung cancer. *Cancer Discov*. 2014; **4**(9): p. 1046–1061.
392. Deshaies, R. J. Protein degradation: prime time for PROTACs. *Nat Chem Biol*. 2015; **11**(9): p. 634–635.
393. An, S., and L. Fu. Small-molecule PROTACs: an emerging and promising approach for the development of targeted therapy drugs. *EBioMedicine*. 2018; **36**: p. 553–562.

394. Burslem, G. M., et al. The advantages of targeted protein degradation over inhibition: an RTK case study. *Cell Chem Biol.* 2018; **25**(1): p. 67–77. e3.
395. Sun, Y., et al. PROTAC-induced BTK degradation as a novel therapy for mutated BTK C481S induced ibrutinib-resistant B-cell malignancies. *Cell Res.* 2018. **28**(7): p. 779–781.
396. Burslem, G. M., et al. Targeting BCR-ABL1 in chronic myeloid leukemia by PROTAC-mediated targeted protein degradation. *Cancer Res.* 2019; **79**(18): p. 4744–4753.
397. Zhang, L., et al. Acquired resistance to BET-PROTACs (proteolysis-targeting chimeras) caused by genomic alterations in core components of E3 ligase complexes. *Mol Cancer Ther.* 2019; **18**(7): p. 1302–1311.
398. Schoepfer, J., et al. Discovery of Asciminib (ABL001), an allosteric inhibitor of the tyrosine Kinase activity of BCR-ABL1. *J Med Chem.* 2018; **61**(18): p. 8120–8135.
399. Wilson, L. J., et al. New perspectives, opportunities, and challenges in exploring the human protein Kinome. *Cancer Res.* 2018; **78**(1): p. 15–29.
400. Jacus, M. O., et al. Pharmacokinetic properties of anticancer agents for the treatment of central nervous system tumors: update of the literature. *Clin Pharmacokinet.* 2016; **55**(3): p. 297–311.
401. Ni, W., W. Chen, and Y. Lu. Emerging findings into molecular mechanism of brain metastasis. *Cancer Med.* 2018; **7**(8): p. 3820–3833.
402. Wang, Q., et al. Evaluation of the MDR-MDCK cell line as a permeability screen for the blood-brain barrier. *Int J Pharm.* 2005; **288**(2): p. 349–359.
403. Johnson, T. W., et al. Discovery of (10R)-7-amino-12-fluoro-2,10,16-trimethyl-15-oxo-10,15,16,17-tetrahydro-2H-8,4-(metheno)pyrazolo[4,3-h][2,5,11]-benzoxadiazacyclotetradecine-3-carbonitrile (PF-06463922), a macrocyclic inhibitor of anaplastic lymphoma kinase (ALK) and c-ros oncogene 1 (ROS1) with preclinical brain exposure and broad-spectrum potency against ALK-resistant mutations. *J Med Chem.* 2014; **57**(11): p. 4720–4744.
404. Yang, Z., et al. AZD3759, a BBB-penetrating EGFR inhibitor for the treatment of EGFR mutant NSCLC with CNS metastases. *Sci Transl Med.* 2016; **8**(368): p. 368ra172.
405. Ballard, P., et al. Preclinical comparison of osimertinib with other EGFR-TKIs in EGFR-Mutant NSCLC brain metastases models, and early evidence of clinical brain metastases activity. *Clin Cancer Res.* 2016; **22**(20): p. 5130–5140.
406. Coombes, R. Cancer drug resistance needs urgent attention, says research chief. *BMJ.* 2019; **365**: p. l1934.
407. Day, D., and L. L. Siu. Approaches to modernize the combination drug development paradigm. *Genome Med.* 2016; **8**(1): p. 115.
408. Celebi, R., et al. In-silico prediction of synergistic anti-cancer drug combinations using multi-omics data. *Sci Rep.* 2019; **9**(1): p. 8949.
409. Mir, S. A., et al. Role of nanotechnology in overcoming the multidrug resistance in cancer therapy: A review. *Molecules.* 2022; **27**(19): p. 6608.
410. Halder, J., et al. Nanotherapeutics approaches to overcome P-glycoprotein-mediated multi-drug resistance in cancer. *Nanomedicine.* 2022; **40**: p. 102494.
411. Pucci, C., C. Martinelli, and G. Ciofani. Innovative approaches for cancer treatment: current perspectives and new challenges. *Ecancermedicalscience.* 2019; **13**: p. 961.

412. Saraswathy, M., and S. Gong. Different strategies to overcome multidrug resistance in cancer. *Biotechnol Adv*. 2013; **31**(8): p. 1397–1407.
413. Zafar, A., et al. Revolutionizing cancer care strategies: immunotherapy, gene therapy, and molecular targeted therapy. *Mol Biol Rep*. 2024; **51**(1): p. 219.
414. Kong, X., et al. Advances of medical nanorobots for future cancer treatments. *J Hematol Oncol*. 2023; **16**(1): p. 74.
415. Elmore, L. W., et al. Blueprint for cancer research: critical gaps and opportunities. *CA Cancer J Clin*. 2021; **71**(2): p. 107–139.

CHAPTER 14

Computational Insights in Metastasis: Exploring the Omics Revolution

Metastasis is responsible for the majority of cancer-related deaths. Understanding the mechanisms driving metastasis is crucial for early diagnosis and the effective clinical treatment of metastatic cancer. Recent advances in multi-omics sequencing technologies, combined with computational approaches, offer significant opportunities to unravel the complexities of cancer metastasis. This review provides an overview of current research on metastatic mechanisms, highlighting the role of multi-omics approaches in uncovering new insights, the development of metastasis process models, and the application of artificial intelligence in addressing metastatic challenges. By integrating genomic, transcriptomic, epigenetic, proteomic, metabolomic, microbiomic, immunomic, as well as single-cell and spatial data, through advanced computational approaches and models, researchers are gaining a more comprehensive understanding of the molecular drivers of metastasis. Continuous research is expected to further elucidate the underlying mechanisms of metastasis, facilitating more precise prognostic tools and therapeutic strategies in the clinical setting.

14.1 Introduction to Challenges in Metastasis Research

Metastasis, the ultimate manifestation of cancer, is responsible for the vast majority of mortality associated with malignant neoplasms [1]. It involves a complex series of biological events, including the intravasation of primary tumor cells into blood vessels or lymph nodes, hematogenous or lymphatic dissemination, extravasation of circulating tumor cells (CTCs) into secondary tissues, and their subsequent progression through stages from dormancy and micrometastasis to macro-metastasis (Fig. 14.1A) [2]. This metastatic cascade is highly inefficient [3]. The success of each step depends on the dynamic, cell-intrinsic reprogramming of cancer cells, which enables them to adapt to the hostile microenvironment through complex interactions and modulations [4, 5]. A deeper understanding of the cellular, molecular, and genetic mechanisms involved in these processes will be crucial for improving therapeutic outcomes for cancer patients [6].

Large-scale, multi-omics sequencing data, coupled with the powerful capabilities of artificial intelligence, have greatly enhanced the ability to unravel the mysteries of cancer metastasis [7–9]. The "7W1H quests" (Fig. 14.1B) outlined below address key concerns regarding cancer metastasis. This review summarizes the current advancements and highlights potential future research within these areas, leveraging computational technologies.

1. "Which" cells possess the ability to initiate metastatic seeding? Cancer cells that initiate and sustain metastasis are referred to as metastasis-initiating cells (MICs) [10]. These cells make up only a small fraction of primary tumor cells [11] but exhibit remarkable adaptability, allowing them to shape their metastatic phenotype. For example, MICs enhance their mobility, invasiveness, and resistance to apoptotic stimuli through dynamic and reversible epithelial-mesenchymal transition (EMT) [12]. Additionally, MICs downregulate MHC-I expression to evade immune detection [13] and uptake lactate and pyruvate to counteract oxidative stress during metastasis [14]. These MICs act as the "seeds" of metastasis, akin to decathlon champions, proficient in all ten events [15]. Identifying the molecular determinants and potential vulnerabilities of

MICs, in comparison to other primary tumor cells and vulnerable CTCs (Fig. 14.1C), could provide valuable insights for developing therapies aimed at preventing and treating metastatic spread.

2. "When" do the MICs start the seeding process? The timing of cancer metastasis is crucial for effective treatment and prevention. However, metastasis often occurs prior to the detection of primary tumors [16], largely due to the limitations of current imaging techniques, which can only detect relatively large tumors (e.g., $>1\,cm^3$). Identifying the genetic drivers in primary tumors and CTCs that contribute to early or late metastatic seeding (Fig. 14.1C) [17] could help stratify patients with aggressive cancer, enabling earlier intervention and more targeted therapeutic strategies.
3. "How" do the MICs leave the primary tumors and reach a secondary tissue? MICs can follow diverse evolutionary pathways to metastasis (Fig. 14.1A) [18]. One pathway involves monoclonal seeding, where a single disseminated tumor cell clone initiates metastatic outgrowth. Alternatively, polyclonal seeding occurs when multiple genetically distinct cancer cell clones give rise to metastases [19]. Given the therapeutic implications of targeting all lethal clones, it is crucial to determine whether monoclonal or polyclonal seeding (Fig. 14.1C) is involved for precise treatment strategies. Furthermore, the lymphatic system plays a pivotal role in some cases of metastasis, known as lymphatic metastasis, while in other cases, MICs spread through the bloodstream, referred to as hematogenous metastasis (Fig. 14.1A) [20, 21]. Distinct mechanisms drive these two evolutionary paths, and understanding them (Fig. 14.1C) may enable the development of more targeted therapies [22, 23].
4. "Where" do the MICs seed in the body? MICs typically seed in specific tissues that provide a particularly hospitable environment, often referred to as seeding in "fertile soil" [15]. Within primary tumors, MICs can induce the formation of microenvironments in distant organs that support cancer cell growth through paracrine signaling. The altered microenvironment is known as the pre-metastasis niche [24]. The nonrandom selection of metastatic organs is termed organotropism [25]. Understanding the molecular mechanisms underlying organotropic metastasis (Fig 14.1C) is crucial for the development of biomarker-based prediction and prognosis, as well as the creation of innovative therapeutic strategies to ultimately improve patient outcomes [26].

5. “Why” do the metastatic tumor cells or masses enter or exit a dormant state? After extravasation into a secondary tissue, some MICs survive in a dormant state through reversible cell cycle arrest, allowing them to evade immune surveillance. In this state, they can alter or unleash functions needed to initiate and expand a lesion when favorable microenvironmental conditions arise [27, 28]. Dormant MICs also evade antiproliferative treatments by remaining inactive [29]. When reactivated, these MICs often accumulate mutations, evolving into multi-therapy-resistant metastasis [30]. In addition to dormant cancer cells, another form of dormancy is tumor mass dormancy, where a balance exists between the proliferation of cancer cells and their death [31]. Understanding dormancy mechanisms (Fig. 14.1C) is crucial, as this period presents a therapeutic opportunity to shift the balance from aggressive to non-aggressive tumor phenotypes.
6. “Whether” micrometastasis has formed in the secondary tissue? Micrometastasis refers to a small number of cancer cells present in the secondary tissue, typically beyond the detection limits of conventional imaging techniques but identifiable through histological or molecular biology methods [32]. Early intervention and targeted eradication of micrometastasis can significantly enhance patient survival [33]. This depends on the early detection of micrometastasis (Fig. 14.1C) and understanding its molecular mechanisms.
7. “What” is the metastatic origin for patients with cancers of unknown primary (CUP)? CUP is defined as a carcinoma or undifferentiated neoplasm for which standard diagnostic methods fail to identify the primary tumor responsible for metastatic seeding [34, 35]. CUP accounts for 2%-5% of new cancers and is associated with a high mortality rate [36]. Its management and treatment largely depend on the accurate determination of the primary site [35], as modern therapeutics are predominantly specific to the primary tumor [37]. Therefore, an approach based on the distinctive features of CUP deserves further study to achieve high accuracy (Fig. 14.1C).
8. “Who” will benefit from treatments targeting metastasis? Treating metastasis remains highly challenging, as the vast majority of patients still die within five years of diagnosis [38]. Identifying patients who are most likely to benefit from current therapies with molecular biomarkers can maximize treatment efficacy while minimizing harm to individuals and communities. For instance, a subset of patients with oligometastasis, characterized by a limited number of metastatic lesions and indolent tumor

growth, can achieve a cure through metastasis-directed localized therapies [39]. Understanding the mechanisms underlying treatment effectiveness and their clinical application is crucial (Fig. 14.1C).

These "7W1H quests" mentioned above are crucial for metastatic clinical diagnosis and treatment. The research related to these questions is summarized below, along with proposed strategies for improvement.

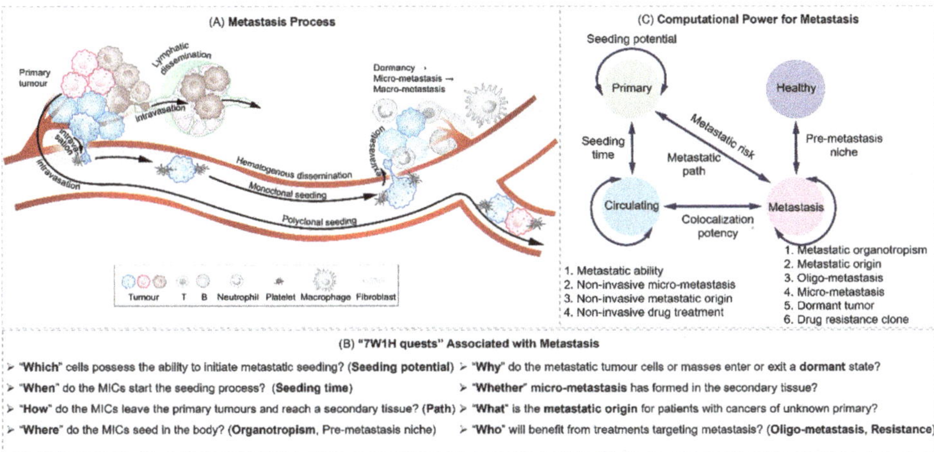

Fig. 14.1 Introduction to challenges in metastasis research. (A) Schematic of the metastasis process. (B) The "7W1H" questions associated with metastasis. (C) Comparison of metastasis-related samples to explore the challenges of metastasis.

14.2 Multi-Omics Approaches to Unravel Metastatic Mechanisms

Recent advancements in high-throughput technologies have enabled comprehensive multi-omics profiling of metastasis-relevant data from primary tumors, metastatic tissues, and liquid biopsies [40–45]. These data include (ct)DNA sequencing (circulating tumor DNA or tumor tissue DNA), RNA sequencing, sc and spatial sequencing, epigenetic sequencing, metabolic mass spectrometry (MS), T-cell receptor (TCR) and B-cell receptor (BCR) sequencing, proteomic MS, microbiome sequencing, and more (Fig. 14.2). By analyzing these multi-omics profiles, researchers can uncover key metastatic mechanisms, such as genetic selection, transcriptomic reprogramming, spatio-cellular evolution, epigenetic rewriting, metabolic adaptation, immune editing, protein diversity, and microbiome dynamics associated with metastasis. The computational approaches for analyzing these data, along with representative results, are outlined in detail below.

1. Genetic selection for metastasis based on (ct)DNA sequencing analysis. The MICs acquire additional genomic alterations to enhance their seeding potential. Identifying genetic mutations selected during metastasis is crucial for developing more effective and personalized therapies [46]. After preprocessing with alignment and co-cleaning workflows using BWA [47] and GATK, DNA sequencing profiles enable the detection of mutations associated with metastasis. These include metastasis-related somatic single nucleotide variants (mSNVs) and indels (mIndels) detected with MuTect2 [48] and VarScan2 [49], copy number variants (mCNVs) analyzed by CNVkit [50], structural variants (mSVs) identified by DELLY [51] and LUMPY [52], mutation signature (mMutSig) delineated using DeconstructSigs [53], driver genes (mDriver) inferred by Mutsig2CV [54], neuroantigens (mNeoantigen) predicted with NetMHCpan [55], and extrachromosomal DNA (mecDNA) derived by AmpliconArchitect [56] and AmpliconClassifier [57] (Fig. 14.2).
   These analyses frequently reveal an increased tumor mutation burden (TMB) [8, 58], low Intratumoral heterogeneity (ITH) [43, 46, 59], elevated chromosomal and microsatellite instability (MSI) [8], significant homologous recombination deficiency (HRD) [46], and enriched loss of heterozygosity [60, 61] in metastases. Specifically, as illustrated in Fig. 14.3, alterations in the tumor suppressor gene TP53 demonstrate metastatic enrichment across multiple cancer types [8, 40, 46, 62]. The androgen receptor (AR), one of the most frequently mutated oncogenes in prostate cancer metastasis, is likely selected in response to androgen-deprivation therapy, facilitating bypass of the AR signaling pathway [62–64]. Focal copy number variations affecting the MYC and PTEN genes, which are observed in a minor subset of primary tumor cells, are present in all CTCs, suggesting a strong selective advantage for metastasis [65]. Additionally, genetic regression of KRAS and SMAD4 mutations is noted in patients with oligometastatic colorectal cancer [66, 67]. These analyses significantly enhance our understanding of the metastatic process from a genomic perspective.
2. Transcriptomic reprogramming for metastasis based on RNA sequencing analysis. The MICs undergo transcriptomic reprogramming to adapt to the distinct stresses during the metastatic cascade and respond to changes in the tumor environment [1, 10, 68]. Understanding these adaptive

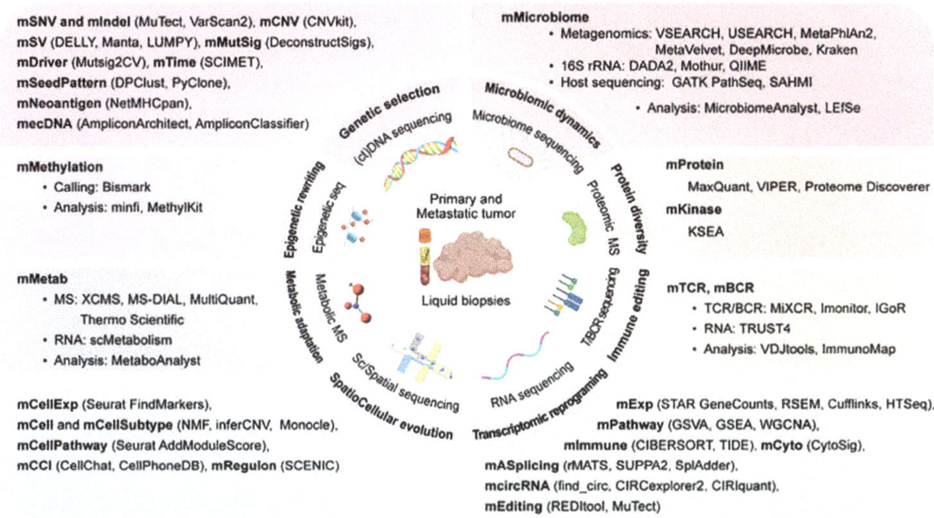

Fig. 14.2 Multi-omics approaches to unravel metastatic mechanisms. Utilizing multi-omics sequencing of primary tumors, metastasis, and liquid biopsies to reveal the cellular and molecular changes during metastasis through diverse computational methods. In this figure, "m" represents metastasis-related features. MS: mass spectrometry.

processes is crucial for evaluating treatment efficacy and developing novel anti-metastatic therapies [69, 70]. After preprocessing with alignment workflow using STAR [71], RNA sequencing profiles allow for the identification of transcriptomic changes during metastasis. These changes include metastasis-related gene expression (mExp) quantified using tools such as STAR GeneCounts [71], HTSeq [72], RSEM [73], and Cufflinks [74], biological pathways (mPathway) analyzed with GSVA [75] and GSEA [76], immune infiltration (mImmune) estimated by CIBERSORT [77] and TIDE [78], cytokine activity modeled by CytoSig (mCyto) [79], alternative splicing detected using rMATS [80], SUPPA2 [81] and SplAdder [82], circular RNAs identified with tools like find_circ [83], CIRCexplorer2 [84], and CIRIquant [85], and RNA editing assessed using mutation detection tools combined with stringent filtering criteria [48, 86] (Fig. 14.2).

As shown in Fig. 14.3, transcriptomic dysregulation occurs at various stages of metastasis. Several circRNAs and long noncoding RNAs have been identified for their potential regulatory roles in the metastasis of different cancers [87–90]. Additionally, overexpression of the macroH2A variant has been shown to induce dormancy in disseminated cancer cells, thereby suppressing metastasis [91]. Differential splicing events in pancreatic ductal adenocarcinoma metastases are enriched with motifs closely resembling the RBFOX2 motif, a potent metastatic suppressor [92]. Furthermore, RNA editing events catalyzed by ADAR or APOBEC3A modulate transcriptomic diversity in regions of primary tumors and can persist within their metastatic sites [93]. Notably, one specific editing event utilizes endoplasmic reticulum stress to promote metastasis in colorectal cancer [94]. These analyses provide valuable insights into the metastatic process from a transcriptomic perspective.

3. Spatio-cellular evolution for metastasis based on sc/spatial sequencing analysis. Recent advances in sc and spatial sequencing technologies have enabled the interrogation of metastasis at unprecedented resolution [95]. With the processed feature-barcode matrix, tools such as Seurat [96] and Scanpy [97] are commonly used to analyze spatial and cellular evolution during metastasis. These include metastasis-related cell-specific gene expression (mCellExp) and biological pathways (mCellPathway) [98], cell–cell interactions associated with metastasis, identified using CellChat [99] and CellPhoneDB [100], cell-specific metastatic regulatory networks (mRegulon) inferred using SCENIC [101], and metastasis-related cell types or subtypes (mCell, mCellSubtype) through clustering [102] or trajectory inference methods [103, 104] (Fig. 14.2).
Advanced technologies are delivering deeper insights into ITH and the role of the local tumor microenvironment in metastasis. For instance, as shown in Fig. 14.3, a subset of cancer cells with high EMT potential, located in close proximity to fibroblasts and endothelial cells, drives peritoneal metastasis in gastric cancer [42]. The stromal cells, as well as immune cells, have demonstrated their essential roles in shaping the metastatic landscape [98, 105]. Furthermore, some traditional views of metastatic mechanisms are being challenged. Notably, metastatic potential is found to peak in a rare, late-hybrid EMT state, rather in a simple binary switch [106]. On the molecular level, KRT17 shows a significant anticorrelation with metastatic capacity, suggesting an unanticipated suppressive role in metastasis [107]. Additionally, ICAM1 is notably upregulated, showing a 60- to 200-fold increase in sc libraries of lung metastases. Depletion of ICAM1 abrogates lung colonization of triple-negative breast cancer cells by inhibiting homotypic tumor cell-tumor cell cluster formation [108]. Further analyses reveal that AXL is highly expressed in dormant myeloma cells [109]. Moreover, PHLDA2 transcript levels are at least twofold

higher in micrometastases than in primary tumors, where its expression is rare [41]. These analyses offer a critical understanding of the metastatic process from a sc and spatial perspective.

4. **Epigenetic rewriting for metastasis based on epigenetic sequencing analysis.** Epigenetic mechanisms, including DNA methylation, play a crucial role in mediating the gene expression of MICs toward metastasis [1]. Bisulfite sequencing can identify metastasis-related methylation sites (mMethylation) that are instrumental in the metastatic process through tools like Bismark [110] (Fig. 14.2).

   Based on these analyses, several epigenetic mechanisms driving metastasis have been uncovered, as shown in Fig. 14.3. For instance, binding sites for stemness- and proliferation-associated transcription factors are specifically hypomethylated in CTC clusters, correlating with increased metastatic potential [111]. Drugs that disrupt these CTC clusters, transforming them into single cells, induce DNA methylation remodeling, thereby suppressing metastasis. Additionally, hypomethylation of metastasizing tumor cells leads to metastatic latency, while methylated tumor cells exhibit low activity and proliferate intravascularly [112]. These analyses provide valuable insights into the regulatory mechanisms underlying the metastatic process from an epigenetic perspective.

5. **Metabolic adaptation for metastasis based on metabolic mass spectrometry (MS) analysis.** The MICs can adapt their metabolism to environmental stresses to survive in circulation and distant organs [1]. The importance of metabolism adaptation during metastatic dissemination guides the analysis of metastasis-related metabolites (mMetab) using tools such as XCMS [113] and MS-DIAL [114] (Fig 14.2).

   These analyses can reveal metastatic mechanisms and suggest potential therapeutic strategies to combat metastasis. For example, as shown in Fig. 14.3, multiple lipids exhibit differential abundance in response to metastatic adaptation [115, 116]. Among them, fatty acids play a critical role in sustaining micrometastases [117] and promoting cellular metastatic properties [87]. Additionally, oxidative phosphorylation is upregulated in lung micrometastases of breast cancer [41] and is enriched in melanoma brain metastases compared to extracranial metastases [118]. Furthermore, dysregulation of nicotinate and nicotinamide metabolism has been observed in lymph node metastasis of esophageal squamous cell carcinoma [119]. These analyses greatly enhance our understanding of the metastatic process from a metabolic perspective.

6. **Immune editing for metastasis based on TCR and BCR sequencing analysis.** The adaptive immune responses co-evolve with the metastatic genomes of MICs [60]. The analysis of TCR and BCR repertoires provides valuable insights into the immune evasion mechanism during metastasis. Using adaptive immune profiling, tools such as MiXCR [120] (Fig. 14.2) can identify metastasis-related TCR/BCR sequences (V(D)J recombination) and clonotypes (mTCR, mBCR), which help to infer immune lineages and potential immune evasion strategies in metastatic progression.
   These analyses provide an insightful view of anti-tumor and anti-metastasis immune responses. For example, metastatic tumors with lower T cell clonality exhibit dysfunction in neoantigen presentation following metastatic colonization [43]. TCR diversities are strongly correlated with tumor-infiltrating lymphocyte densities (Fig. 14.3) [121], which are often reduced in metastases, indicating an immunosuppressed phenotype [118, 122]. The analyses of TCR diversity, combined with neo-antigen and immune cell infiltration data derived from DNA, RNA, sc, and spatial sequencing, offer a relatively comprehensive immune landscape of metastasis.
7. **Protein diversity in metastasis based on proteomic MS analysis.** Proteins, as the actual functional products of cells, directly control the cellular machinery and biological processes that drive diseases, including metastasis [123]. Protein analysis in the context of metastasis provides validation for metastatic mechanisms discovered at the genetic and transcriptomic levels [116] and reveals an additional layer of complexity associated with metastasis [124]. Proteomic MS data are used to identify metastasis-related proteins (mProteins) through tools such as MaxQuant [125] (Fig 14.2). Among these, kinases (mKinase) analyzed using KSEA [126] reveal differential phosphoproteomic patterns that play a crucial role in metastasis. For instance, as shown in Fig. 14.3, increased expression of ANXA5 and DDOST is linked to a reduced time to metastasis in cutaneous squamous cell carcinoma [127]. Plakoglobin, encoded by JUP, is highly overexpressed in CTC clusters compared to single CTCs, enhancing their survival in the bloodstream [128]. These analyses provide valuable insights into metastatic mechanisms from a proteomic perspective.
8. **Microbiome dynamics in metastasis based on microbiome sequencing analysis.** The host and intracellular microbial communities play a crucial role in metastasis by influencing the cellular and molecular characteristics

of tumor immunity, the cytoskeleton, and cell viability [129, 130]. Understanding the role of the microbiome in metastasis is essential for advancing oncology care. To investigate metastasis-related microbiomes (mMicrobiome), metagenomic sequencing, 16S rRNA sequencing, and host tumor sequencing data are aligned to microbiome references using tools such as VSEARCH [131], DADA2 [132], and GATK PathSeq [133] (Fig 14.2).

These analyses can reveal microbiome-related mechanisms underlying metastasis. Specifically, as shown in Fig 14.3, tumour-resident intracellular bacteria, such as *Staphylococcus xylosus, Ligilactobacillus animalis,* and *Spilopsyllus cuniculi*, enhance the survival of CTCs by increasing their resistance to mechanical stress. Disrupting this interaction reduces metastasis [44]. Additionally, *Fusobacterium nucleatum* promotes colorectal cancer metastasis by modulating CARD3 and autophagy [134]. Further, similar analyses could deepen the understanding of metastatic mechanisms from a microbiome perspective.

The analyses of metastatic tumors and liquid biopsies outlined above provide a comprehensive, multi-scale view of the mechanisms underlying metastasis. These approaches can address the six key questions proposed in the first section (Fig. 14.1B), except for the "how and when" questions, which will be discussed below. However, further research is still needed, particularly using multi-region, consecutive, and untreated metastatic samples with minimal interference. With advancements in sequencing technologies and computational approaches, the complexities of metastasis are becoming clearer, and therapies targeting metastasis hold the potential to significantly improve patient outcomes.

14.3 Computational Modeling to Explore Metastatic Seeding and Temporal Dynamics

Recent studies are exploring the metastasis that arises from a sc migration or more complex patterns ("How quest") at specific time points ("When quest") [17, 135]. Addressing these two questions is crucial for understanding the history of metastasis, which can inform treatment strategies. In this section, we summarize the computational methods used to tackle these challenges and propose potential avenues for further improvements.

The simplified model, presented in Fig. 14.4A, illustrates the evolutionary and temporal dynamics of metastatic seeding. The primary tumor originates from a malignant founder cell. Over time (t_p), primary-private clonal mutations accumulate (L_p), leading to the emergence of the most recent common ancestor (P_{MRCA}), which contributes to the expansion of the primary tumor until its clinical detection (T). As the primary tumor grows (P_{size}), metastatic seeding occurs at a specific time before primary detection (t_m). During this process, one or multiple metastatic founder cells (monoclonal or polyclonal seeding), carrying metastasis-private clonal mutations (L_{m1}), initiate the metastatic tumor. Following seeding, metastasis-private clonal mutations continue to accumulate (L_{m2}), with the total mutation burden ($L_{\mathrm{m1}} + L_{\mathrm{m2}} = L_m$) reflecting the evolutionary history of the metastatic tumor. Eventually, the most recent common ancestor of the metastasis (M_{MRCA}) emerges, dominating the metastatic tumor and shaping its progression.

The metastatic seeding pattern and timing are typically inferred from somatic mutation profiling of paired and multiple primary and metastatic tumors within each patient. Using sequencing data, the cancer cell fraction (CCF) for each mutation can be estimated with tools such as ABSOLUTE [136]. The "private" mutations at one site relative to another are operationally defined as those with a smaller CCF at the other site (e.g., <5%) (Fig. 14.4B). Three common approaches are employed to reconstruct metastatic patterns [107]: phylogenetic tree construction using tools like PyClone [137], examining the relationship between L_p, L_m, and primary-metastasis shared mutations (L_s), and analysis of mutation clusters (Fig. 14.4B). As shown in Fig. 14.4C, the structure of the phylogenetic tree provides insight into seeding patterns. For monoclonal seeding, multi-regional metastatic tumor samples cluster together on a single branch. In contrast, polyclonal seeding is characterized by metastatic samples splitting into multiple branches [43]. The relationship between primary-private, metastasis-private, and shared mutations also distinguishes these seeding patterns. Polyclonal seeding typically results in a higher *JSI* value (Eq. 14.1) compared to monoclonal seeding due to a greater proportion of shared mutations and fewer primary-private or metastasis-private clonal mutations [138]. Lastly, in the mutation clusters approach, the presence of mutation clusters that are subclonal in metastases and present in the primary tumour is a hallmark of polyclonal seeding. In monoclonal seeding, such clusters are absent [139, 140].

$$JSI = \frac{L_S}{L_S + L_m + L_P}. \tag{14.1}$$

Regarding metastatic timing (t_m), several approaches are being explored to differentiate early and late metastatic seeding (Fig. 14.4C). In the phylogenetic tree, early seeding is characterized by the divergence of the metastatic lineage before the genetic diversification of the primary tumor, whereas late seeding occurs when divergence happens during the diversification of the primary tumor in patients [17]. In terms of the relationship between primary-private, metastasis-private, and shared mutations, late dissemination is associated with a higher number of clonal mutations in the metastasis, as well as more shared mutations [141], which leads to higher values of *JSI* [43] and *H* (Eq. 14.2) [17]. In the mutation clusters approach, the presence of only clonal mutations in metastasis suggests linear progression from the primary tumour [139], indicating late dissemination. In contrast, early dissemination is characterized by a higher number of subclonal mutation clusters in metastasis. Other methods, such as metastasis-related mutual hazard networks (metMHN), are employed to infer the relative positions of metastatic seeding as well as genotype in trajectories [142]. These approaches aim to quantitatively determine whether metastatic seeding occurs early or late.

$$H = \frac{L_m}{L_p + 1}. \tag{14.2}$$

Additionally, researchers simulate the metastasis process starting from a single primary malignant founder [17, 138, 143] to estimate the specific metastatic time (t_m), (Fig. 14.4D). The spatial growth of tumor is modeled through the expansion of deme subpopulations (N_{cell}, 5-10 K cells) within a defined 3D cubic lattice using a Moore neighborhood (26 neighbors). Each deme is assumed to be well-mixed and grows according to a random neutral branching process with trained birth and death rates for cells in both the primary tumor and metastasis (P_{birth}, P_{death}, M_{birth}, and M_{death}). In the demes, cells with beneficial mutations may experience an increased cell division rate, given by $((1 + P_{selective}) \times P_{birth})$ for the primary tumor and $((1 + M_{selective}) \times M_{birth})$ for the metastasis. At each time step, only the demes located at the tumor surface can grow and divide, with a random empty neighboring site chosen for each new deme generated. Once a deme exceeds the maximum size ($N_{cell} \times 2$), it splits into two offspring demes, sampled from a binomial distribution. During primary tumor growth, one or more tumor cells ($N_{seeding}$) from a random deme at the tumor periphery are randomly chosen to seed the metastasis with a probability of $P_{seeding}$. The metastasis then grows using the similar model as the primary tumor, starting from the disseminated tumor cell(s). The total cell number for the primary tumor at the time of metastatic dissemination (t_m) is denoted by P_{size}. At each cell division, neutral passenger mutations ($P_{selective} = 0$) and advantageous mutations arise in the coding portion of the genome for both the primary tumor and metastasis. These mutations, both neutral ($P_{mutation\_neural}$, $M_{mutation\_neural}$) and advantageous ($P_{mutation\_advantageous}$, $M_{mutation\_advantageous}$), follow a Poisson distribution. The primary tumor and metastasis grow until the tumor reaches a size of approximately 10^9 cells, which aligns with clinical sample sizes that typically range from 4 to 15 cm in maximum diameter. After generating virtual primary and metastatic tumors, multiple regions that are each composed of approximately 10^6 cells are sampled from an octant of the tumor sphere. The parameters governing the dynamics of metastasis are inferred by comparing true variant allele frequencies (VAF) from bulk subpopulations to observed VAF obtained via the simulation model. The time of metastasis is determined by the parameters of P_{size} or t_m. One proposed estimation can be approximated by (Eq. 14.3) [138].

$$t_m = (1 - \frac{L_m}{L_p}\alpha) \times T;\ \alpha = \frac{t_p}{T}. \tag{14.3}$$

The metastatic pattern and timing can provide valuable insights into the stratification and therapeutic targeting of patients with aggressive disease [17]. For instance, survival analysis shows that patients with polyclonal seeding experience rapid progression, whereas monoclonal seeding is associated with attenuated progression [43]. A longer time span between metastatic dissemination and the surgical resection of the primary tumor suggests a higher metastatic risk [17]. Accurately determining the metastatic process is crucial for diagnosis and treatment of cancer patients.

Further efforts are still needed to advance the study of metastasis. More complex metastatic seeding patterns, as illustrated in Fig. 14.4E, include linear, parallel, redirectional, bidirectional, single-source, multi-source, and consecutive seeding. In linear seeding, the metastasis-competent clone arises late (late seeding) during tumorigenesis and disseminates just before the clinical detection of the primary lesion. The metastasis is seeded by the most advanced primary clone or subclone [144]. The parallel seeding model assumes that metastasis-establishing clones disseminate long before the primary tumor is clinically detectable (early seeding) and that genetic alterations necessary for metastatic colonization evolve outside the primary tumor [144]. Re-seeding refers to the possibility of migration from metastatic sites back into the primary tumor [145]. Bidirectional seeding suggests that cancer cell dissemination can

Fig. 14.3 Metastatic mechanisms uncovered through computational analysis of multi-omics data.

occur in both directions between two metastatic sites [107, 146]. Single-source seeding describes metastasis in which all migrations originate from a single anatomical site, either the primary tumor or another specific metastatic site [135]. Multi-source seeding, in contrast, involves metastasis which contains clones originating from different lesions [135]. Consecutive seeding entails metastasis being seeded multiple times by different migrating cells [147]. These complex patterns can further enhance the understanding of the metastasis process.

Moreover, lymph nodes serve as critical hubs for metastatic cell arrest and growth, immune modulation, and secondary dissemination to distant sites [148]. The involvement of lymph nodes dictates two distinct evolutionary routes of metastasis: lymphatic metastasis and hematogenous metastasis [22] (Fig. 14.4F). In lymphatic metastasis, distant metastases are seeded from regional lymph node lesions and cluster in the same clade, bypassing the primary tumor. In contrast, in hematogenous metastasis, distant metastases are directly seeded from primary tumors bypassing regional lymph node tumors. These two routes of metastasis can involve different evolutionary mechanisms and may elicit distinct treatment responses [22, 149], which deserve to be studied further.

14.4 Artificial Intelligence in Supporting Clinical Applications for Metastasis

The ultimate goal of studying metastatic mechanisms is to enable early diagnosis and develop improved therapies for metastasis. In clinical practice, several key issues related to cancer metastasis are of great concern (Fig. 14.5A). These issues are discussed in detail below, with an emphasis on the current research landscape and potential avenues for further improvement through the application of artificial intelligence, leveraging multi-omics data.

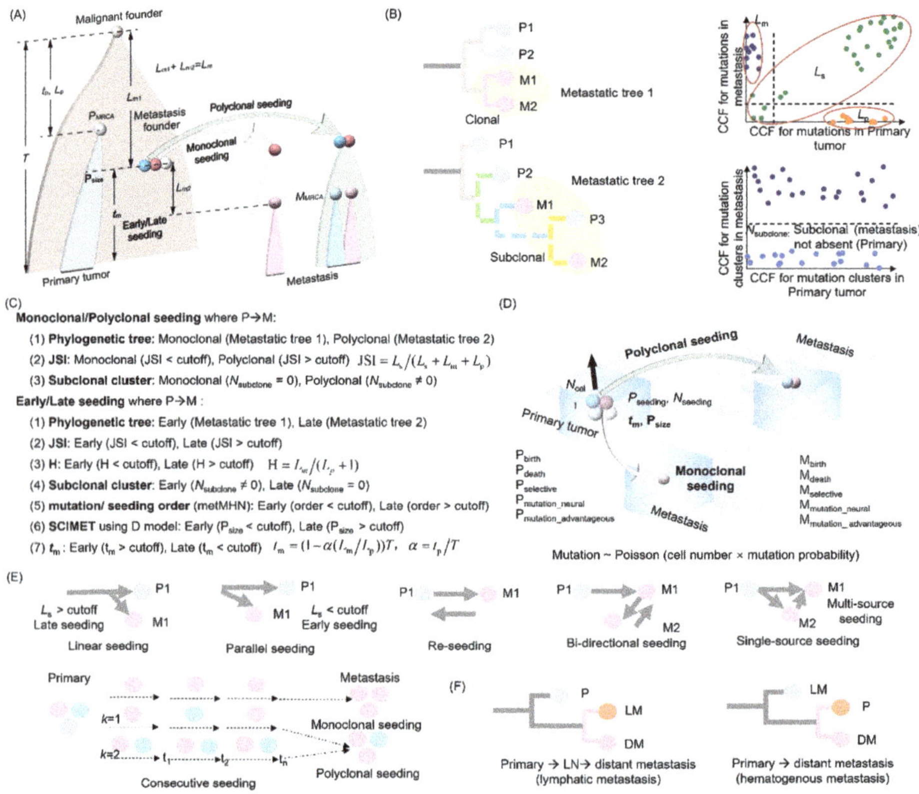

Fig. 14.4 Computational modeling to explore metastatic seeding and temporal dynamics. (A) Mathematical scheme of the metastatic seeding process. (B) Illustration of phylogenetic tree and cancer cell fraction figure showing primary-private mutations, metastasis-private mutations, primary-metastasis shared mutations, and mutation clusters. (C) Approaches to estimate monoclonal or polyclonal seeding, as well as early or late seeding. (D) Mathematical simulation model of the metastasis process. (E) Additional potential complex seeding patterns. (F) Two metastatic evolutionary routes.

1. Estimation of metastasis risk. The nonrandom patterns of cancer metastasis, as outlined above (Fig. 14.3), suggest that primary tumors and CTCs undergo cellular and molecular changes that enable successful metastasis. As a result, pre-metastatic features in primary tumors and bodily fluids can be leveraged to estimate metastatic risk. This is being studied using multi-omics data through various machine learning approaches, including genomics, transcriptomics, epigenomics, metabolomics, proteomics, and microbiomics (Fig. 14.5A). The models proposed in this section can also be utilized for the detection of micrometastases, which are indicative of a high risk of metastasis. Further research could enhance the noninvasive, early, and accurate detection of metastasis by incorporating more detailed characteristics specific to different metastatic stages.

   Specifically, genomics features of tumors are used to train XGBoost models for estimating metastasis likelihood, achieving an area under the receiver operating characteristic curve (AUC) greater than 0.8 for breast and prostate cancers [150]. These features also assess organotropism-based metastasis risk, such as the likelihood of bone and liver metastasis in prostate cancer, using an ordinal regression model [150]. Additionally, DNA methylation signatures provide patient-specific brain metastasis risk probabilities, which are incorporated into a nomogram for lung cancer [151]. Moreover, transcriptomic data from primary and metastatic tumors are increasingly being used to uncover metastatic potential. For example, a six-gene expression panel has demonstrated its ability to identify gastric cancer patients with micrometastasis using a logistic regression model [152]. Additionally, a metastatic potential score can be derived from a deep learning model trained on data from over 13,000 patients [153]. In the case of locoregionally advanced nasopharyngeal carcinoma, nine immune-associated long noncoding RNAs, weighted by their regression coefficients in a penalized Cox model, generated a promising distant metastasis score [154]. These studies highlight the potential of transcriptomics in estimating metastasis risk. For primary lung tumors, transcriptomic features are integrated with genomic data to infer metastatic potential using a combination of logistic regression, random forest (RF), and multilayer perceptron models, with a support vector machine (SVM) terminal layer, achieving an AUC of 0.79 [93].

   Other omics features also play a crucial role in metastasis risk estimation. Specifically, dysregulated metabolites are used to predict the risk of liver and extrahepatic metastasis in pancreatic cancer through a naïve Bayes classifier [155] and to assess the metastatic potential in esophageal squamous cell carcinoma via a partial least square model [119]. In the case of proteomics, data are incorporated into a nomogram to predict brain metastasis in lung cancer patients [156]. Additionally, intratumoral microbiome features are selected to train an RF model for predicting liver metastasis risk in gastrointestinal stromal tumor patients [157]. Collectively, the integration of these diverse omics features holds great promise for improving metastatic diagnosis and enhancing the accuracy of risk predictions.

2. Identification of metastasis origin. Identifying the origin of metastasis is crucial for CUP patients, as the primary tumor responsible for metastatic seeding is often unclear [35]. This uncertainty leads to empirical chemotherapy regimens, resulting in poor survival outcomes [158]. Efforts are now focused on improving the accuracy of metastasis origin identification (Fig. 14.5A) to enhance management strategies and prognosis for CUP patients [159].

   Recent advancements demonstrate the predictive power of genomic data in cancer origin identification. A combination of somatic mutations and copy number alterations in 50 genes achieves 77.7% accuracy for cancer type prediction using a linear SVM [160]. Similarly, genomic alterations, including mutations, indels, focal and broad copy number changes, structural rearrangements, and mutation signatures, trained on a cohort of 7,791 patients, achieve 74.1% accuracy in identifying tumor origins of 11,644 patients [161]. In another study, a multi-class deep neural network trained on mutational types and distributions from 2,436 tumors across 24 cancer types achieves 83% accuracy in metastasis origin prediction [162]. Additionally, 511 somatic driver and passenger mutations, including structural variants, are used to train an ensemble of binary RF classifiers, reaching 89% recall and precision for cancer origin classification in a validation set of 6,082 tumor samples [163].

   DNA methylation signatures also offer promising results. An L1-regularized logistic regression model using DNA methylation fingerprints identifies tumor types for 29 CUP patients, with seven cases later confirmed through pathology [164]. Similarly, microarray DNA methylation data trained on 2,790 samples from 38 cancer types achieve 87% accuracy in predicting cancer origin for 216 CUP patients using RF algorithms [165]. Furthermore, methylation data from cell-free DNA in blood samples have enabled noninvasive tumor type prediction through deconvolution, with improved accuracy for patients with higher ctDNA burden [166].

Transcriptomic data further enhance the prediction of metastasis origin. For example, a neural network accurately identifies primary tumors in 12 CUP patients [167]. Additionally, the expression of 538 genes predicts tissue origin with 92.9% accuracy in 14 metastasis samples using a top-scoring pairs-based majority vote algorithm [168]. In another approach, a 1D-Inception model, utilizing the expression of 817 genes, achieves primary site classification accuracies of 96.7% and 72.5% in two external validation sets [169]. Furthermore, a naive Bayes classifier based on 943 genes successfully diagnoses two clinical CUP cases in the hepatobiliary and pancreatic system [170]. Platelet mRNA profiles also demonstrate 71% accuracy in distinguishing cancer origins across six types when analyzed with an SVM classifier [171]. More recently, a multilayer perceptron classifier, utilizing 214 developmental trajectory signals deconvoluted from bulk tumor transcriptomes based on sc data, achieves a "top-1" prediction concordance of 73% and "top-3" concordance of 90% [172].

Proteins from extracellular vesicles and plasma particles also show diagnostic promise. An RF model using these proteins identifies tumor types in patients with melanoma, colorectal, pancreatic, and lung cancer, demonstrating their diagnostic ability [173]. Collectively, these omics features play a crucial role in improving the management and treatment of CUP patients, providing a more accurate and tailored approach to their care.

3. Stratification of oligometastasis. Targeted localized therapies can cure patients with oligometastasis but show limited effectiveness for those with more widespread metastasis [39]. Identifying patients who would benefit from these specific therapies is crucial for optimizing treatment strategies, minimizing unnecessary treatment costs, and reducing patient discomfort. However, only a limited number of studies have focused on identifying oligometastasis (Fig. 14.5A). For example, a nearest shrunken centroid classification algorithm, trained on 113 gene expressions, is used to identify a subtype of colorectal liver metastases that experience fewer metastatic recurrences or deaths after hepatic resection, achieving an AUC value of 87.5% [174]. This promising result underscores the potential of stratifying patients based on oligometastasis to tailor therapies more effectively. Given the clinical importance of distinguishing oligometastasis from more extensive metastatic disease and the potential oligometastasis biomarkers (Fig. 14.3), further research into better stratification

methods is warranted. Such studies could lead to more precise patient selection for therapy, improving outcomes while reducing the burden of unnecessary treatments.

4. Assessment of metastatic treatment. Personalized cancer therapies require accurate assessment and dynamic tracking of treatment effectiveness (Fig. 14.5A). Liquid biopsy offers a promising approach. For example, CTCs have been shown to predict the benefit of post-surgery adjuvant chemotherapy in metastatic colon cancer [175] and the addition of bevacizumab to first-line letrozole in hormone-receptor-positive (HR+) metastatic breast cancer [176]. Moreover, a reduction in ctDNA is a strong predictor of radiologic response during first-line chemotherapy in metastatic colorectal cancer [177]. Transcriptomics data also demonstrate predictive power. For instance, an 18-gene expression panel can predict prolonged sensitivity to endocrine therapy in HR+/HER2− metastatic breast cancer [178]. Proteomic data further contribute to personalized treatment. In metastatic prostate cancer, patients with AR-V7 protein positive CTCs have improved survival with taxane chemotherapy [179]. Additionally, elastic net regression models using kinase-phosphosubstrate pairs are highly correlated with drug responses to afatinib, gefitinib, and regorafenib [124]. Ongoing research into multi-scale, multi-omics mechanisms, as discussed in Section 14.2, holds promise for improving treatment outcome and tackling drug resistance.
5. Potential avenues for enhancing clinical application in metastasis. These studies highlight significant progress in developing genomics, transcriptomics, epigenetics, metabolomics, proteomics, and microbiomics models for estimating metastasis risk, identifying metastasis origin, stratifying oligometastasis, and assessing metastatic treatments. While these advances offer promising tools for improving metastatic cancer diagnosis and management, further efforts are needed to refine precision therapies in key areas, enhancing their clinical applicability and enabling more effective treatment strategies (Fig. 14.5B).

   First, greater attention must be given to patient variability to improve the reliability of metastasis studies (Fig. 14.5B). Many current metastatic studies overlook the impact of specific treatment regimens. As a result, metastasis-related features identified in these studies may reflect drug resistance rather than true metastatic characteristics. It is also essential to increase the number of patients included in these studies, as a larger cohort will contribute to the development of more robust and clinically relevant models for metastasis.

Second, more detailed and specific features throughout the metastatic seeding cascade are needed for clinical applications in metastasis (Fig. 14.5B). Current features are primarily derived from comparisons between primary tumors and metastases, but more nuanced insights are required. Micrometastasis-specific features in primary tumors and CTCs are better suited for estimating metastasis risk and detecting micrometastases. The metastatic characteristics of CUP patients, in contrast to metastases with a known primary origin, are more relevant for inferring their tumor origins. Additionally, oligometastasis-specific features, distinct from those observed in macro-metastasis, are more appropriate for stratifying oligometastasis. Furthermore, the effective integration of multi-modal omics features, alongside the application of sc and spatial data, could provide more granular insights into metastasis, thereby enhancing diagnosis and treatment in clinical settings.

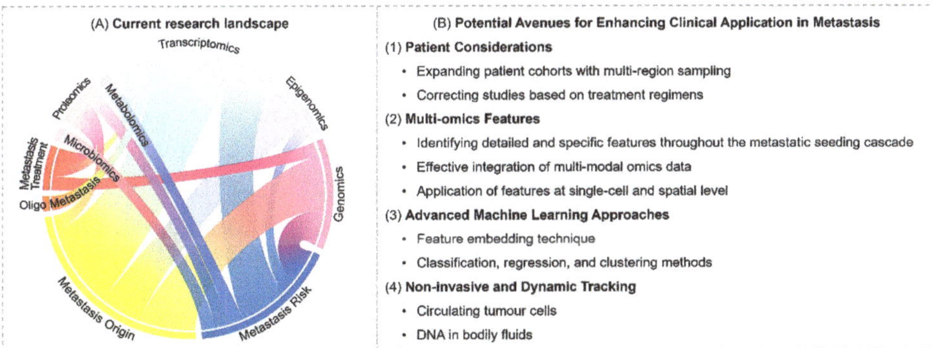

Fig. 14.5 Artificial intelligence in supporting clinical applications for metastasis. (A) Current research landscape addressing metastatic clinical challenges with multi-omics data. (B) Potential avenues for further enhancing clinical applications in metastasis.

Third, advanced machine learning approaches hold significant promise for enhancing metastasis modeling in clinical applications (Fig. 14.5B). Techniques such as Transformer models [180], ensemble learning [181], and graph-based methods [182] have shown considerable potential in capturing complex patterns and improving predictive accuracy. These innovative approaches can enable more accurate detection and predictions, ultimately facilitating early screening, real-time monitoring, and more effective treatment planning for metastatic cancer.

Fourth, CTCs and cell-free DNA in bodily fluids deserve more attention due to their noninvasive nature and ability to dynamically track cancer progression (Fig. 14.5B). These biomarkers offer a promising avenue for real-time monitoring and could play a pivotal role in improving metastatic detection and treatment evaluation.

In conclusion, by addressing these key areas, researchers can significantly enhance the clinical application for metastasis, ultimately leading to more personalized, effective treatments for metastatic cancer patients.

14.5 Conclusion

This review addresses the critical challenges of metastasis and summarizes recent advancements in understanding its various stages from a multi-omics perspective, leveraging computational power. Advances in sequencing technologies have revolutionized cancer research, enabling the identification of cellular and molecular changes during metastasis from the fields of genomics, transcriptomics, epigenomics, metabolomics, proteomics, microbiomics, immunomics, and sc or spatial omics. These innovations are key to discovering metastatic biomarkers, which will help uncover the complex biological mechanisms driving metastasis. Additionally, the application of artificial intelligence, particularly machine learning and deep learning, will enhance the detection and predictive power of metastasis in clinical practice. These ongoing efforts hold great promise for more precise diagnoses, better monitoring of disease progression, and more effective treatments for metastatic cancer.

References

1. Gerstberger, S., Q. Jiang, and K. Ganesh. Metastasis. *Cell*. 2023; **186**(8): p. 1564–1579.
2. Karras, P., et al. Decoding the interplay between genetic and non-genetic drivers of metastasis. *Nature*. 2024; **629**(8012): p. 543–554.
3. Lambert, A. W., Y. Zhang, and R. A. Weinberg. Cell-intrinsic and microenvironmental determinants of metastatic colonization. *Nat Cell Biol*. 2024; **26**(5): p. 687–697.
4. Bhat, G. R., et al. Cancer cell plasticity: from cellular, molecular, and genetic mechanisms to tumor heterogeneity and drug resistance. *Cancer Metastasis Rev*. 2024; **43**(1): p. 197–228.

5. Baumann, Z., P. Auf der Maur, and M. Bentires-Alj. Feed-forward loops between metastatic cancer cells and their microenvironment-the stage of escalation. *EMBO Mol Med.* 2022; **14**(6): p. e14283.
6. Shi, X., et al. Mechanism insights and therapeutic intervention of tumor metastasis: latest developments and perspectives. *Signal Transduct Target Ther.* 2024; **9**(1): p. 192.
7. Garcia-Recio, S., et al. Multiomics in primary and metastatic breast tumors from the AURORA US network finds microenvironment and epigenetic drivers of metastasis. *Nat Cancer.* 2023; **4**(1): p. 128–147.
8. Nguyen, B., et al. Genomic characterization of metastatic patterns from prospective clinical sequencing of 25,000 patients. *Cell.* 2022; **185**(3): p. 563–575. e11.
9. Webster, P. Six ways large language models are changing healthcare. *Nat Med.* 2023; **29**(12): p. 2969–2971.
10. Massagué, J., and K. Ganesh. Metastasis-initiating cells and ecosystems. *Cancer Discov.* 2021; **11**(4): p. 971–994.
11. Ring, A., et al. Biology, vulnerabilities and clinical applications of circulating tumour cells. *Nat Rev Cancer.* 2023; **23**(2): p. 95–111.
12. Mittal, V. Epithelial mesenchymal transition in tumor metastasis. *Annu Rev Pathol.* 2018; **13**: p. 395–412.
13. Haynes, N. M., T. B. Chadwick, and B. S. Parker. The complexity of immune evasion mechanisms throughout the metastatic cascade. *Nat Immunol.* 2024; **25**(10): p. 1793–1808.
14. Bergers, G., and S. M. Fendt. The metabolism of cancer cells during metastasis. *Nat Rev Cancer.* 2021; **21**(3): p. 162–180.
15. Liu, Q., et al. Factors involved in cancer metastasis: a better understanding to "seed and soil" hypothesis. *Mol Cancer.* 2017; **16**(1): p. 176.
16. Hu, Z., and C. Curtis. Looking backward in time to define the chronology of metastasis. *Nat Commun.* 2020; **11**(1): p. 3213.
17. Hu, Z., et al. Quantitative evidence for early metastatic seeding in colorectal cancer. *Nat Genet.* 2019; **51**(7): p. 1113–1122.
18. Alderton, G. K. Metastasis: spreading the seed. *Nat Rev Cancer.* 2015; **15**(5): p. 255.
19. Lee, W. C., et al. Metastasis of cancer: when and how? *Ann Oncol.* 2017; **28**(9): p. 2045–2047.
20. Arbiser, J. L., M. Y. Bonner, and L. C. Gilbert. Targeting the duality of cancer. *NPJ Precis Oncol.* 2017; **1**: p. 23.
21. Wong, S. Y., and R. O. Hynes. Lymphatic or hematogenous dissemination: how does a metastatic tumor cell decide? *Cell Cycle.* 2006; **5**(8): p. 812–817.
22. Lin, M., et al. Evolutionary route of nasopharyngeal carcinoma metastasis and its clinical significance. *Nat Commun.* 2023; **14**(1): p. 610.
23. Calomarde-Rees, L., et al. Risk factors for lymphatic and hematogenous dissemination in patients with stages I to II cutaneous melanoma. *JAMA Dermatol.* 2019; **155**(6): p. 679–687.
24. Wang, Y., et al. Pre-metastatic niche: formation, characteristics and therapeutic implication. *Signal Transduct Target Ther.* 2024; **9**(1): p. 236.
25. Gao, Y., et al. Metastasis organotropism: redefining the congenial soil. *Dev Cell.* 2019; **49**(3): p. 375–391.

26. Chen, W., et al. Organotropism: new insights into molecular mechanisms of breast cancer metastasis. *NPJ Precis Oncol*. 2018; **2**(1): p. 4.
27. Blasco, M. T., I. Espuny, and R. R. Gomis. Ecology and evolution of dormant metastasis. *Trends Cancer*. 2022; **8**(7): p. 570–582.
28. Francescangeli, F., et al. Dormancy, stemness, and therapy resistance: interconnected players in cancer evolution. *Cancer Metastasis Rev*. 2023; **42**(1): p. 197–215.
29. Weston, W. A., and A. R. Barr. A cell cycle centric view of tumour dormancy. *Br J Cancer*. 2023; **129**(10): p. 1535–1545.
30. Talukdar, S., et al. Dormancy and cancer stem cells: an enigma for cancer therapeutic targeting. *Adv Cancer Res*. 2019; **141**: p. 43–84.
31. Endo, H., and M. Inoue. Dormancy in cancer. *Cancer Sci*. 2019; **110**(2): p. 474–480.
32. Liu, Z., et al. Multi-stage mechanisms of tumor metastasis and therapeutic strategies. *Signal Transduct Target Ther*. 2024; **9**(1): p. 270.
33. Janjigian, Y. Y., J. D. Wolchok, and C. E. Ariyan. Eradicating micrometastases with immune checkpoint blockade: Strike while the iron is hot. *Cancer Cell*. 2021; **39**(6): p. 738–742.
34. Kato, S., et al. Cancer of unknown primary in the molecular era. *Trends Cancer*. 2021; **7**(5): p. 465–477.
35. Krämer, A., et al. Cancer of unknown primary: ESMO Clinical Practice Guideline for diagnosis, treatment and follow-up. *Ann Oncol*. 2023; **34**(3): p. 228–246.
36. Olivier, T., et al. Redefining cancer of unknown primary: is precision medicine really shifting the paradigm? *Cancer Treat Rev*. 2021; **97**: p. 102204.
37. Lu, M. Y., et al. AI-based pathology predicts origins for cancers of unknown primary. *Nature*. 2021; **594**(7861): p. 106–110.
38. Ganesh, K., and J. Massagué. Targeting metastatic cancer. *Nat Med*. 2021; **27**(1): p. 34–44.
39. Gutiontov, S. I., et al. (Oligo)metastasis as a spectrum of disease. *Cancer Res*. 2021; **81**(10): p. 2577–2583.
40. Robinson, D. R., et al. Integrative clinical genomics of metastatic cancer. *Nature*. 2017; **548**(7667): p. 297–303.
41. Davis, R. T., et al. Transcriptional diversity and bioenergetic shift in human breast cancer metastasis revealed by single-cell RNA sequencing. *Nat Cell Biol*. 2020; **22**(3): p. 310–320.
42. Peng, H., et al. Single-cell characterization of differentiation trajectories and drug resistance features in gastric cancer with peritoneal metastasis. *Clin Transl Med*. 2024; **14**(10): p. e70054.
43. Sun, Y., et al. Integrated multi-omics profiling to dissect the spatiotemporal evolution of metastatic hepatocellular carcinoma. *Cancer Cell*. 2024; **42**(1): p. 135–156. e17.
44. Fu, A., et al. Tumor-resident intracellular microbiota promotes metastatic colonization in breast cancer. *Cell*. 2022; **185**(8): p. 1356–1372. e26.
45. Magbanua, M. J. M., et al. Expanded genomic profiling of circulating tumor cells in metastatic breast cancer patients to assess biomarker status and biology over time (CALGB 40502 and CALGB 40503, Alliance). *Clin Cancer Res*. 2018; **24**(6): p. 1486–1499.
46. Martínez-Jiménez, F., et al. Pan-cancer whole-genome comparison of primary and metastatic solid tumours. *Nature*. 2023; **618**(7964): p. 333–341.

47. Li, H., and R. Durbin. Fast and accurate short read alignment with Burrows–Wheeler transform. *Bioinformatics*. 2009; **25**(14): p. 1754–1760.
48. Cibulskis, K., et al. Sensitive detection of somatic point mutations in impure and heterogeneous cancer samples. *Nat Biotechnol.* 2013; **31**(3): p. 213–219.
49. Koboldt, D. C., et al. VarScan 2: somatic mutation and copy number alteration discovery in cancer by exome sequencing. *Genome Res*. 2012; **22**(3): p. 568–576.
50. Talevich, E., et al. CNVkit: genome-wide copy number detection and visualization from targeted DNA sequencing. *PLoS Comput Biol.* 2016; **12**(4): p. e1004873.
51. Rausch, T., et al. DELLY: structural variant discovery by integrated paired-end and split-read analysis. *Bioinformatics*. 2012; **28**(18): p. i333–i339.
52. Layer, R. M., et al. LUMPY: a probabilistic framework for structural variant discovery. *Genome Biol.* 2014; **15**(6): p. R84.
53. Rosenthal, R., et al. DeconstructSigs: delineating mutational processes in single tumors distinguishes DNA repair deficiencies and patterns of carcinoma evolution. *Genome Biol.* 2016; **17**: p. 31.
54. Lawrence, M. S., et al. Discovery and saturation analysis of cancer genes across 21 tumour types. *Nature*. 2014; **505**(7484): p. 495–501.
55. Reynisson, B., et al. NetMHCpan-4.1 and NetMHCIIpan-4.0: improved predictions of MHC antigen presentation by concurrent motif deconvolution and integration of MS MHC eluted ligand data. *Nucleic Acids Res*. 2020; **48**(W1): p. W449–W454.
56. Deshpande, V., et al. Exploring the landscape of focal amplifications in cancer using AmpliconArchitect. *Nat Commun*. 2019; **10**(1): p. 392.
57. Luebeck, J., et al. Extrachromosomal DNA in the cancerous transformation of Barrett's oesophagus. *Nature*. 2023; **616**(7958): p. 798–805.
58. Bertucci, F., et al. Genomic characterization of metastatic breast cancers. *Nature*. 2019; **569**(7757): p. 560–564.
59. Reiter, J. G., et al. Minimal functional driver gene heterogeneity among untreated metastases. *Science*. 2018; **361**(6406): p. 1033–1037.
60. De Mattos-Arruda, L., et al. The genomic and immune landscapes of lethal metastatic breast cancer. *Cell Rep*. 2019; **27**(9): p. 2690–2708.e10.
61. McGranahan, N., et al. Allele-specific HLA loss and immune escape in lung cancer evolution. *Cell*. 2017; **171**(6): p. 1259–1271.e11.
62. Birkbak, N. J., and N. McGranahan. Cancer genome evolutionary trajectories in metastasis. *Cancer Cell*. 2020; **37**(1): p. 8–19.
63. Miyamoto, D. T., et al. RNA-Seq of single prostate CTCs implicates noncanonical Wnt signaling in antiandrogen resistance. *Science*. 2015; **349**(6254): p. 1351–1356.
64. Zehir, A., et al. Mutational landscape of metastatic cancer revealed from prospective clinical sequencing of 10,000 patients. *Nat Med*. 2017; **23**(6): p. 703–713.
65. Armenia, J., et al. The long tail of oncogenic drivers in prostate cancer. *Nat Genet*. 2018; **50**(5): p. 645–651.
66. Ottaiano, A., et al. Genetic trajectory and immune microenvironment of lung-specific oligometastatic colorectal cancer. *Cell Death Dis*. 2020; **11**(4): p. 275.
67. Ottaiano, A., et al. Characterization of KRAS mutational regression in oligometastatic patients. *Front Immunol*. 2022; **13**: p. 898561.

68. Wu, S., et al. metsDB: a knowledgebase of cancer metastasis at bulk, single-cell and spatial levels. *Nucleic Acids Res*. 2024. **53**(D1): p. D1427–D1434.
69. Chen, T., et al. S100A6 drives lymphatic metastasis of liver cancer via activation of the RAGE/NF-kB/VEGF-D pathway. *Cancer Lett*. 2024; **587**: p. 216709.
70. Hu, X., et al. Heterogeneous PD-L1 expression in metastases impacts immunotherapy response. *EBioMedicine*. 2023; **97**: p. 104816.
71. Dobin, A., et al. STAR: ultrafast universal RNA-seq aligner. *Bioinformatics*. 2013; **29**(1): p. 15–21.
72. Anders, S., P. T. Pyl, and W. Huber. HTSeq—a Python framework to work with high-throughput sequencing data. *Bioinformatics*. 2015; **31**(2): p. 166–169.
73. Li, B., and C. N. Dewey. RSEM: accurate transcript quantification from RNA-Seq data with or without a reference genome. *BMC Bioinformatics*. 2011; **12**: p. 323.
74. Trapnell, C., et al. Transcript assembly and quantification by RNA-Seq reveals unannotated transcripts and isoform switching during cell differentiation. *Nat Biotechnol*. 2010; **28**(5): p. 511–515.
75. Hänzelmann, S., R. Castelo, and J. Guinney. GSVA: gene set variation analysis for microarray and RNA-seq data. *BMC Bioinformatics*. 2013; **14**: p. 7.
76. Subramanian, A., et al. Gene set enrichment analysis: a knowledge-based approach for interpreting genome-wide expression profiles. *Proc Natl Acad Sci U S A*. 2005; **102**(43): p. 15545–15550.
77. Newman, A. M., et al. Robust enumeration of cell subsets from tissue expression profiles. *Nat Methods*. 2015; **12**(5): p. 453–457.
78. Jiang, P., et al. Signatures of T cell dysfunction and exclusion predict cancer immunotherapy response. *Nat Med*. 2018; **24**(10): p. 1550–1558.
79. Jiang, P., et al. Systematic investigation of cytokine signaling activity at the tissue and single-cell levels. *Nat Methods*. 2021; **18**(10): p. 1181–1191.
80. Shen, S., et al. rMATS: robust and flexible detection of differential alternative splicing from replicate RNA-Seq data. *Proc Natl Acad Sci U S A*. 2014; **111**(51): p. E5593–5601.
81. Trincado, J. L., et al. SUPPA2: fast, accurate, and uncertainty-aware differential splicing analysis across multiple conditions. *Genome Biol*. 2018; **19**(1): p. 40.
82. Kahles, A., et al. SplAdder: identification, quantification and testing of alternative splicing events from RNA-Seq data. *Bioinformatics*. 2016; **32**(12): p. 1840–1847.
83. Memczak, S., et al. Circular RNAs are a large class of animal RNAs with regulatory potency. *Nature*. 2013; **495**(7441): p. 333–338.
84. Zhang, X. O., et al. Diverse alternative back-splicing and alternative splicing landscape of circular RNAs. *Genome Res*. 2016; **26**(9): p. 1277–1287.
85. Zhang, J., et al. Accurate quantification of circular RNAs identifies extensive circular isoform switching events. *Nat Commun*. 2020; **11**(1): p. 90.
86. Picardi, E., and G. Pesole. REDItools: high-throughput RNA editing detection made easy. *Bioinformatics*. 2013; **29**(14): p. 1813–1814.
87. Li, J., et al. A mammalian conserved circular RNA CircLARP1B regulates hepatocellular carcinoma metastasis and lipid metabolism. *Adv Sci (Weinh)*. 2024; **11**(2): p. e2305902.
88. Prensner, J. R., et al. RNA biomarkers associated with metastatic progression in prostate cancer: a multi-institutional high-throughput analysis of SChLAP1. *Lancet Oncol*. 2014; **15**(13): p. 1469–1480.

89. Han, K., et al. CircLONP2 enhances colorectal carcinoma invasion and metastasis through modulating the maturation and exosomal dissemination of microRNA-17. *Mol Cancer.* 2020; **19**(1): p. 60.
90. Hu, Z. Q., et al. Circular RNA sequencing identifies CircASAP1 as a key regulator in hepatocellular carcinoma metastasis. *Hepatology*. 2020; **72**(3): p. 906–922.
91. Sun, D., et al. MacroH2A impedes metastatic growth by enforcing a discrete dormancy program in disseminated cancer cells. *Sci Adv*. 2022; **8**(48): p. eabo0876.
92. Jbara, A., et al. RBFOX2 modulates a metastatic signature of alternative splicing in pancreatic cancer. *Nature*. 2023; **617**(7959): p. 147–153.
93. Martínez-Ruiz, C., et al. Genomic-transcriptomic evolution in lung cancer and metastasis. *Nature*. 2023; **616**(7957): p. 543–552.
94. Wang, S. Y., et al. COPA A-to-I RNA editing hijacks endoplasmic reticulum stress to promote metastasis in colorectal cancer. *Cancer Lett.* 2023; **553**: p. 215995.
95. Klughammer, J., et al. A multi-modal single-cell and spatial expression map of metastatic breast cancer biopsies across clinicopathological features. *Nat Med.* 2024; **30**(11): p. 3236–3249.
96. Hao, Y., et al. Integrated analysis of multimodal single-cell data. *Cell.* 2021; **184**(13): p. 3573–3587. e29.
97. Wolf, F. A., P. Angerer, and F. J. Theis. SCANPY: large-scale single-cell gene expression data analysis. *Genome Biol.* 2018; **19**(1): p. 15.
98. Wu, Y., et al. Spatiotemporal immune landscape of colorectal cancer liver metastasis at single-cell level. *Cancer Discov*. 2022; **12**(1): p. 134–153.
99. Jin, S., et al. Inference and analysis of cell–cell communication using CellChat. *Nat Commun*. 2021; **12**(1): p. 1088.
100. Efremova, M., et al. CellPhoneDB: inferring cell–cell communication from combined expression of multi-subunit ligand-receptor complexes. *Nat Protoc*. 2020; **15**(4): p. 1484–1506.
101. Aibar, S., et al. SCENIC: single-cell regulatory network inference and clustering. *Nat Methods*. 2017; **14**(11): p. 1083–1086.
102. Tickle, T., et al. *inferCNV of the Trinity CTAT Project.* Klarman Cell Observatory, Broad Institute of MIT and Harvard; 2019.
103. Trapnell, C., et al. The dynamics and regulators of cell fate decisions are revealed by pseudotemporal ordering of single cells. *Nat Biotechnol*. 2014; **32**(4): p. 381–386.
104. La Manno, G., et al. RNA velocity of single cells. *Nature*. 2018; **560**(7719): p. 494–498.
105. Zheng, H., et al. PDGFRα(+)ITGA11(+) fibroblasts foster early-stage cancer lymphovascular invasion and lymphatic metastasis via ITGA11-SELE interplay. *Cancer Cell.* 2024; **42**(4): p. 682–700. e12.
106. Simeonov, K. P., et al. Single-cell lineage tracing of metastatic cancer reveals selection of hybrid EMT states. *Cancer Cell.* 2021; **39**(8): p. 1150–1162. e9.
107. Quinn, J. J., et al. Single-cell lineages reveal the rates, routes, and drivers of metastasis in cancer xenografts. *Science*. 2021; **371**(6532): p. eabc1944.
108. Taftaf, R., et al. ICAM1 initiates CTC cluster formation and trans-endothelial migration in lung metastasis of breast cancer. *Nat Commun*. 2021; **12**(1): p. 4867.
109. Khoo, W. H., et al. A niche-dependent myeloid transcriptome signature defines dormant myeloma cells. *Blood*. 2019; **134**(1): p. 30–43.

110. Krueger, F., and S. R., Andrews. Bismark: a flexible aligner and methylation caller for Bisulfite-Seq applications. *Bioinformatics*. 2011; **27**(11): p. 1571–1572.
111. Gkountela, S., et al. Circulating tumor cell clustering shapes DNA methylation to enable metastasis seeding. *Cell*. 2019; **176**(1–2): p. 98–112. e14.
112. Jakab, M., et al. Lung endothelium exploits susceptible tumor cell states to instruct metastatic latency. *Nat Cancer*. 2024; **5**(5): p. 716–730.
113. Smith, C. A., et al. XCMS: processing mass spectrometry data for metabolite profiling using nonlinear peak alignment, matching, and identification. *Anal Chem*. 2006; **78**(3): p. 779–787.
114. Tsugawa, H., et al. MS-DIAL: data-independent MS/MS deconvolution for comprehensive metabolome analysis. *Nat Methods*. 2015; **12**(6): p. 523–526.
115. Jin, X., et al. A metastasis map of human cancer cell lines. *Nature*. 2020; **588**(7837): p. 331–336.
116. Che, G., et al. Sulfotransferase SULT2B1 facilitates colon cancer metastasis by promoting SCD1-mediated lipid metabolism. *Clin Transl Med*. 2024; **14**(2): p. e1587.
117. Ly, M. *Deciphering the biological differences between micro-metastasis and macro-metastasis in medulloblastoma*. University of Toronto (Canada); 2022.
118. Fischer, G. M., et al. Molecular profiling reveals unique immune and metabolic features of melanoma brain metastases. *Cancer Discov*. 2019; **9**(5): p. 628–645.
119. Huang, Q., et al. Multi-omics analysis reveals NNMT as a master metabolic regulator of metastasis in esophageal squamous cell carcinoma. *NPJ Precis Oncol*. 2024; **8**(1): p. 24.
120. Bolotin, D. A., et al. MiXCR: software for comprehensive adaptive immunity profiling. *Nat Methods*. 2015; **12**(5): p. 380–381.
121. Angelova, M., et al. Evolution of metastases in space and time under immune selection. *Cell*. 2018; **175**(3): p. 751–765. e16.
122. Kudo, Y., et al. Suppressed immune microenvironment and repertoire in brain metastases from patients with resected non-small-cell lung cancer. *Ann Oncol*. 2019; **30**(9): p. 1521–1530.
123. Jia, X., et al. Protein translation: biological processes and therapeutic strategies for human diseases. *Signal Transduct Target Ther*. 2024; **9**(1): p. 44.
124. Li, C., et al. Integrated omics of metastatic colorectal cancer. *Cancer Cell*. 2020; **38**(5): p. 734–747. e9.
125. Cox, J., and M. Mann. MaxQuant enables high peptide identification rates, individualized p.p.b.-range mass accuracies and proteome-wide protein quantification. *Nat Biotechnol*. 2008; **26**(12): p. 1367–1372.
126. Wiredja, D. D., M. Koyutürk, and M. R. Chance. The KSEA App: a web-based tool for kinase activity inference from quantitative phosphoproteomics. *Bioinformatics*. 2017; **33**(21): p. 3489–3491.
127. Shapanis, A., et al. Identification of proteins associated with development of metastasis from cutaneous squamous cell carcinomas (cSCCs) via proteomic analysis of primary cSCCs. *Br J Dermatol*. 2021; **184**(4): p. 709–721.
128. Aceto, N., et al. Circulating tumor cell clusters are oligoclonal precursors of breast cancer metastasis. *Cell*. 2014; **158**(5): p. 1110–1122.
129. Battaglia, T. W., et al. A pan-cancer analysis of the microbiome in metastatic cancer. *Cell*. 2024; **187**(9): p. 2324–2335. e19.

130. Gonzalez Lopez, A. Microbial strategist of metastasis. *Nat Rev Microbiol*. 2022; **20**(6): p. 318.
131. Rognes, T., et al. VSEARCH: a versatile open source tool for metagenomics. *PeerJ*. 2016; **4**: p. e2584.
132. Callahan, B. J., et al. DADA2: high-resolution sample inference from Illumina amplicon data. *Nat Methods*. 2016; **13**(7): p. 581–583.
133. Walker, M. A., et al. GATK PathSeq: a customizable computational tool for the discovery and identification of microbial sequences in libraries from eukaryotic hosts. *Bioinformatics*. 2018; **34**(24): p. 4287–4289.
134. Chen, Y., et al. Fusobacterium nucleatum promotes metastasis in colorectal cancer by activating autophagy signaling via the upregulation of CARD3 expression. *Theranostics*. 2020; **10**(1): p. 323–339.
135. El-Kebir, M., G. Satas, and B. J. Raphael. Inferring parsimonious migration histories for metastatic cancers. *Nat Genet*. 2018; **50**(5): p. 718–726.
136. Carter, S. L., et al. Absolute quantification of somatic DNA alterations in human cancer. *Nat Biotechnol*. 2012; **30**(5): p. 413–421.
137. Roth, A., et al. PyClone: statistical inference of clonal population structure in cancer. *Nat Methods*. 2014; **11**(4): p. 396–398.
138. Hu, Z., et al. Multi-cancer analysis of clonality and the timing of systemic spread in paired primary tumors and metastases. *Nat Genet*. 2020; **52**(7): p. 701–708.
139. Gundem, G., et al. The evolutionary history of lethal metastatic prostate cancer. *Nature*. 2015; **520**(7547): p. 353–357.
140. Warner, E. W., et al. Multiregion sampling of de novo metastatic prostate cancer reveals complex polyclonality and augments clinical genotyping. *Nat Cancer*. 2024; **5**(1): p. 114–130.
141. Wang, D., et al. Multiregion sequencing reveals the genetic heterogeneity and evolutionary history of osteosarcoma and matched pulmonary metastases. *Cancer Res*. 2019; **79**(1): p. 7–20.
142. Rupp, K., et al. Modeling metastatic progression from cross-sectional cancer genomics data. *Bioinformatics*. 2024; **40**(Suppl 1): p. i140–i150.
143. Sun, R., et al. Between-region genetic divergence reflects the mode and tempo of tumor evolution. *Nat Genet*. 2017; **49**(7): p. 1015–1024.
144. Turajlic, S., and C. Swanton. Metastasis as an evolutionary process. *Science*. 2016; **352**(6282): p. 169–175.
145. Gerlinger, M. Metastasis seeding cells: lone invaders or mass migrators? *Clin Cancer Res*. 2018; **24**(9): p. 2032–2034.
146. Gui, P., and T. G. Bivona. Evolution of metastasis: new tools and insights. *Trends Cancer*. 2022; **8**(2): p. 98–109.
147. Heyde, A., et al. Consecutive seeding and transfer of genetic diversity in metastasis. *Proc Natl Acad Sci U S A*. 2019; **116**(28): p. 14129–14137.
148. Ji, H., et al. Lymph node metastasis in cancer progression: molecular mechanisms, clinical significance and therapeutic interventions. *Signal Transduct Target Ther*. 2023; **8**(1): p. 367.
149. Reiter, J. G., et al. Lymph node metastases develop through a wider evolutionary bottleneck than distant metastases. *Nat Genet*. 2020; **52**(7): p. 692–700.

150. Jiang, B., et al. Machine learning of genomic features in organotropic metastases stratifies progression risk of primary tumors. *Nat Commun.* 2021; **12**(1): p. 6692.
151. Zuccato, J. A., et al. Prediction of brain metastasis development with DNA methylation signatures. *Nat Med.* 2024. **31**: p. 116–125.
152. Lee, I. S., et al. Transcriptomic profiling identifies a risk stratification signature for predicting peritoneal recurrence and micrometastasis in gastric cancer. *Clin Cancer Res.* 2021; **27**(8): p. 2292–2300.
153. Li, G., et al. Predicting metastatic transcriptomes of patient tumors with deep learning. *Cancer Res.* 2024; **84**(6_Supplement): p. 896–896.
154. Liang, Y. L., et al. A lncRNA signature associated with tumor immune heterogeneity predicts distant metastasis in locoregionally advanced nasopharyngeal carcinoma. *Nat Commun.* 2022; **13**(1): p. 2996.
155. Bojmar, L., et al. Multi-parametric atlas of the pre-metastatic liver for prediction of metastatic outcome in early-stage pancreatic cancer. *Nat Med.* 2024; **30**(8): p. 2170–2180.
156. Zhao, Y., et al. A novel risk signature for predicting brain metastasis in patients with lung adenocarcinoma. *Neuro Oncol.* 2023; **25**(12): p. 2207–2220.
157. Li, Y., et al. Intratumoral microbiome promotes liver metastasis and dampens adjuvant imatinib treatment in gastrointestinal stromal tumor. *Cancer Lett.* 2024; **601**: p. 217149.
158. Moran, S., et al. Precision medicine based on epigenomics: the paradigm of carcinoma of unknown primary. *Nat Rev Clin Oncol.* 2017; **14**(11): p. 682–694.
159. Ma, W., et al. New techniques to identify the tissue of origin for cancer of unknown primary in the era of precision medicine: progress and challenges. *Brief Bioinform.* 2024; **25**(2): p. bbae028.
160. Soh, K. P., et al. Predicting cancer type from tumour DNA signatures. *Genome Med.* 2017; **9**(1): p. 104.
161. Penson, A., et al. Development of genome-derived tumor type prediction to inform clinical cancer care. *JAMA Oncol.* 2020; **6**(1): p. 84–91.
162. Jiao, W., et al. A deep learning system accurately classifies primary and metastatic cancers using passenger mutation patterns. *Nat Commun.* 2020; **11**(1): p. 728.
163. Nguyen, L., A. Van Hoeck, and E. Cuppen. Machine learning-based tissue of origin classification for cancer of unknown primary diagnostics using genome-wide mutation features. *Nat Commun.* 2022; **13**(1): p. 4013.
164. Fernandez, A. F., et al. A DNA methylation fingerprint of 1628 human samples. *Genome Res.* 2012; **22**(2): p. 407–419.
165. Moran, S., et al. Epigenetic profiling to classify cancer of unknown primary: a multicentre, retrospective analysis. *Lancet Oncol.* 2016; **17**(10): p. 1386–1395.
166. Kang, S., et al. CancerLocator: non-invasive cancer diagnosis and tissue-of-origin prediction using methylation profiles of cell-free DNA. *Genome Biol.* 2017; **18**(1): p. 53.
167. Grewal, J. K., et al. Application of a neural network whole transcriptome-based pan-cancer method for diagnosis of primary and metastatic cancers. *JAMA Netw Open.* 2019; **2**(4): p. e192597.
168. Shen, Y., et al. TOD-CUP: a gene expression rank-based majority vote algorithm for tissue origin diagnosis of cancers of unknown primary. *Brief Bioinform.* 2021; **22**(2): p. 2106–2118.

169. Zhao, Y., et al. CUP-AI-Dx: a tool for inferring cancer tissue of origin and molecular subtype using RNA gene-expression data and artificial intelligence. *EBioMedicine*. 2020; **61**: p. 103030.
170. Jiang, W., et al. A naive Bayes algorithm for tissue origin diagnosis (TOD-Bayes) of synchronous multifocal tumors in the hepatobiliary and pancreatic system. *Int J Cancer*. 2018; **142**(2): p. 357–368.
171. Best, M. G., et al. RNA-Seq of tumor-educated platelets enables blood-based pan-cancer, multiclass, and molecular pathway cancer diagnostics. *Cancer Cell*. 2015; **28**(5): p. 666–676.
172. Moiso, E., et al. Developmental deconvolution for classification of cancer origin. *Cancer Discov*. 2022; **12**(11): p. 2566–2585.
173. Hoshino, A., et al. Extracellular vesicle and particle biomarkers define multiple human cancers. *Cell*. 2020; **182**(4): p. 1044–1061. e18.
174. Pitroda, S. P., et al. Integrated molecular subtyping defines a curable oligometastatic state in colorectal liver metastasis. *Nat Commun*. 2018; **9**(1): p. 1793.
175. Satelli, A., et al. Epithelial-mesenchymal transitioned circulating tumor cells capture for detecting tumor progression. *Clin Cancer Res*. 2015; **21**(4): p. 899–906.
176. Magbanua, M. J. M., et al. Clinical significance of circulating tumor cells in hormone receptor-positive metastatic breast cancer patients who received letrozole with or without bevacizumab. *Clin Cancer Res*. 2020; **26**(18): p. 4911–4920.
177. Tie, J., et al. Circulating tumor DNA as an early marker of therapeutic response in patients with metastatic colorectal cancer. *Ann Oncol*. 2015; **26**(8): p. 1715–1722.
178. Sinn, B. V., et al. *SET(ER/PR): a robust 18-gene predictor for sensitivity to endocrine* therapy for metastatic breast cancer. *NPJ Breast Cancer*. 2019; **5**: p. 16.
179. Scher, H. I., et al. Association of AR-V7 on circulating tumor cells as a treatment-specific biomarker with outcomes and survival in castration-resistant prostate cancer. *JAMA Oncol*. 2016; **2**(11): p. 1441–1449.
180. Szałata, A., et al. Transformers in single-cell omics: a review and new perspectives. *Nat Methods*. 2024; **21**(8): p. 1430–1443.
181. Poirion, O. B., et al. DeepProg: an ensemble of deep-learning and machine-learning models for prognosis prediction using multi-omics data. *Genome Med*. 2021; **13**(1): p. 112.
182. Wang, T., et al. MOGONET integrates multi-omics data using graph convolutional networks allowing patient classification and biomarker identification. *Nat Commun*. 2021; **12**(1): p. 3445.

CHAPTER 15

Systems Modeling for Cancer Microenvironment and Regenerative Medicine

Systems modeling or multiscale modeling is the key concept for systems biology, systems medicine, or whole-body digital twin. In this chapter, we will take two applications as examples to describe how to apply systems modeling and agent-based model (ABM) to understand cancer growth and bone regeneration. The ABM model is also the key concept for digital twin.

15.1 Multiscale Modeling for Multiple Myeloma Growth

The multiscale model is depicted in Fig. 15.1. It consists of intracellular level, intercellular level, and tissue level. We will first introduce the intracellular signaling pathways of myeloma-initiating cells (MyICs) or stem-like cells and bone marrow stromal cells (BMSCs) which may dominate multiple myeloma (MM) cell's biomechanical phenotype at intercellular level, MM development and disease prognosis of tissue level, cell-cell interactions via lineage model, and MM cancer growth model. The basic idea is that cell-to-cell interactions are the pivot chain linking intracellular level features of MIC and BMSC to intracellular biomechanical phenotype switch of MIC, BMSC, progenitor cell (PC), and MM. On the contrary, MIC-BMSC interaction triggers immediate intracellular signaling events, which in turn determine the growth and differentiation patterns of MyICs, shape the midterm MM growth profile, and determine the long-term effects of relapse possibility at the tissue level.

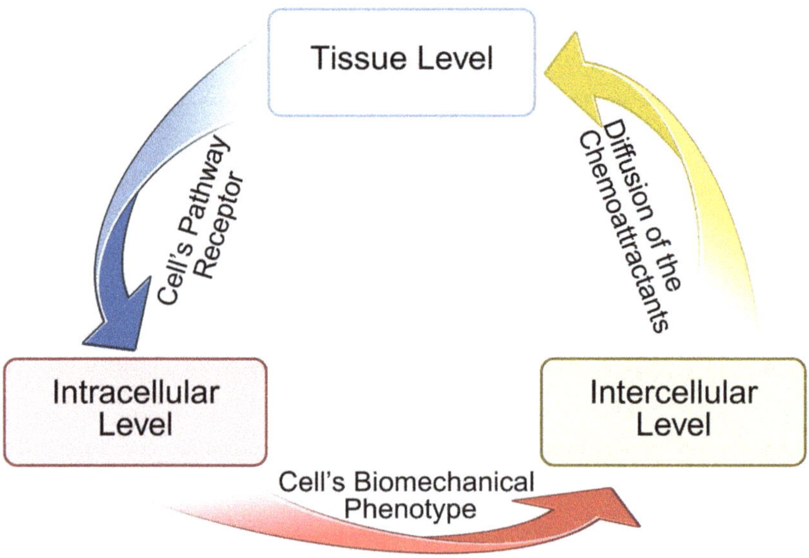

Fig. 15.1 Multiscale model of MM growth.

15.1.1 Biomechanical Studies at Intracellular Level

15.1.1.1 Myeloma BMSCs are stiffer and more contractile than normal BMSCs

We have established six lines of BMSCs from myeloma patients and age-matched controls using their bone marrow samples as previously described [1], two of which have been engineered with red fluorescent protein (RFP). Then a cyto-indenter was used to measure the stiffness in BMSCs at a single-cell level by modifying the protocol reported previously [2]. The mean stiffness (represented by Young's modulus) of myeloma BMSCs (443 ± 184 Pa) was significantly greater than that of normal BMSCs (181 ± 79 Pa). We further evaluated if the differences in stiffness in individual BMSCs may lead to changes in biomechanical properties at the tissue level using a gel contraction assay as described in previous studies [3, 4]. We discovered that the myeloma BMSCs contract the gels much more than the normal BMSCs (Fig. 15.2, Student's t-test), indicating that myeloma BMSCs exert higher contraction forces than normal BMSCs.

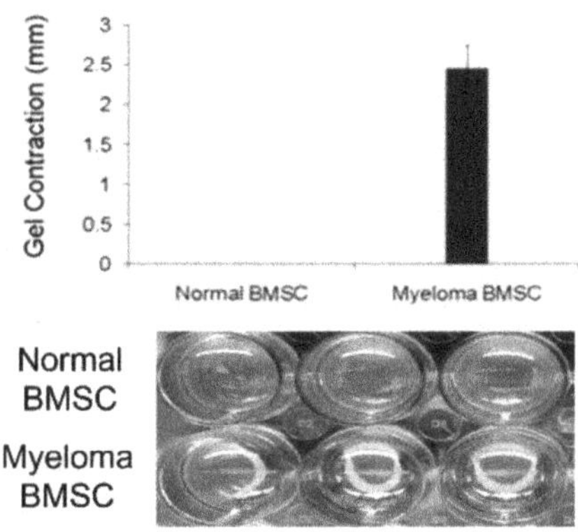

Fig. 15.2 Myeloma BMSCs show more contraction of collagen gels than normal BMSCs.

15.1.1.2 MyICs form more colonies, adhere more tightly, and become more resistant to drugs when they are co-cultured with myeloma BMSCs or stiffer hydrogels

To preliminarily evaluate the interaction between the MyICs and their niche, MyICs were recognized by their high expression of the ATP-binding cassette family of membrane transporters, leading to an efflux of the DNA-binding dye Hoechst 33342, resulting in the "side population (SP) phenotype" characteristic of MyICs by flow cytometry assay [5–7]. It is found that the single MyICs co-cultured with myeloma BMSCs form colonies more frequently than with control BMSCs (24/132 or 18.1% vs. 11/128 or 8.6%, chi-square, $p = 0.0235$). Furthermore, the serially re-plated assay indicated that the MyICs had greater clonogenic expansion with the support of myeloma BMSCs than with normal BMSCs in the third re-plating.

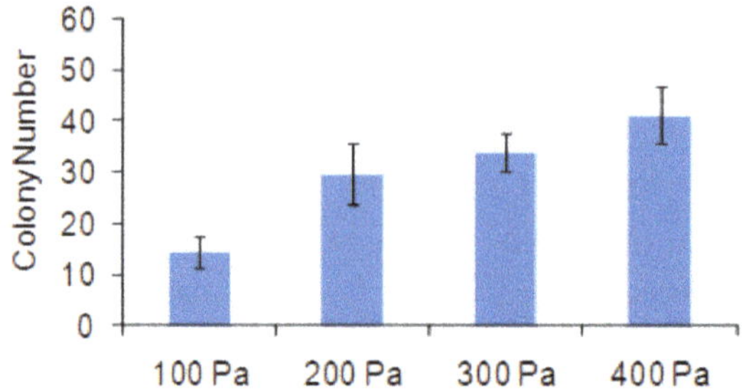

Fig. 15.3 SP cells (MyICs) form more colonies on the stiffer hydrogels.

To examine if the increased stiffness of myeloma BMSCs contributes to the increase of colony formation, SP cells were sorted into wells coated with different stiffnesses of hydrogels using the Extracel™ Hydrogel Kit (Glycosan Biosystem, Salt Lake City, UT, USA) as previously reported [8, 9]. The results indicated that the MyICs formed more colonies when they were cultured on the hydrogels with higher stiffness (Fig. 15.3). Furthermore, SP cells of the RPMI 8226 line were sorted into a six-well plate coated by hydrogels with different stiffnesses (200 and 400 Pa, respectively) and were cultured for one week in Iscove's Modified Dulbecco's Medium (IMDM) with 10% Fetal bovine serum (FBS). Afterward, the cells were harvested for SP cell analysis by flow cytometry. As shown in Fig. 15.4, purified SP cells co-cultured on stiffer hydrogel maintained a higher percentage of SP cells, indicating that stiffer hydrogel supports SP cell expansion (self-renewal).

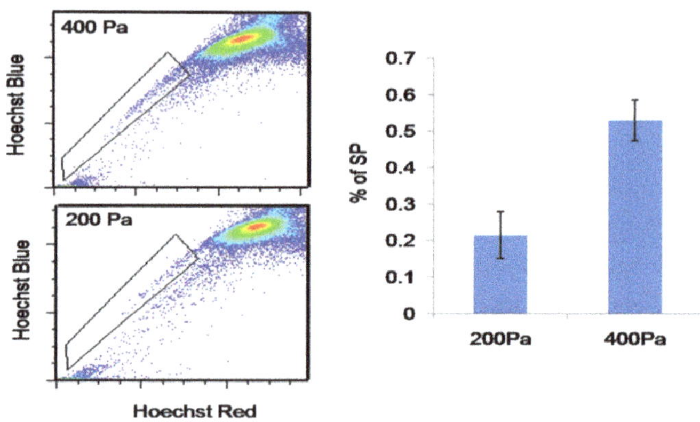

Fig. 15.4 Stiffer hydrogel maintains a higher percentage of SP cells.

It was postulated that the increased stiffness and contractile force of the myeloma BMSCs may promote more MIC attachment than normal BMSCs.

SP cells of the RPMI 8226 cells were cultured on myeloma or control BMSCs for 24 hours, and then shear stress was applied by shaking on a rotator for 20 minutes. The SP cells that remained adherent to BMSCs were then counted, and the percentages of cells that remained adherent were calculated. The results indicated that the myeloma BMSCs were able to hold about two times more SP cells than the normal BMSCs upon the given shear stress. Then, using hydrogels, a gradual increase in the percentage of adherent SP cells was observed as gel stiffness increased from 100 Pa to 300 or 400 Pa, with cell numbers rising from 5 to 40 (data not shown). These results suggest that the biomechanical properties of myeloma BMSCs contribute to maintaining MyICs within the stem cell niches. In addition, the viability of SP cells gradually increases as the stiffness of hydrogel increases when SP cells are treated with bortezomib, suggesting that biomechanical properties of BMSCs contribute to the drug resistance of MyICs (data not shown). Besides the canonical mechanisms on how the cell proliferation is regulated by substrate stiffness, we found that the SDF-1 (stromal cell-derived factor-1)/CXCR4 (CXC chemokine receptor type 4) autocrine signaling has been reported to promote proliferation and survival of bone marrow cells (CD34+ cells [10] and myeloid progenitors [11]), possibly via the phosphatidylinositol 3-kinase/protein kinase B (PI3K-Akt) axis [10]. It is showed that the expression of CXCR4 in SPs also significantly increased when cultured on stiffer surfaces [12], implying that the microenvironmental biomechanical properties may also modulate SP self-renewal by modulating the SDF1/CXCR4. We will also examine the role of the SDF1/CXCR4 autocrine in the proliferation induced by the stiffness of microenvironment.

15.1.1.3 SDF1/CXCR4 pathway may play a role in regulating the biomechanical phenotype of myeloma BMSCs and the fate of MyICs

To study the mechanism underlying the increased stiffness of myeloma BMSCs, RNA was extracted from myeloma and control BMSCs, and their gene expression profiles were analyzed. Differentially expressed genes (DEGs) were then imported into ingenuity pathway analysis (IPA) for gene network analysis. It was found that genes with functions related to the SDF1/CXCR4 pathway were significantly enriched in myeloma BMSCs. In addition, it was found that SP cells had much higher SDF1 and CXCR4 expression and secreted much more SDF1 than the mature myeloma cells (non-SP cells) (Fig. 15.5). Of interest, SDF1 significantly increased the gel contraction of BMSCs (Fig. 15.6). Furthermore, the treatment of AMD3100 inhibited the adherence of SP cells to myeloma BMSCs and decreased the colony formation of SP cells seeding on myeloma BMSCs (data not shown). These findings suggest that the SDF1/CXCR4 pathway plays a role in regulating the biomechanical phenotype of BMSCs and may impact the fates of MyICs via changes in biomechanical phenotype of BMSCs.

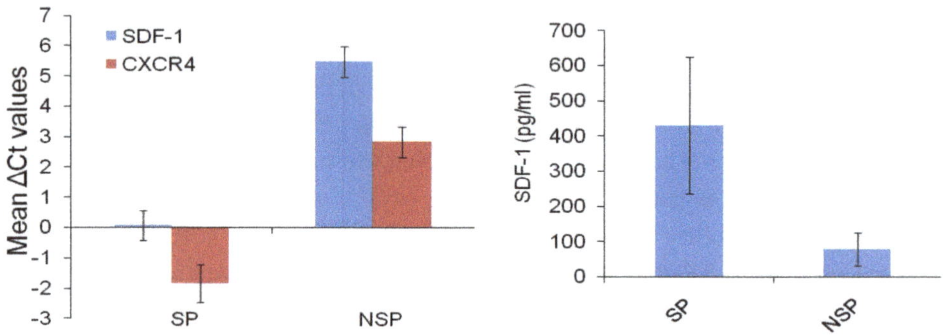

Fig. 15.5 SP cells had higher SDF1 and CXCR4 expression and secreted more SDF1 than the mature myeloma cells (non-SP cells).

15.1.1.4 Intracellular level: RPPA reveals stiffness-triggered MIC signaling pathway

In order to study how stiffness regulates MyICs and BMSCs fate and how MyICs and BMSCs communicate, our team conducted a reverse-phase protein array (RPPA) experiment [13–15] based on which we successfully determined Tumor necrosis factor alpha (TNF-α)-induced myelodysplastic syndrome (MDS) stem cell behavior and signaling pathway changes recently [15]. MIC preliminary cells were lysed and proteins were prepared for RPPA assays according to protocols provided by the phosphoproteomics core at MDACC. The RPPA platform, established at the MD Anderson Cancer Center (MDACC), provides a robust and highly accurate functional proteomics assay, and its coverage of protein species is quickly expanding with the fast growth of the validated antibody library [16–18]. Although a number of papers [19–22] have been published using the technology to interrogate signaling networks, no informatics tools are available to model the time-lapse RPPA data. Therefore, a systems biology tool (p38 mitogen-activated protein kinases (P38-MAPK) pathway modeling [14]) was developed to analyze the time-lapse RPPA data, which will be used in this project.

Hydrogel surfaces (Young's modulus of 200 and 400 Pa) were used to mimic the stiffness of MIC stroma under physiological and pathological conditions, respectively. MIC cells, sorted according to SP staining (Hoechst 33342) from other MM cells, were seeded onto hydrogel surfaces and collected at 0, 30, and 60 min and overnight for RPPA assay.

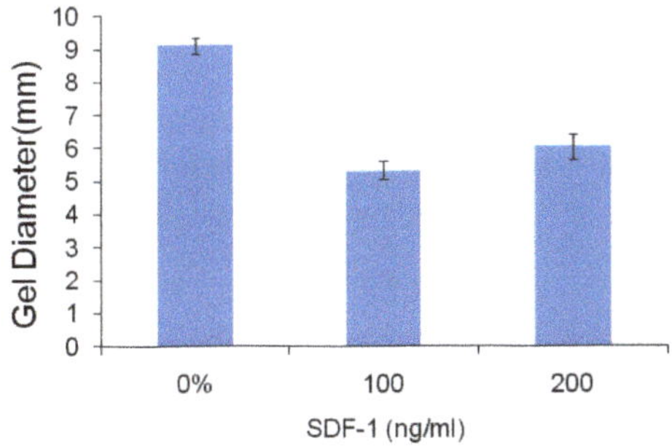

Fig. 15.6 SDF1 increased the gel contraction of BMSCs.

Analysis of RPPA data (Fig. 15.7, upper panel) showed that stiffness-triggered signaling pathways can be disentangled from other effects. The introduction of hydrogel surfaces initiated both common and stiffness-specific phenotypical changes in MIC, and such changes are either transient or sustained. Statistical analysis (two-way ANOVA) suggested that stiffness difference dominates early-stage signaling and alters key pathways in the long term. Indeed, within the first 60 minutes, 65.4% of all significantly varied functional proteins were stiffness-specific, compared with less than 25% after one day. It was reasoned that microenvironment stiffness cues were quickly recognized by MIC via transient signaling messengers (e.g., the Akt-pT308 and pErk1, Fig. 15.7, right column) and finally modulated cell behaviors and fates in terms of late-stage sustainable protein expression patterns (such as the upregulation of BCL2, cyclin D, and pP90RSK, Fig. 15.7, right column).

RPPA data allows us to study how the stiffness regulates MIC fate in a global view. To study such a signaling pathway, a generic signaling pathway network was generated using bioinformatics tools and databases (IPA® by Ingenuity Systems®) (Fig. 15.7, lower-left plot). For clarity, only a portion of the nonspecific pathways is shown in gray. The final MIC-specific signaling pathways are displayed at normal size, with red font indicating upregulated functional proteins on the stiffer surface, green indicating downregulated ones, and gray representing proteins not available in the current RPPA antibody library. This generic signaling pathway has over 80 proteins. The optimization problem involves how to get our MIC-specific signaling pathway by fitting the stiffness-specific RPPA data to the generic pathway map. We built 126 ODEs with over 90 parameters for this generic signaling pathway following the mass action law.

It is not feasible to solve such a complex system directly. A modulated factor graph approach can solve the optimization problem. First, our team construct

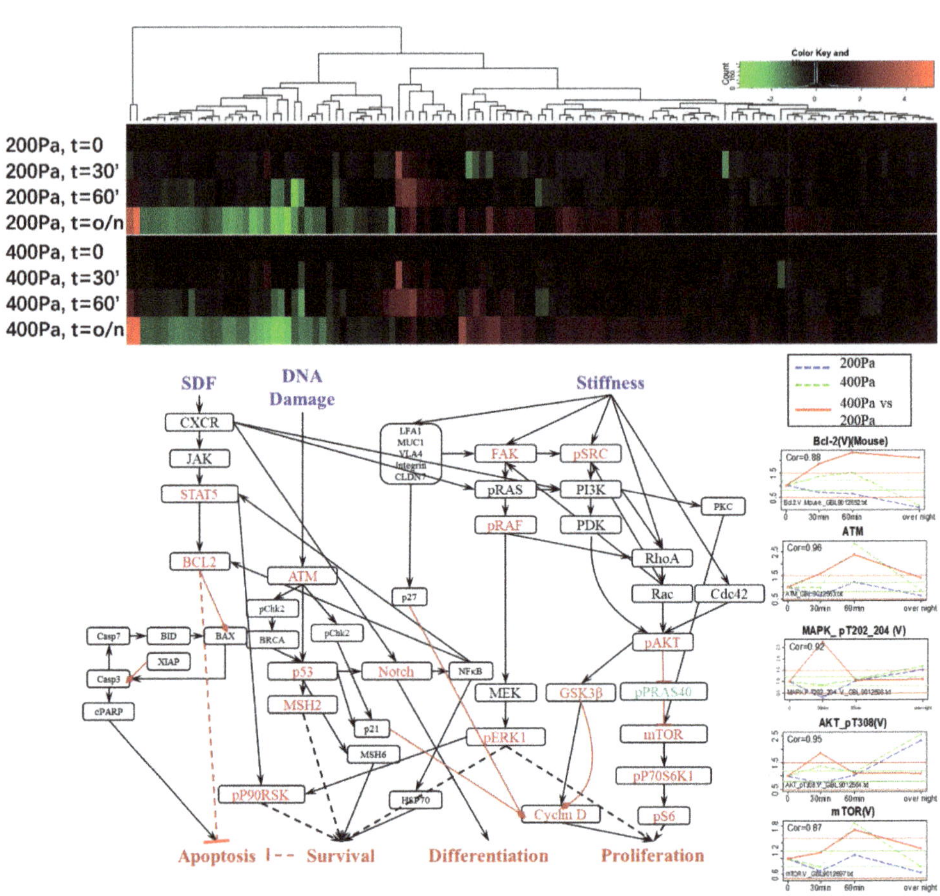

Fig. 15.7 Heatmaps of RPPA (upper) show common and stiffness-specific signaling patterns. The dynamics of protein expression and activity after seeding are similar on both 200 and 400 Pa surfaces. The stiffness-specific MIC signaling pathway (lower) is inferred from RPPA data using BP algorithm.

the mathematical model (126 ODEs) of the signaling pathway using mass action equations. Second, we represent the ODEs using a hybrid functional Petri net. Third, because there are many proteins and connections in the generic pathway network, it is natural to decompose them into smaller components and estimate the parameters for each component separately. Guided by some heuristic rules, the Petri net is decomposed into smaller components that can be efficiently estimated. Fourth, we estimate the parameters of each component via the particle swarm optimization algorithm [23] step by step. The results of the fourth step gave us the initial values for further refining the parameters from a global view. In order to do global optimization with low computational complexity, we adopted the factor graph approach and belief propagation (BP) algorithm [24] in the fifth step. Finally, the optimal parameters of the specific network were obtained. The major reason for using the BP algorithm is that it is, in essence, a dynamic programming algorithm which performs a more global search than strategies such as gradient descent.

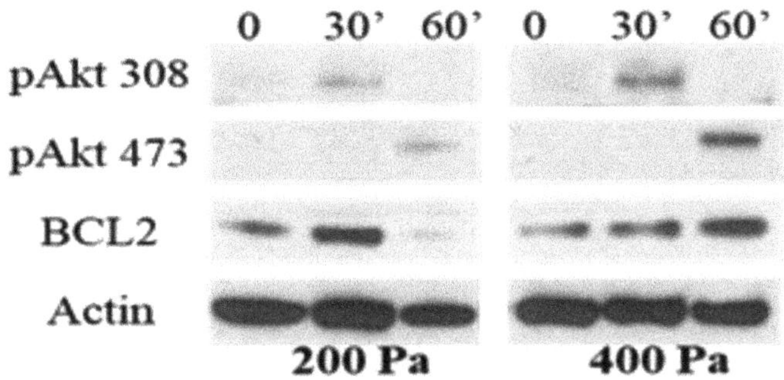

Fig. 15.8 Validation of key proteins in the MIC-specific signaling pathways.

The MIC-specific stiffness signaling pathway was established based on estimated ODE parameters of the generic pathway network. The rationale is that extremely small parameter values imply nonsignificant signaling transduction, which should be trimmed. All such "trivial edges" in the generic pathway network were deleted, as well as the resultant "orphan nodes" (i.e., proteins that no longer interact with other network components), and finally a simplified MIC signaling pathway was delineated (Fig. 15.7, lower-left). It showed that stiffer niches triggered Focal Adhesion Kinase (FAK)/Rapidly Accelerated Fibrosarcoma (RAF)/Src proto-oncogene (SRC) pathways, which branched to mitogen-activated protein kinase (MEK)/extracellular-signal-regulated kinase (ERK) and Akt/mTOR pathways and promoted MIC proliferation and survival. Meanwhile, the expression levels of SDF1 and CXCR4 were upregulated, which enhanced the SDF1/CXCR4 autocrine system and caused upregulation of apoptosis inhibitor BCL2. The Notch pathway was also activated by SDF1/CXCR4 axis, which drove MIC self-renewal. The enhanced ATM/p53 pathway promoted DNA repair and MIC survival.

All parameters were individually and slightly altered (1% from their estimated values), and the parameter sensitivity (defined as the resulting changes in system output) was achieved from simulation. Key proteins defined by parameter sensitivities were validated with Western blotting and real-time polymerase chain reaction (some are shown in Fig. 15.8). In summary, RPPA provides an efficient way to MIC signaling induced by the stiffness from BMSC. Some nodes in the signaling pathway imply a possible hidden pathway.

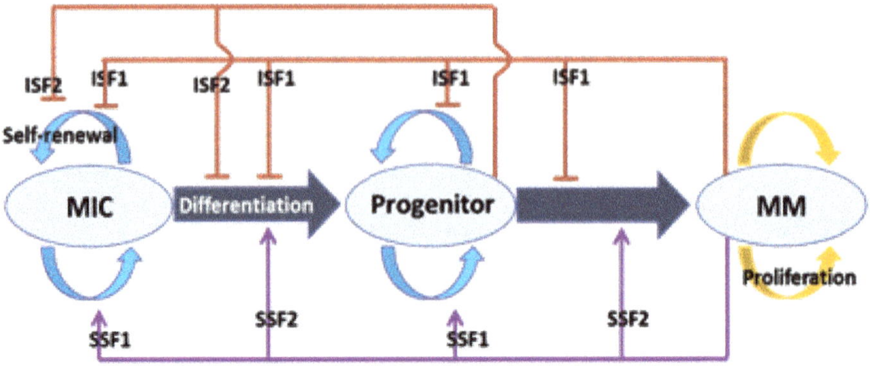

Fig. 15.9 Intercellular level: MIC lineage model.

15.1.2 Lineage Model of MIC-PC-MM Interaction at Intercellular Level

Our team modeled the cell-cell interactions among MyICs, PCs, and MM cells. Figure 15.9 gives the basic description of the hypothesis at the intercellular level: we assumed the MM cells secrete stimulation factor Stimulation Secretion Factor 1/2 (SSF1/2) to promote the proliferation and differentiation of MyICs and PCs while secreting inhibition factor Inhibition Secretion Factor 1/2 (ISF1/2) to inhibit the proliferation and differentiation of MyICs and PCs. Similarly, we set the assumptions for the PCs. The question was how to identify those factors and their functions at

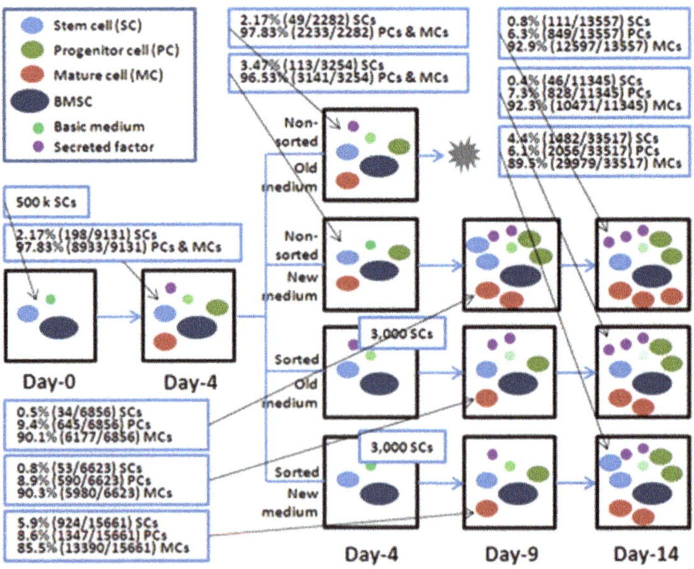

Fig. 15.10 Culture diagram and experimental data.

the cellular level. We designed biological experiments to obtain the sizes of each cell population and developed a novel lineage model to answer these questions.

1. Cellular assay experimental design for sorting different types of MM-related cells. Hoechst 33342 and SP analysis and cell culture: Hoechst staining was performed according to the protocol outlined by the Goodell' lab at the Baylor College of Medicine. Briefly, RPMI8226 cells were harvested by centrifugation and washed in PBS and then resuspended at 1×10^6/ml in Dulbecco's modified Eagle medium (DMEM) supplemented with 10 mM HEPES (Invitrogen) and 2% Fetal bovine serum (FBS) with Hoechst 33342 dye at a final concentration of 10 μg/ml. After incubation at 37°C for 60 minutes, the cells were centrifuged and re-suspended in cold Hanks' balanced salt solution (HBSS) buffers with 2 μg/ml propidium iodide (PI) to exclude dead cells. Samples were then put on ice before sorting. Hoechst dye is then excited with a UV laser at 350 nm, and signals were measured with Hoechst blue and Hoechst red filters. SP cells were sorted out by cell sorter and used for further experiments. The collected SP cells were then cultured as shown below.

$$\frac{dX_1}{dt} = (2f_1 - 1)\mu_1 X_1 - d_{SC} X_1$$

$$\frac{dX_2}{dt} = 2(1 - f_1)\mu_1 X_1 + (2f_2 - 1)\mu_2 X_2 - d_{PC} X_2$$

$$\cdots\cdots$$

$$\frac{dX_{n-1}}{dt} = 2(1 - f_{n-2})\mu_{n-2} X_{n-2} + (2f_{n-1} - 1)\mu_{n-1} X_{n-1} - d_{PC} X_{n-1}$$

$$\frac{dX_n}{dt} = (1 - f_{n-1})\mu_{n-1} X_{n-1} - d_{MC} X_n$$

Fig. 15.11 The lineage model of MyICs, PCs, and MCs.

As summarized in Fig. 15.10, the SP cells were sorted and cultured in DMEM (10% FBS) for four days, and then 80% of the cells were re-sorted for the SP cells. The remaining 20% of the cells were then separated into two groups: one "non-sorted old medium" (1/2 of the old medium and 1/2 of fresh DMEM with 10% FBS) and the other "non-sorted new medium" (fresh DMEM with 10% FBS). The resorted SP cells were divided into the sorted-old medium or sorted-new medium culture as above for further SP analysis on day 9 and day 14, with medium change every 3 days. Before the SP analysis, the total cells were stained with CD138-PE as a marker for cell differentiation from MyICs to progenitors (CD138– non-SP cells) or mature myeloma cells (CD138+ non-SP cells). The results indicated that when the sorted SP cells were cultured for four days, most of the progenies were committed progenitor or MM cells, with only 2.17% of cells remaining stem cells. In addition, the stem cell population continued to decline on day 9 to 0.8% with replacement of culture medium on day 4, while when continuing to culture the cells with old medium, most cells died on day 9. Similarly, the re-sorted myeloma stem cells showed a significant decline in myeloma stem cells with the old medium compared to the replaced new medium on day 4. These findings suggest that the differentiated cells secrete inhibitor factors to MyICs and mature MM cells.

A

$$\frac{d[SSF1]}{dt} = sr1 \cdot X$$
$$\frac{d[SSF2]}{dt} = sr2 \cdot X$$
$$\frac{d[ISF1]}{dt} = sr3 \cdot X$$
$$\frac{d[ISF2]}{dt} = sr4 \cdot \left(\frac{[ISF1]}{Ls + [ISF1]}\right) \cdot \sum X$$

B

$$f_i = f_{max} \exp\left[-\frac{(i-1)^2}{2D_{SR}^2}\right] \cdot \frac{1+[SSF1]^{k1}}{1+[ISF2]^{k4}+[SSF1]^{k1}}$$
$$\mu_i = \mu_{max} \exp\left[-\frac{(i-n_{max})^2}{2D_{GR}^2}\right] \cdot \frac{t^{kt}}{\tau_D^{kt}+t^{kt}} \cdot \frac{1+[SSF2]^{k2}}{1+[ISF1]^{k3}+[SSF2]^{k2}}$$
$$i = 1, \ldots, n-1;$$

Fig. 15.12 Mathematical model for the secreted factor system: (A) illustrate how the factors are secreted from different cell types and (B) tell us how the major parameters in the MM cancer cell models are determined by the secreted factors.

2. MIC lineage model: As it has been described in Fig. 15.9, the MyICs can self-renew or differentiate from PCs, and the PCs also can self-renew or differentiate to mature plasma cells, and the mature cells can proliferate themselves. In addition, apoptosis will occur in all three kinds of cancer cells during MIC development. Several compartments were considered during the division of PCs as modeled in previous studies [25–28]. As we can see from the symbols in Fig. 15.10, $X1$ represents the number of MyICs, $X2$ to X_{n-1} represent the number of PCs in each compartment among multiple compartments for the process of progenitor development, and Xn represents the number of MM cells, respectively. The symbol f_i represents the probability of self-renewal of the stem cells or PCs in the ith compartment; and u_i represents the proliferation rate of these three kinds of cells in the ith compartment; dSC, dPC, and dMC represent the apoptosis or degradation rates of these three kinds of cells, respectively. Therefore, by referring to previous studies [25–27], an ODE system for

the cellular growth model for these three cell types was constructed as the equations in Fig. 15.11. To determine the parameters in this model, the parameters were linked to the corresponding secreted factors. (Fig. 15.12A). The secreted factors were modeled using the simple linear function referring to [26]. The parameters sr_1 through sr_4 represent the secretion rates for different factors. Combining the secreted factor model with some a priori knowledge for the MIC self-renewal and differentiation, we can determine the self-renewal probabilities, proliferation rates, and death rates during the development of the three MM cell types using the Hill-type functions in Fig. 15.12B. Dynamic data from the experiment in Fig. 15.10 were used as a training set to estimate parameters using a genetic algorithm followed by the Nelder-Mead simplex algorithm [29]. All parameters were identifiable according to coefficients of variation of 1,000 bootstrap experiments. Parameter sensitivity analysis shows our model is stable.

The experimental data of cell population numbers on day 14 for three separate culture conditions were used to validate the model. The validation results (Fig. 15.13, left on day 14) showed that the simulations fitted the experimental data well with approximately 90% accuracy, which demonstrates the accuracy of the model parameter estimation and the reliability of prediction from the in silico model. The predicted results also showed that stem cell and PC numbers rapidly plateaued and began to decline within a week. Increased frequency of culture manipulations at later time points (day 9) improved stem cell expansion while having no effect on the output of total cells or PCs. We also noticed that cell density affected the concentration of secreted factor and the switch between the promotion and inhibition of tumor growth. Stem (MIC) and PC output can, thus, be specifically manipulated by dynamically perturbing intercellular feedback, and the culture decline was due to increased inhibitory secreted factor (ISF1 and ISF2) feedback from the accumulating differentiating cells. In general, the in silico system demonstrated that intercellular feedback signaling regulates cell population dynamics.

The prediction suggested that ISF1 and SSF1 were similar to the profiles of IL-6 and TGFβ, and ISF2 (Fig. 15.13, right) yields a different profile from the other three secreted factors. As expected from the parameter sensitivity analysis, the inhibitor factors always had higher concentration values than the simulator factors that contributed to the

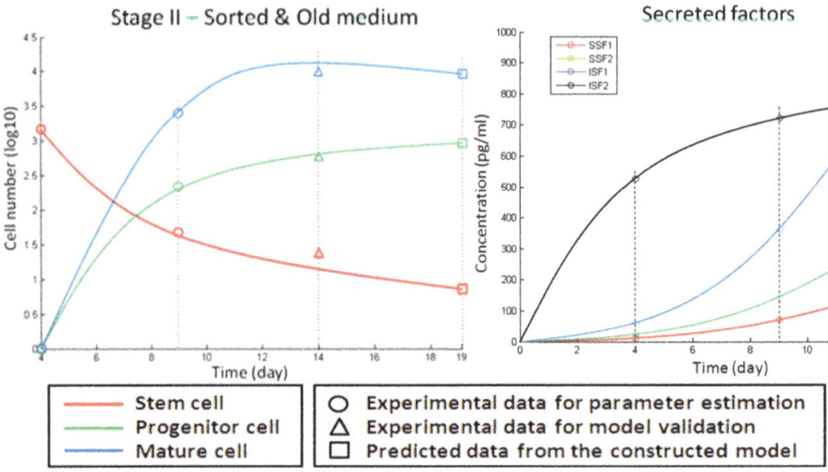

Fig. 15.13 (left) The cell population dynamics and (right) the estimated profiles of the secreted factors.

culture decline in the experiment. In summary, intercellular feedback model of MIC growth predicatively simulated the dynamic characteristics of MyICs, PCs, and MM cells in vitro. The obtained information can be further integrated into ABM for MM growth studies at the tissue level.

15.1.3 Agent-Based Model at Tissue Level

A 3D multiscale ABM was developed to study the role of tumor–stroma interactions in MM tumorigenesis. The system was classified into three levels: the intracellular, the intercellular, and the tissue level. Their relations were conceptually defined as "interfaces" among these levels. As indicated in Fig. 15.1: (a) Cell biomechanical phenotypes were defined as the interface between the intracellular and the intercellular level. The signaling pathway at the intracellular level determined BMSC biomechanical phenotypes, and cell-cell interactions at the intercellular level regulated and depended on the biomechanical phenotypes of BMSCs. (b) We denoted the diffusion of the cytokines as the interface between the intercellular and the tissue level. On the one hand, as the behavior of the intercellular level, cells secreted cytokines into the interstitial space. On the other hand, secreted cytokines diffused at the tissue level according to their gradient of concentrations. (c) Cell's pathway receptors were defined as the interface between the tissue and the intracellular level. The paracrine or autocrine of cytokines on the tissue level was sensed by cellular receptors and triggered intracellular signaling pathways to determine cell biomechanical phenotypes.

The ABM integrates events at multiple spatial and temporal scales.

(1) Spatial scales: Intracellular signaling pathways are encapsulated into each cell to determine either the BMSC intercellular biomechanical phenotype (cell stiffness) or tumor cell behaviors (migration or proliferation). And then, cancer cells compete for the best location in the 3D extracellular matrix to migrate or proliferate regarding the changes in BMSC cell's stiffness (Fig. 15.3) and sensitivity of chemoattractant (SDF1) (Fig. 15.5). In turn, chemoattractant cues at the tissue level will trigger cell's intracellular signaling pathways inside agents via the agent interfaces (receptors) (Fig. 15.7), and the resultant feedbacks are the changes of either agents' properties (change of cell stiffness or sensitivity to outside stimuli) or behaviors (secretion of cytokines, proliferation, differentiation, apoptosis, or migration). (2) Temporal scales: The model covers minute-to-hour-level signaling pathway dynamics (RPPA, Fig. 15.8); day-level cell division, apoptosis (lineage model, Fig. 15.9), and local migration (ABM model, Fig. 15.14); week-level drug responses; month-level tumor growth; and year-level disease prognosis.

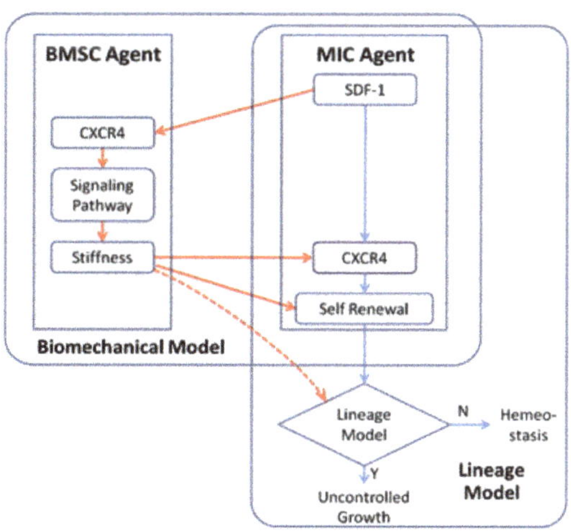

Fig. 15.14 A sketch of ABM model.

Functions of the agents:

We defined four types of agents in the model. The temporal-spatial information (time t and position p) was denoted in subscripts. (1) BMSC agent, which was for modeling the 3D reticular network formed by BMSCs. Our preliminary results showed that the stiffness of the BMSC is determined by myosin II, so the change rate of BMSC stiffness was described by a myosin-triggered Hill-type function (Eq. 15.1). (2) Myeloma stem cell agent, which was used to model myeloma stem cells. MIC proliferation, differentiation, and apoptosis were governed by intracellular signaling pathways shown in Fig. 15.7, and therefore, their rates were also described by Hill functions in similar forms as Fig. 15.12B. (3) PC agent, which was used to model progenitor cells. PC's differentiation rate, self-renewal rate, and death rate were determined from the lineage model (Fig. 15.12B). (4) MM cell agent, which was used to model mature MM cells. Again, MM's self-renewal rate and death rate were determined from the lineage model (Fig. 15.12B). Meanwhile, MM agents secreted lineage-specific cytokines (SSF1/2 and ISF1/2, please refer to the lineage model in Fig. 15.9).

$$Stiff\big|_{t,p} = Stiff\big|_{a,p}\frac{\left(pMY\big|_{t,p}/K_x\right)^n}{1+\left(pMY\big|_{t,p}/K_x\right)^n}. \tag{15.1}$$

Cell-cell interaction. Interactions between BMSCs and MyICs (Fig. 15.14) were: (a) MyICs secreted SDF1 into the 3D extracellular matrix. Described by Eq. 15.2, SDF1 signaling is autocrined for MyICs and paracrined for BMSCs. As a response to extracellular SDF1 signals, the BMSCs become stiffer (Eq. 15.1); meanwhile, MyICs prefer to attach to stiffer BMSCs. When attached, stiffer BMSCs boost proliferation and self-renewal of MyICs and thus promote MIC expansion, which drives more sustainable MM growth. As indicated by the lineage model (Fig. 15.9), MM cells secreted cytokines to promote or inhibit the proliferate rate of PC and MM cells. PC and MM cells preferred to attach to stiffer BMSCs, and such attachment to stiffer BMSCs in turn accelerated the self-renewal and proliferation of PC as well as the proliferation of MM. In summary, the cell-cell interactions among BMSC, MIC, PC, and MM cells stimulated the MM cancer expansion.

$$\begin{cases} L\big|_{t,p} = L\big|_{t-1,p} & p=0 \\ L\big|_{t,p} = L\big|_{t-1,p} + S\big|_{t,p} - AC\big|_{t,p} & p=1. \\ L\big|_{t,p} = L\big|_{t-1,p} - PC\big|_{t,p} & p=2 \end{cases} \tag{15.2}$$

Implementation of tumor growth function in ABM model. Our previous research [30–36] demonstrated that cancer cells not only favored the least resistance, the most permissive, and the most attractive neighborhood position to migrate or proliferate, but also most of their trajectories were dependent on the external factors, such as SDF1 concentration, gel concentration, and cell-cell adhesion.

$$R_p = a \cdot E_p + b \cdot M_p + c \cdot Stiff_p \tag{15.3}$$

$$\frac{\partial u}{\partial t} = k\left(\frac{\partial^2 u}{\partial x^2} + \frac{\partial^2 u}{\partial y^2} + \frac{\partial^2 u}{\partial z^2}\right) + Q - DG. \tag{15.4}$$

Therefore, a pilot tumor growth function (Eq. 15.3) was developed to rank the value of NEIGHBORHOOD POSITIONs, where RA_p was the ranking value of the location (p); E_p and M_p were the concentrations of SDF1 and MatriGel at the location (p); $Stiff_{t,p}$ is the stiffness value of BMSC at location (p); a, b, and c were the coefficients of E_p, M_p, and $Stiff_{t,p}$, respectively. At each time step, the cancer cells (MIC, PC, and MM) chose such a NEIGHBORHOOD POSITION with the highest ranking value to migrate or proliferate. Here, we treat each cell type as one agent in Figs. 15.14 and 15.15).

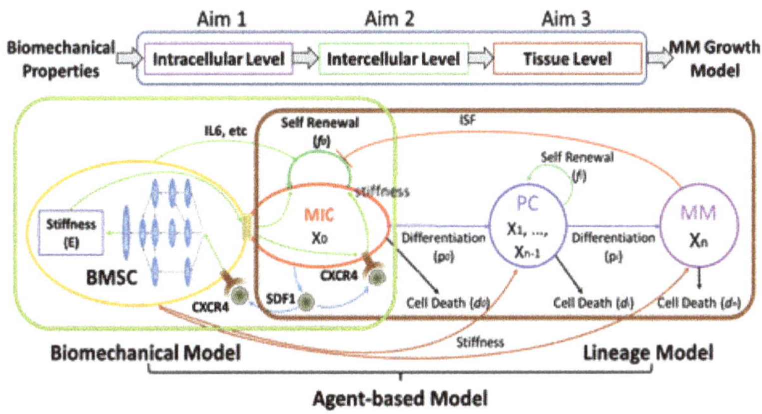

Fig. 15.15 Multiscale modeling of effects of the BMSC biomechanical phenotypes on MM growth.

Diffusion of the chemoattractants and cytokines. Chemoattractants (SDF1) and cytokines (ISF1 and SSF1) diffused at the tissue level as Eq. 15.4. Here, u describes the concentration of chemoattractants and cytokines, k is the diffusivity of chemoattractants and cytokines, Q is the source of chemoattractants and cytokines at each point, and DG is the degrading rate of chemoattractants and cytokines. The Von-Neumann condition was used for the boundary condition. The diffusion model was studied in a recent work [31, 34].

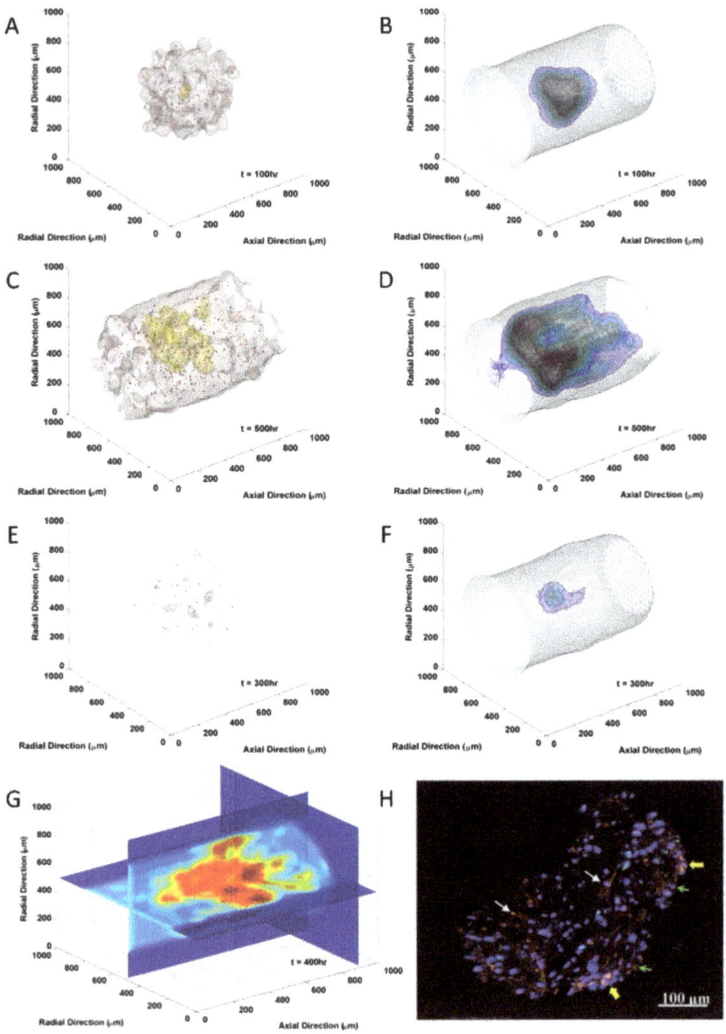

Fig. 15.16 Simulated 3D MM.

Using a graphics processing unit (GPU) to speed up the computing:

GPU was employed to speed up the multiscale agent model at the intracellular and tissue levels by paralleling numerical ODE and PDE solvers for signaling pathway and the diffusion model (Eq. 15.4). Our recent publications [37, 38] demonstrated that a GPU-based parallel PDE solver can increase the performance by 50 times compared to a sequential PDE solver.

Validation for the multiscale ABM: In order to validate the 3D ABM myeloma growth model, a 3D culture scaffold was designed per the manufacturer's guidelines (BD™ PuraMatrix™). Briefly, MyICs along with myeloma RFP-BMSCs or with control RFP-BMSCs were sorted into a Matrigel/sucrose (BD™ PuraMatrix™ 100 Pa) to a final Matrigel concentration of 0.15%, then the mixture was quickly transferred into cell culture inserts (pore size: 0.4 μm; pore density, $2 \times 10^6/cm^2$) and immersed into 24-well plates where the contact with culture medium triggered gelation, and a 2-mm-thick cell-encapsulated scaffold was formed in each insert. By using a gel softer than BMSCs, the microenvironment stiffness that influences MIC fates is mainly due to BMSC contraction. Cell proliferation and apoptosis were marked by BrdU and Annexin V staining. The 3D growth of myeloma colonies was reconstructed from confocal microscopic images at time points of interest (such as 1, 2, 4, 7, and 14 days) to validate ABM predictions. To validate the effectiveness of ABM, we compared simulated results with experimental results from spatial and temporal scales, respectively. On the one hand, ABM employed various colors to label the BMSC, MIC, PC, and MM cells and described the volume and population of each type of cell from spatial scales (Fig. 15.16). On the other hand, we used the particle swarm optimization algorithm [23] to optimize the parameters of the signaling pathway by comparing with RPPA experimental data (Fig. 15.16) at the minute-to-hour temporal level. Moreover, we employed a lineage model (Fig. 15.12) to determine the proliferation rates of PC and MM cells as well as tumor growth function to decide cell migration at the daily temporal level. After that, the minute-to-hour and day-level information was integrated into the ABM model to simulate MM cancer progression at the month level. In pilot simulations, the tumor cell population predicted by the model was found to be qualitatively consistent with the experimental results. Especially, during a specific time period (around 4 days), simulated root-mean-squared speed and displacement effectiveness [39–41] were around 80%, similar to experimental results. The discrepancy between simulated and experimental results was explained as the current tumor growth function (Eq. 15.3), which only considered the impact of chemoattractant, MatriGel, and the stiffness of myeloma BMSCs on the MIC, PC, and MM, but not the well-known spatial structure of stem cell niches, which is important for stem cell behaviors.

The simulation results of the ABM of myeloma growth under different biophysical microenvironments were compared with experimental results to validate the model. The simulated MIC populations as well as the total tumor sizes under the Non-Myeloma-Bone Marrow Stromal Cell (NMBSC) and Myeloma-Bone Marrow Stromal Cell (MBMSC) after 4 weeks of growth were consistent with experimental results. Tumor also showed consistent Bortezomib (BZM) dose responses after 48-hour treatment in simulation and in experiments. Please refer to our publication [42] for more details.

15.2 Multiscale Modeling for Multiple Myeloma Growth

15.2.1 Overall Strategies in Bone Regeneration with Systems Modeling

Bone tissue engineering [43–45] is a complex biological and medical event that includes the scaffolds themselves, cellular and signaling cues, neovascularization, and interactions between them [46]. Many in vitro and/or in vivo experimental studies have been conducted to determine strategies promoting osteogenesis and angiogenesis [46, 47]. However, due to the complexity of the process of bone regeneration within scaffolds, such coordinated processes involved in bone regeneration have often been examined piecemeal rather than systematically.

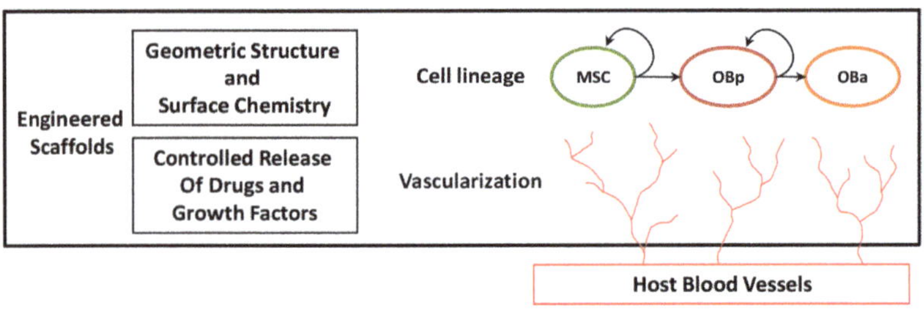

Fig. 15.17 Overview of the bone regeneration system. Engineered scaffolds provide specific geometric structures and surface chemistries as well as accurately release drugs and growth factors to promote both bone regeneration (from the MSCs toward a healthy and well-functional osteoblast lineage) and vascularization between the scaffolds and the host blood vessel system. MSC, bone marrow–derived mesenchymal stem cell; OBp, preosteoblast; OBa, (matured) osteoblast.

One previous study [48] described a multiscale systems biological model that linked the intracellular and intercellular signaling pathways with dynamic cellular properties to study the combination of effects and optimal doses of cytokines on bone remodeling based on the intracellular signaling pathway. Therein, we mainly focused on the responses of cells and bone mass to cytokines (including BMP2, Wnt ligands, and TGF-β) after they were released from the scaffold, but the explicit kinetics of growth factor (GF) release was not studied.

Some drug release models [49–52] provide insights into the mass transport and chemical processes involved in GF delivery systems; however, none of them has considered the responses of cell activity and angiogenesis to cytokine concentration dynamics. A computational systems model of bone regeneration within a 3D porous biodegradable Tricalcium Phosphate scaffold (TCP) can be developed by simulating osteogenesis in response to GF release, based on molecular mechanisms and incorporating angiogenesis and nutrient transportation. The overall schematic of the system is illustrated in Fig. 15.17. First, the 3D bone

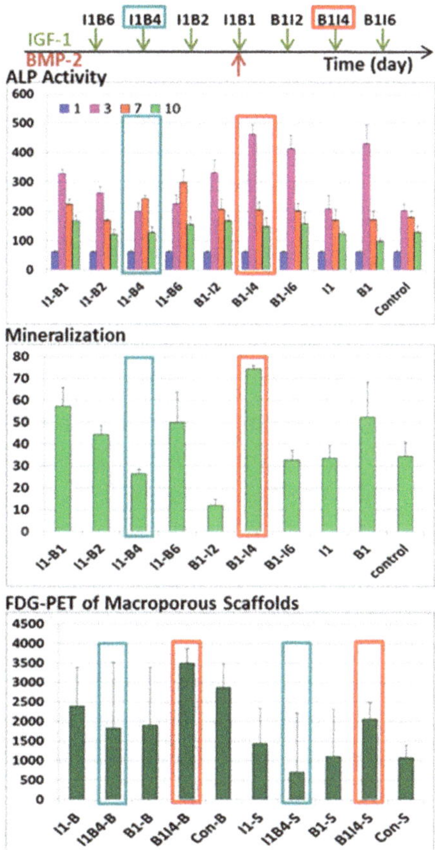

Fig. 15.18 B1I4 temporal combination (highlighted in the red box) induced the greatest differentiation at the early stage (ALP activity) and the best late bone mass regeneration (mineralization). I1B4 condition (green box) and other conditions showed poor osteoblastic differentiation and bone formation. The same trends were also observed by long-term (day 35) cell activities on β-TCP scaffolds of various pore sizes assayed by F^{18} FDG-uptake using PET-CT. –B, big pore; -S, small pore; BMP-2 and IGF-1, 50 ng/mL.

regeneration process will be reconstructed, followed by an investigation of the effects of pore size and porosity on the release of growth factors (GFs), angiogenesis, and bone formation. Finally, it will be examined whether the combination of GFs, such as Bone Morphogenetic Protein 2 (BMP2) and Insulin-like Growth Factor 1 (IGF1), promotes angiogenesis and bone formation more effectively than single GFs. Experimental data in vitro and in vivo will be employed to estimate the system parameters and optimize as well as validate the model.

Bone repair and regeneration involves multiple cell lineages and key GFs in a spatial-temporal way. It is highly desired to create a novel scaffold with controlled GF release that will mimic a physiological cascade to accelerate bone regeneration. We and others showed that delivery of GFs in a properly controlled manner significantly promoted bone formation in vitro and in vivo [53–56]. BMP2 and IGF1-like drugs (Increlex [mecasermin]) have been approved by the FDA for clinical use. We anticipate that this project would, for the first time, predict a biological cascade by the temporal combination of GFs and the development of novel bone grafts, paving the way for clinical trials of multiple GF combinations and bone scaffolds.

15.2.2 Systematically Inferring and Modeling the Molecular Mechanisms of the BMP2/IGF1 Temporal Combination System

The combination of BMP2 and IGF1 was widely investigated in dual cytokine delivery systems. BMP2 is an osteoinductive factor that can potently induce osteoblast differentiation in vitro and in vivo [57–59] and has been approved by the FDA for clinical use, while IGF1 is a mitogenic factor affecting the growth of adult cells and has been used to induce osteoblast proliferation and growth toward local osseointegration [60, 61]. Previous studies revealed that sequential delivery of BMP2 and IGF1 in a controlled manner significantly enhanced osteoblastic differentiation compared to single cytokine treatment and simultaneous release of both factors [53, 54, 62]. Recently, it was further specified that treatment of MSCs with BMP2 on day 1 followed by IGF1 on day 4 (termed the B1I4 bone induction condition) resulted in optimal osteoblast formation compared to other temporal combinations. These studies suggested that BMP2 and IGF1 function cooperatively in a time-dependent manner. However, the underlying molecular mechanisms are still unknown. A systems biology approach was proposed to address these challenges by integrating experimental data and a multi-temporal-scale model at the systems level. The involved signaling pathways were inferred from molecular data through bioinformatics analysis. Based on this systems biology framework, the hypothesis was formulated that the dynamics of cell phenotypes (differentiation and proliferation) define the synergism of temporal combinations of growth factors (GFs) during MSC-initiated bone regeneration. The priming of BMP2 induces the differentiation of the MSC toward a temporary IGF1-sensitive phenotype of preosteoblast whose sensitivity reaches the peak on day 4, which explains the optimal IGF1 boosting time; in contrast, the priming of the naive MSCs with IGF1 will induce a nonosteoblastic phenotype, which shows less sensitivity to BMP2 and undermined bone generation potential.

15.2.2.1 B1I4 is the optimal temporal combination of the BMP2/IGF1 system

A detailed screening of temporal combinations of the BMP2/IGF1 system (Fig. 15.18) revealed that the B1I4 condition (red box) induced the strongest differentiation of murine MSC-like cells (W-20-17) in vitro at an early stage, which is marked by alkaline phosphatase (ALP) activity, and the best late bone mass generation (mineralization assayed by Alizarin Red S staining). In contrast, the I1B4 condition (green box) showed significantly diminished bone differentiation and function. In addition, consistent temporal combination effects were observed from long-term in vitro 3D assays of cell activities in macroporous TCP scaffolds of various pore sizes (from 300 to 800 μm), indicated by F18 FDG-uptakes using positron emission tomography–computed tomography (PET-CT).

15.2.2.2 Systematic screening of the dynamics of multiscale system features

Our team proposed a multi-temporal-scale working hypothesis: (1) At a short time scale (minutes to hours), the growth factor cues trigger cell-type-specific signaling pathway networks and modulate the activities of transcription factors and regulators; which, (2) at a medium time scale (days to weeks), gradually alter the gene expression patterns of the cells and redefine the cell phenotypes, cell functionalities, and signaling pathway topologies; (3) at a long time scale (weeks to months), both the composition of the cell population of osteogenic lineage and the phenotypes of each cell type reflect the effects of different temporal combinations of growth factors, and different bone regeneration efficiencies are thus observed. Per this hypothesis, Our team designed a series of coordinated experiments to comprehensively screen system features at multiple temporal scales and to reveal the underlying mechanisms of the temporal combination effects of the B1I4 condition, using other conditions such as I1B4, I1, B1, and I1B1 as controls.

Fig. 15.19 summarizes the flowchart of the systems biology approach, from systematic experiments, to the descriptions of the multi-temporal-scale bio-events by comprehensive data processing and analysis, to the signaling mechanism inference based on the experimental data at the molecular level and the multiscale modeling [63]. For short-time scale features, GF-induced

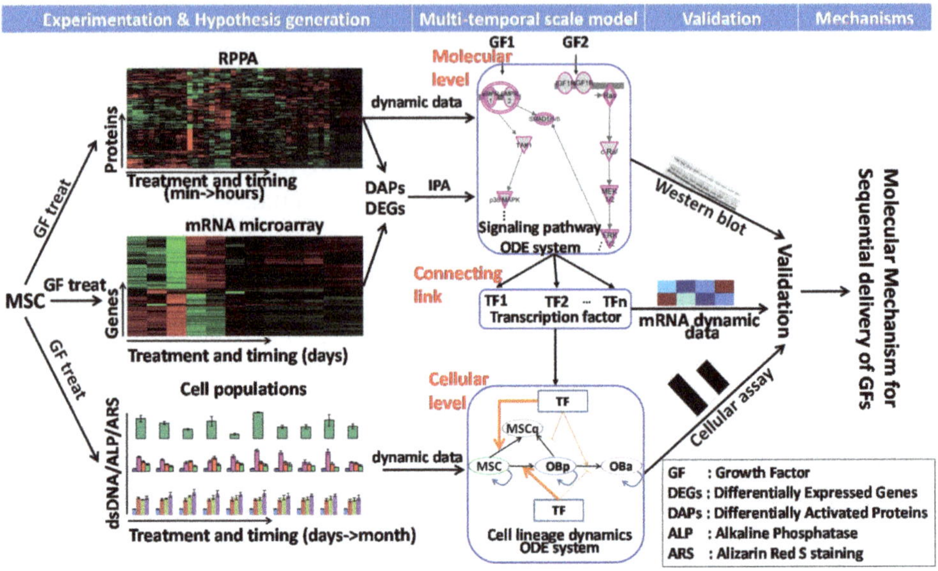

Fig. 15.19 Flowchart of the coordinated systems biology approach to exploring the molecular mechanism of sequential GF treatment and the optimal combination of GFs.

immediate signaling profiles within a few hours were assayed by RPPA and Western blot; for medium-to-long time scale features, mRNA expression patterns and MSC early differentiation were assayed within 10 days of GF delivery by mRNA microarrays and ALP activity, respectively; for long-term features, osteoblast formation was assayed for 28 days by matrix calcium staining [64] for cells cultured on TCP surfaces and by F18 FDG PET-CT imaging for cells cultured in macroporous scaffolds. In addition, the total cell population was monitored by the double-stranded DNA (dsDNA) level. The linear composition, that is, the percentage of MSCs, preosteoblasts (OBp), and matured osteoblasts (OBa), was inferred from gene expression patterns and functional assays [65]. The mRNA microarray and RPPA results showed significant differences between the B1I4 and other conditions from short to long time scale, implying that the working hypothesis was valid and the key dynamic features from molecular to cellular to function level were well captured.

15.2.2.3 Inferring the dynamic signaling pathway and revealing the molecular mechanisms

We inferred the generic signaling pathway related to MSC differentiation and proliferation based on our mRNA microarray and RPPA data. First, we extracted the DEGs under each treatment condition by setting an intensity variation threshold θg=1 after log base 2 transformation on original microarray data. Likewise, we identified the differentially activated proteins (DAPs) from the RPPA data, with a fold change threshold θp = 1.5. This procedure yielded a batch of molecules (genes/proteins) that probably associate with responses of the cells to various cytokine treatments. These molecules were then uploaded to IPA (http://www.ingenuity.com) for upstream analysis, through which we obtained a collection of upstream regulators (UTRs). All of the obtained DEGs, DAPs, and their UTRs were then mapped to the canonical cell signaling pathways of IPA. The eventual generic pathway (Fig. 15.20) was determined by integrating the pathways that were well overlapped with IPA canonical pathways, especially the BMP2 and IGF1-initiated signaling cascades. Then we refined the generic pathway to a simple one by extending our recently developed modularized factor graph (MFG) method [66] to optimize the network structure and interaction rates between genes. After obtaining the refined signaling pathways, we will construct a multi-temporal scale model to simulate the cell lineage process under GF treatment, as detailed in the following section.

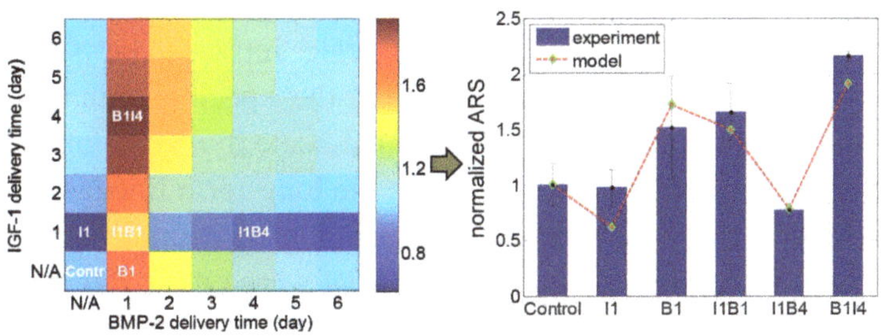

Fig. 15.20 Model prediction and experiment results of osteoblast formation for various treatments. Conditions covered by experiments are denoted in the grids (left) and compared to the experimental data (right). Cell mass was measured by Alizarin red S staining, detecting calcium deposition on day 28.

15.2.2.4 Multi-time-scale modeling of the temporal combination effects on osteogenesis

Our team previously developed a multiscale model for synergy prediction of combined GF treatment [48]. Recently, we extended this model to investigate the effects of specific temporal combinations of GFs on MSC lineage commitment (Fig. 15.21). The temporal dynamics of the multiscale signaling events, the resultant gene expression changes, and the dynamics of cell activities such as proliferation, differentiation, and bone generation were simulated using ODEs.

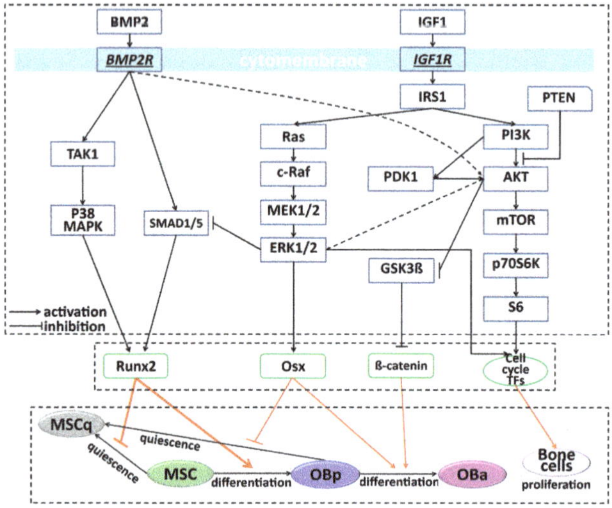

Fig. 15.21 Molecular signaling pathways triggered by BMP2 and IGF1. This generic pathway was inferred from our molecular (mRNA, RPPA) and cellular (dsDNA, ALP) data by mathematical modeling and bioinformatics analysis.

As shown in Fig. 15.19, short-time-scale intracellular signaling was described using Michealis-Menten law and the specific signaling network inferred from the previous step. Medium-time-scale dynamics of gene expression patterns and cell behaviors, as well as the long-term lineage composition changes and bone mass generation, were simulated using a separated ODE system. Experimental data listed in Fig. 15.19 were employed for parameter estimation using a genetic algorithm [53, 67–69]. Fig. 15.21 shows that the model prediction results successfully recapitulated the experimental observations that the B1I4 condition induced optimal long-term mineralization. Moreover, other conditions not covered by experiments can be predicted by the model. Based on this model, an optimal dosage combination of GFs could also be identified. The output of this model will serve as an important input for the integrative 3D multiscale model of bone regeneration within the β-TCP scaffold.

15.2.3 Multiscale Modeling of Bone Regeneration in Engineered Scaffolds and Systems Validation

Our team define a translational solution for bone regeneration as a comprehensive package including (1) the engineered and fine-tunable bone scaffold for individualized medicine, (2) a bio-model of the underlying mechanisms that

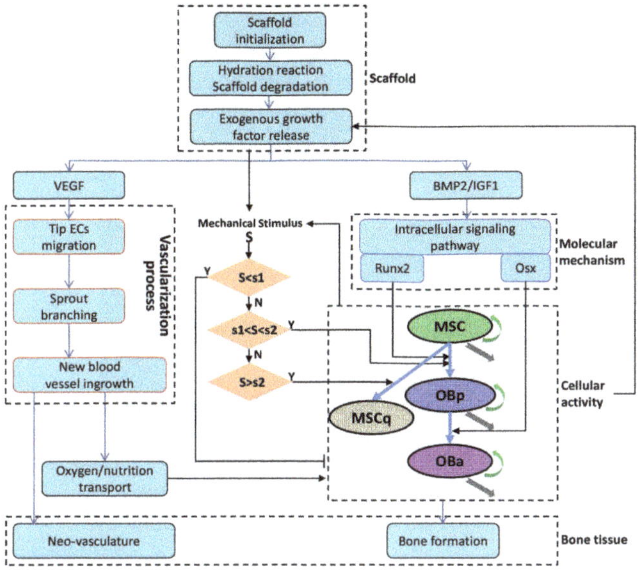

Fig. 15.22 Computational framework of 3D multiscale bone regeneration within a GF-loaded porous biodegradable β-TCP scaffold. VEGF is added in the HUVEC cell medium instead of loading it into the scaffold directly.

offers prediction power, addresses safety concerns, and promises clinical success, and (3) a predictive 3D multiscale computational model for designing optimization and clinical application. We will first introduce the model construction and simulation results and then present the experimental results.

15.2.3.1 Data-guided 3D multiscale model

Sun et al. [67] recently developed a 3D multiscale ABM for bone regeneration within porous biodegradable β-TCP scaffolds. In this model, each cell (e.g., MSC, OBp, OBa, or endothelial cell) was treated as an individual agent. The intracellular signaling events of each agent were described by ODEs, and the distribution and dynamics of GFs were simulated by partial differential equations (PDEs). Thus, the continuum reaction-diffusion sub-model for exogenous GF release and the agent-based sub-model for cell response were integrated in our model. Sun et al. applied a set of reaction–diffusion equations to model the process of GF release from the porous biodegradable scaffold as well as nutrient transportation by the vasculature and employed an ABM [70] to simulate activities of each osteoblast and endothelial cell. This model contains four biophysical scales: scaffold, molecular, cellular, and bone tissue scales (Fig. 15.22), as detailed in the following subsections.

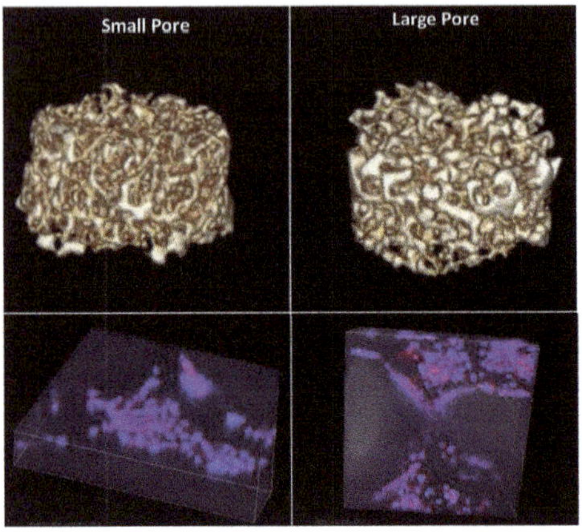

Fig. 15.23 3D image used for model parameter estimation and validation. Top: micro-CT images of scaffolds with small (450 μm) and large (700 μm) pore sizes. Bottom: cell spatial distribution and morphology in these scaffolds.

1. **Scaffold scale: Scaffold degradation and growth factor release.** GFs encapsulated into gelatin microspheres or chitosan hydrogel were loaded into porous biodegradable β-TCP scaffolds. After being implanted into defective bone, calcium phosphate, as well as gel and microspheres, will degrade through hydration reactions and network breakage [71–73]. The diffusion of extracellular liquids such as water and the disintegration of the calcium phosphate network via hydrolysis are described below using the standard second-order rate expression [50]. Meanwhile, GFs were released from the degrading β-TCP scaffold and continuously diffused within the scaffold pores. The diffusivity of each GF varies over regions of β-TCP matrix and scaffold pores, initial loading, degradation rate, and uptake rate by bone cells, which indicates important feedback of cells to the cytokine release system. These processes were explicitly depicted by a group of reaction-diffusion equations and solved by the finite difference method. Since we did not simulate the interaction between scaffolds and host bone, homogenous Neumann boundary conditions were imposed for the scaffold degradation and GF diffusion equations, with details provided in our publication [67]. Parameter estimation: Scaffold pore geometry and distribution were estimated from micro-CT 3D images (Fig. 15.23, top); β-TCP degradation rates were measured by Attenuated Total Reflectance - Fourier Transform Infrared Spectroscopy (ATR-FTIR) as we previously did [74]; GF releasing was measured by ELISA as we previously proposed [53].

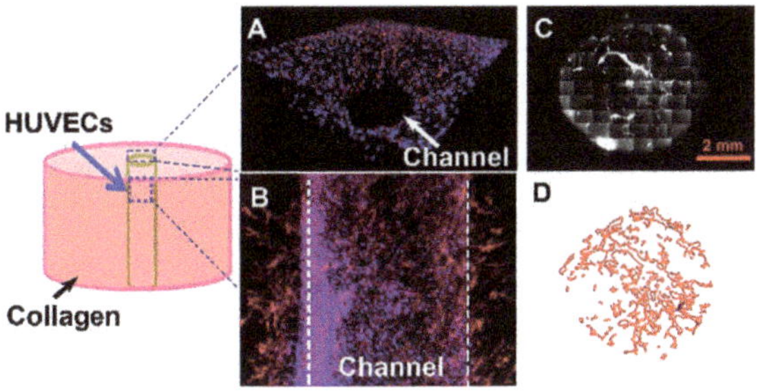

Fig. 15.24 Confocal images of a human CD31-expressing HUVECs endothelialized lumen in collagen (A and B). The fluorescent image (C) and skeleton outline (D) of the microcapillary networks in the gel.

2. **Molecular scale: Signaling pathway.**
   Runx2 and Osx are two critical transcription factors for osteoblastic differentiation and bone regeneration [75]. The molecular regulatory mechanisms involved in the intracellular signaling pathway were modeled using a system of ODEs [53]. The binding and/or dissociation reactions in the intracellular signaling pathway were assumed to be much faster than cellular phenotype switches and cellular population changes [27, 76], so the quasi-steady state of the intracellular signaling pathway was used in this session to model the signal transduction from extracellular GFs to intracellular proteins and transcription factors. Parameter estimation: Signaling pathway kinetics parameters were estimated from molecular data (RPPA, mRNA, and Western blot) as described earlier.

3. **Cellular scale: Cell activities.**
   Migration: Cells migrate along the gradient of the concentration of GFs and oxygen as described below. Different cell types responsd to these environmental molecules at distinct rates [77]. Differentiation: Activated Runx2 and Osx play different roles in stages of osteoblastic lineage [75, 78]. The probabilities of MSC differentiating into OBp and OBp into OBa were related to the expression levels of activated Runx2 and Osx. Proliferation: MSC, OBp, and OBa can proliferate with different probabilities [79, 79]. Several GFs, such as IGF1, can regulate the proliferation of bone cells. Apoptosis: The probability of apoptosis in osteoblasts was related to the concentration of oxygen and ambient space availability. Mechanical stress: Mechanical stimuli are one of the factors that influence tissue differentiation [80, 81]. The mechanical stimulus regulating cell differentiation was defined as $S = \gamma/a + \nu/b$, where γ was the tetrahedral

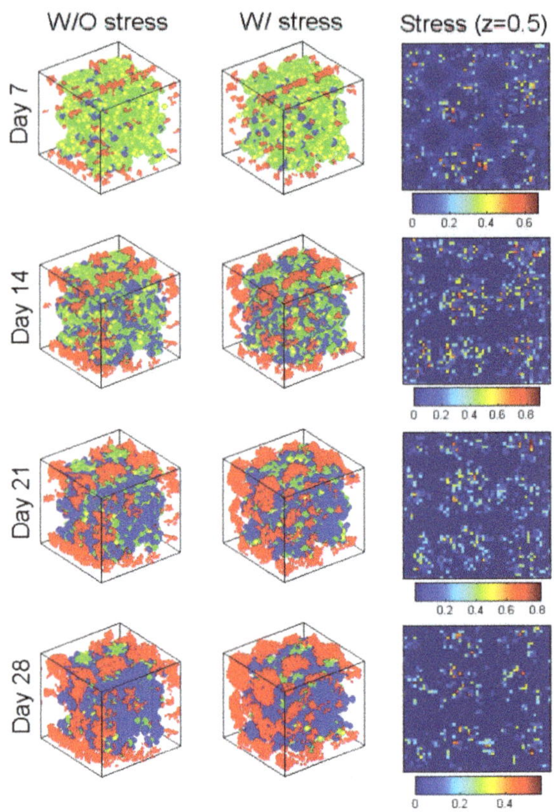

Fig. 15.25 Dynamics of MSC lineage commitment and neovascularization with and without considering the mechanical stress within the scaffold over simulation time. The stress scenario yielded an 8.05% decrease in bone formation but a 9.24% increase in neo-vessels compared to the scenario without considering mechanical stress. The fluid diffusion equation was solved by the finite element method with adaptive meshing technique. The mechanical stress was calculated based on the tetrahedral shear strain and fluid velocity as proposed in literature. Column 3 shows the stress profiles at the $z = 0.5$ horizontal section. Yellow: MSCs; Green: preosteoblasts; Blue: matured osteoblasts; Red: neo-vasculatures.

shear strain, v was the fluid velocity, and a and b were empirical coefficients adapted from the literature [80, 82]. Since the displacement in a small time interval could be assumed small, the Green-Lagrange strain tensor relates the strains γ to the displacement vector u by a simple formula [83]: $\gamma = (\nabla \mathbf{u} + \nabla \mathbf{u}^T)/2$. The displacement vector u was determined by the deformation of cell/vessel structure within one time step of simulation. Parameter estimation: Cell spatial distribution assayed by 3D confocal microscopy (Fig. 15.24 bottom) was used to estimate migration

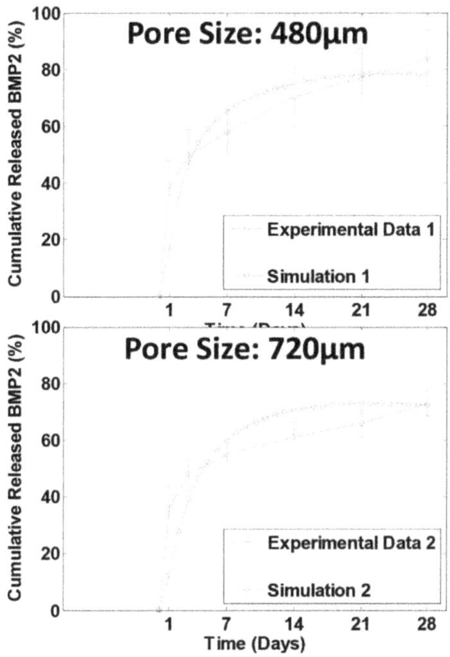

Fig. 15.26 Predictions of BMP2 cumulative released from scaffolds with small (480 μm, top) and large pore size (720 μm, bottom), compared with experimental data.

parameters, population changes (dsDNA) for proliferation and apoptosis rates, and ALP activity assay for differentiation rate.

4. **Bone tissue scale: Angiogenesis, oxygen transportation, and bone formation.**

We assume that the motion of an individual endothelial cell located at the tip of a capillary sprout will govern the motion of the whole sprout, and chemotaxis in response to VEGF gradients will guide the motion of the endothelial cells at the capillary sprout tip [84]. We define the probability of migration of endothelial cells based on the VEGF gradient. A series of rules will be set for sprout migration, branching, and vessel growth. Meanwhile, oxygen can be transported by the neovasculature to osteoblasts within scaffold pores, which is related to oxygen diffusivity, the vessel permeability for oxygen, blood oxygen concentration, and cells' uptake rate of oxygen. Media were perfused into an endothelialized lumen embedded in collagen, and extensive sprouting and microvascularization were observed in vitro. (Fig. 15.24 A and B). In addition, the microcapillary networks in a 6-mm gel were quantified at 83 endothelial branches, with an average branch length of 106 μm. The largest network had about 258 branch points (Fig. 15.24 C and D). Parameter estimation: Confocal imaging of capillary topology and growth rates was used for estimating angiogenesis-related parameters such as sprouting rate, branches, and capillary tip growth rate. The cumulative number of osteoblasts was validated by bone mass formation at designated time points, estimated from the micro-CT image.

15.2.3.2 Simulation results in comparison with experimental data

Sun et al. [67] developed a 3D multiscale systems model to reconstruct the bone regeneration process within the porous, GF-, and MSC-loaded scaffold. The main purpose was to investigate how geometrical properties (e.g., pore size and porosity) of the scaffold will influence GF release, angiogenesis process, and eventual bone formation. This studies indicated that, by tuning the geometrical parameters of the scaffold, the GF release profile could be controlled, and consequently, the eventual bone regeneration efficiency could be manipulated. This implied that the proposed systems approach is promising in providing guidance for the design of GF delivery system and scaffolds. Here, we summarize the major findings obtained through these efforts as below:

1. **Baseline simulation of 3D vascularized bone regeneration with and without mechanical stress.**

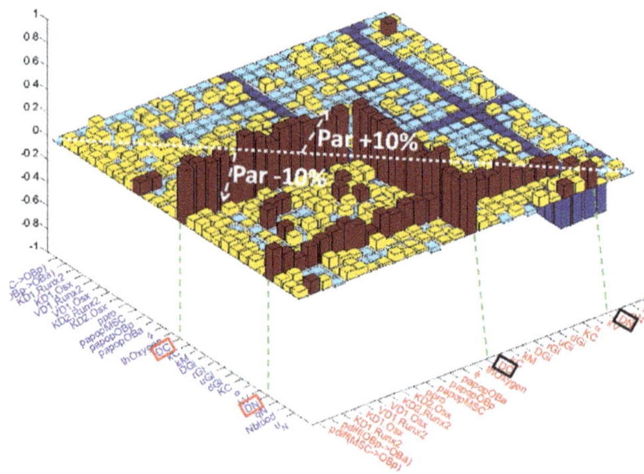

Fig. 15.27 Sensitivity analysis on 28 parameters used in the 3D multiscale model for bone cell formation. Shown are outcome variations upon perturbations of single or double parameters by ±10% (up or down as denoted) from their benchmark values. Color coding for the magnitude of outcome variation: red: >10% increase; yellow: 0%–10% increase; light blue: 0%–10% decrease; blue: >10% decrease.

Sun et al. reconstructed the system of vascularized bone regeneration within the 3D porous β-TCP scaffold, which involved the coupled processes of scaffold degradation, exogenous GF release, angiogenesis, MSC differentiation, and cell growth within porous scaffolds over time. Here, They present a typical simulation of 3D multiscale model with and without considering the mechanical stimulus within the scaffold (Fig. 15.25). The first column refers to the scenario without considering the mechanical stress, with details provided in our published work [67]. The second column shows their recently achieved advancement on the original model by taking into account the mechanical stimuli during the bone regeneration process. The third column shows a sectional profile of mechanical stress distribution within the scaffold. This investigation demonstrated the critical role of mechanical stimulus in regulating the cell differentiation and neovascularization processes.

2. **Tailoring growth factor release profile by controlling scaffold pore size.**
The simulation suggested that the GF release rate could be tuned by pore sizes; specifically, the cumulative BMP2 release from scaffolds with smaller pores (480 μm) was faster than that in larger pore sizes (720 μm), which is well consistent with the experimental results (cells were seeded into BMP2-coated scaffolds and the released BMP2 at series time points up to 28 days was measured by ELISA) (Fig. 15.26). The good agreement between the simulation prediction and experimental data provided an important validation of our model.

3. **Effects of porosity on angiogenesis and bone formation.**
Scaffold porosity was calculated as the ratio of void pore volume versus total scaffold volume. The simulations explored how different pore sizes and porosities affect the bone regeneration procedure. The optimal pore size for both angiogenesis and bone formation was 540 μm [67]. Sun et al. also examined the correlations between the scaffold porosity and the number of active osteoblasts, endothelial cells. Overall, the numbers of endothelial cells and active osteoblasts increased with scaffold porosity, showing a stronger dependence of angiogenesis and bone formation on scaffold porosity than on pore size. These results were in accordance with previous reports [64, 85–87] and experimental studies [87–89].

4. **Effects of combining growth factors.**
Sun et al. also studied the effect of different GF combinations released from the scaffold on angiogenesis and bone formation [67]. Their results

indicated that the combination of Wnt, BMP2, and VEGF promoted the most efficient growth of active osteoblasts. Angiogenesis was promoted significantly when exogenous VEGF was introduced. After 8 weeks, bone formation was greatest when all three GFs were combined. In addition, bone formation was faster with a higher BMP2 dose. Therefore, these results suggested that when growth factors are combined, a high ratio of BMP2 could accelerate the bone formation efficiency.

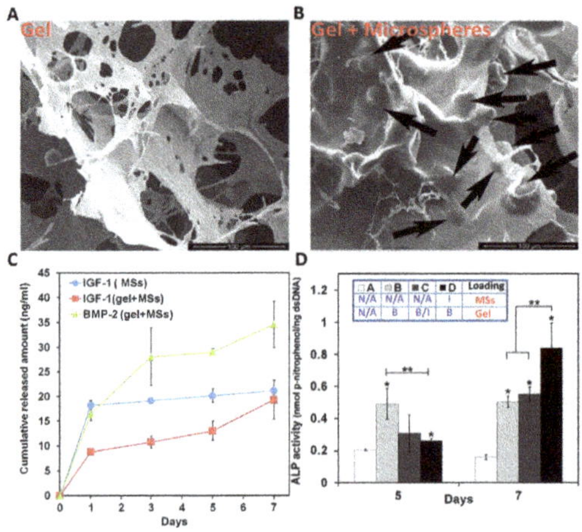

Fig. 15.28 Sequential release of BMP2 and IGF1 from the materials over a week period (C) and their effect on ALP activity of MSCs (D). SEM microscopy of chitosan gel before (A) and after (B) gelatin microspheres embedding is also illustrated.

5. **Sensitivity and uncertainty analysis on model parameters.**
The earlier work [67] primarily focused on the rationality and feasibility of the 3D multiscale modeling of MSC lineage within GF-preloaded scaffolds and therefore did not emphasize the sensitivity and uncertainty of involved parameters to the model output. Recently, a thorough sensitivity and uncertainty analysis was conducted on all parameters involved in cell lineage progression, scaffold degradation, and fluid diffusion processes. As shown in Fig. 15.27, one or two parameters were perturbed at a time by 10% from their benchmark values, and the resulting variation in model output (bone cell mass on day 28) was examined. Generally, the variation of model output was bounded by 10% (yellow and light blue) upon a 10% perturbation on the input parameter(s), indicating that this system model was quite robust with these model parameters. The bone cell mass was very sensitive to the diffusivity of water (DC)

and diffusivity of oxygen (DN), indicating their important roles in regulating osteogenesis within scaffolds. These two parameters need to be adjusted in a more elaborate way when designing the scaffold. An uncertainty analysis on the whole system was further performed, and it was shown that the coefficients remained consistently bounded within 0.05 throughout the simulation period, suggesting that the uncertainty was controllable.

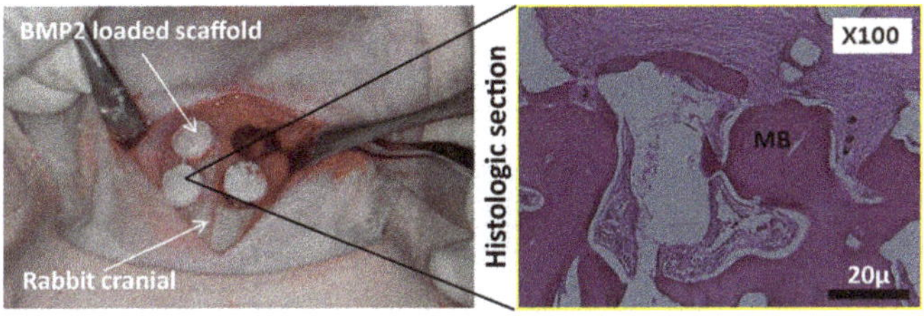

Fig. 15.29 BMP2-loaded scaffolds promoted bone repair *in vivo*. MB: newly formed mineralized bone at 8 weeks.

15.2.4 In Vitro and in Vivo Experiments

Our team recently achieved a series of important in vivo and in vitro experimental results toward effective and successful scaffold design. The results proved that it is feasible to integrate three critical functional modules (GF delivery system, vessel grafting, and stem cell preloading) into a scaffold to enable clinically promising bone regeneration. These achievements provided us with a great opportunity to estimate the model parameters and to validate the model predictions.

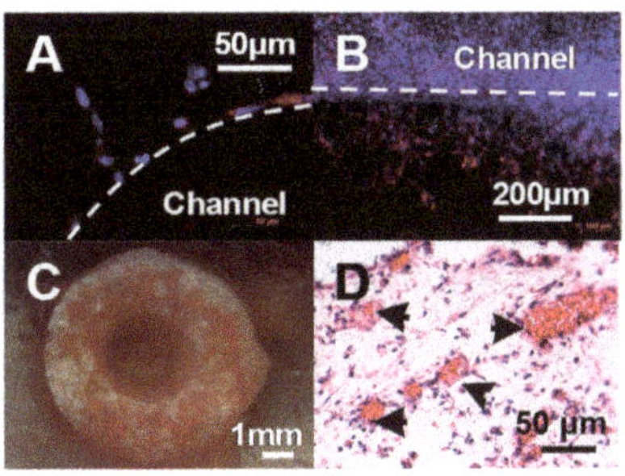

Fig. 15.30 Vascularization from pre-grafted macroporous TCP scaffolds. Details (A, confocal microscopy) and overview (B) of sprouts from the embedded channel. (C) Subcutaneous implants. (D) H&E staining of paraffin sections of the explants.

1. **Sequential GF delivery system design.**
   Kim et al. [53] recently developed a dual drug delivery system by embedding crosslinked gelatin microspheres into chitosan gel (Fig. 15.28 A and B). By varying the crosslinking degree of the gelatin microspheres and GF types, a wide range of growth factor releasing profiles were achieved (Fig. 15.28C) and promoted distinct bone differentiation (Fig. 15.28D). These results were used to calibrate and validate the scaffold scale in our model. It should be noted that the molecular experiments regarding microarray and RPPA have been presented in the previous section, which were used for validating the molecular scale.

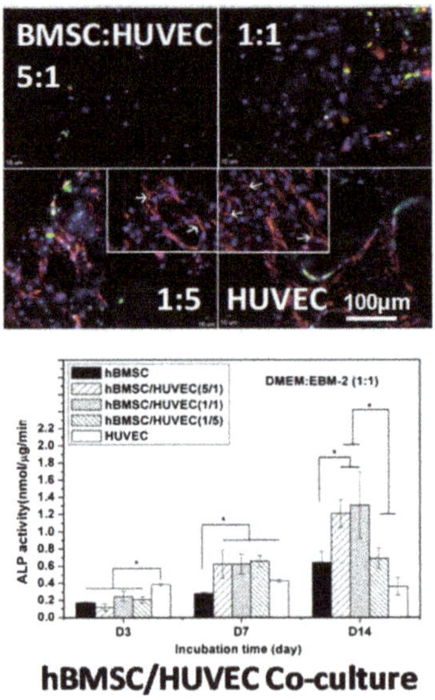

Fig. 15.31 MSCs promoted *in vivo* vascularization.

2. **Bone repair in in vivo β-TCP scaffolds.**
   Lim et al. [55] evaluated the role of BMP2-loaded β-TCP scaffolds on bone defect repair in rabbit cranials. Defects with BMP2-impregnated scaffolds showed significantly enhanced bone formation compared to non-impregnated scaffolds at 8 weeks (Fig. 15.29). These results were used to tune and validate the cellular scale.

3. **Vascularization in in vivo β-TCP scaffolds.**
   Kang et al. [90] devised the vessel-grafted macroporous β-TCP scaffold by in vitro guiding vessel-like structures on solid rods and transferring them into β-TCP scaffolds using EDT technique; the pregrafted Human umbilical

vein endothelial cell (HUVEC) in artificial vessels successfully sprouted in vivo and formed vascular topology at day 14 (Fig. 15.30 A and B), which successfully connected with peripheral blood vessels (Fig. 15.30C) and showed blood perfusion across the scaffold over time (Fig. 15.30D).

4. **Preloaded MSCs and HUVECs facilitating vascularization.** Kang et al. [91] investigated the effects of preexistence of MSC-derived osteogenic lineage on vascularization. Co-seeding experiments revealed the optimal MSC to endothelial cell ratio for bone regeneration (Fig. 15.31). The results presented here can be used to estimate parameters in the tissue (including neo-vessels) scale and also for validation purposes.

In summary, we introduced the ABM for tumor growth modeling and bone regeneration in this chapter. This approach has been actively applied to different areas in biomedical research.

References

1. Corre, J., et al. Bone marrow mesenchymal stem cells are abnormal in multiple myeloma. *Leukemia*. 2007; **21**(5): p. 1079–1088.
2. Shin, D. and K. Athanasiou. Cytoindentation for obtaining cell biomechanical properties. *J Orthop Res*. 1999; **17**(6): p. 880–890.
3. Park, D.W., et al. A well-defined in vitro three-dimensional culture of human endometrium and its applicability to endometrial cancer invasion. *Cancer Lett*. 2003; **195**(2): p. 185–192.
4. Park, D.W., et al. Localization of matrix metalloproteinases on endometrial cancer cell invasion in vitro. *Gynecol Oncol*. 2001; **82**(3): p. 442–449.
5. Goodell, M.A., et al. Isolation and functional properties of murine hematopoietic stem cells that are replicating in vivo. *J Exp Med*. 1996; **183**(4): p. 1797–1806.
6. Gottesman, M.M., T. Fojo, and S.E. Bates. Multidrug resistance in cancer: role of ATP-dependent transporters. *Nat Rev Cancer*. 2002; **2**(1): p. 48–58.
7. Hirschmann-Jax, C., et al. A distinct "side population" of cells with high drug efflux capacity in human tumor cells. *Proc Natl Acad Sci U S A*. 2004; **101**(39): p. 14228–14233.
8. Hanjaya-Putra, D., et al. Vascular endothelial growth factor and substrate mechanics regulate in vitro tubulogenesis of endothelial progenitor cells. *J Cell Mol Med*. 2010; **14**(10): p. 2436–2447.
9. Vanderhooft, J.L., et al. Rheological properties of cross-linked hyaluronan-gelatin hydrogels for tissue engineering. *Macromol Biosci*. 2009; **9**(1): p. 20–28.
10. Lataillade, J.-J., et al. Stromal cell-derived factor 1 regulates primitive hematopoiesis by suppressing apoptosis and by promoting G0/G1 transition in CD34+ cells: evidence for an autocrine/paracrine mechanism. *Blood*. 2002; **99**(4): p. 1117–1129.
11. Broxmeyer, H.E., et al. Stromal cell-derived factor-1/CXCL12 directly enhances survival/antiapoptosis of myeloid progenitor cells through CXCR4 and G{alpha}i proteins and enhances engraftment of competitive, repopulating stem cells. *J Leukoc Biol*. 2003; **73**(5): p. 630–638.

12. Feng, Y., et al. Unique biomechanical interactions between myeloma cells and bone marrow stroma cells. *Prog Biophys Mol Biol.* 2010; **103**(1): p. 148–156.
13. Peng, H., et al. Drug inhibition profile prediction for NFκB pathway in multiple myeloma. *PLoS ONE.* 2011; **6**(3): p. e14750.
14. Peng, H., et al. Prediction of treatment efficacy for prostate cancer using a mathematical model. *Sci Rep.* 2016; **6**(1): p. 21599.
15. Peng, H.M., J.G.Wen, C.C. Chang and X.B. Zhou. Systematic modeling study on mechanism of p38 MAPK activation in MDS. In: *The Eighteenth International Conference on Intelligent Systems for Molecular Biology* (*ISMB 2010*). 2010: Boston, USA.
16. Coombes, K., and et al. Object-oriented microarray and proteomics analysis library (oompa). http://bioinformatics.mdanderson.org/software.html, 2007.
17. Hu, J., et al. Non-parametric quantification of protein lysate arrays. *Bioinformatics.* 2007; **23**(15): p. 1986–1994.
18. Neeley, E.S., et al. Variable slope normalization of reverse phase protein arrays. *Bioinformatics.* 2009; **25**(11): p. 1384–1389.
19. Tibes, R., et al. Reverse phase protein array: validation of a novel proteomic technology and utility for analysis of primary leukemia specimens and hematopoietic stem cells. *Mol Cancer Ther.* 2006; **5**(10): p. 2512–2521.
20. Park, E.S., et al. Integrative analysis of proteomic signatures, mutations, and drug responsiveness in the NCI 60 cancer cell line set. *Mol Cancer Ther.* 2010; **9**(2): p. 257–267.
21. Tsavachidou-Fenner, D., et al. Gene and protein expression markers of response to combined antiangiogenic and epidermal growth factor targeted therapy in renal cell carcinoma. *Ann Oncol.* 2010; **21**(8): p. 1599–1606.
22. Iadevaia, S., et al. Identification of optimal drug combinations targeting cellular networks: integrating phospho-proteomics and computational network analysis. *Cancer Res.* **70**(17): p. 6704–6714.
23. Zhan, Z.H., et al. Adaptive particle swarm optimization. *IEEE Trans Syst Man Cybern B Cybern.* 2009; **39**(6): p. 1362–1381.
24. Kschischang, F., B.J. Frey and H.A. Loeliger. Factor graphs and the sum-product algorithm. *IEEE Trans Inf Theory.* 2001; **47**(2): p. 498–519.
25. Ganguly, R. and I.K. Puri. Mathematical model for the cancer stem cell hypothesis. *Cell Prolif.* 2006; **39**: p. 3–14.
26. Kirouac, D.C., et al. Cell-cell interaction networks regulate blood stem and progenitor cell fate. *Mol Syst Biol.* 2009; **5**: p. 293.
27. Pivonka, P., et al. Model structure and control of bone remodeling: a theoretical study. *Bone.* 2008; **43**(2): p. 249–263.
28. Lander, A.D., et al. Cell lineages and the logic of proliferative control. *PLoS Biol.* 2009; **7**(1): p. e15.
29. Coope, I.D.a.P., C.J. Positive bases in numerical optimization. *Comput Optim Appl.* 2002; **21**(2): p. 169–176.
30. Zhang, L., et al. Multiscale agent-based modeling of ovarian cancer progression under the stimulation of the STAT 3 pathway. *Int J Data Min Bioinform.* 2014; **9**(3): p. 235–253.
31. Zhang, L., et al. Multiscale agent-based cancer modeling. *J Math Biol.* 2009; **58**(4–5): p. 545–559.

32. Zhang, L., et al. Simulating brain tumor heterogeneity with a multiscale agent-based model: linking molecular signatures, phenotypes and expansion rate. *Math Comput Model.* 2009; **49**(1–2): p. 307–319.
33. Zhang, L., L.L. Chen, and T.S. Deisboeck. Multi-scale, multi-resolution brain cancer modeling. *Math Comput Simul.* 2009; **79**(7): p. 2021–2035.
34. Zhang, L., C.A. Athale, and T.S. Deisboeck. Development of a three-dimensional multiscale agent-based tumor model: simulating gene-protein interaction profiles, cell phenotypes and multicellular patterns in brain cancer. *J Theor Biol.* 2007; **244**(1): p. 96–107.
35. Kim, H.D., et al. Epidermal growth factor-induced enhancement of glioblastoma cell migration in 3D arises from an intrinsic increase in speed but an extrinsic matrix- and proteolysis-dependent increase in persistence. *Mol Biol Cell.* 2008; **19**(10): p. 4249–4259.
36. Deisboeck, T.S., et al. Pattern of self-organization in tumour systems: complex growth dynamics in a novel brain tumour spheroid model. *Cell Prolif.* 2001; **34**(2): p. 115–134.
37. Jiang, B., et al. Employing graphics processing unit technology, alternating direction implicit method and domain decomposition to speed up the numerical diffusion solver for the biomedical engineering research. *Int J Numer Methods Biomed Eng.* 2011; **27**(11): p. 1829–1849.
38. Jiang, B., et al. Accelerate numerical diffusion solver of 2D multiscale and multi-resolution agent-based brain cancer model by employing graphics processing unit technology. In: *The 2011 International Conference on Biocomp.* 2011; Las Vegas, NV.
39. Zygourakis, K. Quantification and regulation of cell migration. *Tissue Eng.* 1996; **2**(1): p. 1–16.
40. Ware, M.F., A. Wells, and D.A. Lauffenburger. Epidermal growth factor alters fibroblast migration speed and directional persistence reciprocally and in a matrix-dependent manner. *J Cell Sci.* 1998; **111**(Pt 16): p. 2423–2432.
41. Costa Lda, F., L.C. Cintra, and D. Schubert. An integrated approach to the characterization of cell movement. *Cytometry A.* 2005; **68**(2): p. 92–100.
42. Su, J., et al. Targeting the biophysical properties of the myeloma initiating cell niches: a pharmaceutical synergism analysis using multi-scale agent-based modeling. *PLoS One.* 2014; **9**(1): p. e85059.
43. Trautvetter, W., et al. Tissue-engineered polymer-based periosteal bone grafts for maxillary sinus augmentation: five-year clinical results. *J Oral Maxillofac Surg.* 2011; **69**(11): p. 2753–2762.
44. Petite, H., et al. Tissue-engineered bone regeneration. *Nat Biotechnol.* 2000; **18**(9): p. 959–963.
45. Dragoo, J.L., et al. Tissue-engineered bone from BMP-2-transduced stem cells derived from human fat. *Plast Reconstr Surg.* 2005; **115**(6): p. 1665–1673.
46. Nguyen, L.H., et al. Vascularized bone tissue engineering: approaches for potential improvement. *Tissue Eng Part B Rev.* 2009; **15**(3): p. 353–370.
47. Lovett, M., et al. Vascularization strategies for tissue engineering. *Tissue Eng Part B Rev.* 2009; **15**(3): p. 353–370.
48. Sun, X., et al. Cytokine combination therapy prediction for bone remodeling in tissue engineering based on the intracellular signaling pathway. *Biomaterials.* 2012; **33**(33): p. 8265–8276.

49. Arifin, D.Y., L.Y. Lee, and C.H. Wang. Mathematical modeling and simulation of drug release from microspheres: Implications to drug delivery systems. *Adv Drug Deliv Rev.* 2006; **58**(12): p. 1274–1325.
50. Rothstein, S.N., W.J. Federspiel, and S.R. Little. A unified mathematical model for the prediction of controlled release from surface and bulk eroding polymer matrices. *Biomaterials.* 2009; **30**(8): p. 1657–1664.
51. Siepmann, J. and A. Gopferich. Mathematical modeling of bioerodible, polymeric drug delivery systems. *Adv Drug Deliv Rev.* 2001; **48**(2): p. 229–247.
52. Tzafriri, A. Mathematical modeling of diffusion-mediated release from bulk degrading matrices. *J Control Release.* 2000; **63**(1): p. 69–79.
53. Kim, S., et al. Sequential delivery of BMP-2 and IGF-1 using a chitosan gel with gelatin microspheres enhances early osteoblastic differentiation. *Acta Biomater.* 2012; **8**(5): p. 1768–1777.
54. Raiche, A.T. and D.A. Puleo. In vitro effects of combined and sequential delivery of two bone growth factors. *Biomaterials.* 2004; **25**(4): p. 677–685.
55. Lim, H.P., et al. The effect of rhBMP-2 and PRP delivery by biodegradable beta-tricalcium phosphate scaffolds on new bone formation in a non-through rabbit cranial defect model. *J Mater Sci Mater Med.* 2013; **24**(8): p. 1895–1903.
56. Yang, J., et al. Graded porous β-TCP scaffolds enhance bone regeneration in mandible augmentation. *J Craniofac Surg.* 2015; **in press**.
57. Kim, S., et al. In vitro evaluation of an injectable chitosan gel for sustained local delivery of BMP-2 for osteoblastic differentiation. *J Biomed Mater Res B Appl Biomater.* 2011; **99B**(2): p. 380–390.
58. Groeneveld, E.H.J. and E.H. Burger. Bone morphogenetic proteins in human bone regeneration. *Eur J Endocrinol.* 2000; **142**(1): p. 9–21.
59. Kang, Y., et al. Creation of bony microenvironment with CaP and cell-derived ECM to enhance human bone-marrow MSC behavior and delivery of BMP-2. *Biomaterials.* 2011; **32**(26): p. 6119–6130.
60. Meinel, L., et al. Localized insulin-like growth factor I delivery to enhance new bone formation. *Bone.* 2003; **33**(4): p. 660–672.
61. Jaklenec, A., et al. Sequential release of bioactive IGF-I and TGF-beta 1 from PLGA microsphere-based scaffolds. *Biomater.* 2008; **29**(10): p. 1518–1525.
62. Strobel, C., et al. Sequential release kinetics of two (gentamicin and BMP-2) or three (gentamicin, IGF-I and BMP-2) substances from a one-component polymeric coating on implants. *J Control Release.* 2011; **156**(1): p. 37–45.
63. Tan, H., et al. A systems biology approach to studying the molecular mechanisms of osteoblastic differentiation under cytokine combination treatment. *NPJ Regen Med.* 2017; **2**(1): p. 5.
64. Roy, T.D., et al. Performance of degradable composite bone repair products made via three-dimensional fabrication techniques. *J Biomed Mater Res A.* 2003; **66**(2): p. 283–291.
65. Ju, W.J., et al. Defining cell-type specificity at the transcriptional level in human disease. *Genome Res.* 2013; **23**(11): p. 1862–1873.
66. Peng, H., et al. Characterization of p38 MAPK isoforms for drug resistance study using systems biology approach. *Bioinformatics.* 2014; **30**(13): p. 1899–1907.

67. Sun, X., et al. Modeling vascularized bone regeneration within a porous biodegradable CaP scaffold loaded with growth factors. *Biomater.* 2013; **34**(21): p. 4971–4981.
68. Peng, H., et al. A systematic modeling study on the pathogenic role of p38 MAPK activation in myelodysplastic syndromes. *Mol Biosyst.* 2012; **8**(4): p. 1366–1374.
69. Peng, H., et al. Drug inhibition profile prediction for NFκB pathway in multiple myeloma. *PLoS One.* 2011; **6**(3): p. e14750.
70. Sun, X., et al. Multi-scale agent-based brain cancer modeling and prediction of TKI treatment response: Incorporating EGFR signaling pathway and angiogenesis. *BMC Bioinformatics.* 2012; **13**(1): p. 218.
71. Wang, Q., et al. Degradation kinetics of calcium polyphosphate bioceramic: an experimental and theoretical study. *Mater Res.* 2009; **12**: p. 495–501.
72. Fernández, E., et al. Calcium phosphate bone cements for clinical applications. Part I: solution chemistry. *J Mater Sci Mater Med.* 1999; **10**(3): p. 169–176.
73. Lu, J., et al. The biodegradation mechanism of calcium phosphate biomaterials in bone. *J Biomed Mater Res.* 2002; **63**(4): p. 408–412.
74. Kang, Y., et al. The osteogenic differentiation of human bone marrow MSCs on HUVEC-derived ECM and beta-TCP scaffold. *Biomaterials.* 2012; **33**(29): p. 6998–7007.
75. Gordeladze, J.O., et al. From stem cells to bone: phenotype acquisition, stabilization, and tissue engineering in animal models. *ILAR J.* 2009; **51**(1): p. 42–61.
76. Lee, J.M., et al. Dynamic analysis of integrated signaling, metabolic, and regulatory networks. *PLoS Comput Biol.* 2008; **4**(5): p. e1000086.
77. Sandino, C., et al. Simulation of angiogenesis and cell differentiation in a CaP scaffold subjected to compressive strains using a lattice modeling approach. *Biomaterials.* 2009; **31**(8): p. 2446–2452.
78. Komori, T. Regulation of bone development and maintenance by Runx2. *Front Biosci.* 2008; **13**: p. 898–903.
79. Isaksson, H., et al. A mechano-regulatory bone-healing model incorporating cell-phenotype specific activity. *J Theor Biol.* 2008; **252**(2): p. 230–246.
80. Sandino, C., et al. Simulation of angiogenesis and cell differentiation in a CaP scaffold subjected to compressive strains using a lattice modeling approach. *Biomaterials.* 2010; **31**(8): p. 2446–2452.
81. Checa, S. and P.J. Prendergast. Effect of cell seeding and mechanical loading on vascularization and tissue formation inside a scaffold: a mechano-biological model using a lattice approach to simulate cell activity. *J Biomech.* 2010; **43**(5): p. 961–968.
82. Checa, S. and P.J. Prendergast. A mechanobiological model for tissue differentiation that includes angiogenesis: a lattice-based modeling approach. *Ann Biomed Eng.* 2009; **37**(1): p. 129–145.
83. Mollemans, W., et al. Predicting soft tissue deformations for a maxillofacial surgery planning system: from computational strategies to a complete clinical validation. *Med Image Anal.* 2007; **11**(3): p. 282–301.
84. Artel, A., et al. An agent-based model for the investigation of neovascularization within porous scaffolds. *Tissue Eng Part A.* 2011; **17**(17–18): p. 2133–2141.
85. Story, B.J., et al. In vivo performance of a modified CSTi dental implant coating. *Int J Oral Maxillofac Implants.* 1998; **13**: p. 749–757.

86. Lewandrowski, K.U., et al. Developing porosity of poly (propylene glycol-co-fumaric acid) bone graft substitutes and the effect on osteointegration: a preliminary histology study in rats. *J Biomater Sci Polym Ed*. 2000; **11**(8): p. 879–889.
87. Karageorgiou, V. and D. Kaplan. Porosity of 3D biomaterial scaffolds and osteogenesis. *Biomaterials*. 2005; **26**(27): p. 5474–5491.
88. Kujala, S., et al. Effect of porosity on the osteointegration and bone ingrowth of a weight-bearing nickel-titanium bone graft substitute. *Biomaterials*. 2003; **24**(25): p. 4691–4697.
89. Fisher, J.P., et al. Soft and hard tissue response to photocrosslinked poly (propylene fumarate) scaffolds in a rabbit model. *J Biomed Mater Res*. 2001; **59**(3): p. 547–556.
90. Kang, Y., et al. Engineering a vascularized collagen-β-tricalcium phosphate graft using an electrochemical approach. *Acta Biomater*. 2015; **11**: p. 449–458.
91. Kang, Y., et al. Osteogenic and angiogenic potentials of monocultured and co-cultured human-bone-marrow-derived mesenchymal stem cells and human-umbilical-vein endothelial cells on three-dimensional porous beta-tricalcium phosphate scaffold. *Acta Biomater*. 2013; **9**(1): p. 4906–4915.

CHAPTER 16

AI and Synthetic Biology

16.1 What Is Synthetic Biology?

Synthetic biology is an interdisciplinary field that focuses on the application of engineering principles to biology. Its primary aim is to design and construct novel biological devices, parts, or systems that do not naturally occur or to modify existing biological systems for practical applications [1]. The ultimate goal of synthetic biology is to create living systems capable of processing information, manipulating chemicals, producing materials and energy, providing food, and enhancing human health. Researchers and companies worldwide are leveraging the power of nature to address challenges across diverse fields such as medicine, manufacturing, and agriculture.

The history of genetic manipulation stretches back thousands of years, rooted in the practices of farmers and agriculturalists who employed selective breeding to enhance desirable traits in crops and livestock. This long tradition laid the groundwork for modern genetics, which began to take shape in the 1850s and 1860s with Gregor Mendel's experiments on pea plants. Mendel's work established the foundational principles of genetic inheritance, now known as the laws of Mendelian inheritance.

Synthetic biology represents a more recent advancement, emerging just a few decades ago. It shifts the genetic manipulation paradigm by enabling direct DNA editing and incorporating engineering principles. Synthetic biology has traditionally been both costly and time-consuming. However, advancements at the intersection of genomics and digitization are transforming the field. The digital revolution's ability to manipulate data efficiently and affordably has created unprecedented economies of scale, accelerated innovation, and reduced both the cost and duration of experiments. A striking example is the evolution of DNA sequencing costs. In 2001, sequencing the first human genome required over a decade of work and around $100 million. Today, that same process can be completed in a single day for approximately $600, according to the National Human Genome Research Institute. As a result, synthetic biology is becoming a transformative force in science, providing deeper insights into biological systems and opening new frontiers for reengineering the natural world and driving innovation [2]. For example, white blood cells could be engineered to target and eliminate specific mutations, such as those found in cancer. Custom-designed organisms might be developed to capture atmospheric carbon, contributing to climate change mitigation. Similarly, advancements in agricultural biotechnology could yield more nutritious crops, enhance food security, and lower production costs.

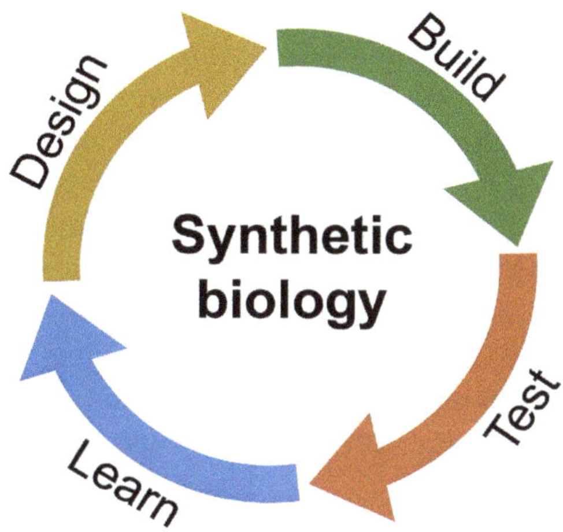

Fig. 16.1 Design–build–test–learn (DBTL) cycle.

Synthetic biologists apply engineering design principles to harness the predictability of engineering in controlling complex biological systems. Key principles include the use of standardized genetic components and the iterative design-build-test-learn (DBTL) cycle, which is tailored to achieve specific outcomes (Fig. 16.1). In synthetic biology, the DBTL cycle is adapted to encompass the following four stages:

Design: Formulate a hypothesis for a DNA sequence or set of cellular modifications that could achieve the desired outcome.

Build: Execute the design by synthesizing the DNA sequence and successfully introducing it into a biological system, typically by transforming it into a host cell.

Test: Collect data to evaluate how well the resulting phenotype aligns with the design goal, while also assessing any unintended effects or off-target outcomes.

Learn: Analyze the test results to uncover insights that guide the next iteration more efficiently than a random trial-and-error approach. This stage often involves diagnosing failures caused by unexpected off-target effects. Artificial intelligence (AI) can play a crucial role here by optimizing subsequent designs and minimizing the number of DBTL cycles needed to achieve the target outcome.

16.2 How Is AI Used for Synthetic Biology?

Artificial intelligence (AI) leverages machines and software to carry out tasks traditionally associated with human intelligence, including language comprehension, pattern recognition, decision-making, and problem-solving [3, 4]. When applied to complex datasets, AI excels at identifying subtle patterns, extracting valuable features from noisy data, and providing precise predictions [3, 5]. These capabilities make it exceptionally well suited for synthetic biology at various levels of biological complexity, including genomics (the study of the DNA sequence), transcriptomics (the analysis of gene expression), proteomics (the study of proteins), and metabolomics (the study of tiny molecules) (Fig. 16.2). Indeed, innovative AI algorithms have

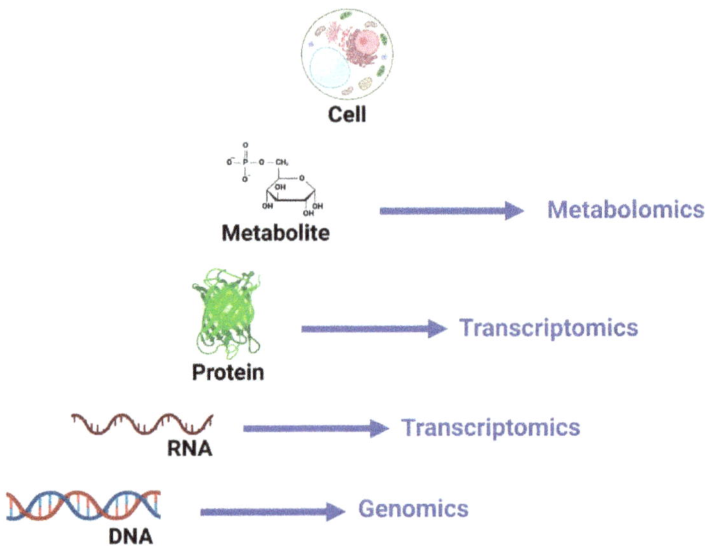

Fig. 16.2 The genetic code is transcribed into RNA, which is then translated into proteins (central dogma of biology). Proteins enable a variety of reactions, among them the transformation of metabolites (chemical species) into other metabolites. Genomics, transcriptomics, proteomics, and metabolomics are all examples of omics data.

been developed to generate new and useful genes, speed analysis of possible applications, and conduct (low-cost, low-risk, and rapid) virtual testing. The power of the partnership between synthetic biology and AI is based on three elements:

Data: Even the simplest organisms have more than 100,000 base pairs of DNA, while complex life forms, such as humans, have more than three billion. Each of those DNA pairs is a data point that can be used by machine learning (ML) algorithms for training and to create outputs.

Hidden patterns: The massive variety of DNA combinations leads to complexity and optionality that is difficult for the human mind to comprehend, but ML is adept at uncovering variations, patterns, and interconnections (and can be trained to identify the most promising candidates).

Scalability: Laboratory testing by humans is characterized by lengthy and often wasteful trial and error, whereas virtual and automated testing by AI can be exponentially quicker, cheaper, and safer. That in turn makes synthetic biology scalable.

16.3 AI and Genomics Synthetic Biology-Genome Editing

16.3.1 Introduction

With developments in synthetic biology, "engineering biology" has emerged through standardization and platformization based on hierarchical, orthogonal, and modularized biological systems. Genome editing is necessary to manufacture and design synthetic cells with desired functions using bioparts obtained from sequence databases. Among various genome-editing tools, the clustered regularly interspaced short palindromic repeats (CRISPR) and CRISPR-associated protein 9 (Cas9) system show advantages that enable researchers to manipulate genomic elements efficiently and precisely.

Initially reported in the prokaryotic genome as part of an antiviral defense mechanism against bacteriophages, the CRISPR/Cas9 system was later recognized as a revolutionary genome-editing tool enabling insertion, removal, and deletion of existing genes with high precision and specificity. CRISPR-Cas9 uses a single-guide RNA (sgRNA) and the Cas9 nuclease to target and cleave DNA. sgRNA is a guide sequence that recognizes a target sequence in the genome. The sgRNA is made up of two functional domains: the gRNA sequence and the tracrRNA sequence. The gRNA sequence is 20 nucleotides long and pairs with the target sequence. Cas9 nuclease acts like a pair of scissors to cut the double strands of DNA (Fig. 16.3). CRISPR has quickly become a cornerstone in biological research and biotechnology due to its simplicity, efficiency, and versatility. Over the past decade, scientists have utilized CRISPR technology as a readily adaptable tool to probe biological function, dissect genetic interactions, and inform strategies to combat human diseases and engineer crops. CRISPR holds immense potential for a wide range of applications, including correcting genetic defects, engineering disease-resistant crops, and advancing treatments for genetic disorders and cancer. However, its widespread adoption brings ethical challenges and concerns, such as off-target effects, unintended consequences, and ensuring equitable access to genetic modification technologies. Meanwhile, AI has become a transformative force in gene editing, revolutionizing the approach and execution of genetic modifications [6].

AI and deep learning (DL) algorithms are transforming CRISPR systems by automating key processes such as gRNA activity prediction, Cas9 variant optimization, and anti-CRISPR protein identification [7]. These advancements are streamlining CRISPR-based research and applications, significantly reducing the time and resources required [7]. Furthermore, the integration of AI with CRISPR technology is enabling more precise and efficient genome editing.

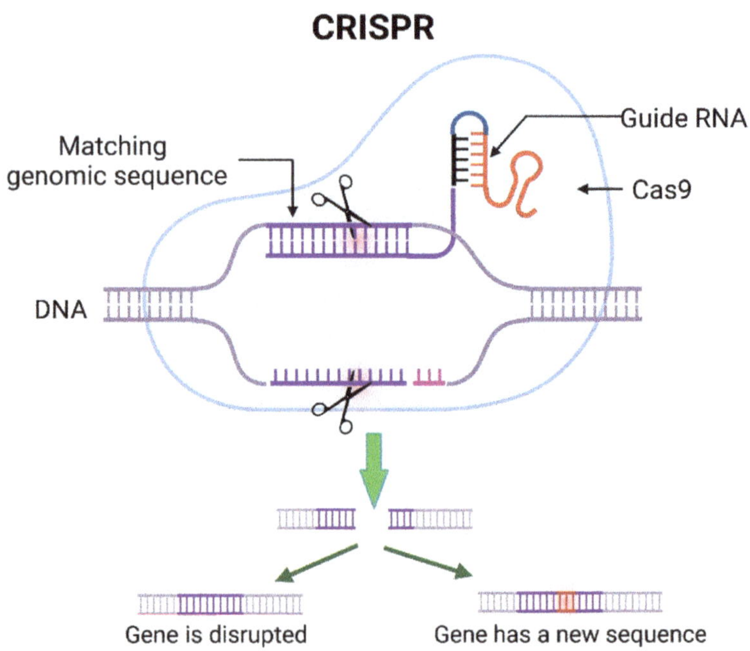

Fig. 16.3 Clustered regularly interspaced short palindromic repeats (CRISPR).

DL models can predict the activity and specificity of various Cas9 variants with high accuracy, facilitating the selection of the most suitable enzyme for specific applications [7, 8]. This predictive capability plays a vital role in minimizing off-target effects while improving the safety and effectiveness of CRISPR-based therapies.

16.3.2 AI for CRISPR

The cleavage efficiency of guide RNAs (gRNAs) in the CRISPR-Cas9 system is influenced by numerous features, including protospacer adjacent motifs (PAMs), gRNA sequence motifs, nucleotide composition, position-specific nucleotide patterns, secondary structure, and epigenetic factors [8–10]. These features can also contribute to off-target effects, a significant concern in CRISPR applications. To optimize gRNA design for high on-target efficacy while minimizing off-target activity, researchers have developed advanced computational tools and algorithms. These tools predict gRNA performance and improve the precision and reliability of CRISPR-based gene editing [11].

Despite these advances, gRNA efficacy is determined by a complex interplay of factors, including gRNA properties, target sequence characteristics, experimental conditions, and cellular environment. ML approaches have emerged as powerful solutions for navigating these complexities. By training on large datasets, ML algorithms can predict on-target and off-target effects with high accuracy. These methods generally fall into three categories: (a) regression-based approaches that estimate the quantitative efficacy of gRNAs [12–16], (b) classification-based methods that categorize gRNAs as effective or ineffective [16–19], and (c) ensemble-based methods that combine predictions from multiple models to enhance robustness [20, 21].

A critical distinction among these ML methods lies in their feature selection and representation of target sites. Advanced ML methods have paved the way for integrating DL into CRISPR applications. DL models, such as artificial neural networks (ANNs), have demonstrated remarkable precision in predicting target sites. These models typically encode gRNA-DNA sequences into a 23-nucleotide matrix, which serves as input. The process involves multiple layers:

1. Convolutional layers extract features from the input matrix using various filter sizes.
2. Batch normalization stabilizes learning and minimizes overfitting.
3. Pooling layers refine the extracted features by filtering and condensing the data.

4. Dense layers interconnect neurons to enable complex pattern recognition.
5. The final output layer predicts whether a sequence corresponds to an on-target or off-target site [9].

AI-driven machine and deep learning (MDL) approaches have revolutionized gRNA design by leveraging global gene-editing datasets. These algorithms predict gRNA activity and specificity with unprecedented accuracy, offering a more efficient and cost-effective alternative to experimental methods [22]. Several MDL-based tools have been developed to enhance on-target efficacy, including CRISPRater [14], CRISPRScan [13], DeepCRISPR [23], DeepCas9 [24], WU-CRISPR [18], SgRNA Scorer [25], CRISPRpred [26], CNN-SVR [27], and C-RNNCrispr [28]. These tools represent a significant advancement in computational gene editing, streamlining gRNA design and improving the efficiency and specificity of CRISPR-Cas9 experiments [9].

16.4 AI and Transcriptomics Synthetic Biology-mRNA Optimization

16.4.1 Introduction

In the past, synthetic biologists have primarily used DNA as the molecule of choice for designing synthetic systems. As a result, synthetic DNA has been instrumental in the development of artificial genes, gene regulatory networks, and even entire genomes, enabling scientists to study complex biological processes and even create synthetic organisms. However, with the rise of RNA therapeutics, there has been a growing interest in developing synthetic systems exploiting the unique attributes of RNA molecules. Over recent years, computational metrics that can be used as proxies for different messenger RNA (mRNA) features have been proposed together with optimization algorithms. Recent advances in AI and the generation of high-throughput experimental data pushed our boundaries further in mRNA sequence design.

mRNA medicine

mRNA is a transient carrier that transfers genetic information from DNA to ribosomes, where that information can be translated into proteins. The fundamental structure of mRNA medicine consists of an encapsulated mRNA strand, comprising an open reading frame (ORF) flanked by 5′ and 3′ untranslated regions (UTRs) (Fig.16.4). The core principle behind mRNA therapies is to deliver the transcript of interest, encoding one or more genes, into the host cell

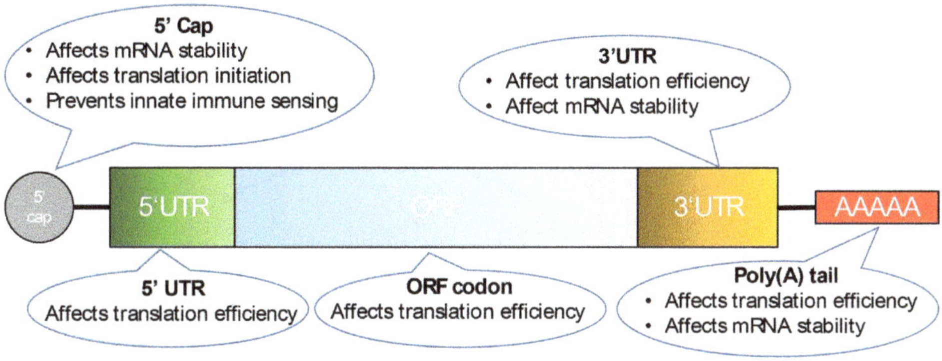

Fig. 16.4 Structure of mRNA and the function of each element.

cytoplasm, where expression generates translated protein(s) within the membrane, secreted or intracellularly located (Fig. 16.5).

The first successful experimental injections of mRNA that resulted in translated proteins were performed in murine muscle cells in vitro in the early 1990s [29]. This encouraged the development of mRNA as a new concept of vaccination [30–32]. However, it was not until critical modifications of the mRNA leading to increased stability and translational rate were introduced that the field of mRNA vaccination started its real expansion [33]. The first clinical phase 1 studies using modified mRNA vaccines in LNP were against influenza virus H10 and H7 hemagglutinin (HA) during the years between 2015 and 2018, resulting in 100% seroconversion [34, 35]. The success of COVID-19 mRNA vaccines has proven that RNA technology is safe and effective for commercial production. mRNA therapies have become the focus of molecular medicine research. Various mRNA applications have reached major milestones at high speed in the immuno-oncology field. The competitive advantage of mRNA vaccines is that the vaccine production and purification processes are similar, despite the encoded antigens. This feature gives prospects of using harmonized protocols for different mRNA vaccines [36]. Production of new mRNA vaccines can thereby be initiated immediately once the genomic sequence of a target antigen has been identified. This can result in reduced costs and faster production of new vaccines against emerging infectious diseases, as exemplified by the mRNA vaccines against SARS-CoV-2 [37].

The challenge in studying mRNA medicines

Eukaryotic gene expression is regulated at the translation level by several key elements, including the cap structure, 5′UTR, codon usage, 3′UTR, and poly(A) tail.

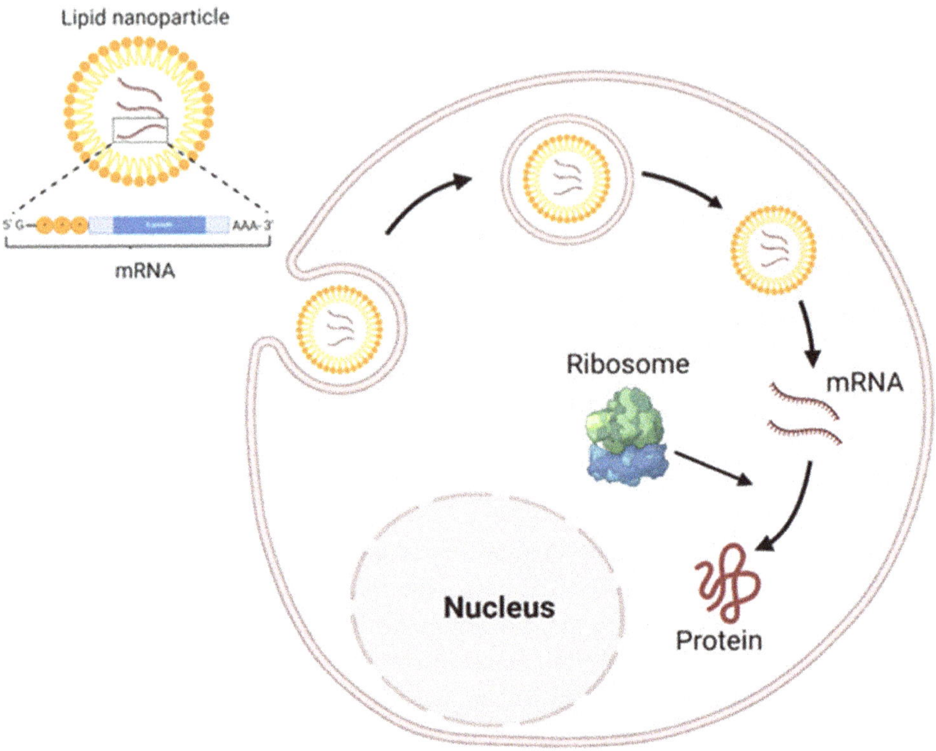

Fig. 16.5 mRNA medicine. mRNA is encapsulated within a lipid nanoparticle and delivered to the target cell. Then the target cell's own machinery recognizes the mRNA and translates it, restoring or augmenting protein function to treat or prevent disease.

The UTRs of genes play a crucial role in post-transcriptional regulation, influencing mRNA stability, cytoplasmic export, subcellular localization, and translation efficiency, all of which determine the overall protein yield [38]. The 5′UTR significantly affects translational efficiency by assisting ribosome binding near the start codon. Additionally, certain RNA elements within the 5′UTR, such as internal ribosome entry sites (IRES), upstream start codon (AUG), or upstream open reading frames (uORF), can alter the secondary structure and substantially influence the translation rates. Moreover, the 5′UTR may contain sequences that serve as binding sites for regulatory proteins [39]. Similarly, the 3′UTR plays a key role in determining mRNA fate through interactions with RNA-binding proteins and microRNAs (miRNAs), which help regulate mRNA localization, stability, and translation [40, 41]. Codon usage also impacts protein levels, as the genetic code is redundant, with most amino acids (AAs) being encoded by multiple codons. However, not all synonymous codons are functionally equivalent, some stabilize mRNA, while others reduce its stability. Codon usage affects cellular protein production and organismal fitness, as observed in bacteria, eukaryotic microorganisms, and human cell lines [42–45]. What's more, during translation, mRNA forms intramolecular interactions between distant elements, with the 5′UTR, ORF, and 3′UTR working together. For instance, the 5′UTR and the upstream region of the ORF collectively create an optimal context for translation initiation [46], and there are reported physical interactions between ORF and 3′UTR [47]. Additionally, base-pairing between 5′ and 3′UTR has been shown to regulate human genes [48]. These interdependencies make optimizing mRNA very challenging.

Thus far, AI-based mRNA optimization approaches have been developed for 5′UTR, 3′UTR, and ORF codon, and most of them only focus on a single element. Here, we focus on the integration optimization approach developed by our team.

16.4.2 AI and Codon Optimization

Recombinant protein production in synthetic biology heavily relies on heterologous expression, where codon optimization plays a pivotal role. Traditional optimization methods primarily focus on biological metrics, but Fu et al. introduced an innovative DL-based approach. Their method leverages the concept of "codon boxes," which recode DNA sequences into codon box sequences, disregarding base order. This transformation reframes codon optimization as a sequence annotation problem, aligning codon boxes with their corresponding AAs. To demonstrate their approach, they trained codon optimization models for *Escherichia coli* using bidirectional long-short-term memory conditional random fields (BiLSTM-CRF). This DL framework effectively captures the distributional characteristics of DNA. Their method was evaluated through protein expression experiments, including tests on *Plasmodium falciparum* candidate vaccine proteins and polymerase acidic protein. These experiments compared original sequences, sequences optimized with conventional methods, and those optimized using their model. Results revealed that the DL-based method significantly enhances protein expression, establishing it as both efficient and competitive [49].

RiboCode, a DL-based method, advances protein expression by modeling the intricate relationship between mRNA codon sequences and their translation efficiency using large-scale Ribo-seq data. Its prediction model demonstrates robust performance, with optimization strategies highlighting a complex interplay between sequence features and translational outcomes. In vitro experiments revealed up to a 72-fold increase in protein expression, surpassing previous methods. RiboCode also enables cell-type-specific expression and maintains consistent performance across various mRNA formats, including unmodified, m1Ψ-modified, and circular mRNA. In vivo, optimized influenza virus HA mRNA elicited approximately tenfold stronger neutralizing antibody responses in mice, while optimized nerve growth factor (NGF) mRNA provided equivalent neuroprotection for retinal ganglion cells at just one-fifth the dose in an optic nerve crush mouse model [50].

16.4.3 AI and 5′ UTR Optimization

Sample et al. integrated polysome profiling of a library comprising 280,000 randomized 5′ UTRs with DL to develop a predictive model linking human 5′UTR sequences to translation efficiency. Leveraging this model alongside a genetic algorithm, they designed new 5′UTRs that precisely control ribosome loading levels, enabling fine-tuning of sequences for optimal protein expression. Additionally, they demonstrated that this approach is adaptable to chemically modified RNA, a critical advancement for applications in mRNA therapeutics and synthetic biology [51].

Hair et al. employed DL to design 5′UTRs optimized for efficient mRNA translation. Using polysome profiling, they analyzed fully or partially randomized 5′UTR libraries across three cell types and observed a high correlation in UTR performance between cell types. They trained predictive models on their datasets and utilized gradient descent and generative neural networks to guide the design of high-performing 5′UTRs. Experimentally, they tested the designed 5′UTRs with mRNA encoding megaTAL™ gene-editing enzymes targeting two different genes in two distinct cell lines, finding that the designed 5′UTRs effectively supported robust gene-editing activity [52].

16.4.4 AI and 3′UTR Optimization

Flynn et al. developed a high-throughput massively parallel reporter assay (MPRA) that, in a single experiment, measures the half-life of tens of thousands of unique mRNA sequences containing designed 3′UTRs. Over multiple design-build-test iterations, they have accumulated 180,000 unique measurements of mRNA stability covering unique genomic and synthetic 3′UTRs, representing the largest such dataset of sequences. They trained highly accurate ML models to map from 3′UTR sequences to mRNA stability, and used them to guide the design of synthetic 3′UTRs that increase mRNA stability in cell lines. Finally, they validated the function of several ML-designed 3′UTRs in mouse models, resulting in up to twofold more protein production over time and 30–100-fold higher protein output at later time points compared to a commonly used benchmark. These results highlight the potential of ML-driven sequence design for mRNA therapeutics [53].

16.4.5 AI and Integrated Optimization for Codon, 5′UTR, and 3′UTR

Since the 5′UTR, ORF, and 3′UTR interact and work together during translation, much like a symphony orchestra, and therefore, optimizing each "player" independently may disrupt this coordination. For instance, the 5′UTR and the upstream region of the ORF collectively create an optimal context for translation initiation [46], and there are reported physical interactions between the ORF and 3′UTR [47].

To address these limitations, we have developed a novel DL-based integrated optimization approach, integrated deep-learning-based mRNA optimization (iDRO) [54]. Unlike other benchmark optimization methods, our approach not only optimizes the codon usage of the ORF but also generates 5′UTR and 3′UTR sequences based on the ORF. We split mRNA into three subregions (5′UTR, ORF, 3′UTR) and applied different models on different subregions. We then developed a new pipeline for the optimization of mRNA sequences as shown in Fig. 16.6. Based on the composition of mRNA, our algorithm consists of two parts, ORF optimization and UTRs generation, which help iDRO focus on the biological rules within each component of mRNA. For ORF optimization, we adapted BiLSTM-CRF [55] to arrange codons. Inspired by the multilingual bidirectional and auto regressive transformers (mBART) [56] model in NLP research, we developed RNA-BART for UTR generation, with a stronger ability to capture mRNA information.

Data collection for preliminary study

To address the long-sequence problem, we divided the mRNA optimizing task into two subtasks: ORF optimization and UTR optimization. The NCBI human gene dataset and an R library called "GenomicFeatures" are our training data sources. Without further explanation, we only choose the UTRs whose length is between 50 and 500. This R library can link to the dataset UCSC.hg19.knownGene [57] and export the data into a ".csv" file. This ".csv" file has three columns, including 5′UTR, CDS sequence, and 3′UTR, and contains 63,691 samples. After that, we used Pandas, a Python library, to select data. We chose the data that UTRs are between 50 nt and 500 nt and CDS length is less than 2,500 nt. Finally, we obtained 17,029 training data corresponding to 17,029 genes, respectively. We randomly split data into the training set and validation set.

Network structure

iDRO is shown in Fig. 16.6A. It consists of one word2vec layer, six BiLSTM layers, one CRF layer, one tokenize layer, eight transformer layers, and one softmax layer. The input is the AA sequence of the target protein, and the output is the optimized mRNA sequence for humans. In Fig. 16.6B, there are two main stages. ORF optimization aims to find the best codon arrangement for an AA sequence. In which AAs will first input to the word2vec layer to get a word vector that embeds context information for each AA. Then, these vectors are fed into BiLSTM-CRF to get optimized ORF. For further enhancement, optimized ORF will be split into several short sequences (tokens) by the tokenize layer and fed to RNA-BART to generate UTRs.

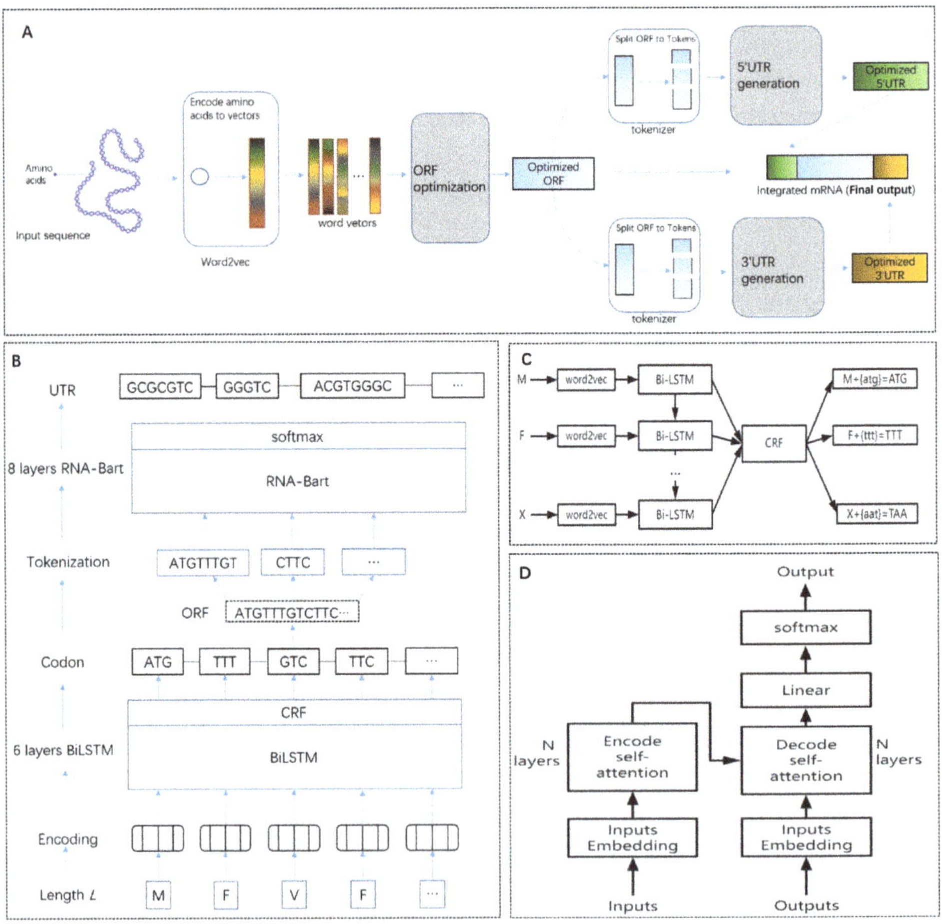

Fig. 16.6 iDRO pipeline and architectural details. (A) iDRO pipeline (B), network structure of iDRO. Arrows show the information flow. (C) Structure of BiLSTM + CRF. (D) Structure of RNA-BART.

RNA-BART

Compared with the original mBART, RNA-BART (Fig. 16.6B and 16.6D) has fewer layers, because transformer-based model has redundant parameters and can be simplified [58, 59]. Inspired by mBART in NLP, RNA-BART consists of two-part training. We first covered 15% of tokens in UTR and ORF sequences and trained an autoregressive model with a maximum length of 512 in the pre-training stage. This allows the model to successfully encode UTR and ORF sequences. In the fine-tuning part, we kept the autoregressive model and fine-tuned it to the machine translation model. Unlike mBART, we deliberately used a small learning rate to focus on the feature of UTR and ORF to avoid confusion since both these two kinds of sequences are made up of "ATCG."

ORF optimization

Codons are not randomly selected but follow some codon rules embedded in cellular environmental information [60]. We introduced a DL method to accurately unravel the rules of codon selection. Based on the injective relation between codon and AA, ORF optimization can be regarded as a name entity recognition (NER) task in natural language processing (NLP). In NLP, the goal of NER is to recognize the category (people, location, etc.) for each word. In this task, we recognized the best codon (category) for each AA. Here, we employed the BiLSTM-CRF network to optimize ORF (Fig. 16.6C). The input of BiLSTM-CRF is the AA sequence, and the output is the optimized ORF. In the training phase, we set the human's ORF as the ground truth, because we assume it is the most suitable way to encode AA in humans. In Fig. 16.6C, each AA is first fed into a word2vec layer, which is pre-trained with all the AA sequences we collected. At this step, each AA is encoded to a word vector that carries some context information. After that, the word vectors are input to the BiLSTM-CRF network. For the output part, the codon box paradigm was used to reduce the tag class number and improve performance (please refer to our recent work [54]). Codons can be categorized into 20 classes and the class is named codon box. Given the combination of AA and a codon box, there is only one corresponding codon. Instead of outputting the exact codon, this model outputs the codon box and then used the combination of codon box and AA to determine the codon. This algorithm used four layers of BiLSTM. The hidden layer dimension was 200; the learning rate was 0.003; the batch size was 32; the dropout rate was 0.5. The accuracy on the test dataset and training dataset can reach 0.63 and 0.71, respectively, that is comparable with performance in bacterial [61].

UTR generation

The UTR generation can be framed as a machine translation task, where the goal is to identify the most relevant sentence (or sequence) based on the input. Here, we aim to find the most relevant UTR sequence given an ORF. Therefore, machine translation serves as an appropriate model for UTR generation. We trained RNA-BART for this purpose, with its structure shown in Fig. 16.6D. To manage the length of input sequences, we used tokenization, a common technique in NLP. Tokenization splits the mRNA sequence into short sequences based on their frequency in the corpus. For example, "ATGGACTTTC" would be tokenized into "ATGGAC," "TTTC" (two tokens) instead of being split into individual nucleotides ("A," "T," "G," etc.). In addition to tokenization, we employed two other preprocessing methods to distinguish ORF from UTR: adding special tokens and converting the ORF into codon boxes. Special tokens (which do not appear naturally in ORF nor UTR) were inserted before the input and target sequences to categorize each sequence type. Specifically, we added "<ORF>" before ORF and "<5UTR>" before 5′UTR. Furthermore, the ORF was converted to codon boxes, meaning the input sequence was represented as codon boxes rather than the ORF itself. This further differentiates OFR from UTR, and by combining the codon box with the AA sequence, the model can generate the most suitable UTR for a given ORF.

Numerical results

The training for UTR optimizing was conducted on two 2080ti GPUs over 10 epochs. The learning rate was set at 0.00005, with 8 layers and a hidden dimension of 512. Training for 5′UTR optimization took 4 hours and 43 minutes, while 3′UTR optimization took 2 hours and 32 minutes. The data was randomly split into training, validation, and test sets following 8:1:1, and cross-validation was employed to evaluate iDRO. The results are shown in Table 16.1.

Table 16.1 Jaccard and Rouge scores for different pre-methods.

Area	Preprocessing method	Jaccard	Rouge
5′UTR	Only tokenize	0.692	0.633
	Special tokens	0.694	0.701
	Codon box	0.786	0.742
	Codon box + special tokens	**0.811**	**0.783**

Area	Preprocessing method	Jaccard	Rouge
3′UTR	Only tokenize	0.738	0.692
	Special tokens	0.0774	0.799
	Codon box	0.703	0.72
	Codon box + special tokens	**0.798**	**0.856**

Note: Bold indicates that the preprocessing method using Codon box + special tokens has the best performance compared with other methods.

Our objective was to generate UTR sequences that closely resemble human endogenous UTR patterns. We used the Jaccard score [62] and Rouge [63] as the primary metrics for sequence comparison, both of which are commonly used in machine translation to measure sequence similarity. Higher scores indicate a closer match between predicted sequences and human UTRs. Without further preprocessing, the average Jaccard and Rough scores for 5′UTR generation were 0.692 and 0.633, respectively, and for 3′UTR generation, 0.738 and 0.692. These results were suboptimal and might not support high translation efficiency. To identify the best preprocessing method, we compared the UTR sequences generated by previous preprocessing methods (as detailed in our work [55]) to the original UTR sequences. We found that the combination of codon box with special tokens yielded the highest score for both 5′UTR and 3′UTR prediction: 0.811 and 0.783 for 5′UTR, and 0.798 and 0.856 for 3′UTR. These results substantially outperformed the tokenize-only approach.

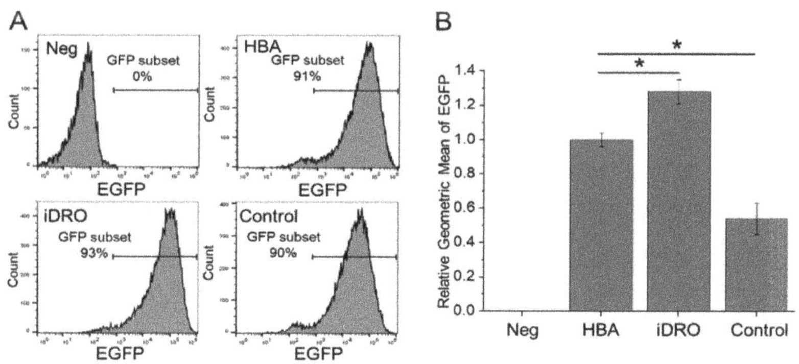

Fig. 16.7 Flow cytometry analysis of HEK-293 cells transfected with encoding EGFP mRNA. (A) EGFP gene expression in HEK-293 cells transfected with encoding EGFP mRNA. (B) Relative EGFP fluorescence density in EGFP-positive cells. Neg: no transfection. HBA: EGFP gene flanked by 5′UTR/3′UTR from HBA. Model: 5′UTR/3′UTR generated from our model. Control: random 5′UTR/3′UTR.

Experimental validation

We used the EGFP gene as an example to validate our method. The EGFP AA sequence was input into iDRO, and the optimal UTR combination (including codon optimization and special tokens for 5′UTR and 3′UTR) was compared with the globin UTR combination. To highlight the importance of preprocessing, a tokenize-only model was used as a control. We generated EGFP mRNA sequences with optimized codon and 5′UTR/3′UTR as designed in this study. These mRNAs were transfected into HEK 293 cells, resulting in efficient translation into functional protein. Flow cytometry analysis showed that ≥90% of cells expressed EGFP (Fig. 16.7A). We then analyzed the EGFP fluorescence density in each group. The results indicated that cells transfected with mRNA containing the de novo designed 5′UTR/3′UTR exhibited more than twice the fluorescence intensity of the control group. Notably, the de novo designed 5′UTR/3′UTR outperformed the widely used HBA 5′UTR/3′UTR in mRNA vaccines [64] (Fig. 16.7B).

16.5 AI and Proteomics Synthetic Biology-Protein Design

Proteins are essential molecules that drive nearly every biological process, serving as the foundation of life. Composed of linear chains of AAs, their specific sequence and chemical properties dictate how these chains fold into intricate 3D structures, which in turn determine their functions. The complexity of protein folding is influenced by factors such as AA interactions, molecular environment, and post-translational modifications, making it nearly impossible for humans to predict a protein's 3D structure from its sequence alone.

Historically, experimental methods like X-ray crystallography have been the gold standard for resolving protein structures. This technique, which relies on crystallizing proteins and analyzing X-ray diffraction patterns, delivers highly detailed atomic-level insights. However, it is expensive, time-consuming, and requires significant expertise. Despite these limitations, thousands of protein structures have been determined through such methods, creating a rich dataset that has catalyzed advancements in computational approaches to protein structure prediction.

One breakthrough in this domain is DeepMind's AlphaFold, which demonstrated the power of ML to accurately predict protein structures based solely on AA sequences. The subsequent AlphaFold2 (AF2) system refined this capability, leveraging a dataset of over 170,000 experimentally resolved structures to achieve unprecedented accuracy. These advancements have transformed structural biology, enabling faster and more accessible protein modeling.

In parallel, researchers have explored whether AI can reverse this process, designing protein sequences to achieve specific functions. Natural proteins exhibit a vast array of chemical properties, inspiring efforts in synthetic biology to engineer artificial proteins for targeted applications. Traditional methods, such as phage display [65] and directed evolution [66], have been used to identify and enhance proteins with desirable traits. However, these techniques are often constrained by high costs, limited scalability, and reliance on natural proteins as templates.

To address these challenges, computational protein design has emerged as a powerful alternative, enabling the generation of custom protein sequences tailored to specific needs. Recent advancements in this field have accelerated progress, with state-of-the-art computational methods now rivaling traditional experimental approaches in efficiency and precision. These innovations pave the way for the creation of novel proteins with transformative potential in medicine, biotechnology, and beyond [66].

16.5.1 Introduction

Generative models

DL has transformed protein design, enabling the creation of powerful generative models that explore and innovate beyond the constraints of natural proteins. These generative approaches fall into two primary categories: sequence-based and structure-based models.

Recent advancements in structure-based design have been driven by diffusion models, algorithms originally developed for text-to-image generation [67–69]. These models, often enhanced by graph neural networks (GNNs), excel at modeling the intricate relationships between protein structures and their properties. On the other hand, sequence-based design has leveraged large language models (LLMs) [70], which draw on NLP techniques to analyze and generate protein sequences with unprecedented accuracy and versatility.

Both approaches extend the boundaries of protein design by generalizing beyond the "protein space" of natural sequences and structures. This capability allows researchers to explore entirely new regions of possibilities, crafting novel proteins with desired functions or features [71, 72]. Furthermore, these models are increasingly accessible through intuitive interfaces. Researchers can now specify target properties using user-friendly programming languages [73] or even natural language interfaces [74], making the design process more efficient and approachable. By integrating DL into protein design, generative models are unlocking new avenues for discovery, advancing synthetic biology, and paving the way for transformative applications in biotechnology and medicine.

Protein structure prediction

Protein structure prediction algorithms, including AlphaFold [75], RoseTTAFold [76, 77], RoseTTAFoldAA [78], and ESMFold [79], have revolutionized de novo protein design by addressing two critical challenges. First, these algorithms provide in silico metrics to evaluate and score proposed designs, significantly enhancing the success of experimental validation. For example, Bennett et al. [80] demonstrated that applying metrics such as the predicted local distance difference Test (pLDDT), a per-residue confidence measure, and the root mean squared deviation (RMSD) between the designed atomic coordinates and those predicted by structure algorithms led to a tenfold improvement in experimental outcomes. By using these metrics to filter designs, researchers can identify candidates with a higher likelihood of stability and functionality. Second, these algorithms generate extensive structural data, which serves as valuable training material for other models. For instance, Hsu et al. [81] leveraged a dataset of 12 million protein structures predicted by AF2 to train ESM-IF, a model that translates structural features into sequence designs. This abundance of high-quality data enables the development of more sophisticated tools that bridge the gap between protein structure and sequence. The integration of structure prediction algorithms into protein design workflows not only accelerates the discovery of novel proteins but also improves the reliability of experimental outcomes, marking a transformative step forward in synthetic biology and bioengineering.

16.5.2 Sequence-Based Protein Design

LLMs have emerged as transformative tools for generating plausible protein sequences by learning the "rules" of protein language. Trained to infer the identity of missing AAs based on their sequence context, these models adjust their parameters to improve accuracy through iterative comparison of predictions with true sequences [82]. When tasked with generating novel sequences, LLMs construct proteins step by step, yielding designs that are unique yet biologically realistic [66].

LLM-generated proteins often show low sequence similarity to natural proteins while maintaining physicochemical properties consistent with biological functionality. For instance, Verkuil et al. [71] demonstrated the potential of LLMs in protein design, creating proteins with intricate structural features like hydrogen bond networks. Of 228 designs tested, 152 were experimentally validated, including 35 with no significant sequence similarity to known proteins and the rest exhibiting a median sequence identity of 27% to their closest natural counterparts. Similarly, ProGEN [82] achieved experimental success rates comparable to both random and natural control proteins. ProtGPT2 [83] further demonstrated the ability of LLMs to generate proteins with minimal structural resemblance to existing ones, highlighting their capacity to explore uncharted regions of protein space.

Despite their successes, LLMs face notable challenges. They have yet to achieve the design of entirely novel protein structures with unprecedented catalytic mechanisms or functions, a domain where rational design has traditionally excelled [84, 85]. Additionally, methods like Markov chain and Monte Carlo (MCMC) often suffer from inefficiencies, such as getting stuck in local minima and requiring numerous iterations. To overcome these limitations, diffusion models have emerged as a promising alternative in sequence-based protein design [86, 87]. Unlike MCMC, diffusion models refine sequences iteratively, offering greater efficiency and flexibility. A key innovation in this space is EvoDiff [88], the first sequence diffusion model trained on the entire UniProt database. EvoDiff's framework successfully generated intrinsically disordered protein regions, a task beyond the reach of structure-based techniques due to the lack of stable, singular structural representations. By bridging gaps in protein design and exploring previously unattainable regions of protein space, LLMs and diffusion models are driving innovation in synthetic biology and protein engineering, while addressing longstanding challenges in the field.

16.5.3 Structure-Based Protein Design

Structure-based design methods aim to create new protein structures by linking desired properties to structural outputs. In this context, diffusion models, advanced generative models, have emerged as particularly promising tools for protein structure generation.

Diffusion models simulate a stepwise generative process akin to "drawing" proteins. During training, experimental protein structures are incrementally corrupted by adding Gaussian noise over multiple steps. The model learns to reverse this process, reconstructing the original structure through iterative denoising. This is accomplished by minimizing a loss function that captures the difference between the predicted and actual structures at each step [67–69]. During generation, diffusion models begin with random Gaussian noise and iteratively refine it into realistic, novel protein structures. This approach allows for de novo protein generation, bypassing the constraints of requiring an existing structural input [66]. Early applications of diffusion models have demonstrated their ability to generate diverse protein backbones, from individual monomers to complex assemblies. For instance, Chroma [74] generated proteins with fragment distributions closely mirroring those found in natural proteins. Another notable example, RFDiffusion, used pretrained weights from RoseTTAFold to directly predict protein structures, successfully producing alpha-beta barrel architectures validated through CryoEM. It is important to note that diffusion models typically generate only backbone coordinates. A subsequent structure-to-sequence step is necessary to translate these outputs into viable protein sequences, integrating the generative structural framework into the broader protein design pipeline.

Structure-based protein design continues to evolve rapidly, with diffusion models leading the charge in redefining what is possible in protein engineering. By bridging structural prediction, generative modeling, and sequence optimization, these approaches are expanding the horizons of synthetic biology and computational design.

16.6 AI and Metabolomics Synthetic Biology-Analytical Detection, Data Preprocessing, Biomarker Discovery, Predictive Modeling, and Multi-Omics Data Integration

16.6.1 Introduction

Metabolomics is a cutting-edge field focused on analyzing the metabolome, which encompasses the small molecule metabolites and their interactions within a biological system [89, 90]. These metabolites include essential compounds such as sugars, AAs, lipids, and other organic molecules critical for cellular processes [91]. Metabolic reactions are generally classified into two main types: catabolism, which involves breaking down molecules to produce energy, and anabolism, which involves synthesizing compounds necessary for cellular function [92]. Additionally, metabolic pathways play a vital role in the detoxification, deactivation, and elimination of foreign substances [92]. By simultaneously measuring thousands of metabolite levels and compositions, metabolomics offers a comprehensive tool to explore metabolic pathways, biochemical processes, and physiological changes occurring within organisms [93, 94].

Metabolomics generates extensive datasets, primarily using two key analytical technologies: nuclear magnetic resonance (NMR) spectroscopy and mass spectrometry (MS). MS is often paired with liquid or gas chromatography (LC-MS or GC-MS) to enhance detection capabilities [95]. These methods enable the identification of hundreds to thousands of metabolites or metabolic features within a single biological sample, facilitating large-scale exploration of the metabolome. Leveraging these expansive datasets, metabolomics provides a comprehensive view of metabolic profiles, offering valuable insights into the biochemical processes that respond to internal and external stimuli [95–98].

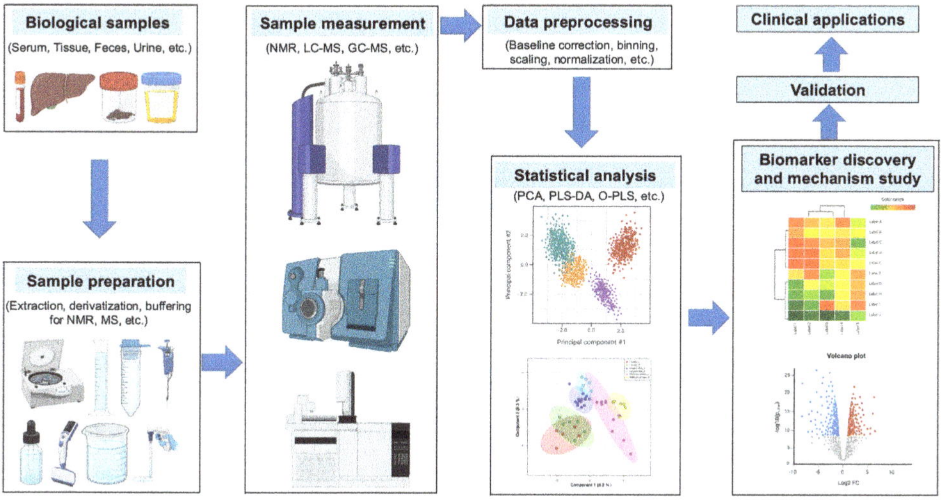

Fig. 16.8 A general workflow in metabolomics.

Fig. 16.8 illustrates the general workflow for metabolomics, a comparative science that requires samples from two or more groups under varying endogenous and/or exogenous conditions [95, 99]. A variety of biological samples, including blood, urine, tissue, and cells, can be used. These samples undergo either untargeted (broad screening) or targeted (focused on specific metabolites) metabolomics analysis. Analytical platforms for metabolomics necessitate distinct sample preparation protocols to extract metabolites effectively and enhance detection sensitivity [100–102]. Following data acquisition, raw data undergoes processing, including background correction, peak picking, deconvolution, and integration. Metabolite identification is achieved by comparing experimental data to chemical standards, reference libraries, or databases. Data pretreatment steps such as baseline correction, binning, scaling, and normalization ensure data consistency. Statistical modeling employs both univariate and multivariate approaches to identify significant metabolites. These metabolites are subsequently mapped to metabolic pathways and integrated with other omics data to reconstruct molecular networks.

An optimal set of metabolites, genes, transcripts, or proteins can be identified to effectively differentiate samples under various conditions [103]. To ensure the robustness of findings, both internal and external cross-validation are essential in metabolomics studies [104]. Additionally, targeted measurement can quantify metabolite concentrations with precision [105]. Prior to statistical analysis, thorough validation and rigorous data quality control are necessary to enhance study reliability.

Applying AI models on metabolomics data suits well because of the large datasets and complex patterns among them. The integration of AI in metabolomics has greatly improved analytical detection, data extraction and preprocessing, and statistical modeling [106, 107].

16.6.2 AI in Analytical Detection

MS datasets generated through LC-MS and GC-MS analyses often include abundances for both annotated metabolites and unidentified compounds [108]. AI has emerged as a powerful tool for addressing the challenge of identifying these unknowns by comparing experimental mass spectra and retention times (RTs) against authentic chemical standards and existing database entries [109]. For instance, Jang et al. developed AI-screener of illicit drugs and analogues (AI-SIDA), a standalone software platform designed to efficiently screen unknown erectile dysfunction (ED) drugs and their analogues [110]. Similarly, the OngLai algorithm was introduced to identify homologous series using RDKit [111], while another approach utilized entropy-based AI to classify stilbenoid compounds [112].

Beyond annotation, AI algorithms have shown remarkable promise in predicting compound structures without prior references [113]. A notable example is AF2, created by DeepMind, which accurately predicts the 3D structures of chemical compounds at atomic resolution. These AI-driven methodologies not only streamline metabolite identification but also hold potential for discovering novel compounds that remain unexplored [114].

AI has also been instrumental in analyzing complex spectral data. Zhang et al., for example, developed the DeepSpectra model, which leverages raw spectroscopic data to enhance quantitative analysis [115]. DeepSpectra demonstrated superior performance compared to traditional linear and nonlinear calibration methods by minimizing artifacts, preserving critical patterns, and delivering improved repeatability and accuracy. Collectively, these advancements highlight AI's transformative role in metabolomics and compound discovery [99].

16.6.3 AI in Data Extraction and Preprocessing

The primary objective of data extraction in metabolomics is to transform complex, high-dimensional raw data into a format suitable for downstream analysis and interpretation [107, 116]. Convolutional neural networks (CNNs) have demonstrated exceptional performance in tasks such as processing, selecting, and integrating chromatographic peaks, particularly those with strong local correlations. Compared to traditional methods, CNNs improve classification accuracy by an average of 6%-24% [117, 118]. Additionally, CNNs exhibit reduced reliance on data preprocessing techniques and excel at resolving overlapping peaks in complex chromatograms [119, 120]. For instance, the Serrano group introduced an advanced ANN model inspired by physics-informed neural networks (PINNs), which significantly enhances peak selection and effectively quantifies analytes with overlapping chromatographic peaks using single-detector instruments [121].

In untargeted metabolomics, incorporating RT alongside the mass-to-charge ratio (m/z) enhances metabolite identification [122, 123]. While RT knowledge improves peak picking accuracy, predicting RT for unknown metabolites remains challenging. To address this, various ML approaches have been developed. For example, the Fiehn group integrated five ML algorithms into the Retip R package, reducing candidate structures by 68% during isomer searches in mouse plasma samples [124]. In the initial stages of RT prediction, molecular structural information is typically encoded into vector formats using methods like quantitative structure–retention relationship (QSRR) models or molecular fingerprints [125–127]. Liapikos et al. developed an ML model employing four regression algorithms, Bayesian ridge regression, extreme gradient boosting, and support vector regression (SVR) with linear and nonlinear kernels, leveraging molecular descriptors to represent the physical, chemical, and structural properties of molecules mathematically [128].

Bouwmeester et al. further advanced the field in 2019 by comparing various ML methods for RT prediction in liquid chromatography (LC). Their study extracted 151 features from simplified molecular input line entry system (SMILES) notation to train seven models, encompassing both linear and nonlinear techniques. Among these, stand-alone models like ANN and support vector machines (SVM) exhibited high predictive accuracy. Notably, ensemble methods combining multiple ML tools outperformed individual models, highlighting the advantage of integrative approaches [129].

Systematic errors, such as batch differences, longitudinal drifts, and instrument variations, are common challenges in metabolomics data analysis. To address these issues, Kehoe et al. developed a sparse SVM model specifically designed to mitigate batch effects prevalent in large-scale metabolomics studies [130]. Similarly, Shen et al. introduced an SVR method to minimize intra- and inter-batch variations while integrating multiple data batches. Their approach significantly enhanced downstream multivariate statistical analyses, improving classification and prediction accuracy [131]. ML methods, with their capacity to handle nonlinear data, are particularly effective in capturing complex patterns and eliminating systematic errors. For instance, Fan et al. proposed the systematic error removal using random forest (SERRF) method, which demonstrated superior performance over 15 commonly used normalization techniques by effectively reducing systematic variation and emphasizing biological variance [132].

16.6.4 AI in Statistical Modeling

Statistical modeling of preprocessed metabolomics data aims to unravel complex metabolic interactions, discover novel biomarkers, and deepen our understanding of biological processes [133–135]. Selecting the right AI algorithms, fine-tuning model configurations, and designing new computational approaches require a thorough understanding of the unique characteristics of metabolomics data. Key challenges include [99]

Curse of dimensionality [136]: Metabolomics experiments often measure hundreds to thousands of metabolites, but the high cost of these studies typically limits the number of samples. This imbalance, where the number of variables far exceeds the number of samples, makes the data prone to overfitting.

Complex relationships between metabolites [137]: Metabolites are interlinked through a network of biochemical pathways and reactions, resulting in varying levels of correlation among their concentrations. These dependencies can complicate analysis.

High variability [138]: Metabolite levels can fluctuate significantly due to intrinsic and extrinsic factors, such as biological responses to stimuli. Additionally, individual differences and experimental variability may obscure subtle metabolic changes of interest. Covariates such as age, sex, and demographic factors also influence metabolite levels and should be accounted for during feature selection and predictive modeling.

Ambiguity in metabolic causality: Metabolic changes may either drive or result from phenotypic alterations. With the wide array of metabolites involved, determining which should be treated as independent or dependent variables presents a significant challenge.

To address these challenges, various strategies have been developed, focusing on feature extraction, feature selection, and robust modeling and evaluation techniques. These approaches aim to enhance the accuracy and interpretability of metabolomics data analysis across diverse applications.

16.6.4.1 Feature selection

Feature selection focuses on identifying the most relevant and informative features while reducing the dimensionality of metabolomics data [139]. By isolating key metabolites associated with a specific phenotype or condition, supervised ML algorithms play a pivotal role in this process [140]. Feature selection strategies can be broadly categorized into four main approaches [99]:

Filter methods: These methods rely on univariate and multivariate statistical tests to evaluate features based on their individual relevance. Their simplicity and efficiency make them popular in metabolomics studies [141–143]. However, because they do not account for feature interactions, the selected metabolites may not always be the best predictors for distinguishing phenotypes.

Wrapper methods: These approaches address the limitations of filter methods by incorporating an evaluation step that builds predictive models using the selected features. The feature subset is iteratively refined to optimize model performance. For instance, Gaul et al. used recursive feature elimination (RFE) with an SVM to develop a 16-metabolite predictive model for diagnosing early stage serous epithelial ovarian cancer [144].

Embedded methods: These methods integrate feature selection into the model training process. Algorithms like decision trees, random forests (RF), and least absolute shrinkage and selection operator (LASSO) simultaneously identify relevant features and construct predictive models [145]. For example, LASSO was employed to generate lipid profiles distinguishing molecular subtypes of breast cancer [146].

Hybrid methods: Combining multiple feature selection approaches can address scenarios where individual methods alone fall short. For example, Bifarin et al. used filter techniques to reduce an initial feature set of 7,147 NMR and MS features to 128 based on t-test p-values, fold changes, and correlation tests. They then applied wrapper methods, including RF-RFE and partial least squares (PLS), to refine this subset further, ultimately identifying the top 20 features for predicting renal cell carcinoma status [147].

16.6.4.2 Predictive modeling

AI offers distinct advantages over traditional statistical methods for modeling complex, nonlinear relationships in metabolomics data. The choice of method depends on the study type, data size, and analysis goals.

For targeted metabolomics studies, where datasets are typically small and interpretability is critical, models that provide clear insights into feature–response relationships, such as decision trees or conventional statistical regression, are often preferred. Conversely, in untargeted metabolomics studies, which involve larger datasets and prioritize predictive performance over interpretability, advanced "black box" models, including DL, are more suitable. Model selection is an iterative process. Initially, a range of algorithms, from simple linear regression to more complex methods like RFs, SVMs, and DL, is tested to establish baseline performance. Based on these results, the selection process is refined by tuning hyperparameters and reassessing model outcomes. This systematic approach ensures that the chosen method aligns with study objectives, such as phenotype classification or quantitative outcome prediction.

One limitation of DL in metabolomics has been the need for large training datasets [148]. To address this, Choudhury et al. employed generative adversarial networks (GANs) to generate synthetic metabolomics data closely resembling real profiles [149]. By augmenting a small dataset with a substantial synthetic dataset, they trained the reconstruction of kinetic models using deep learning (REKINDLE) model. This approach enabled REKINDLE to explore a wide parameter space, incorporating sufficient diversity to enhance model robustness and expand its applicability.

References

1. Benner, S. A., and A. M. Sismour. Synthetic biology. *Nat Rev Genet.* 2005; **6**(7): p. 533–543.
2. Garner, K. L. Principles of synthetic biology. *Essays Biochem.* 2021; **65**(5): p. 791–811.
3. Li, R. C., S. M. Asch, and N. H. Shah. Developing a delivery science for artificial intelligence in healthcare. *NPJ Digit Med.* 2020; **3**(1): p. 107.
4. Tien, J. M. Internet of things, real-time decision making, and artificial intelligence. *Ann Data Sci.* 2017; **4**: p. 149–178.
5. Barnes, E. A., et al. Indicator patterns of forced change learned by an artificial neural network. *J Adv Model Earth Syst.* 2020; **12**(9): p. e2020MS002195.
6. Dara, M., et al. Convergence of CRISPR and artificial intelligence: a paradigm shift in biotechnology. *Hum Gene.* 2024; **41**: p. 201297.
7. Lee, M. Deep learning in CRISPR-Cas systems: a review of recent studies. *Front Bioeng Biotechnol.* 2023; **11**: p. 1226182.
8. Niazi, S. K. The coming of age of AI/ML in drug discovery, development, clinical testing, and manufacturing: The FDA Perspectives. *Drug Des, Devel Ther.* 2023; **17**: p. 2691–2725.
9. Bhat, A. A., et al. Integration of CRISPR/Cas9 with artificial intelligence for improved cancer therapeutics. *J Transl Med.* 2022; **20**(1): p. 534.

10. Konstantakos, V., et al. CRISPR–Cas9 gRNA efficiency prediction: an overview of predictive tools and the role of deep learning. *Nucleic Acids Res*. 2022; **50**(7): p. 3616–3637.
11. Henry, V. J., et al. OMICtools: an informative directory for multi-omic data analysis. *Database*. 2014; **2014**: p. bau069.
12. Xu, H., et al. Sequence determinants of improved CRISPR sgRNA design. *Genome Res*. 2015; **25**(8): p. 1147–1157.
13. Moreno-Mateos, M. A., et al. CRISPRscan: designing highly efficient sgRNAs for CRISPR-Cas9 targeting in vivo. *Nat Methods*. 2015; **12**(10): p. 982–988.
14. Labuhn, M., et al. Refined sgRNA efficacy prediction improves large-and small-scale CRISPR–Cas9 applications. *Nucleic Acids Res*. 2018; **46**(3): p. 1375–1385.
15. Doench, J. G., et al. Optimized sgRNA design to maximize activity and minimize off-target effects of CRISPR-Cas9. *Nat Biotechnol*. 2016; **34**(2): p. 184–191.
16. Wilson, L. O., et al. High activity target-site identification using phenotypic independent CRISPR-Cas9 core functionality. *CRISPR J*. 2018; **1**(2): p. 182–190.
17. Doench, J. G., et al. Rational design of highly active sgRNAs for CRISPR-Cas9–mediated gene inactivation. *Nat Biotechnol*. 2014; **32**(12): p. 1262–1267.
18. Wong, N., W. Liu, and X. Wang. WU-CRISPR: characteristics of functional guide RNAs for the CRISPR/Cas9 system. *Genome Biol*. 2015; **16**: p. 1–8.
19. Chari, R., et al. Unraveling CRISPR-Cas9 genome engineering parameters via a library-on-library approach. *Nat Methods*. 2015; **12**(9): p. 823–826.
20. Peng, H., et al. CRISPR/Cas9 cleavage efficiency regression through boosting algorithms and Markov sequence profiling. *Bioinformatics*. 2018; **34**(18): p. 3069–3077.
21. Hiranniramol, K., et al. Generalizable sgRNA design for improved CRISPR/Cas9 editing efficiency. *Bioinformatics*. 2020; **36**(9): p. 2684–2689.
22. Chuai, G. H., Q.-L. Wang, and Q. Liu. In silico meets in vivo: towards computational CRISPR-based sgRNA design. *Trends Biotechnol*. 2017; **35**(1): p. 12–21.
23. Chuai, G., et al. DeepCRISPR: optimized CRISPR guide RNA design by deep learning. *Genome Biol*. 2018; **19**: p. 1–18.
24. Xue, L., et al. Prediction of CRISPR sgRNA activity using a deep convolutional neural network. *J Chem Inf Model*. 2018; **59**(1): p. 615–624.
25. Chari, R., et al. sgRNA Scorer 2.0: a species-independent model to predict CRISPR/Cas9 activity. *ACS Synth Biol*. 2017; **6**(5): p. 902–904.
26. Rahman, M. K., and M. S. Rahman. CRISPRpred: a flexible and efficient tool for sgRNAs on-target activity prediction in CRISPR/Cas9 systems. *PloS One*. 2017; **12**(8): p. e0181943.
27. Zhang, G., Z. Dai, and X. Dai. A novel hybrid CNN-SVR for CRISPR/Cas9 guide RNA activity prediction. *Front Genet*. 2020; **10**: p. 1303.
28. Zhang, G., Z. Dai, and X. Dai. C-RNNCrispr: prediction of CRISPR/Cas9 sgRNA activity using convolutional and recurrent neural networks. *Comput Struct Biotechnol J*. 2020; **18**: p. 344–354.
29. Wolff, J. A., et al. Direct gene transfer into mouse muscle in vivo. *Science*. 1990; **247**(4949): p. 1465–1468.
30. Ulmer, J. B., et al. Heterologous protection against influenza by injection of DNA encoding a viral protein. *Science*. 1993; **259**(5102): p. 1745–1749.
31. Zhou, X., et al. Self-replicating Semliki Forest virus RNA as recombinant vaccine. *Vaccine*. 1994; **12**(16): p. 1510–1514.

32. Martinon, F., et al. Induction of virus-specific cytotoxic T lymphocytes in vivo by liposome-entrapped mRNA. *Eur J Immunol*. 1993; **23**(7): p. 1719–1722.
33. Karikó, K., et al. Incorporation of pseudouridine into mRNA yields superior nonimmunogenic vector with increased translational capacity and biological stability. *Mol Ther*. 2008; **16**(11): p. 1833–1840.
34. Bahl, K., et al. Preclinical and clinical demonstration of immunogenicity by mRNA vaccines against H10N8 and H7N9 influenza viruses. *Mol Ther*. 2017; **25**(6): p. 1316–1327.
35. Feldman, R. A., et al. mRNA vaccines against H10N8 and H7N9 influenza viruses of pandemic potential are immunogenic and well tolerated in healthy adults in phase 1 randomized clinical trials. *Vaccine*. 2019; **37**(25): p. 3326–3334.
36. Schlake, T., et al. Developing mRNA-vaccine technologies. *RNA Biol*. 2012; **9**(11): p. 1319–1330.
37. Cagigi, A., and K. Loré. Immune responses induced by mRNA vaccination in mice, monkeys and humans. *Vaccines*. 2021; **9**(1): p. 61.
38. Moore, M. J. From birth to death: the complex lives of eukaryotic mRNAs. *Science*. 2005; **309**(5740): p. 1514–1518.
39. Sultana, N., et al. Optimization of 5′ untranslated region of modified mRNA for use in cardiac or hepatic ischemic injury. *Mol Ther-Methods Clin Dev*. 2020; **17**: p. 622–633.
40. Mayr, C. Regulation by 3′-untranslated regions. *Annu Rev Genet*. 2017; **51**: p. 171–194.
41. Bazzini, A. A., M. T. Lee, and A. J. Giraldez. Ribosome profiling shows that miR-430 reduces translation before causing mRNA decay in zebrafish. *Science*. 2012; **336**(6078): p. 233–237.
42. Hershberg, R., and D. A. Petrov. Selection on codon bias. *Annu Rev Genet*. 2008; **42**: p. 287–299.
43. Plotkin, J. B., and G. Kudla. Synonymous but not the same: the causes and consequences of codon bias. *Nat Rev Genet*. 2011; **12**(1): p. 32–42.
44. Kudla, G., et al. Coding-sequence determinants of gene expression in *Escherichia coli*. *Science*. 2009; **324**(5924): p. 255–258.
45. Gustafsson, C., S. Govindarajan, and J. Minshull. Codon bias and heterologous protein expression. *Trends Biotechnol*. 2004; **22**(7): p. 346–353.
46. Zur, H., and T. Tuller. New universal rules of eukaryotic translation initiation fidelity. *PLoS Comput Biol*. 2013; **9**(7): p. e1003136.
47. Eldad, N., Y. Yosefzon, and Y. Arava. Identification and characterization of extensive intramolecular associations between 3′-UTRs and their ORFs. *Nucleic Acids Res*. 2008; **36**(21): p. 6728–6738.
48. Chen, J., and M. B. Kastan. 5′–3′-UTR interactions regulate p53 mRNA translation and provide a target for modulating p53 induction after DNA damage. *Genes Dev*. 2010; **24**(19): p. 2146–2156.
49. Fu, H., et al. Codon optimization with deep learning to enhance protein expression. *Sci Rep*. 2020; **10**(1): p. 17617.
50. Li, Y., et al. Deep generative optimization of mRNA codon sequences for enhanced protein production and therapeutic efficacy. *bioRxiv*. 2024: p. 2024.09.06.611590.
51. Sample, P. J., et al. Human 5′ UTR design and variant effect prediction from a massively parallel translation assay. *Nat Biotechnol*. 2019; **37**(7): p. 803–809.

52. Castillo-Hair, S., et al. Optimizing 5'UTRs for mRNA-delivered gene editing using deep learning. *Nat Commun*. 2024; **15**(1): p. 5284.
53. Morrow, A. K., et al. ML-driven design of 3' UTRs for mRNA stability. *bioRxiv*. 2024: p. 2024.10.07.616676.
54. Gong, H., et al. Integrated mRNA sequence optimization using deep learning. *Brief Bioinform*. 2023;**24**(1): p. bbad001.
55. Huang, Z., W. Xu, and K. Yu. Bidirectional LSTM-CRF models for sequence tagging. 2015.
56. Bettini, E., and M. Locci. SARS-CoV-2 mRNA vaccines: immunological mechanism and beyond. *Vaccines*. 2021; **9**(2): p. 147.
57. Jiang, T., et al. Tumor neoantigens: from basic research to clinical applications. *J Hematol Oncol*. 2019; **12**(1): p. 1–13.
58. Jiao, X., et al. Tinybert: distilling bert for natural language understanding. 2019. arXiv:1909.10351.
59. Lan, Z., et al. Albert: a lite bert for self-supervised learning of language representations. 2019. arXiv:1909.11942.
60. Hershberg, R., and D. A. Petrov. Selection on codon bias. *Annu Rev Genet*. 2008; **42**: p. 287–299.
61. Fu, H., et al. Codon optimization with deep learning to enhance protein expression. *Sci Rep*. 2020; **10**(1): p. 1–9.
62. Real, R., and J. M. Vargas. The probabilistic basis of Jaccard's index of similarity. *Syst Biol*. 1996; **45**(3): p. 380–385.
63. Lin, C.-Y. Rouge: A package for automatic evaluation of summaries. In *Text summarization branches out*. Basel, Switzerland: MDPI; 2004.
64. Zhuang, X., et al. mRNA vaccines encoding the HA protein of influenza A H1N1 virus delivered by cationic lipid nanoparticles induce protective immune responses in mice. *Vaccines*. 2020; **8**(1): p. 123.
65. Valldorf, B., et al. Antibody display technologies: selecting the cream of the crop. *Biol Chem*. 2022; **403**(5-6): p. 455–477.
66. Winnifrith, A., C. Outeiral, and B. L. Hie. *Generative artificial intelligence for de novo* protein design. *Curr Opin Struct Biol*. 2024; **86**: p. 102794.
67. Sohl-Dickstein, J., et al. Deep unsupervised learning using nonequilibrium thermodynamics. In: *International conference on machine learning*. 2015. PMLR.
68. Ho, J., A. Jain, and P. Abbeel. Denoising diffusion probabilistic models. *Adv Neural Inf Process Syst*. 2020; **33**: p. 6840–6851.
69. Song, Y., et al. Score-based generative modeling through stochastic differential equations. arXiv preprint arXiv:2011.13456, 2020.
70. Chowdhery, A., et al. Palm: Scaling language modeling with pathways. *J Mach Learn Res*. 2023; **24**(240): p. 1–113.
71. Verkuil, R., et al. Language models generalize beyond natural proteins. *bioRxiv*. 2022: p. 2022.12. 21.521521.
72. Watson, J. L., et al. De novo design of protein structure and function with RFdiffusion. *Nature*. 2023; **620**(7976): p. 1089–1100.
73. Hie, B., et al. A high-level programming language for generative protein design. *bioRxiv*. 2022: p. 2022.12. 21.521526.

74. Ingraham, J. B., et al. Illuminating protein space with a programmable generative model. *Nature*. 2023; **623**(7989): p. 1070–1078.
75. Jumper, J., et al. Highly accurate protein structure prediction with AlphaFold. *Nature*. 2021; **596**(7873): p. 583–589.
76. Baek, M., et al. Accurate prediction of protein structures and interactions using a three-track neural network. *Science*. 2021; **373**(6557): p. 871–876.
77. Baek, M., et al. Efficient and accurate prediction of protein structure using RoseTTAFold2. *bioRxiv*. 2023: p. 2023.05. 24.542179.
78. Krishna, R., et al. Generalized biomolecular modeling and design with RoseTTAFold All-Atom. *Science*. 2024; **384**(6693): p. eadl2528.
79. Lin, Z., et al. Evolutionary-scale prediction of atomic-level protein structure with a language model. *Science*. 2023; **379**(6637): p. 1123–1130.
80. Bennett, N. R., et al. Improving de novo protein binder design with deep learning. *Nat Commun*. 2023; **14**(1): p. 2625.
81. Hsu, C., et al. Learning inverse folding from millions of predicted structures. In: *International conference on machine learning*. 2022. Baltimore, Maryland, USA: PMLR.
82. Madani, A., et al. Large language models generate functional protein sequences across diverse families. *Nat Biotechnol*. 2023; **41**(8): p. 1099–1106.
83. Ferruz, N., S. Schmidt, and B. Höcker. ProtGPT2 is a deep unsupervised language model for protein design. *Nat Commun*. 2022; **13**(1): p. 4348.
84. Kaplan, J., and W. DeGrado. De novo design of catalytic proteins. *Proc Natl Acad Sci*. 2004; **101**(32): p. 11566–11570.
85. Chen, Z., et al. De novo design of protein logic gates. *Science*. 2020; **368**(6486): p. 78–84.
86. Hoogeboom, E., et al. Autoregressive diffusion models. arXiv preprint arXiv:2110.02037, 2021.
87. Austin, J., et al. Structured denoising diffusion models in discrete state-spaces. *Adv Neural Inf Process Syst*. 2021; **34**: p. 17981–17993.
88. Alamdari, S., et al. Protein generation with evolutionary diffusion: sequence is all you need. *bioRxiv 2023*. Google Scholar.
89. Muthubharathi, B. C., T. Gowripriya, and K. Balamurugan. Metabolomics: small molecules that matter more. *Mol Omics*. 2021; **17**(2): p. 210–229.
90. Gu, H., et al. Principal component directed partial least squares analysis for combining NMR and MS data in metabolomics: application to the detection of breast cancer. *Anal Chim Acta*. 2010; **686**(1-2): p. 57.
91. Yang, P. L. Metabolomics and lipidomics: yet more ways your health is influenced by fat. In *Viral pathogenesis*. 2016, Elsevier. p. 181–198.
92. Chandel, N. S. Basics of metabolic reactions. *Cold Spring Harb Perspect Biol*. 2021; **13**(8): p. 1–11.
93. Schilling, C. H., et al. Metabolic pathway analysis: basic concepts and scientific applications in the post-genomic era. *Biotechnol Prog*. 1999; **15**(3): p. 296–303.
94. Johnson, C. H., et al. Bioinformatics: the next frontier of metabolomics. *Anal Chem*. 2015; **87**(1): p. 147–156.
95. Zhang, A., et al. Modern analytical techniques in metabolomics analysis. *Analyst*. 2012; **137**(2): p. 293–300.

96. Pearce, E. L., et al. Fueling immunity: insights into metabolism and lymphocyte function. *Science*. 2013; **342**(6155): p. 1242454.
97. Zhang, J., et al. Metabolomics study of esophageal adenocarcinoma. *J Thorac Cardiovasc Surg*. 2011; **141**(2): p. 469–475. e4.
98. Gibney, M. J., et al. Metabolomics in human nutrition: opportunities and challenges. *Am J Clin Nutr*. 2005; **82**(3): p. 497–503.
99. Chi, J., et al. Artificial intelligence in metabolomics: a current review. *TrAC Trends Anal Chem*. 2024; **178**: p. 117852.
100. Dettmer, K., P. A. Aronov, and B. D. Hammock. Mass spectrometry-based metabolomics. *Mass Spectrom Rev*. 2007; **26**(1): p. 51–78.
101. Wishart, D. Metabolomics: the principles and potential applications to transplantation. *Am J Transplant*. 2005; **5**(12): p. 2814–2820.
102. Want, E. J., et al. From exogenous to endogenous: the inevitable imprint of mass spectrometry in metabolomics. *J Proteome Res*. 2007; **6**(2): p. 459–468.
103. Chen, Y., E.-M. Li, and L.-Y. Xu. Guide to metabolomics analysis: a bioinformatics workflow. *Metabolites*. 2022; **12**(4): p. 357.
104. Masutin, V., C. Kersch, and S. Schmitz-Spanke. A systematic review: metabolomics-based identification of altered metabolites and pathways in the skin caused by internal and external factors. *Exp Dermatol*. 2022; **31**(5): p. 700–714.
105. Ribbenstedt, A., H. Ziarrusta, and J. P. Benskin. Development, characterization and comparisons of targeted and non-targeted metabolomics methods. *PLoS One*. 2018; **13**(11): p. e0207082.
106. Odenkirk, M. T., D. M. Reif, and E. S. Baker. Multiomic big data analysis challenges: increasing confidence in the interpretation of artificial intelligence assessments. *Anal Chem*. 2021; **93**(22): p. 7763–7773.
107. Liebal, U. W., et al. Machine learning applications for mass spectrometry-based metabolomics. *Metabolites*. 2020; **10**(6): p. 243.
108. Petrick, L. M., and N. Shomron. AI/ML-driven advances in untargeted metabolomics and exposomics for biomedical applications. *Cell Rep Phys Sci*. 2022; **3**(7): p. 100978.
109. Giese, S. H., et al. Retention time prediction using neural networks increases identifications in crosslinking mass spectrometry. *Nat Commun*. 2021; **12**(1): p. 3237.
110. Jang, I., et al. LC–MS/MS software for screening unknown erectile dysfunction drugs and analogues: artificial neural network classification, peak-count scoring, simple similarity search, and hybrid similarity search algorithms. *Anal Chem*. 2019; **91**(14): p. 9119–9128.
111. Lai, A., et al. An algorithm to classify homologous series within compound datasets. *J Cheminform*. 2022; **14**(1): p. 85.
112. Turi, K. N., et al. A review of metabolomics approaches and their application in identifying causal pathways of childhood asthma. *J Allergy Clin Immunol*. 2018; **141**(4): p. 1191–1201.
113. Yang, Z., et al. AlphaFold2 and its applications in the fields of biology and medicine. *Signal Transduct Target Ther*. 2023; **8**(1): p. 115.
114. Frusciante, L., et al. Artificial intelligence approaches in drug discovery: towards the laboratory of the future. *Curr Top Med Chem*. 2022; **22**(26): p. 2176–2189.

115. Zhang, X., et al. DeepSpectra: An end-to-end deep learning approach for quantitative spectral analysis. *Anal Chim Acta.* 2019; **1058**: p. 48–57.
116. Boccard, J., J. L. Veuthey, and S. Rudaz. Knowledge discovery in metabolomics: an overview of MS data handling. *J Sep Sci.* 2010; **33**(3): p. 290–304.
117. Su, K.-M., et al. Intelligent geochemical interpretation of mass chromatograms: based on convolution neural network. *Pet Sci.* 2024; **21**(2): p. 752–764.
118. Acquarelli, J., et al. Convolutional neural networks for vibrational spectroscopic data analysis. *Anal Chim Acta.* 2017; **954**: p. 22–31.
119. Bjerrum, E. J., M. Glahder, and T. Skov. Data augmentation of spectral data for convolutional neural network (CNN) based deep chemometrics. arXiv preprint arXiv:1710.01927, 2017.
120. Kensert, A., et al. Convolutional neural network for automated peak detection in reversed-phase liquid chromatography. *J Chromatogr A.* 2022; **1672**: p. 463005.
121. Hervás, C., et al. Improving the quantification of highly overlapping chromatographic peaks by using product unit neural networks modeled by an evolutionary algorithm. *J Chem Inf Model.* 2005; **45**(4): p. 894–903.
122. Zeki, Ö. C., et al. Integration of GC–MS and LC–MS for untargeted metabolomics profiling. *J Pharm Biomed Anal.* 2020; **190**: p. 113509.
123. Choi, E., et al. Machine learning liquid chromatography retention time prediction model augments the dansylation strategy for metabolite analysis of urine samples. *J Chromatogr A.* 2023; **1705**: p. 464167.
124. Bonini, P., et al. Retip: retention time prediction for compound annotation in untargeted metabolomics. *Anal Chem.* 2020; **92**(11): p. 7515–7522.
125. Wolfer, A. M., et al. UPLC–MS retention time prediction: a machine learning approach to metabolite identification in untargeted profiling. *Metabolomics.* 2016; **12**(1): p. 8.
126. Creek, D. J., et al. Toward global metabolomics analysis with hydrophilic interaction liquid chromatography–mass spectrometry: improved metabolite identification by retention time prediction. *Anal Chem.* 2011; **83**(22): p. 8703–8710.
127. Berry, M. W., A. Mohamed, and B. W. Yap. *Supervised and unsupervised learning for data science.* Springer; 2019.
128. Liapikos, T., et al. Quantitative structure retention relationship (QSRR) modelling for analytes' retention prediction in LC-HRMS by applying different machine learning algorithms and evaluating their performance. *J Chromatogr B.* 2022; **1191**: p. 123132.
129. Bouwmeester, R., L. Martens, and S. Degroeve. Comprehensive and empirical evaluation of machine learning algorithms for small molecule LC retention time prediction. *Anal Chem.* 2019; **91**(5): p. 3694–3703.
130. Kehoe, E. R., et al. Biomarker selection and a prospective metabolite-based machine learning diagnostic for lyme disease. *Sci Rep.* 2022; **12**(1): p. 1478.
131. Shen, X., et al. Normalization and integration of large-scale metabolomics data using support vector regression. *Metabolomics.* 2016; **12**: p. 1–12.
132. Fan, S., et al. Systematic error removal using random forest for normalizing large-scale untargeted lipidomics data. *Anal Chem.* 2019; **91**(5): p. 3590–3596.
133. Biswas, N., and S. Chakrabarti. Artificial intelligence (AI)-based systems biology approaches in multi-omics data analysis of cancer. *Front Oncol.* 2020; **10**: p. 588221.

134. Helmy, M., D. Smith, and K. Selvarajoo. Systems biology approaches integrated with artificial intelligence for optimized metabolic engineering. *Metab Eng Commun*. 2020; **11**: p. e00149.
135. Edison, A. S., et al. NMR: unique strengths that enhance modern metabolomics research. *Anal Chem*. 2020; **93**(1): p. 478–499.
136. Berisha, V., et al. Digital medicine and the curse of dimensionality. *NPJ Digit Med*. 2021; **4**(1): p. 153.
137. Cuperlovic-Culf, M. Machine learning methods for analysis of metabolic data and metabolic pathway modeling. *Metabolites*. 2018; **8**(1): p. 4.
138. Miyazawa, T., et al. Artificial intelligence in food science and nutrition: a narrative review. *Nutr Rev*. 2022; **80**(12): p. 2288–2300.
139. Grissa, D., et al. Feature selection methods for early predictive biomarker discovery using untargeted metabolomic data. *Front Mol Biosci*. 2016; **3**: p. 30.
140. Barberis, E., et al. Precision medicine approaches with metabolomics and artificial intelligence. *Int J Mol Sci*. 2022; **23**(19): p. 11269.
141. Xie, Y., et al. Early lung cancer diagnostic biomarker discovery by machine learning methods. *Transl Oncol*. 2021; **14**(1): p. 100907.
142. He, Z., Z. Liu, and L. Gong. Biomarker identification and pathway analysis of rheumatoid arthritis based on metabolomics in combination with ingenuity pathway analysis. *Proteomics*. 2021; **21**(11-12): p. 2100037.
143. Mehmood, T., S. Sæbø, and K. H. Liland. Comparison of variable selection methods in partial least squares regression. *J Chemom*. 2020; **34**(6): p. e3226.
144. Gaul, D. A., et al. Highly-accurate metabolomic detection of early-stage ovarian cancer. *Sci Rep*. 2015; **5**(1): p. 16351.
145. Vidyasagar, M. Identifying predictive features in drug response using machine learning: opportunities and challenges. *Annu Rev Pharmacol Toxicol*. 2015; **55**(1): p. 15–34.
146. Santoro, A. L., et al. In situ DESI-MSI lipidomic profiles of breast cancer molecular subtypes and precursor lesions. *Cancer Res*. 2020; **80**(6): p. 1246–1257.
147. Bifarin, O. O., et al. Machine learning-enabled renal cell carcinoma status prediction using multiplatform urine-based metabolomics. *J Proteome Res*. 2021; **20**(7): p. 3629–3641.
148. Zampieri, G., et al. Machine and deep learning meet genome-scale metabolic modeling. *PLoS Comput Biol*. 2019; **15**(7): p. e1007084.
149. Choudhury, S., et al. Reconstructing kinetic models for dynamical studies of metabolism using generative adversarial networks. *Nat Mach Intell*. 2022; **4**(8): p. 710–719.

CHAPTER 17

Small Molecular Generation

17.1 Introduction

The discovery of new molecules is pivotal for addressing some of the most pressing challenges in chemistry and engineering, from combating complex diseases to mitigating the impacts of climate change. Traditionally, molecular discovery has relied heavily on human expertise to propose, synthesize, and experimentally evaluate candidates. While this approach has led to many groundbreaking advances, it is inherently resource-intensive, both in terms of cost and time. Furthermore, the vastness of chemical space, estimated to range from 10^{23} to 10^{60} theoretically feasible compounds, renders traditional exploration methods capable of sampling only a minuscule fraction of potential candidates [1, 2].

To overcome these limitations, generative models have emerged as a transformative tool, leveraging recent advances in artificial intelligence (AI) and deep learning. Generative models allow for the efficient exploration of chemical space, significantly accelerating the design process while enhancing diversity and optimization. Recent progress in this field has led to a variety of model architectures and molecular representations, ranging from one-dimensional string-based methods (e.g., SMILES [3] and SELFIES [4]) to graph-based and 3D models that explicitly account for molecular geometry [5].

In particular, 3D generative models have gained traction due to their ability to encode and optimize spatial properties critical for structure-based applications, such as drug-protein interactions. Unlike earlier models that relied on indirect representations of 3D structure, modern approaches directly generate 3D molecular geometries, offering advantages in accuracy, efficiency, and scalability. These advancements are particularly relevant in drug design, where small molecules must satisfy complex biochemical and pharmacological constraints [6, 7], and in protein engineering, where the goal often involves designing entirely novel biomolecules [8].

Despite these advancements, the application of generative models in practical scenarios remains in its infancy. Many studies focus on optimizing computational metrics, such as drug-likeness or synthetic accessibility, with fewer examples of experimental validation for real-world applications. Moreover, challenges persist in integrating generative approaches with existing drug discovery pipelines, particularly in addressing the scalability of models, handling multi-objective optimization, and ensuring the interpretability of results.

This chapter provides an overview of the current landscape in molecular generative modeling, emphasizing both its transformative potential and the challenges that remain. Through a comprehensive description and a detailed explanation of specific generative model algorithms, our aim is to provide readers with a thorough understanding of how AI accelerates and enhances discovery processes across various fields, including AI-assisted drug design.

17.2 AI Accelerated Drug Discovery

Drug discovery is a complex, resource-intensive, and multiphase endeavor aimed at identifying molecules with the potential to address unmet medical needs. Traditionally, this process spans several stages, each requiring significant time and expertise. In recent years, AI has emerged as a transformative tool, significantly accelerating various aspects of drug discovery and development. This section primarily introduces the traditional drug discovery process and the AI-powered acceleration of drug discovery, aiming to help readers understand the relationship between them.

The Traditional Drug Discovery Process

1. Target identification and validation: The process begins with identifying a biological target, such as a protein or receptor, that plays a critical role in the pathophysiology of a disease. Validation ensures that modulating this target has therapeutic potential. This step relies on techniques such as genomics, proteomics, and functional studies.
2. Lead compound identification: Following target validation, the next step is to identify lead compounds, molecules that show preliminary activity against the target. Traditional methods include: High-throughput screening (HTS): Large chemical libraries are experimentally screened for activity; and virtual screening: Computational techniques identify candidates by simulating molecular interactions.
3. Lead optimization: Promising lead compounds undergo structural modifications to enhance their efficacy, specificity, and drug-like properties while minimizing toxicity. This stage involves iterative testing and refinement cycles.
4. Preclinical testing: Candidates are tested in vitro (in cells) and in vivo (in animal models) to assess safety, pharmacokinetics, and pharmacodynamics. Toxicological studies determine dosage ranges and potential side effects.

5. Clinical trials: Compounds that pass preclinical evaluations enter clinical trials, conducted in three phases: Phase I: Evaluates safety and tolerability in healthy volunteers. Phase II: Tests efficacy and optimal dosing in a small patient cohort. Phase III: Confirms efficacy and safety in a large, diverse patient population.
6. Regulatory approval and post-market surveillance: Successful candidates are submitted for regulatory approval, followed by ongoing safety monitoring after-market release.

While effective, this traditional pipeline is costly and time-consuming, often taking 10-15 years and billions of dollars to bring a single drug to market [9, 10]. Moreover, the process explores only a fraction of the vast chemical space of potential compounds.

Ai-Powered Acceleration of Drug Discovery

AI is revolutionizing drug discovery by addressing bottlenecks, improving accuracy, and significantly reducing costs and timelines. Its applications span the entire drug discovery process.

1. Target identification and validation: AI leverages large datasets from genomics, proteomics, and clinical records to identify novel drug targets. Machine learning models can predict target-disease associations by analyzing complex biological networks. Additionally, AI-driven systems help prioritize targets based on druggability and therapeutic relevance.
2. Compound screening and design: AI excels in virtual screening by simulating molecular interactions on a massive scale, identifying promising candidates faster than traditional HTS. Generative models, such as variational autoencoders (VAEs) and reinforcement learning frameworks, enable de novo molecule design, producing novel compounds optimized for specific properties.
3. Lead optimization: AI-driven predictive models streamline lead optimization by estimating key properties, such as binding affinity, solubility, and toxicity, based on molecular structure. These predictions enable rapid prioritization of candidates for synthesis and testing, reducing experimental iterations.
4. Preclinical and clinical testing: AI models analyze preclinical data to predict human pharmacokinetics and toxicity, helping to prioritize candidates with higher likelihoods of clinical success. In clinical trials, AI aids in

patient recruitment, stratification, and monitoring, ensuring more efficient trial designs and improved success rates.

5. Exploration of chemical space: The vastness of chemical space, estimated to include 10^{60} possible compounds, presents a formidable challenge for traditional methods. AI-driven algorithms, such as deep generative models, explore this space more effectively, identifying candidates that might otherwise remain undiscovered.
6. Real-world data integration: AI systems integrate data from multiple sources, including real-world evidence, electronic health records, and omics studies, to refine hypotheses and adaptively guide the discovery process. This integration enhances decision-making at every stage of development.

17.3 Process of Small Molecule Generation

This section outlines the workflow for small molecule generation, as illustrated in Fig. 17.1.

Compared to the high consistency between training and prediction in discriminative models, the training and generation (sampling) processes of generative models are less aligned, and this is also true for molecular generation.

During the training phase of generative models for molecular generation, inputs typically include molecules and noise, and sometimes additional conditional constraints, such as the highest occupied molecular orbital (HOMO) energy, the lowest unoccupied molecular orbital (LUMO) energy, the HOMO-LUMO gap (HLG), relative energy (E_{rel}), adiabatic ionization potential (IP), and adiabatic electron affinity (EA) [11]. In such cases, the loss function is often related to probabilities or probability densities, or their variant representations. Specific examples for different generative models will be discussed in subsequent sections.

In the sampling phase, which corresponds to the practical generation of molecules for applications such as AI-accelerated drug discovery (described in the previous section), the model requires only noise input (e.g., Gaussian noise). The model processes this input to generate new small molecules. It is worth noting that this generation process can either be a single-step procedure or a multistep iterative process, depending on the model architecture.

Finally, it is important to emphasize that effective training and sampling in generative models rely on the preprocessing of input data. Generative models

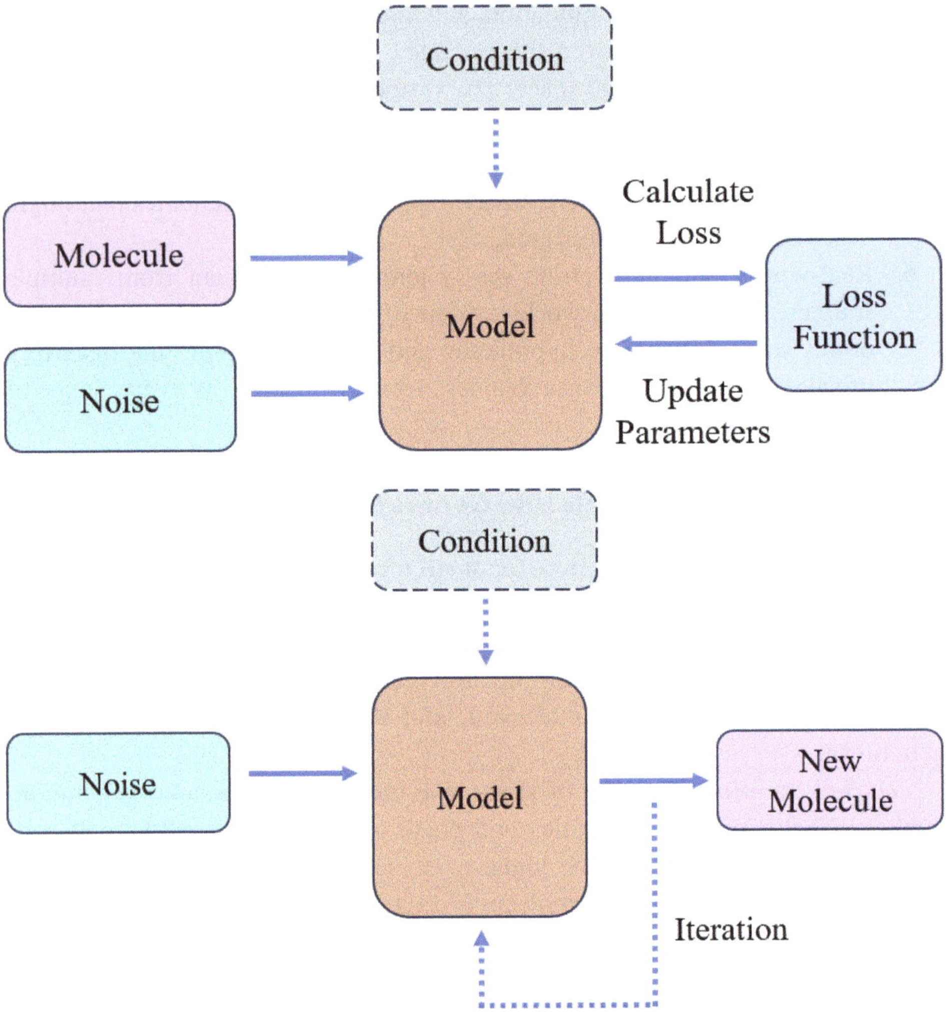

Fig. 17.1 Molecule generation model. (a) Training stage. (b) Generating (sampling) stage. Processes indicated by dashed lines or dashed boxes represent steps that are optional.

operate on preprocessed representations of molecular data. For molecular generation tasks, the input data are typically represented as follows:

A molecule consisting of N atoms and M bonds is described using:

An $N \times d$ one-hot vector representing atom types, an $N \times 3$ matrix representing atomic coordinates, a $2 \times M$ matrix specifying bond connections, and an $M \times k$ one-hot vector encoding bond types.

Noise is represented in a similar format, but its values are sampled from a noise distribution rather than from the distribution of real molecular data.

This standardized representation ensures that the model can effectively process and generate molecular data.

17.4 Generative AI Models

Generative models are designed to create new data samples that mimic the characteristics of existing data by learning underlying patterns and distributions. Traditional approaches include generative adversarial networks (GANs), which utilize adversarial training between a generator and a discriminator; VAEs, which rely on probabilistic encoding and decoding of data; and flow-based models, which map simple distributions to complex ones through invertible transformations. More recently, diffusion models have emerged as powerful alternatives, leveraging iterative denoising processes to generate high-quality samples. These methods collectively form the foundation for various applications in data generation and synthesis.

Variational Autoencoders

VAEs are a type of generative model that combines probabilistic reasoning with neural networks to generate new data samples [12]. Unlike traditional autoencoders, which encode data into a deterministic latent space, VAEs encode data into a probabilistic latent space, allowing for controlled sampling and variability in the generated outputs.

As shown in Fig. 17.2, VAEs consist of two main components. Encoder maps input data to a latent space by outputting a distribution (typically Gaussian), represented by a mean vector and a variance vector, and the Decoder reconstructs data samples by sampling from the latent distribution and mapping these samples back to the original data space.

Formally, the *encoder* can be represented $q_\phi(z \mid x) = \mathcal{N}\left(z; \mu_\phi(x), \sigma_\phi^2(x) I\right)$, where each input x is mapped to a latent space characterized by a mean $\mu_\phi(x)$ and variance $\sigma_\phi^2(x)$. These parameters define a Gaussian distribution in the latent space.

In the *decoding process*, the decoder is expressed as $p_\theta(x \mid z)$, where z represents a sample drawn from the latent Gaussian distribution $\mathcal{N}\left(\mu_\phi(x), \sigma_\phi^2(x) I\right)$. The decoder subsequently maps z back to a reconstructed data point x in the original data space.

The VAE loss function combines two components: the reconstruction loss and the Kullback–Leibler (KL) divergence loss. These two terms are designed to balance the fidelity of data reconstruction and the smoothness of the latent space distribution.

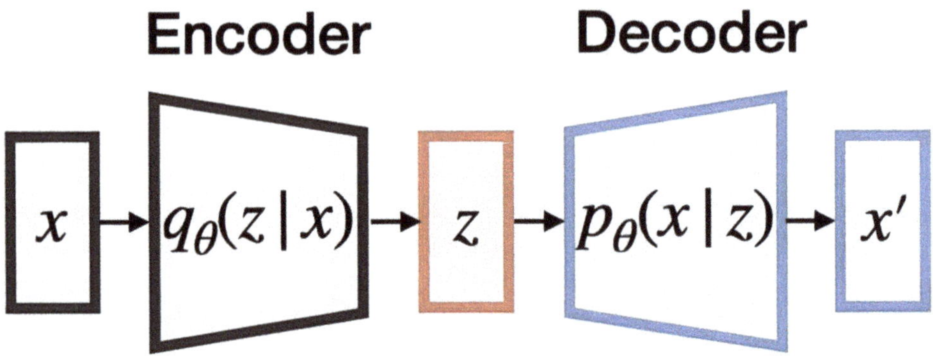

Fig. 17.2 Variational autoencoders [13].

The reconstruction loss measures how well the decoder reconstructs the input data x based on the latent variable z sampled from the encoder's distribution $q_\phi(z \mid x)$. It evaluates the expected negative log-likelihood of the data given the latent variables, expressed as

$$\mathcal{L}_{\text{recon}} = -\mathbb{E}_{q_\phi(z|x)}\left[\log p_\theta(x \mid z)\right] = -\int q_\phi(z \mid x) \log p_\theta(x \mid z)\, dz. \quad (17.1)$$

This term ensures that the decoder generates samples that are as close as possible to the original input.

The KL divergence term measures the difference between the encoder's learned distribution $q_\phi(z \mid x)$ and a predefined prior distribution $\mathcal{N}(0, I)$, typically a standard Gaussian. The KL divergence is given by:

$$\mathcal{L}_{\text{KL}} = D_{\text{KL}}\left(q_\phi(z \mid x) \,\|\, \mathcal{N}(0, I)\right) = \frac{1}{2} \sum_{k=1}^{K} \left(1 + \log\left(\sigma_k^2\right) - \mu_k^2 - \sigma_k^2\right). \quad (17.2)$$

Here, μ_k and σ_k^2 are the mean and variance of the latent space's k-th dimension, respectively, as produced by the encoder. The KL divergence regularizes the latent space by encouraging $q_\phi(z \mid x)$ to remain close to the prior $\mathcal{N}(0, I)$, ensuring smoothness and continuity.

The total loss function for the VAE is the weighted sum of the reconstruction loss and the KL divergence, expressed as

$$\mathcal{L} = \mathcal{L}_{\text{recon}} + \beta \mathcal{L}_{\text{KL}}. \quad (17.3)$$

Here, β is a hyperparameter that controls the trade-off between reconstruction accuracy and the regularization imposed on the latent space. A higher β value places more emphasis on the KL divergence, encouraging stronger regularization.

Generative Adversarial Networks

GANs [14] are a class of deep learning models consisting of two neural networks, the generator (G) and the discriminator (D), that are trained simultaneously in a competitive setting. This adversarial framework aims to improve both networks iteratively, creating a dynamic optimization process, as shown in Fig. 17.3.

The generator G takes a random noise vector z sampled from a prior distribution $p_z(z)$ (e.g., a Gaussian distribution) and maps it to a synthetic data sample $G(z)$ that aims to resemble the real training data distribution. Conversely, the discriminator D is designed to distinguish between real samples from the training data and synthetic samples generated by G. The discriminator outputs a probability $D(x)$, indicating how likely a given sample x is to be real.

The training process involves optimizing a minimax objective function, expressed as

$$\min_G \max_D \mathbb{E}_{x \sim p_{\text{data}}} [\log D(x)] + \mathbb{E}_{z \sim p_z(z)} [\log(1 - D(G(z)))]. \quad (17.4)$$

Here, $\mathbb{E}_{x \sim p_{\text{data}}} [\log D(x)]$ represents the expected log-probability that the discriminator correctly classifies a real data sample x. $\mathbb{E}_{z \sim p_z(z)} [\log(1 - D(G(z)))]$ denotes the expected log-probability that the discriminator correctly identifies a generated sample $G(z)$ as fake.

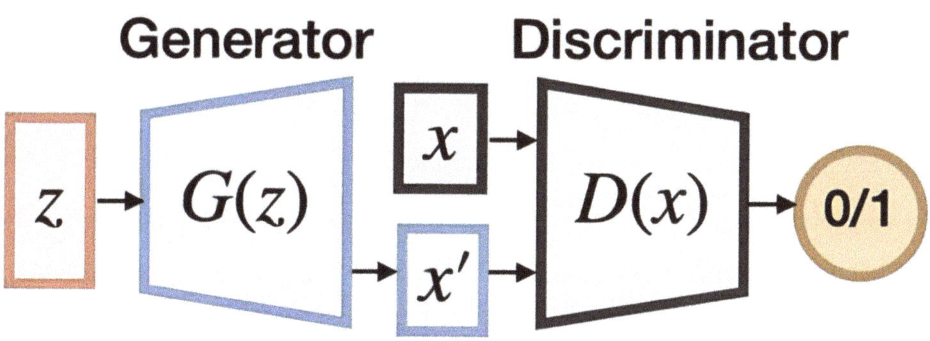

Fig. 17.3 Generative adversarial networks [13].

The discriminator attempts to maximize this objective, improving its ability to distinguish between real and synthetic data. Simultaneously, the generator seeks to minimize the objective by producing synthetic samples that can "fool" the discriminator into classifying them as real. This competition forms a zero-sum game where the generator improves its outputs as the discriminator becomes more accurate, and vice versa.

At equilibrium, the generator produces samples indistinguishable from real data, and the discriminator is no longer able to make meaningful predictions, outputting a probability of 0.5 for all inputs.

Flow-Based Models

Flow-based generative models [15] are a type of deep learning framework that generates data by transforming a simple latent distribution, such as a Gaussian distribution $z_0 \sim p_0(z_0)$, into a more complex target distribution $x \sim p(x)$, as shown in Fig. 17.4. This transformation is achieved through an invertible function f, parameterized by θ, which maps the latent variable z to the data space x. The relationship can be expressed as

$$x = f(z;\theta) \quad \text{and} \quad z = f^{-1}(x;\theta). \tag{17.5}$$

Since f is invertible, the mapping between x and z is bijective, ensuring that the dimensionality of x and z is the same. Typically, f is composed of a sequence of simpler functions, represented as $f(x) = f_K \circ f_{K-1} \circ \cdots \circ f_1(x)$, enabling the model to represent highly complex data distributions by combining multiple transformations.

An essential property of flow-based models is that they allow the probability density $p(x)$ to be computed explicitly. Using the change of variable formula, the log-likelihood of a data point x can be expressed as

$$\log p(x) = \log p_0(z) + \log\left|\det\frac{\partial f}{\partial z}\right|. \tag{17.6}$$

Here, $\log p_0(z)$ represents the log-probability of the latent variable z under the base distribution. $\log\left|\det\frac{\partial f}{\partial z}\right|$ is the log-determinant of the Jacobian matrix of f, accounting for the transformation's effect on the distribution.

This explicit density formulation allows flow-based models to maximize the likelihood of observing the data directly, providing a principled way to train the parameters θ. By adjusting θ, the model learns to transform the simple base distribution into the desired target distribution.

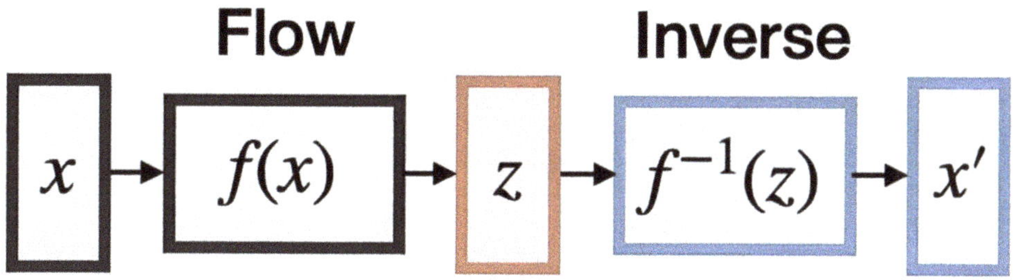

Fig. 17.4 Flow-based models [13].

Diffusion Models

Diffusion models [16] are a class of generative models that progressively learn to transform noise into meaningful data by simulating a forward and reverse process. The forward process incrementally corrupts data with Gaussian noise over a series of time steps, while the reverse process learns to remove the noise to reconstruct the original data distribution, as shown in Fig. 17.5.

In the *forward process*, data x_t at a time step t is perturbed into x_{t+1} by adding Gaussian noise in a Markov chain. This step can be expressed as

$$x_{t+1} = \sqrt{1-\beta_t}\,x_t + \sqrt{\beta_t}\,\epsilon,\ \epsilon \sim \mathcal{N}(0,I). \tag{17.7}$$

Here, β_t is the noise schedule, which determines the variance of the noise added at each step. ϵ represents Gaussian noise sampled from a standard normal distribution. As t progresses, the variance β_t increases, resulting in x_t converging to a nearly isotropic Gaussian distribution $\mathcal{N}(0,I)$.

The *reverse process* attempts to denoise the data, progressively reversing the noise introduced in the forward process. It learns a parameterized denoising function $f_\theta(x_t, t)$ to approximate the original data distribution. Each reverse step is represented as

$$x_{t-1} = f_\theta\left(x_t, t\right) + \sqrt{\beta_t}\,\epsilon,\ \epsilon \sim \mathcal{N}(0,I). \tag{17.8}$$

The denoising function $f_\theta(x_t, t)$ is typically implemented using a neural network, parameterized by θ, and trained to predict the denoised data.

To train the diffusion model, the added noise at each step is approximated and compared to the true noise. The loss function minimizes the mean squared error between the true noise ϵ and the noise predicted by the model, $\epsilon_\theta(x_t, t)$:

$$L_t = \mathbb{E}_{t\sim U(1,T), x_0, \epsilon_t}\left[\| \epsilon_t - \epsilon_\theta(x_t, t)\|^2\right]. \tag{17.9}$$

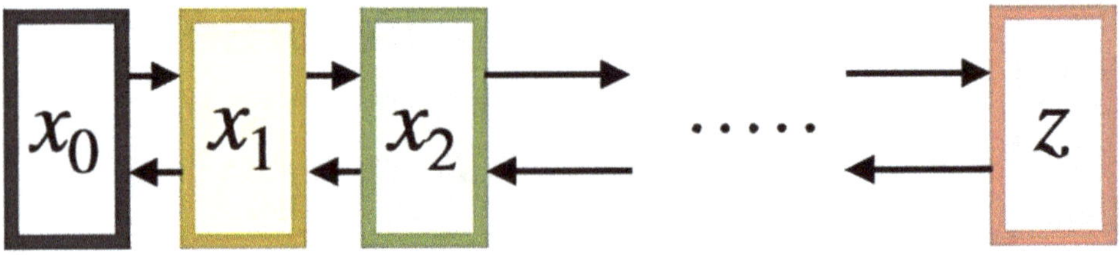

Fig. 17.5 Diffusion models [13].

Here, $t \sim U(1, T)$ indicates that the time step t is sampled uniformly from the range $\{1, \ldots, T\}$. $\epsilon_\theta(x_t, t)$ represents the model's predicted noise for data corrupted at step t. T is the total number of diffusion steps.

Once trained, the model uses the learned denoising function to iteratively sample from the noise distribution and reconstruct data through the reverse process.

17.5 Small Molecule Generation

In the third section of this chapter, we demonstrated the ease of utilizing molecular generative models. Through the discussion in Section 17.4, readers were introduced to the core principles of the most widely used generative models. In this section, we aim to deepen the reader's understanding of the molecular generation task by presenting two practical examples of small molecule generative models.

Task

Molecule generation focuses on designing novel chemical compounds, often with the goal of aiding drug discovery. The molecules produced by such methods are expected to satisfy several key criteria: they must be valid (i.e., chemically feasible), stable (retaining structural integrity), and unique (not redundant or duplicative). Beyond these basic requirements, molecules are typically optimized for their pharmaceutical applicability, a broad term encompassing the binding affinity of molecules to biological targets, as well as other drug-like properties.

The field of molecular generation can be categorized into distinct approaches based on their intended purpose: (1) *Target-agnostic* molecule generation aims to produce sets of chemically valid molecules without focusing on their interactions with specific biological targets. The objective is to explore diverse regions of chemical space, irrespective of biological context. (2) *Target-aware* molecule generation (also referred to as ligand generation) involves designing molecules that are intended to bind to specific protein structures. This approach incorporates biological and pharmacological considerations, focusing on optimizing molecular properties for interactions with a defined target.

To train and evaluate generative models for molecule generation, molecular data can be represented in various formats depending on the requirements of the task: (1) 1D representations: Simplified notations, such as Simplified Molecular Input Line Entry System (SMILES), encode molecules as strings of characters. (2) 2D representations: Graph-based formats represent molecules using nodes for atoms and edges for bonds, providing a two-dimensional view of connectivity. (3) 3D representations: Point clouds or embeddings are used to capture the spatial arrangement of atoms, which is essential for modeling molecular interactions and properties.

Target-Agnostic Molecule Generation

The task of target-agnostic molecule generation involves generating novel, chemically valid, and stable molecules without considering specific biological targets. This approach aims to explore chemical space to produce new molecular structures that satisfy these general constraints.

Datasets: To ensure the generation of valid and stable molecules, models require training on comprehensive datasets of chemically viable molecules. Two widely used datasets for this task are: (1) Quantum Machines 9 (QM9) [17]. A curated dataset containing small, stable molecules extracted from the broader GDB-17 chemical database. QM9 is commonly used for tasks requiring high data quality and interpretability. (2) GEOM-Drug (Geometric Ensemble of Molecules) [18]. A dataset of more complex and drug-like molecules, used to evaluate model scalability beyond the simpler structures in QM9.

Metrics: Evaluating models for target-agnostic molecule generation often involves metrics that assess molecular validity, stability, and diversity. The most common metrics include: (1) Atom stability. The proportion of atoms in generated molecules that satisfy valency constraints. (2) Molecule stability. The percentage of molecules in which all atoms are stable. (3) Validity. The fraction of generated molecules that are chemically valid is typically assessed using cheminformatics tools like RDKit. (4) Uniqueness. The proportion of generated molecules that are distinct from one another. (5) Novelty. The percentage of generated molecules that do not overlap with the training dataset.

(6) Quantitative estimate of drug-likeness (QED) [19]. A metric combining various molecular properties to estimate a molecule's likelihood of being suitable for pharmaceutical purposes.

It is important to note that the novelty metric is sometimes omitted, particularly for datasets like QM9, which encompass the majority of plausible molecules within a predefined space of constraints. Consequently, novel molecules often violate constraints, making novelty less reliable as a standalone metric. Similarly, while QED is a well-established metric, many generative models focus exclusively on validity and stability, with QED evaluation reserved for specific pharmaceutical tasks [20].

In addition to unconditional molecule generation, models may be evaluated on their ability to generate molecules conditioned on specific properties. For such tasks, a property classifier is trained on a portion of the dataset (e.g., QM9) to predict target properties. The generative model is then trained on the remaining data and evaluated by comparing the generated molecules' properties to the desired targets. Key molecular properties often considered include: (1) Polarizability. The molecule's tendency to acquire a dipole moment under an external electric field, measured in cubic Bohr radius. (2) HOMO energy. The energy of the HOMO, measured in millielectron volts (meV). (3) LUMO energy. The energy of the LUMO, measured in meV. (4) HLG. The energy difference between the HOMO and LUMO orbitals, measured in meV. (5) Dipole moment. The molecule's dipole moment, measured in Debye (D). (6) Molar heat capacity. Heat capacity at 298.15 K, measured in calories per Kelvin per mole (cal/(mol·K)).

Target-Aware Molecule Generation

Target-aware molecule generation focuses on designing molecules that interact specifically with given biological targets, such as proteins, with the goal of maximizing binding affinity and other desirable interactions. The input typically includes target-specific information, such as amino acid sequences in ligand-based drug design (LBDD) or 3D protein structures in structure-based drug design (SBDD) [7]. These inputs guide the generation of molecules tailored to exhibit high affinity and compatibility with the target.

Datasets: To train and evaluate models for target-aware molecule generation, several widely used datasets are available: (1) CrossDocked2020 [21]. This dataset contains ligand–protein complexes generated through cross-docking, where similar binding sites (referred to as "pockets") are grouped into clusters. CrossDocked2020 supports the creation of large-scale data, significantly increasing the number of ligand–protein pairs available for training compared to other datasets. (2) ZINC20 [22]. A fully enumerated dataset of chemical compounds that serves as a source of potential ligands, offering diverse molecular structures for generative tasks. (3) Binding Mother of All Databases (MOAD) [23]. A curated subset of protein-ligand complexes derived from the Protein Data Bank (PDB) [24]. This dataset provides experimentally verified binding interactions, making it a valuable benchmark for model evaluation.

Metrics: Models for target-aware molecule generation are assessed using multiple metrics to evaluate their ability to generate molecules with high binding affinity, synthetic feasibility, and diversity. Key metrics include: (1) Vina score [25]. A docking-based scoring function that calculates a weighted sum of atomic interactions, useful for evaluating the binding potential of generated molecules. (2) Vina energy [25]. The binding energy predicted by the docking platform reflects the strength of the interaction between the molecule and the target protein. (3) High affinity percentage. The fraction of generated molecules that exhibit a lower binding energy (better affinity) compared to a reference molecule. (4) QED. A composite metric is used to assess the pharmaceutical potential of generated molecules. This metric is also used in target-agnostic tasks. (5) Synthetic accessibility score (SAScore) [26]. A formula-based measure of how easily a molecule can be synthesized in a laboratory setting. (6) Diversity. The diversity of generated molecules for a specific binding site is calculated using Tanimoto similarity between molecular fingerprints (e.g., Morgan fingerprints). Greater diversity allows for a broader exploration of chemical space.

Example 1: A Target-Agnostic Molecule Generation Model-EquiFM

Motivation and contribution

While diffusion models have proven effective in generating complex geometries, they are often hindered by unstable probability dynamics and inefficient sampling processes, which impact both training stability and inference speed [27, 28]. To overcome these issues, Equivariant Flow Matching (EquiFM) is introduced as a novel framework that combines equivariant modeling with stabilized probabilistic dynamics [29].

The framework introduces an equivariant optimal transport prior to stabilize the generation of atomic coordinates, ensuring smoother transitions in the probabilistic space and improving training stability. To further address the inconsistency between categorical and continuous modalities, EquiFM adopts a hybrid probability path that distinguishes the generative processes of different components while maintaining computational efficiency. Additionally, the model employs a continuous normalizing flow (CNF), parameterized using an ordinary differential equation (ODE) solver, which enables faster sampling compared to conventional stochastic approaches.

Empirical results demonstrate that EquiFM achieves significant improvements in 3D molecular generation benchmarks, including enhanced structural quality, stability, and alignment of generated molecules. Notably, the model achieves a 4.75× speedup in sampling time on average, offering superior efficiency without compromising generation performance. These advancements make EquiFM a promising approach for addressing the challenges of 3D molecular generation with improved stability, controllability, and computational efficiency.

Preliminary

The core of EquiFM is essentially a variant of the previously discussed flow-based generative models; however, its formulation appears slightly different. Therefore, a brief introduction to the concept of simulation-free flow matching is provided here as prerequisite knowledge [30].

Flow matching is introduced as an efficient, simulation-free method to train continuous flow models for generative tasks. It provides an alternative to traditional CNFs, which are often challenging to train due to their reliance on numerical simulations of ODEs [31]. The flow matching framework defines a probability path over time, transforming a prior distribution p_1 (e.g., a Gaussian distribution) into a target distribution q. This is achieved by learning a neural network $v_\theta(x, t)$ that approximates a target vector field $u_t(x)$. The transformation is governed by the following ODE:

$$\frac{d}{dt}\psi_t(x) = v_t(\psi_t(x)), \psi_1(x) = x, \tag{17.10}$$

where ψ_t represents the time-dependent transformation of the data, and v_t is the time-dependent vector field. To avoid the numerical challenges of CNFs, flow matching directly learns the vector field $u_t(x)$ that generates the desired probability path.

The flow matching objective $\mathscr{L}_{\text{FM}}$ minimizes the difference between the learned vector field $v_\theta(x, t)$ and the target field $u_t(x)$, expressed as

$$\mathscr{L}_{\text{FM}}(\theta) = \mathbb{E}_{t, p_t(x)}\left[\| v_\theta(x, t) - u_t(x) \|^2\right]. \tag{17.11}$$

Since directly accessing $u_t(x)$ and $p_t(x)$ can be infeasible, the vector field and probability path are often derived from a conditional formulation, where $u_t(x \mid x_0)$

and the conditional probability $p_t(x|x_0)$ are more easily defined. The marginal form of $p_t(x)$ can then be recovered from a mixture of these conditional probabilities. This reformulation leads to the conditional flow matching (CFM) objective, expressed as

$$\mathcal{L}_{\mathrm{CFM}}(\theta) = \mathbb{E}_{t,q(x_0),p_t(x|x_0)}\left[\| v_\theta(x,t) - u_t(x \mid x_0) \|^2\right]. \tag{17.12}$$

The CFM objective is tractable for optimization and enables efficient training. For inference, ODE solvers can integrate the learned vector field $v_\theta(x,t)$ to generate new samples.

Two specific probability paths are discussed to demonstrate different strategies for flow matching. The first is the conditional optimal transport (OT) path, which approximates the data distribution $p_0(x \mid x_0)$ with a peaked Gaussian centered at x_0 and small variance. The probability path is expressed as

$$p_t(x \mid x_0) = \mathcal{N}\left((1-t)x_0, (\sigma_{\min} + (1-\sigma_{\min})t)^2 I\right), \tag{17.13}$$

and the corresponding vector field is derived as

$$u_t(x \mid x_0) = -x_0 + (1-\sigma_{\min})x_1. \tag{17.14}$$

The OT path learns the transformation direction from noise to data along a straight-line trajectory, which ensures geometric smoothness in the learned probability dynamics.

The second approach is the variance preserving (VP) path, which uses a variance schedule to control the noise dynamics. The VP path follows:

$$p_t(x \mid x_0) = \mathcal{N}\left(\alpha_t x_0, (1-\alpha_t^2) I\right), \tag{17.15}$$

where α_t is a noise schedule function. The corresponding vector field can be expressed as

$$u_t(x \mid x_0) = \frac{\alpha_t'}{\alpha_t}(\alpha_t x_0 - x). \tag{17.16}$$

The objective for VP CFM then becomes

$$\mathcal{L}_{\mathrm{CFM}}^{\mathrm{VP}}(\theta) = \mathbb{E}_{t,q(x_0),p_t(x|x_0)}\left[\| v_\theta(x,t) - \frac{\alpha_t'}{\alpha_t}(\alpha_t x_0 - x) \|^2\right]. \tag{17.17}$$

The VP path is flexible in controlling the noise-to-data transformation dynamics through the choice of different noise schedules. This flexibility allows for fine-tuning the information flow within the generative process.

Methodology

The framework of EquiFM for 3D molecular generation models a molecule as a tuple $g = \langle x, h \rangle$, where $x = (x^1, \ldots, x^N) \in \mathbb{R}^{N\times3}$ represents the atomic coordinates, and $h = (h^1, \ldots, h^N) \in \mathbb{R}^{N\times d}$ represents node features such as atom types or charges. These features are modeled in the zero center-of-mass (CoM) space, ensuring that the average of all atomic coordinates satisfies $\frac{1}{N}\sum_{i=1}^{N} x^i = 0$. This ensures translation invariance throughout the modeling process.

To address the need for equivariant modeling in geometric domains, the framework enforces that both the prior distribution $p_1(x, h)$ and the parameterized vector field $v_\theta(x, h, t)$ respect SE(3) equivariance properties. Specifically, the vector field must satisfy the following property under any rotational transformation R:

$$v_\theta(Rx, h, t) = R v_\theta(x, h, t), \tag{17.18}$$

where Rx denotes the rotated coordinates, and $v_\theta(x, h, t)$ is the parameterized vector field. This ensures that the probability path $p_t(x, h)$, derived from v_θ, is **invariant** to rotational transformations.

The hybrid probability modeling approach is introduced to handle the generation of both continuous variables x (atomic coordinates) and discrete variables h (atomic types or charges) independently. This allows separate probability paths for each variable. For the hybrid probability path, the CFM objective is defined as

$$\mathcal{L}_{\text{CFM}}(\theta) = \mathbb{E}_{t, q(g_0), p_t(g|g_0)}\left[\| v_\theta^x(g, t) - u_t^x(g|g_0)\|^2 + \| v_\theta^h(g, t) - u_t^h(g|g_0)\|^2\right], \tag{17.19}$$

where $g = \langle x, h \rangle$, and $u_t^x(g|g_0)$ and $u_t^h(g|g_0)$ represent the target vector fields for the continuous and discrete components, respectively.

For modeling atomic coordinates x, the equivariant optimal transport (EOT) path is employed. Given two-point clouds $z = (z^1, \ldots, z^N)$ and $y = (y^1, \ldots, y^N)$

in $\mathbb{R}^{N\times 3}$, the EOT minimizes the distance between the paired points while accounting for SE(3) equivariance. The optimal transport plan is defined as

$$\pi^*, R^* = \arg\min_{\pi, R} \sum_{i=1}^{N} \| Rz^{\pi(i)} - y^i \|_2^2. \tag{17.20}$$

where π is a permutation of indices, R is a rotation matrix in $\mathbb{R}^{3\times 3}$, and both π and R satisfy the Zero CoM constraint. The resulting EOT map transforms coordinates x from the prior distribution p_1 to the target distribution p_0. The EOT-based probability path is then defined as

$$p_t(x) = [\psi_t^{\mathrm{EOT}}]_* p_1, \psi_t^{\mathrm{EOT}}(x) = \left(\sigma_{\min} + \left(1 - \sigma_{\min}\right)t\right) \cdot \pi^*(R^* x) + (1-t)x_0, \tag{17.21}$$

where $\sigma_{\min}$ controls the variance at the start of the trajectory, x_0 is the sampled point cloud from the target distribution, and $t \in [0,1]$ parameterizes the path.

Based on the aforementioned methods, the following presents a detailed algorithmic framework for EquiFM.

Algorithm 1 Training Algorithm of EquiFM

1. **Input**: geometric data distribution $p_{\mathbf{g}}$, $\mathbf{g} = \langle \mathbf{x}, \mathbf{h} \rangle$
2. **Initial**: vector field network v_θ, minimum constant variance $\sigma_{\min}$
3. **while** θ have not converged do
4. $t \sim \mathrm{U}(0,1), \epsilon \sim \mathcal{N}(0, \boldsymbol{I}), g_0 \sim p_{\mathbf{g}}$
5. Subtract center of gravity from ϵ_x in $\epsilon = [\epsilon_x, \epsilon_h]$
6. Obtaining the Equivariant Optimal Transport plan $\pi^*, \mathbf{R}^*$
7. $x_t = (\sigma_{\min} + (1 - \sigma_{\min})t)\pi^*(\mathbf{R}^* \epsilon_x) + (1-t)x_0$
8. $h_t = \alpha_t h_0 + (1 - \alpha_t^2)\epsilon_h, \alpha_t = e^{-\frac{1}{2}T(t)}$
9. $\mathcal{L}_{\mathrm{EquiFM}} = \| v_\theta^x(\langle x_t, h_t \rangle, t) - (-x_0 + (1 - \sigma_{min})\epsilon_x) \|^2 + \| v_\theta^h(\langle x_t, h_t \rangle, t) - \frac{\alpha_t'}{1 - \alpha_t^2}(\alpha_t h_t - h_0) \|^2$
10. **end while**
11. **return** v_θ

Algorithm 2 Sampling Algorithm of EquiFM

1. Input: vector field model v_θ
2. $g_1 \sim \mathcal{N}(0, \boldsymbol{I})$
3. Subtract center of gravity from x_1 in $g_1 = [x_1, h_1]$
4. $g_0 = \text{ODESolve}(g_1, v_\theta, 1, 0)$
5. {During the ODE solving, always subtract the center of gravity from v_θ^x}
6. sample $\hat{p}(x_0, h_0 | g_0)$
7. return $\langle x_0, h_0 \rangle$

Using the algorithm, the model can be easily trained and used to generate new molecules. It should be noted that the evaluation of molecular properties has been omitted here, as the evaluation metrics were introduced in earlier parts.

Example 2: A Target-Aware Molecule Generation Model-TargetDiff

Motivation and contribution

Existing approaches to target-aware molecule generation face notable challenges. First, models relying on voxelization suffer from scalability issues as the spatial resolution and protein pocket-size increase [32]. Additionally, methods based on autoregressive sampling introduce unrealistic assumptions, such as imposing a fixed atom generation order, which limits the structural diversity of the generated molecules and introduces bias [33]. These approaches often lack effective mechanisms to globally model physical interactions in 3D space, leading to poor performance on large-scale or highly specific tasks. Furthermore, such methods typically require separate post-processing steps or external tools to evaluate the binding quality of generated molecules [25].

To address these limitations, a 3D diffusion-based framework named TargetDiff is proposed [34, 35]. This framework integrates SE(3)-equivariant graph neural networks and diffusion modeling to enable the joint generation of both atomic coordinates and atom types while ensuring invariance to global transformations like rotation and translation. Unlike existing autoregressive methods, TargetDiff introduces a non-autoregressive process in which noise is gradually reduced during generation, aligning both discrete (atom types) and continuous (atomic positions) features with protein structural constraints. This approach leverages the context provided by protein atoms to predict molecular interactions more effectively.

One notable advantage of this framework is its ability to extract meaningful molecular features during the generation process. By feeding generated molecules into the model, TargetDiff can estimate binding affinities without additional retraining. These extracted features provide a robust signal for downstream tasks such as ranking candidate molecules or screening them for drug-like properties. Empirical evaluations on datasets such as CrossDocked2020 demonstrate that TargetDiff produces molecules with more realistic 3D structures and higher binding affinities compared to existing methods.

Methodology

Since one of the major contributions of TargetDiff is the application of the original diffusion model, as previously introduced to molecule generation, the preliminary to diffusion models has been omitted here.

In this framework, $\mathcal{P}$ represents the protein binding site, which includes both the 3D coordinates of the protein atoms and their associated features (e.g., atom types). Similarly, $\mathcal{M}$ denotes the ligand molecule being generated, where $\mathcal{M} = \{x, v\}$ combines x, the atomic coordinates, and v, the categorical atom types of the ligand.

To address the limitations of autoregressive sampling models, which often introduce unnatural generation order and scalability issues, TargetDiff employs a non-autoregressive approach based on diffusion models. The framework incorporates SE(3)-equivariance to ensure invariance to global transformations such as translation and rotation. This property is critical for accurately modeling 3D spatial interactions between atoms in the protein-ligand complex.

The generative process is formalized as a forward and reverse diffusion procedure. In the *forward process*, Gaussian noise is gradually added to the molecular data, transforming the data distribution into a standard Gaussian distribution over time. At each diffusion step, the joint distribution of ligand atomic coordinates and atom types, conditioned on the protein binding site $\mathcal{P}$, is expressed as [36]:

$$q(\mathcal{M}_t | \mathcal{M}_{t-1}, \mathcal{P}) = \mathcal{N}(x_t; \sqrt{1-\beta_t} x_{t-1}, \beta_t I) \cdot \mathcal{C}(v_t | (1-\beta_t) v_{t-1} + \beta_t / K), \quad (17.22)$$

where $\mathcal{N}$ represents the Gaussian distribution over atomic coordinates, $\mathcal{C}$ is the categorical distribution over atom types, x_t represents the atomic coordinates of the ligand at step t, v_t represents the atom types, β_t is the variance schedule controlling the noise, and K is the number of atom categories. The protein binding site $\mathcal{P}$ provides the conditioning context throughout the process.

In the *reverse process*, the model learns to gradually denoise the molecular data, recovering both the atomic coordinates and atom types step by step. The reverse probability distribution is parameterized using an SE(3)-equivariant neural network, which models the interaction between protein atoms and the generated ligand molecules. For the reverse step, the predicted posterior distributions for atomic coordinates and atom types are derived using Bayes' theorem and can be expressed in closed form.

The SE(3)-equivariant neural network alternates between updating the atom embeddings h_i and atomic coordinates x_i, incorporating pairwise distances and feature similarities in its computations. The equations governing this process are defined as [37]:

$$\begin{aligned} h_i^{l+1} &= h_i^l + \sum_{j\neq i} f_h\left(d_{ij}^l, h_i^l, h_j^l, e_{ij}; \theta_h\right),\ x_i^{l+1} \\ &= x_i^l + \sum_{j\neq i} (x_i^l - x_j^l) f_x(d_{ij}^l, h_i^l, h_j^l, e_{ij}; \theta_x) \cdot 1_{\text{mol}} \end{aligned}, \tag{17.23}$$

where $d_{ij} = \| x_i - x_j \|_2$ is the Euclidean distance between atoms i and j, h_i and h_j are their respective embeddings, and e_{ij} encodes additional interaction features between the two atoms.

The training objective of TargetDiff is formulated as a VAE, optimizing the variational bound on negative log likelihood. For the continuous atomic coordinates, the KL divergence between the predicted posterior distribution and the target Gaussian is computed as

$$\mathcal{L}_{t-1}^{(x)} = \frac{1}{2\sigma_t^2} \| \tilde{\mu}_t(x_t, x_0) - \mu_\theta((x_t, v_t), \text{t}, \mathcal{P}) \|^2 + C, \tag{17.24}$$

where μ_θ and $\tilde{\mu}_t$ are the predicted and ground truth means, respectively. It is important to note that, in practice, the loss function can be reparameterized, and predicting the mean of the posterior distribution, predicting the noise, or predicting the original sampling data are equivalent approaches. For the discrete atom types, the loss is computed as the KL divergence between the categorical distributions:

$$\mathcal{L}_{t-1}^{(v)} = \sum_k c(v_t, v_0)_k \log \frac{c(v_t, v_0)_k}{c(v_t, \hat{v}_0)_k}. \tag{17.25}$$

The final loss combines the weighted sums of these two components, balancing the contributions of atomic coordinates and atom types. This integrated training and generation process ensures that the output molecules align closely with the protein targets in both 3D structure and chemical properties, providing a robust solution for target-aware molecular design.

Based on the aforementioned methods, the following presents a detailed algorithmic framework for TargetDiff ($\alpha_t = 1 - \beta_t$ and $\overline{\alpha}_t = \Pi_{s=1}^{t} \alpha_s$).

Algorithm 3 Training Procedure of TargetDiff

Input: Protein-ligand binding dataset $\{\mathcal{P}, \mathcal{M}\}_{i=1}^{N}$, neural network ϕ_θ

1. **while** ϕ_θ not converge **do**
2. Sample diffusion time $t \in \mathcal{U}(0, \ldots, T)$
3. Move the complex to make **CoM** of protein atoms zero
4. Perturb x_0 to obtain $x_t = \sqrt{\overline{\alpha}_t} x_0 + \sqrt{1 - \overline{\alpha}_t} \epsilon, \epsilon \sim \mathcal{N}(0, I)$
5. Perturb v_0 to obtain v_t:

$$\log c = \log\left(\overline{\alpha}_t v_0 + (1 - \overline{\alpha}_t)/K\right)$$

$$v_t = \text{one\_hot}\,(\arg\max_i [g_i + \log c_i]), g \sim \text{Gumbel}(0,1)$$

6. Predict $[\hat{x}_0, \hat{v}_0]$ from $[x_t, v_t]$ with ϕ_θ: $[\hat{x}_0, \hat{v}_0] = \phi_\theta([x_t, v_t], t, \mathcal{P})$
7. Compute the posterior atom types $c(v_t, v_0)$ and $c(v_t, \hat{v}_0)$
8. Compute the unweighted MSE loss on atom coordinates and the KL loss on

   posterior atom types: $L = \| x_0 - \hat{x}_0 \|^2 + \alpha \text{KL}(c(v_t, v_0) \,\|\, c(v_t, \hat{v}_0))$
9. Update θ by minimizing L
10. **end while**

Algorithm 4 Sampling Procedure of TargetDiff

Input: The protein binding site $\mathcal{P}$, the learned model ϕ_θ.

Output: Generated ligand molecule $\mathcal{M}$ that binds to the protein pocket.

1. Sample the number of atoms in $\mathcal{M}$ based on a prior distribution conditioned on pocket size

2. Move **CoM** of protein atoms to zero
3. Sample initial molecular atom coordinates x_T and atom types v_T:

$$x_T \sim \mathcal{N}(0,I);\ v_T = \text{one}_{\text{hot}}\left(\arg\max_i g_i\right),\ \textit{where}\ g \sim \text{Gumbel}(0,1)$$

4. **for** $t \in T, T-1, \ldots, 1$ **do**
5. Predict $[\hat{x}_0, \hat{v}_0]$ from $[x_t, v_t]$ with ϕ_θ: $[\hat{x}_0, \hat{v}_0] = \phi_\theta([x_t, v_t], t, \mathcal{P})$
6. Sample x_{t-1} from the posterior $p_\theta(x_{t-1} | x_t, \hat{x}_0)$
7. Sample v_{t-1} from the posterior $p_\theta(v_{t-1} | v_t, \hat{v}_0)$
8. **end for**

17.6 Conclusion

This chapter provides a brief introduction to AI-assisted drug discovery, with a more comprehensive overview of one specific aspect, small molecule generation. Furthermore, it offers a detailed explanation of two popular molecular generation algorithms to facilitate a deeper understanding at the algorithmic level. Naturally, as a relatively new field, current methods for small molecule generation face certain limitations, such as challenges in efficiency, effectiveness, and interpretability, as well as underexplored opportunities in leveraging domain knowledge. Overall, interdisciplinary research at the intersection of AI and other fields is expected to remain a key focus for an extended period, and advancements in AI will undoubtedly drive progress in domains such as drug discovery and beyond.

References

1. Polishchuk, P.G., T.I. Madzhidov, and A. Varnek. Estimation of the size of drug-like chemical space based on GDB-17 data. *J Comput Aided Mol Des*. 2013; **27**(7): p. 675–679.
2. Reymond, J.-L. and M. Awale. Exploring chemical space for drug discovery using the chemical universe database. *ACS Chem Neurosci*. 2012; **3**(9): p. 649–657.
3. Weininger, D. SMILES, a chemical language and information system. 1. Introduction to methodology and encoding rules. *J Chem Inf Comput Sci*. 1988; **28**(1): p. 31–36.
4. Krenn, M., et al. Self-referencing embedded strings (SELFIES): a 100% robust molecular string representation. *Mach Learn Sci Technol*. 2020; **1**(4): p. 045024.
5. Xie, W., et al. Advances and challenges in de novo drug design using three-dimensional deep generative models. *J Chem Inf Model*. 2022; **62**(10): p. 2269–2279.

6. Wang, M., et al. Deep learning approaches for de novo drug design: an overview. *Curr Opin Struct Biol.* 2022; **72**: p. 135–144.
7. Thomas, M., A. Bender, and C. de Graaf. Integrating structure-based approaches in generative molecular design. *Curr Opin Struct Biol.* 2023; **79**: p. 102559.
8. Ding, W., K. Nakai, and H. Gong. Protein design via deep learning. *Brief Bioinform.* 2022; **23**(3): p. bbac102.
9. Horvath, C. Comparison of preclinical development programs for small molecules (drugs/pharmaceuticals) and large molecules (biologics/biopharmaceuticals): studies, timing, materials, and costs. *Pharm Sci Encycl.* 2010: p. 1–35. https://doi.org/10.1002/9780470571224.pse166
10. DiMasi, J.A., H.G. Grabowski, and R.W. Hansen. Innovation in the pharmaceutical industry: new estimates of R&D costs. *J Health Econ.* 2016; **47**: p. 20–33.
11. Weiss, T., et al. Guided diffusion for inverse molecular design. *Nat Comput Sci.* 2023; **3**(10): p. 873–882.
12. Kingma, D.P. Auto-encoding variational bayes. *arXiv preprint arXiv:1312.6114.* 2013.
13. Tang, X., et al. A survey of generative AI for de novo drug design: new frontiers in molecule and protein generation. *Brief Bioinform.* 2024; **25**(4): p. bbae338.
14. Goodfellow, I., et al. Generative adversarial nets. *Adv Neural Inf Process Syst.* 2014; **27**: p. 2672–2680.
15. Rezende, D. and S. Mohamed. Variational inference with normalizing flows. *Int Conf Mach Learn.* 2015; **37**: p. 1530–1538.
16. Yang, L., et al. Diffusion models: a comprehensive survey of methods and applications. *ACM Comput Surv.* 2023; **56**(4): p. 1–39.
17. Ramakrishnan, R., et al. Quantum chemistry structures and properties of 134 kilo molecules. *Sci Data.* 2014; **1**(1): p. 1–7.
18. Axelrod, S. and R. Gomez-Bombarelli. GEOM, energy-annotated molecular conformations for property prediction and molecular generation. *Sci Data.* 2022; **9**(1): p. 185.
19. Bickerton, G.R., et al. Quantifying the chemical beauty of drugs. *Nat Chem.* 2012; **4**(2): p. 90–98.
20. Vignac, C. and P. Frossard. Top-n: equivariant set and graph generation without exchangeability. *arXiv preprint arXiv:2110.02096.* 2021.
21. Francoeur, P.G., et al. Three-dimensional convolutional neural networks and a cross-docked data set for structure-based drug design. *J Chem Inf Model.* 2020; **60**(9): p. 4200–4215.
22. Irwin, J.J., et al. ZINC20—a free ultralarge-scale chemical database for ligand discovery. *J Chem Inf Model.* 2020; **60**(12): p. 6065–6073.
23. Hu, L., et al. Binding MOAD (mother of all databases). *Proteins.* 2005; **60**(3): p. 333–340.
24. Berman, H.M., et al. The protein data bank. *Nucleic Acids Res.* 2000; **28**(1): p. 235–242.
25. Trott, O. and A.J. Olson. AutoDock Vina: improving the speed and accuracy of docking with a new scoring function, efficient optimization, and multithreading. *J Comput Chem.* 2010; **31**(2): p. 455–461.
26. Ertl, P. and A. Schuffenhauer. Estimation of synthetic accessibility score of drug-like molecules based on molecular complexity and fragment contributions. *J Cheminform.* 2009; **1**: p. 1–11.

27. Razavi, A., A. Van den Oord, and O. Vinyals. Generating diverse high-fidelity images with vq-vae-2. *Adv Neural Inf Process Syst*. 2019; **32**: p. 14837–14847.
28. Hoogeboom, E., et al. Equivariant diffusion for molecule generation in 3d. *Int Conf Mach Learn*. 2022; **162**: p. 8867–8887.
29. Song, Y., et al. Equivariant flow matching with hybrid probability transport for 3d molecule generation. *Adv Neural Inf Process Syst*. 2023; **36**: p. 549–568.
30. Lipman, Y., et al. Flow matching for generative modeling. *arXiv preprint arXiv:2210.02747*. 2022.
31. Chen, R.T.Q., et al. Neural ordinary differential equations. *Adv Neural Inf Process Syst*. 2018; **31**: p. 6572–6583.
32. Ragoza, M., T. Masuda, and D.R. Koes. Generating 3D molecules conditional on receptor binding sites with deep generative models. *Chem Sci*. 2022; **13**(9): p. 2701–2713.
33. Peng, X., et al. Pocket2mol: efficient molecular sampling based on 3d protein pockets. *Int Conf Mach Learn*. 2022; **162**: p. 17644–17655.
34. Ho, J., A. Jain, and P. Abbeel. Denoising diffusion probabilistic models. *Adv Neural Inf Process Syst*. 2020; **33**: p. 6840–6851.
35. Guan, J., et al. 3d equivariant diffusion for target-aware molecule generation and affinity prediction. *arXiv preprint arXiv:2303.03543*. 2023.
36. Hoogeboom, E., et al. Argmax flows and multinomial diffusion: learning categorical distributions. *Adv Neural Inf Process Syst*. 2021; **34**: p. 12454–12465.
37. Satorras, V.G., E. Hoogeboom, and M. Welling. E(n) equivariant graph neural networks. *Int Conf Mach Learn*. 2021; **139**: p. 9323–9332.

Index

www.ingramcontent.com/pod-product-compliance
Ingram Content Group UK Ltd.
Pitfield, Milton Keynes, MK11 3LW, UK
UKHW020207190926
8624IPUK00034BA/174